Edited by
Satoshi Horikoshi
Nick Serpone

Microwaves in Catalysis

Related Titles

Pombeiro, A.J. (ed.)

Advances in Organometallic Chemistry and Catalysis

The Silver/Gold Jublilee International Conferenceon Organometallic Chemistry Celebratory Book

2014
Print ISBN: 978-1-118-51014-8; also available in electronic formats

Sheldon, R.A., Arends, I., Hanefeld, U.

Green Chemistry and Catalysis

Second Edition

2014
Print ISBN: 978-3-527-32947-2; also available in electronic formats

Li, C., Liu, Y. (eds.)

Bridging Heterogeneous and Homogeneous Catalysis

Concepts, Strategies, and Applications

2014
Print ISBN: 978-3-527-33583-1; also available in electronic formats

Wilson, K., Lee, A.F. (eds.)

Heterogeneous Catalysts for Clean Technology

Spectroscopy, Design, and Monitoring

2014
Print ISBN: 978-3-527-33213-7; also available in electronic formats

Wirth, T. (ed.)

Microreactors in Organic Chemistry and Catalysis

Second Edition

2013
Print ISBN: 978-3-527-33299-1; also available in electronic formats

Zhang, W., Cue, B.W. (eds.)

Green Techniques for Organic Synthesis and Medicinal Chemistry

2012
Print ISBN: 978-0-470-71151-4; also available in electronic formats

Li, C. (ed.)

Handbook of Green Chemistry - Green Processes

Volume 7 - Green Synthesis

2012
Print ISBN: 978-3-527-32602-0; also available in electronic formats

Hoz, A.d., Loupy, A. (eds.)

Microwaves in Organic Synthesis

Third Edition

2012
Print ISBN: 978-3-527-33116-1; also available in electronic formats

Kappe, C.O., Stadler, A., Dallinger, D.

Microwaves in Organic and Medicinal Chemistry

Second Edition

2012
Print ISBN: 978-3-527-33185-7; also available in electronic formats

Horikoshi, S. and Serpone, N.

Microwaves in Nanoparticle Synthesis

Fundamentals and Applications

2013
Wiley-VCH Verlag GmbH, Weinheim

Edited by
Satoshi Horikoshi
Nick Serpone

Microwaves in Catalysis

Methodology and Applications

WILEY-VCH
Verlag GmbH & Co. KGaA

The Editors

Prof. Satoshi Horikoshi
Sophia University
Department of Materials and Life Sciences
Kioicho 7-1
102-8554 Tokyo
Japan

Prof. Nick Serpone
Universita di Pavia
Dipartimento di Chimica
Via Taramelli 10
27100 Pavia
Italy

Library of Congress Card No.: applied for

British Library Cataloguing-in-Publication Data
A catalogue record for this book is available from the British Library.

Bibliographic information published by the Deutsche Nationalbibliothek
The Deutsche Nationalbibliothek lists this publication in the Deutsche Nationalbibliografie; detailed bibliographic data are available on the Internet at <http://dnb.d-nb.de>.

Print ISBN: 978-3-527-33815-3
ePDF ISBN: 978-3-527-68812-8
ePub ISBN: 978-3-527-68814-2
Mobi ISBN: 978-3-527-68813-5
oBook ISBN: 978-3-527-68811-1

Cover Design Adam-Design, Weinheim, Germany
Typesetting SPi Global, Chennai, India
Printing and Binding Markono Print Media Pte Ltd, Singapore

Printed on acid-free paper

Contents

List of Contributors

Rick Arneil Desabille Arancon
University of Cordoba
Organic Chemistry Department
Faculty of Sciences
Rabanales Campus
Marie Curie Building (C-3)
Ctra Nnal IV-A
Km 396
E14014 Cordoba
Spain

Yusuke Asakuma
University of Hyogo
Department of Mechanical and
Systems Engineering
2167 Shosha
Himeji, Hyogo
671-2280
Japan

Faysal Benaskar
University of Warwick
School of Engineering
Energy Intensified Reactor
Engineering Laboratory
Coventry
CV4 7AL
UK

Diego Carnaroglio
University of Turin
Department of Drug Science and
Technology and
NIS – Centre for Nanostructured
Interfaces and Surfaces
Via P. Giuria 9
10125 Turin
Italy

Hyejin Cho
University of Massachusetts
Boston
Department of Chemistry
100 Morrissey Boulevard
Boston
MA 02125
USA

Giancarlo Cravotto
University of Turin
Department of Drug Science and
Technology and
NIS – Centre for Nanostructured
Interfaces and Surfaces
Via P. Giuria 9
10125 Turin
Italy

Volker Hessel
Eindhoven University of Technology
Chemical Reactor Engineering
Department of Chemical Engineering and Chemistry
De Rondom 70
5612 AP, Eindhoven
The Netherlands

Takenori Hirano
Catalyst Research Inc.
1474 Ohkubo
Takaono-cho
Izumi-shi
Kagoshima 899-0401
Japan

Satoshi Horikoshi
Sophia University
Department of Materials and Life Sciences
Faculty of Science and Technology
7-1 Kioicho, Chiyodaku
Tokyo, 102-8554
Japan

Leonid M. Kustov
Russian Academy of Sciences
N.D. Zelinsky Institute of Organic Chemistry
Leninsky Prospect 47
Moscow 119991
Russia

and

Moscow State University
Chemistry Department
Leninskie Gory 1
Building 3
Moscow 119992
Russia

Rafael Luque
University of Cordoba
Organic Chemistry Department
Faculty of Sciences
Rabanales Campus
Marie Curie Building (C-3)
Ctra Nnal IV-A
Km 396
E14014 Cordoba
Spain

Mitsuhiro Matsuzawa
Hitachi, Ltd.
Hitachi Research Laboratory
832-2, Horiguchi
Hitachinaka
Ibaraki 312-0034
Japan

Shigeru Mineki
Tokyo University of Science
Faculty of Science and Technology
Department of Applied Biological Science
264 Yamazaki, Noda
Chiba 305-8506
Japan

Tomohiko Mitani
Kyoto University
Research Institute for Sustainable Humanosphere
Gokasho, Uji
Kyoto 611-0011
Japan

Shokichi Ohuchi
Kyushu Institute of Technology
Graduate School of Life Science and System Engineering
680-4, Kawazu
Iizuka
Fukuoka 820-8502
Japan

Ayumu Onda
Kochi University
Faculty of Science
Research Laboratory of
Hydrothermal Chemistry
Akebono-cho 2-5-1
Kochi 780-8520
Japan

Narendra Patil
Eindhoven University of
Technology
Chemical Reactor Engineering
Department of Chemical
Engineering and Chemistry
De Rondom 70
5612 AP, Eindhoven
The Netherlands

Volker Rebrov
University of Warwick
School of Engineering
Energy Intensified Reactor
Engineering Laboratory
Coventry
CV4 7AL
UK

Laura Rinaldi
University of Turin
Department of Drug Science and
Technology and
NIS – Centre for Nanostructured
Interfaces and Surfaces
Via P. Giuria 9, 10125 Turin
Italy

and

University of Turin
Dipartimento di Biotecnologie
Molecolari e Scienze per la Salute
Via Quarello 15
10135 Turin
Italy

Antonio Angel Romero
University of Cordoba
Organic Chemistry Department
Faculty of Sciences
Rabanales Campus
Marie Curie Building (C-3)
Ctra Nnal IV-A
Km 396
E14014 Cordoba
Spain

Christian Schäfer
University of Massachusetts
Boston
Department of Chemistry
100 Morrissey Boulevard
Boston
MA 02125
USA

Jaap Schouten
Eindhoven University of
Technology
Chemical Reactor Engineering
Department of Chemical
Engineering and Chemistry
De Rondom 70
5612 AP, Eindhoven
The Netherlands

Nick Serpone
University of Pavia
PhotoGreen Laboratory
Department of Chemistry
12 Taramelli Street
27100 Pavia
Italy

A. E. Stiegman
Florida State University
Department of Chemistry and Biochemistry
95 Chieftan Way
Tallahassee
FL 32312
USA

Shigenori Togashi
Hitachi, Ltd.
Hitachi Research Laboratory
832-2, Horiguchi
Hitachinaka
Ibaraki 312-0034
Japan

Béla Török
University of Massachusetts Boston
Department of Chemistry
100 Morrissey Boulevard
Boston
MA 02125
USA

Shuntaro Tsubaki
Tokyo Institute of Technology
Department of Applied Chemistry
Graduate School of Science and Technology
2-12-1, Ookayama
Meguro
Tokyo 152-8552
Japan

Tadaharu Ueda
Kochi University
Department of Applied Science
Akebono-cho 2-5-1
Kochi 780-8520
Japan

Takashi Watanabe
Kyoto University
Research Institute for Sustainable Humanosphere
Gokasho, Uji
Kyoto 611-0011
Japan

Takeo Yoshimura
Tokyo University of Science
Faculty of Science and Technology
Department of Applied Biological Science
264 Yamazaki
Noda
Chiba 305-8506
Japan

Preface

From the early days of *Homo Sapiens*, as we know it today civilization developed mostly by burning wood for cooking and keeping warm, which was subsequently followed by burning coal and other fossil fuels. The thermal energy released from the combustion process is used to drive chemical reactions and materials sintering, which ultimately led to the birth of such fields as hydrodynamics, thermodynamics, and photochemistry, with the latter through the use of sunlight. These fields are connected through chemical engineering and materials processing. Unfortunately, the generation of heat by the combustion of wood and fossil fuels inevitably generates greenhouse gases such as carbon dioxide (CO_2), responsible in part for global warming. The energy released from the combustion process cannot be sustained for very long, so that further development of energy sources has relied heavily on the energy of the atom through nuclear reactions that have been shown to have serious consequences. For the long term, to sustain life and further development, mankind will have to rely on the energy from electromagnetic waves (radio waves and light).

Microwave radiation has now been used for more than 20 years as a heat source for materials synthesis and sintering. For this reason, the energy embodied in the microwaves can be absorbed by substances directly; energy saving is one of the advantages in the use of microwave radiation.

The chemistry of the twentieth century was essentially developed through the use of substrates that acted as catalysts. For example, at the beginning of the twentieth century, whatever soil nitrogen (mostly nitrates) was available proved insufficient for intensive production of crops. Accordingly, several commercial processes to produce nitrogen fertilizers were developed with the best known process being the Haber–Bosch process (discovered by Fritz Haber in 1906 and industrial production by Carl Bosch in 1913) that converted air nitrogen into ammonia using iron-based catalysts. The challenges of the twenty-first century will be, among others, the development of processes that achieve lower greenhouse gas (e.g., CO_2) emissions and discover novel energy-saving catalyzed reactions. A natural choice in such efforts would be the combination of catalysis and microwave radiation.

The main purpose of this book is to give an overview into the preparation of catalysts and explore catalytic reactions using the microwave method. As such, the book is divided into three principal topics: (i) introduction to microwave chemistry and microwave materials processing; (ii) fundamentals of microwave chemistry and microwave-driven catalysis; and (iii) applications of microwaves that include the preparation of catalysts, microwave flow methodology, hybrid microwave method with other techniques such as organic reactions, hydrogenation and fuel formation, oil refining, biomass and wastes, and environmental remediation. To the extent that the data available in the literature were enormous, we have had to make the difficult choice to include only the more relevant and up-to-date literature. We are grateful to all the contributing authors who have answered our call, and thank the Wiley-VCH editorial staff for their thorough and professional assistance. The data presented would not have been possible without the fruitful collaboration of many university and industrial researchers, and not least without the cooperation of students whose names appear in many of the publications cited. We are indeed very grateful for their effort.

We hope this book becomes a starting point for researchers in other fields to become interested in pursuing microwave chemistry, in general, and microwave-driven catalyzed reactions, in particular.

April 2015
Tokyo

Satoshi Horikoshi
Nick Serpone

1
General Introduction to Microwave Chemistry

Satoshi Horikoshi and Nick Serpone

1.1
Electromagnetic Waves and Dielectric Materials

The common characteristics of electromagnetic waves (light and radio waves) are that they (i) have no mass, (ii) are free to move at the speed of light, (iii) cannot collide with each other, (iv) possess significant energy, and (v) are reflected on a metal surface. By contrast, when microwaves (MWs) are used to irradiate dielectric materials, various phenomena occur according to the nature of the electromagnetic waves. The influence of electromagnetic waves on dielectric materials in various ranges of the electromagnetic spectrum is summarized in Figure 1.1 [1], which also shows the various analytical instruments and their usefulness in the ranges indicated.

Atomic species: If a dielectric is placed in an alternating electric field of the electromagnetic waves, the positions of the atomic nucleus (protons) and of the electrons that constitute the dielectric tend to follow the course of the electric field. The spatial relationship of a proton and an electron in an atom becomes distorted when subjected to the electric field (referred to as *electronic polarization*). Since the electric dipole that reflects the electric deviation (a proton and an electron) inside an atom is very small, a resonance phenomenon will occur with light of short wavelengths such as X-rays and ultraviolet rays.

Molecular species: The distortion of the electric charge by electromagnetic waves also occurs in a molecule with an electric dipole moment and in a salt. For example, in the structure of NaCl salt, composed of Na^+ and Cl^- ions, distortion of the structure arises when subjected to an alternating electric field (referred to as *ionic polarization*). An electric dipole moment of a molecule is large compared with that of an atom, so that resonance occurs in the infrared domain, that is, at the longer wavelengths. The magnitude of ionic polarization is dictated by the interionic distance and the types of ions involved. Moreover, changes in bond lengths in an organic molecule occur through stretching and bending vibrations. These

Microwaves in Catalysis: Methodology and Applications, First Edition.
Edited by Satoshi Horikoshi and Nick Serpone.

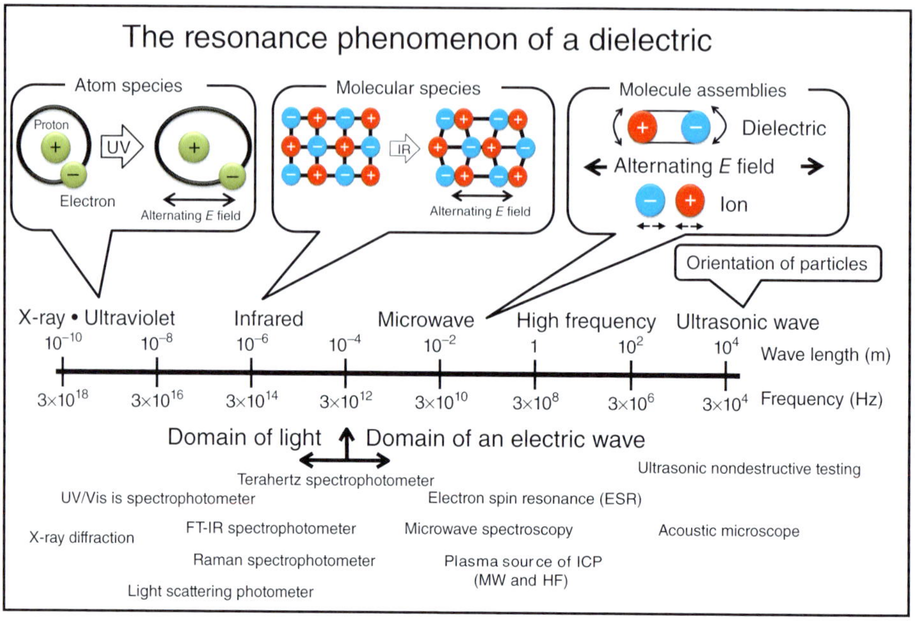

Figure 1.1 Resonance of dielectric to electromagnetic waves and positioning of the analytical equipment. Reproduced from Ref. [1] Copyright 2013 by Wiley-VCH Verlag GmbH.

phenomena occur at various wavelength regions absorbed, depending on the chemical structure; they can be observed by means of an infrared absorption spectrophotometer.

Molecular assemblies: Orientation polarization of a dielectric possessing an electric dipole takes place on interaction with the electromagnetic waves in the microwave range. Moreover, as for the case of ions in a solution, Joule heating takes place by space–charge polarization. For example, when microwave heating an electrolyte/water solution, dielectric heating and Joule heating occur simultaneously compared with pure water, and thus the exothermic efficiency becomes remarkably high. Accordingly, dielectric heating by orientation polarization of water and resistance heating by the Joule process are enhanced in electrolyte/water media.

1.2
Microwave Heating

Microwave radiation is electromagnetic radiation spanning the frequency range 30 GHz–300 MHz (i.e., from a wavelength of 1 m to 1 cm). It is used widely in communications and in heating processes, especially in the heating of foodstuff. Historically, the powerful interaction of microwaves with materials was discovered in 1946 from the melting of chocolate, a process attributed to microwave heating, whereas the first commercial microwave oven was developed in 1952 by the Raytheon Company [2]. This discovery is frequently taken up as an example of serendipity. Before the discovery of microwaves, high-frequency induction heating was commonly used. The patent of dielectric heating by means of high-frequency induction was issued in 1933 [3]. In the 1970s, the microwave generator was re-engineered by Japanese scientists into a domestic microwave oven (using a simple, reliable, and inexpensive magnetron), which allowed for simple food and mass processing [4]. Domestic and industrial microwave ovens generally operate at a frequency of 2.45 GHz corresponding to a wavelength of 12.24 cm and an energy of 1.02×10^{-5} eV, or around 900 MHz with a longer wavelength of 37.2 cm, which can provide up to 100 kW in larger process heating applications [5]. Since then, microwave chemistry has become a rapidly developing branch in numerous fields of research, in industry, and in its applications.

A molecule is formed when atoms combine to share one or more of their electrons. This rearrangement of electrons may cause an imbalance in charge distribution, thereby creating a permanent electric dipole moment. These moments are oriented in a random manner in the absence of an electric field so that no polarization exists. Electric dipole moments in a polar molecule undergo group movement so that a mutual electric dipole may be left without. When a microwave is applied to this state, an electric field has an influence on the electric dipole (Figure 1.2). The microwaves' electric field (*E* field) will exercise a torque (*N*) on the electric dipole, and the dipole will consequently rotate to align itself with the electric field thus causing orientation polarization to occur [6]. If the field changes direction, the torque will also change. The orientation polarization changes by a vibration in a microwave electric field. A time difference is caused between the frequency of the microwave electric field and the electric dipole of molecules. At the general frequency of 2.45 GHz, the microwaves vibrate 2 450 000 000 times/s; a molecular cluster cannot follow this vibration through the power chain with the surrounding molecules. This delay changes to heat energy (kinetic energy) as loss of the electromagnetic wave energy. The friction accompanying the orientation of the dipole will contribute to dielectric losses. The dipole rotation causes a variation in both ε_r' (the relative dielectric constant) and ε_r'' (the relative dielectric loss) at the relaxation frequency, which usually occurs in the microwave region. Water is an example of a substance that exhibits a strong orientation polarization.

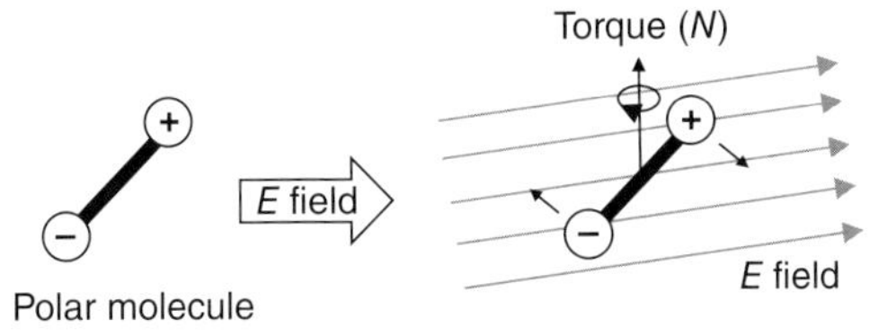

Figure 1.2 Image of dipole rotation for polar molecule in an electric field (*E* field).

By contrast, a solid substance with a partial electric charge (a dielectric material) is also an insulator. The slight distortion of the atomic position (lattice point) in a structure lattice, and the lattice strain of a crystal of solid substances cannot follow the changing time of the microwave electric field. Microwave heating of a solid substance progresses by these phenomena [7]. On the other hand, the heating of solid substances with a magnetic dipole moment occurs by the microwaves' magnetic field component. The heating process is the same phenomenon as by the microwaves' electric field [8]. Generation of heat by magnetic loss heating is expected only in magnetic (solid) materials. Joule heating progresses by the interaction of an electric or a magnetic field with solid particles of activated carbon and with solids possessing conductivity-like metallic properties [9].

1.3 The Various Types of Microwave Heating Phenomena

Polarization phenomena are generated regardless of the polar or nonpolar nature of the molecules. On the other hand, alignment polarization (molecular alignment) occurs only when the polarity of molecules is affected by the electric field because of their permanent dipoles. Thus, the role of microwaves (and radio waves) is not simply as a heating source. With regard to the latter, we must recognize three types of heating phenomena caused by microwaves: (i) conduction loss heating, (ii) dielectric heating, and (iii) magnetic loss heating. In this regard, the thermal energy P produced per unit volume originating from microwave radiation is given by Eq. (1.1) [10]:

$$P = \frac{1}{2}\sigma|E|^2 + \pi f \varepsilon_0 \varepsilon_r'' |E|^2 + \pi f \mu_0 \mu_r'' |H|^2 \tag{1.1}$$

where $|E|$ and $|H|$ denote the strength of the microwaves' electric and magnetic fields, respectively; σ is the electrical conductivity; f is the frequency of the microwaves; ε_0 is the permittivity in vacuum; ε_r'' is the relative dielectric loss factor; μ_0 is the magnetic permeability in vacuum; and μ_r'' is the relative magnetic loss.

The first term in Eq. (1.1) expresses *conduction loss heating*; the second term denotes *dielectric loss heating*, whereas *magnetic loss heating* is given by the third term. Microwave heating of solutions is governed mostly by dielectric loss heating, whereas conduction loss heating involves mostly, but not exclusively (see later), solid materials. Therefore, microwave heating of materials is dictated by the electrical, dielectric, and magnetic properties. Moreover, when a change of frequency occurs, their behavior also changes. It is possible to design catalyzed reactions based on these properties and build innovative microwave chemistry processes. A catalyzed reaction occurring on a solid catalyst is expected to show characteristic differences relative to the solution bulk because the reacting substrates in the liquid medium may be adsorbed onto the solid's surface. Temperature rise in systems involving solely gaseous molecules is negligible because the molecular density is rather small.

1.3.1 Conduction Loss Heating (Eddy Current Loss and Joule Loss)

The hydroxide ion is a typical ionic species with both ionic and dipolar characteristics. The conductive loss effect can become larger than dipolar relaxation in solutions that contain large quantities of ionic salts. Conduction losses tend to be less at ambient temperatures in the case of solids but can change substantially with an increase in temperature. A typical example is alumina (Al_2O_3) whose dielectric losses are negligibly small ($\sim 10^{-3}$) at ambient temperature, but can reach fusion levels in very short times (minutes) in a microwave cavity. This effect originates from a strong increase in conduction losses associated with the thermal activation of electrons, which migrate from the oxygen's 2p valence band to the aluminum's 3s, 3p conduction band. Moreover, conduction losses in solids are usually enhanced by defects in materials. These sharply reduce the energy needed to generate electrons and holes in the conduction and valence bands, respectively. In the particular case of carbon black powder, conduction losses are rather high, so much so that this material is often used as an impurity additive to induce losses within solids whose dielectric losses are too small.

1.3.2 Dielectric Heating

Nonconductive materials can be heated by the microwaves' electric field through molecular rotation of the molecules' electric dipoles, with the latter aligning themselves with the field. As the field alternates, the molecular dipoles reverse direction (vectorially) and accelerate the motion of individual molecules or atoms, thereby generating heat through friction of the molecules as they rotate against each other. Dielectric heating can therefore be applied to the heating of solutions. By contrast, heating of electroconductive solids, such as metals, implicates ohmic losses in the material in virtue of the current flow induced by the oscillating electric field.

The dipolar polarization mechanism only concerns polar compounds, for example, water, methanol, ethanol, which possess a permanent dipole moment [4]. With regard to the frequency of the applied electric field, the dipoles can behave threefold: (i) at low frequencies, the dipoles keep in phase with the electric field and the temperature of the irradiated medium hardly rises; (ii) at high frequencies, the response time of the dipoles is too long to follow very rapid oscillations of the electric field and no energy is absorbed; and (iii) at the microwave frequencies, the dipole rotation lags behind the changes of the electric field and results in the heating of the irradiated material.

Dipolar polarization, P_d, occurs on a timescale of the order of those associated with microwaves. Hence, when a dielectric is subjected to an external electric field of strength E, the polarization is related to the intrinsic properties of the material through the relation expressed by Eq. (1.2) [11]:

$$P_{\mathrm{d}} = \varepsilon_0(\varepsilon_{\mathrm{r}} - 1)E \tag{1.2}$$

where ε_0 is the permittivity in free space and ε_{r} is the relative permittivity of the material. Given that in reality the permittivity is a complex quantity, ε^* (equal to the product $\varepsilon \cdot \varepsilon_{\mathrm{r}}$ and expressed as $\varepsilon^* = \varepsilon' + i\varepsilon''$) [12], then the loss tangent ($\tan\delta$) is commonly used to describe the interaction of a dielectric with the microwaves. The ability of a material to convert microwave energy to thermal energy ($\tan\delta$) depends on the dielectric properties of the material and is defined by Eq. (1.3):

$$\tan\delta = \frac{\varepsilon_{\mathrm{r}}''}{\varepsilon_{\mathrm{r}}'} \tag{1.3}$$

Dielectric parameters of common organic substances typically used in industry have been determined at various temperatures at 10 °C intervals using a conventional plate heater. The temperatures of the solutions were measured with an optical fiber thermometer. The temperature-dependent $\tan\delta$ of these substrates at 20, 50, and 90 °C are summarized in Table 1.1.

1.3.3 Magnetic Loss Heating (Eddy Current Loss and Hysteresis Loss Heating)

In metal oxides such as ferrites and other magnetic materials, magnetic losses occur in the microwave region. These losses are different from hysteresis or Eddy current losses because they are induced by domain wall and electron spin resonance (ESR). Such materials are typically placed at positions of magnetic field maxima for optimal absorption of the microwave energy. Transition metal oxides (e.g., iron, nickel, and cobalt oxides) possess high magnetic losses and can thus be used as added loss impurities or additives to induce losses within those solids for which dielectric loss is too small. For instance, induction heating by the microwaves' magnetic field is expected to occur in catalyzed reactions involving solids, an example being the rapid induction heating of magnetite (Fe_3O_4) but not hematite (Fe_2O_3) as the latter is not a magnetic material [13].

1.3.4 Penetration Depth of Microwaves

Even though the dielectric loss of a solvent may be high, the heating efficiency is sometimes low. This may be caused by the shallow penetration depth of the microwaves. For example, though solar light is absorbed by sea water, it is nonetheless dark at greater depths. A similar phenomenon occurs with microwaves in that as the microwave energy is changed into heat it decreases. Therefore, if a large size reactor is used, the microwaves may not heat the center of the solution. The penetration depth D_{p} (in cgs units) is the depth at which microwaves pervade into the material and the power flux has fallen to

Table 1.1 Temperature-dependent dielectric loss factor (tan δ) of common organic substances at temperatures of 20, 50, and 80 °C.

Organic substance	tan δ at 20 °C	Organic substance	tan δ at 50 °C	Organic substance	tan δ at 80 °C
Ethylene glycol	1.0137	1-Propanol	0.8832	1-Butanol	0.6943
Ethanol	0.9499	2-Propanol	0.8707	1-Propanol	0.6237
1-Propanol	0.7097	Glyciline	0.8130	1-Pentanol	0.5995
2-Propanol	0.6660	Ethanol	0.8027	α-Lonone	0.5339
Amino methoxyethanol	0.6527	2-Butanol	0.7089	1-Hexanol	0.4998
Methanol	0.5537	Ethylene glycol	0.6994	2-Propanol	0.4792
β-Lonone	0.5125	1-Butanol	0.6835	2-Butanol	0.4659
1-Butanol	0.4992	Amino methoxyethanol	0.6047	Ethanol	0.4450
Tetrahydrofurfuryl alcohol	0.4608	α-Lonone	0.5345	β-Lonone	0.4441
Glyciline	0.4442	β-Lonone	0.5237	Amino methoxyethanol	0.4240
2-Butanol	0.4338	1-Pentanol	0.4968	1-Heptanol	0.4164
Ethyl lactate	0.4171	Methanol	0.3958	Ethylene glycol	0.3606
α-Lonone	0.4071	1-Hexanol	0.3773	Benzyl salicylate	0.3136
o-Methyl phenol	0.4038	*o*-Methyl phenol	0.3698	Benzyl benzoate	0.2902
Butyl benzoate	0.3637	Tetrahydroxyfurfuryl alcohol	0.3350	*o*-Methyl phenol	0.2692
p-Methyl phenol	0.3383	Benzyl benzoate	0.3279	Butyl benzoate	0.2204
Benzyl benzoate	0.3052	1-Heptanol	0.2850	2-Butanone	0.1980
1-Pentanol	0.3036	Diethylhydroxylamine	0.2837	Tetrahydroxyfurfuryl alcohol	0.1909
Diethylhydroxylamine	0.3025	Butyl benzoate	0.2818	*p*-Methyl phenol	0.1865
Ethyl benzoate	0.2860	Benzyl salicylate	0.2647	Diethylhydroxylamine	0.1704
Dimethyl sulfoxide	0.2576	*p*-Methyl phenol	0.2579	Ethyl benzoate	0.1549
o-Dichlorobenzene	0.2451	Ethyl lactate	0.2479	Methyl benzoate	0.1446
Methyl benzoate	0.2411	Acetic acid	0.2403	Ethyl lactate	0.1319
Acetic acid	0.2409	Ethyl benzoate	0.2196	*o*-Dichlorobenzene	0.1289
1-Hexanol	0.2325	*o*-Dichlorobenzene	0.1839	Dimethyl sulfoxide	0.1282
1-Heptanol	0.1763	Dimethyl sulfoxide	0.1837	Bromobenzene	0.0821
Benzyl salicylate	0.1666	Methyl benzoate	0.1793	*m*-Dichlorobenzene	0.0812
N,*N*-Dimethylformamide	0.1370	Bromobenzene	0.1060	Octane	0.0763

(*continued overleaf*)

Table 1.1 (*Continued*)

Organic substance	tan δ at 20 °C	Organic substance	tan δ at 50 °C	Organic substance	tan δ at 80 °C
m-Dichlorobenzene	0.1339	*m*-Dichlorobenzene	0.1039	Propionic acid	0.0733
Bromobenzene	0.1264	*N*,*N*-Dimethylformamide	0.0971	*o*-Xylene	0.0725
Chlorobenzene	0.0704	Nonane	0.0714	*N*,*N*-Dimethylformamide	0.0688
2-Chlorobutane	0.0654	*m*-Xylene	0.0688	Chlorobenzene	0.0476
Butyl acetate	0.0639	*p*-Xylene	0.0670	Butyl acetate	0.0428
m-Xylene	0.0498	2-Chlorobutane	0.0666	Propyl benzene	0.0282
p-Xylene	0.0475	Propionic acid	0.0650	Acetonitrile	0.0128
o-Xylene	0.0469	Chlorobenzene	0.0628	Propyl acetate	0.0075
Propionic acid	0.0466	*o*-Xylene	0.0610	Benzylamine	0.0059
1-Bromopropane	0.0439	Octane	0.0535	Benzene	0
Nonane	0.0435	Butyl acetate	0.0469	Toluene	0
2-Butanone	0.0366	1-Bromopropane	0.0276	Triethylamine	0
Acetonitrile	0.0329	Acetonitrile	0.0217	Methanol	Over b.p.
2-Bromopropane	0.0303	2-Butanone	0.0213	Glyciline	Over b.p.
Benzylamine	0.0280	2-Bromopropane	0.0189	Acetic acid	Over b.p.
Acetone	0.0256	Acetone	0.0181	2-Chlorobutane	Over b.p.
Propylacetate	0.0256	Propyl benzene	0.0181	*m*-Xylene	Over b.p.
Octane	0.0197	Benzylamine	0.0141	*p*-Xylene	Over b.p.
Ethyl acetate	0.0106	Propyl acetate	0.0136	1-Bromopropane	Over b.p.
Propyl benzene	0.0095	Tetrahydrofuran	0.0064	Nonane	Over b.p.
Tetrahydrofuran	0.0057	Ethyl acetate	0.0045	2-Bromopropane	Over b.p.
Benzene	0	Benzene	0	Acetone	Over b.p.
Dichloromethane	0	*n*-Hexane	0	Ethyl acetate	Over b.p.
Monopropylamine	0	Toluene	0	Tetrahydrofuran	Over b.p.
n-Hexane	0	Triethylamine	0	Dichloromethane	Over b.p.
Toluene	0	Dichloromethane	Over b.p.	Monopropylamine	Over b.p.
Triethylamine	0	Monopropylamine	Over b.p.	*n*-Hexane	Over b.p.

$1/e$ (=36.8%) of its surface value. That is, it denotes the depth at which the power density of the microwaves is reduced to $1/e$ of its initial value; D_p can be estimated by Eq. (1.4) [7]:

$$D_p = \frac{\lambda}{4\pi}\left[\frac{2}{\varepsilon'\left(\sqrt{1+(\varepsilon r''/\varepsilon r')^2}-1\right)}\right]^{1/2} \tag{1.4}$$

where λ is the wavelength of the radiation, $\lambda_{(2.45\,\text{GHz})} = 12.24\,\text{cm}$. Since the dielectric constant and the dielectric loss are components of this equation, we can expect that the penetration depth will change with an increase in temperature. For example, since the penetration depth at 25 °C is about 1.8 cm for water, even if we used a reactor 10 cm in diameter, the microwaves will not reach the center of the reactor. However, at 50 °C, the penetration depth is 3.1 cm, and at 90 °C it is 5.4 cm so that now the microwaves do reach the center of the water sample [14]. The penetration depth of the microwaves into almost all nonpolar solvents is very deep compared with polar solvents. On the other hand, if ions are added to water, the penetration depth then becomes remarkably shallow. For example, when NaCl (0.25 M) is added to pure water, the penetration depth into this saline solution changes from 1.8 cm (pure water) to 0.5 cm at ambient temperature [15]. Moreover, even if this electrolyte/water system was to be heated, the penetration depth would hardly change (0.4 cm at 90 °C). Therefore, in the case of solutions containing ions, since microwave heating does not occur, temperature hot spots will occur only at the surface of the reactor and it will be necessary to carry out vigorous stirring to control these. Hence, it is necessary to examine the heating efficiency and the penetration depth of microwaves into a substance to achieve optimal microwave heating.

1.4
Fields of Applications with Microwave Heating

To the extent that microwaves are electromagnetic waves, they move at the speed of light. Since a sample is directly heated by microwaves, the time of heat conduction is shortened considerably relative to conventional heating, and this regardless of however complex the reactor design might be because reactor design has no influence on reaction times. Accordingly, microwaves have become an important heat source in a wide set of applications superseding the more traditional thermal methods. Microwave energy has been widely developed and applied to almost every field of chemistry. Figure 1.3 summarizes some of the applications where microwaves have become a common source of thermal energy, for example, in such fields as attenuation of environmental pollution, in the medical field, in printing, in paints, in foodstuff, in the fabrication of thin films, in agriculture, and in the drying of wood [16].

Microwave heating in application

Environmental
- ✓ Sintering of asbestos
- ✓ Decomposition of dioxin and PCB
- ✓ Oil recovery from plastics
- ✓ Process treatment of waste oil
- ✓ Processing of medical wastes
- ✓ Processing of radioactive contaminated substances
- ✓ Degradation of chlorofluorocarbons
- ✓ Decomposition of VOCs
- ✓ Exhaust gas treatment
- ✓ Incineration of garbage
- ✓ Electrodeless lamps
- ✓ Biomass
- ✓ Oxidization of soot
- ✓ Enhancement of activity of photocatalysts
- ✓ Recycling of scrap tires
- ✓ Solidification of waste plastics

Medical uses
- ✓ Hyperthermia
- ✓ Sterilization
- ✓ Muscular warming
- ✓ Cutting of blood vessels

Usage for foodstuffs
- ✓ Sterilization
- ✓ Processing
- ✓ Defrosting
- ✓ High-speed cooking
- ✓ Reduced-pressure drying
- ✓ Activation of enzyme and yeast

Ink and paint
- ✓ Drying of printing ink
- ✓ Drying of paints

Films and paper
- ✓ Dryness of book cover
- ✓ Selective heating of pastes
- ✓ Sintering of electronic wiring
- ✓ Film curating

- ✓ **Agriculture**
- ✓ Sterilization of soil
- ✓ Drying of wood
- ✓ Extermination of noxious insects in wood
- ✓ Extraction and degradation of contaminants

Usage in Wood treatments
- ✓ Adhesion processing
- ✓ Bending
- ✓ Drying of wood

Microwave chemistry and material processing

Organic chemistry
- ✓ Rapid syntheses
- ✓ Solid-phase syntheses
- ✓ Non solvent and non catalytic processes
- ✓ Organometallic complexes
- ✓ Combinatorial chemistry

Analytical chemistry
- ✓ Carbonization processing
- ✓ Acid and alkali treatment
- ✓ High-speed concentration, extraction and degradation

Biochemistry
- ✓ Enzyme reactions
- ✓ PCR
- ✓ Heating brain of rat
- ✓ DNA dyeing

Polymer
- ✓ Vulcanization and firing of rubber
- ✓ Higher selectivity in polymer syntheses
- ✓ Heat processing of plastics
- ✓ Size control of polymers

Catalytic chemistry
- ✓ Heterogeneous catalysis (reduction reactions, etc)
- ✓ Homogeneous catalysis (Suzuki coupling, and others)
- ✓ Syntheses of catalysts and catalyst supports (e.g., zeolites)

Inorganic materials and metal chemistry
- ✓ Functional material composition
- ✓ Composition of quality nanomaterials
- ✓ Hydrothermal syntheses
- ✓ Particulate coatings
- ✓ Control of crystallinity
- ✓ Syntheses of nitrides
- ✓ Drying of refractory products
- ✓ Calcination of ceramics and formation of interfacial junctions
- ✓ Microwave-assisted iron manufacturing
- ✓ Metallic powder metallurgy
- ✓ Carbonization
- ✓ High-speed treatments of lightweight fire-resistant building materials
- ✓ Heat treatments of specialty glasses
- ✓ Syntheses of artificial zeolites
- ✓ Drying of extruded molding ceramics
- ✓ Syntheses of artificial bones

Photochemistry
- ✓ Electrodeless lamps
- ✓ UV hardening of transparent coatings
- ✓ Photochemical syntheses

Figure 1.3 Fields of applications with microwave heating. Reproduced from Ref. [16]. Copyright 2013 by Wiley-VCH Verlag GmbH.

Not least, microwaves have also been used for other purposes such as when used as a plasma source in the treatment of semiconductor surfaces or in the fabrication of industrial diamonds. Moreover, ESR spectroscopy used in chemistry makes use of 9 GHz microwaves. Also, the microwave radiation used in military communications has found a domestic niche in the widely used microwave ovens. The technology has developed into various other fields, one of which that is relevant here is in microwave chemistry, an area that has witnessed unparalleled growth in the last decade. The applications of microwaves in the chemical field are also summarized in Figure 1.3, the major ones being organic chemistry, analytical chemistry, biochemistry, polymer chemistry, catalysis, photochemistry, and inorganic chemistry of materials.

1.5 Microwaves in Solid Material Processing

A study carried out on the calcination and sintering of ceramics by microwaves was reported as early as 1972 [17]. Fundamental and application studies soon followed around 1980. The early industrial use of microwaves as a heat source was to remove water from inorganic materials. It was soon discovered [18], however, that ceramics could also be heated with microwaves (latter half of the 1970s), which led to the effective sintering of ceramic powders. The conventional heating methods for such tasks are heat conduction and radiation heating. With regard to microwave irradiation, the sample itself generates heat. Therefore, a sintered compact product is obtained at low temperature for a short time.

A symposium held at a conference on materials science by the Materials Research Society (MRS) revealed that materials processing by microwaves was a topic that drew a great deal of interest [19]. Indications from the conference also showed that microwaves could be used not only to determine the quantity of water in ceramics but also as a technique to detect and analyze formation of fissures in ceramics and metals [19]. The basic interactions between the microwaves and materials were also widely discussed. Various kinds of ceramics have been treated by the microwave method [20]. In recent years, the calcination of special ceramics with a specific microwave method has also been reported [21].

Since the 1990s, several reactions have been carried out in the presence of solid catalysts for a potential application of microwave radiation in materials synthesis: for example, the synthesis of acetylene by oxidization of methane using particulate metal catalysts [22]. In contrast to microwave heating, when a conventional heating method was used, catalyst activity stopped when the reaction was only about 25% complete, whereas with microwave heating the reaction proceeded to about 90% completion. Research into the dechlorination of chlorinated substrates with heterogeneous metal catalysts was also examined in the initial stages (1990s) of utilization of microwave radiation. The influence of both the microwave frequency and pulsed microwave irradiation was considered; optimal reaction conditions were determined in the presence of a solid catalyst [23]. Of particular interest were catalysts that absorbed microwave radiation, and not least was the notion of differential heating in the quality of a reaction group that involved formation of hot spots on the catalyst surface. Accordingly, the next section examines selective heating of a heterogeneous catalyst (metallic or organic catalyst) by the microwaves.

1.6 Microwaves in Organic Syntheses

Two pioneering papers that made use of microwaves in organic synthesis were published nearly three decades ago in 1986. Gedye and coworkers [24] carried out organic syntheses using a Teflon reactor and a commercial/domestic microwave oven; this resulted in a remarkable decrease in reaction time compared to conventional heating methods. Giguere and coworkers [25] pointed out that reactor, solvent, and temperature management were important factors in microwave-assisted organic syntheses. Research into such application of microwaves increased but slowly and gradually. By contrast, the number of publications in microwave chemistry has increased dramatically since 2000 [26], due in large part to the availability of commercial microwave devices intended for organic synthesis that led many chemists to convert from the traditional microwave oven to these newly fabricated devices. Safety, reproducibility, control of microwave output power, temperature, and pressure were some of the important and attractive features of such devices. Currently, microwave organic synthesis systems have become fully automated by combining with robot technology. Such systems are playing an active part in such fields as combinatorial chemistry. This has also led research of microwave chemistry from the academic laboratory level to the industrial level where microwaves are being used as a heating tool in organic syntheses.

The biggest attraction of microwaves in this field is the dramatic enhancement of reaction rates. An example is the Suzuki–Miyaura crossing coupling reaction, which when performed with microwaves occurs in a reaction time of about 50 s, whereas when using the traditional heating methods the reaction times are of several hours [27]. Similar findings have been reported in various other organic reactions. Several synthetic examples of solvent-free and noncatalytic reactions have also been reported. Not surprisingly then, the use of microwaves in chemical syntheses is particularly attractive in the field of Green Chemistry. Many excellent books are now available that describe some of the details of microwave-assisted organic syntheses [28].

1.7 Microwave Chemical Equipment

The inner intricacies in commercial microwave chemical equipment cannot be judged by looking at the steel case. The equipment consists of various parts to apply microwave radiation to a sample. The various parts included in microwave chemical equipment are explained below. An applicator is needed to apply microwaves to a sample either in single-mode (mono-mode) operation or in multi-mode operation. In single-mode operation, the sample is placed at the maximum location of the standing wave. The resonance of the microwaves in the single-mode applicator can be adjusted with the iris and the plunger. Heating of the sample solution in a reactor can be achieved in the single-mode microwave apparatus of Figure 1.4a either at positions of maximal electric field (*E* field; position (i)) density or at the maximal magnetic field (*H* field; position (ii)) density within the waveguide. It is not possible to change the strength of the electric field and the magnetic field under multi-mode operation. On the other hand, the sample size is subject to restrictions. One of the advantageous features of the multi-mode applicator is that it can irradiate large samples. To apply microwaves uniformly, a stirrer and/or a turntable are used. Therefore, the heating efficiency of a sample is lower than under single-mode operation. The multi-mode applicator is used in a domestic microwave oven.

The composition of a simple microwave chemical equipment consists of an applicator connected to the microwave generator (magnetron, power source, and control unit) by a waveguide. Low-cost microwave commercial ovens consist of the same components. On the other hand, when performing highly precise experiments, it is possible to add additional parts to the equipment. For example, the parts used in a microwave chemical reactor of a single-mode system are displayed in the schematic of Figure 1.4a, whereas the actual system is pictured in Figure 1.4b. An iris, a three-stub tuner, a power monitor, and an isolator are connected between the microwave generator and the single-mode applicator. A short plunger is also connected to the single-mode applicator. The impedance between the equipment and the sample can be adjusted with the short plunger and the three-stub tuner. The higher efficient microwave radiation can be applied to a sample with impedance matching. On the other hand, an isolator is used to

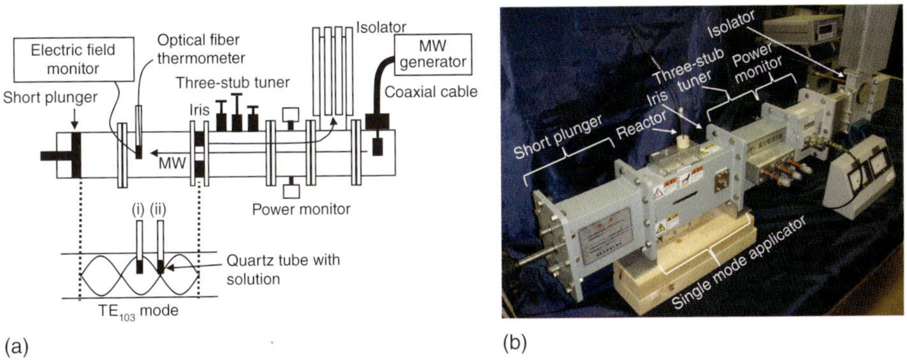

Figure 1.4 (a) Details of the experimental setup and positioning of the samples in the single-mode microwave resonator; (i) maximal position of the electric field (*E* field) density and (ii) maximal position of the magnetic field (*H* field) density. (b) Photograph of the single-mode microwave resonator and the 2.45 GHz semiconductor microwave generator.

protect the equipment from absorbing the reflection wave. The power monitor is used to monitor the incident wave and the reflection wave.

A waveguide is a tube made of metal. An optical fiber is a useful tool currently used to transmit light to its destination. On the other hand, in the domain of microwaves, the waveguide consisting of a metal tube is used to transmit the microwaves. The cross-sectional size of the waveguide tube (Figure 1.5) changes according to the frequency (wavelength) of the microwave radiation generated by the magnetron or by the semiconductor generator [14].

A magnetron generator is generally used in commercial microwave organic synthesis apparatuses, regardless of whether it operates in the single-mode or the multi-mode method. Since high reproducibility has been achieved in recent years in single-mode apparatuses, organic syntheses carried out using single-mode equipment have attracted much attention. As a follow-up on this, a semiconductor generator began to be used in order to obtain high resonance inside the single-mode applicator. However, a disadvantage of the semiconductor generator has been the price and the maximum power available. Nonetheless, semiconductor generators with maximum power of 500 W and with highly accurate microwave frequencies have recently become available at a price similar to the costs of magnetrons. A further spread in the use of semiconductor microwave generators should reduce the price even further. The performance of the electrical appliance was improved remarkably by switching to a transistor semiconductor from a vacuum tube. The spread of small cellular phones depends on the development of a microwave semiconductor oscillator. This course is also having an influence on microwave chemical equipment. The features of the semiconductor and magnetron generators are described later.

The frequency distribution of the microwave radiation from the magnetron generator in microwave cooking ovens is reported in Figure 1.6a, which shows that the frequency of the microwaves is distributed over a large frequency range from 2.25

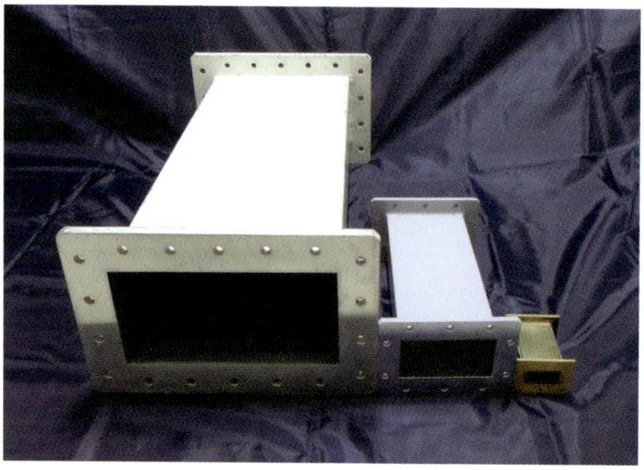

Figure 1.5 Actual photograph displaying the different sizes (and sectional areas) of the waveguides used for 915 MHz microwaves (left), 2.45 GHz microwaves (center), and 5.8 GHz microwaves (right). Reproduced from Ref. [14]. Copyright 2012 by Wiley-VCH Verlag GmbH.

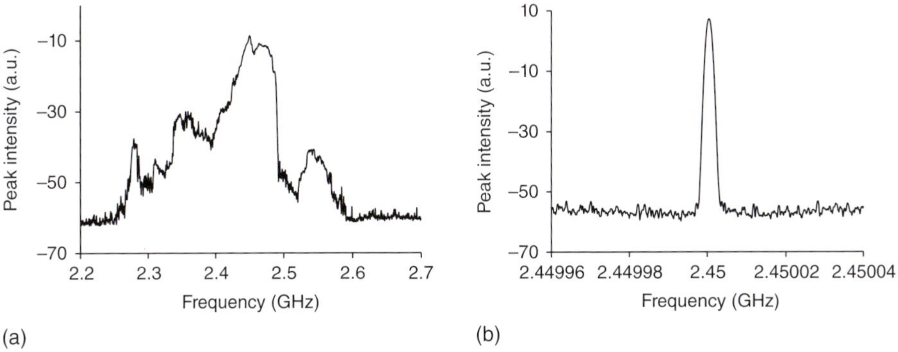

Figure 1.6 Frequency spectral distribution of the 2.45 GHz microwave radiation emitted from the (a) magnetron generator and (b) the semiconductor generator. Reproduced from Ref. [29]. Copyright 2011 by the American Chemical Society.

to 2.60 GHz [29]. The distribution of the microwave frequency changes depending on the characteristics of the microwave generator equipment. Note that narrower frequency waves can be generated with a magnetron of high accuracy. By contrast, the semiconductor generator produces microwaves only within a very narrow frequency range of 2.4500 ± 0.0025 GHz (Figure 1.6b). In the case of the magnetron generator, the microwave input power distribution reflects the widely dispersed frequency distribution. Therefore, the output of the actual 2.45 GHz microwaves is smaller than the input power. On the other hand, when using a semiconductor microwave generator, the heating can progress efficiently because the microwave input power is concentrated at the 2.45000 GHz frequency. The microwaves can resonate better in a single-mode apparatus.

More importantly, a different advantage of a semiconductor microwave generator is the compact size of the equipment. For example, the power supply of the magnetron necessary for a 1-kW input power weighs 6 Kg, whereas the

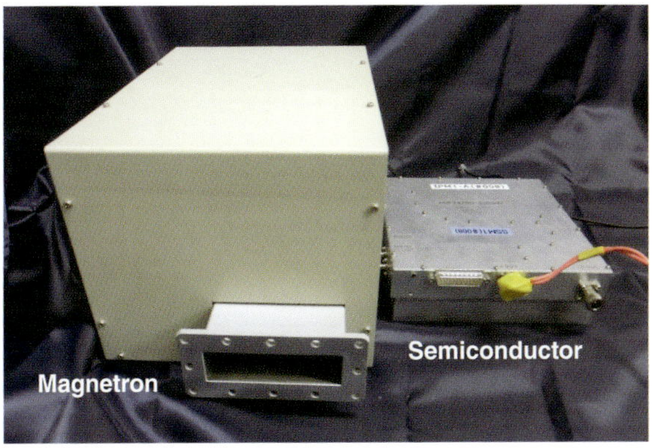

Figure 1.7 Photograph of a GaN semiconductor generator (right) and a magnetron generator (left).

power supply of the semiconductor generator necessary for a 1-kW input power weighs only 0.5 Kg. A photograph of a magnetron generator and a semiconductor generator is displayed in Figure 1.7. Clearly, the semiconductor transmitter is smaller. Microwaves from a semiconductor generator are distributed through a coaxial cable, not a waveguide. Therefore, a flexible coaxial cable can be used for connecting a chemical reactor to a semiconductor generator. It is also possible to install a semiconductor generator in a small available space. Also important, the lifetime of a semiconductor generator is very long (20 years) compared to a magnetron generator (about 500–1000 hrs). For the input, microwave radiation of hundreds of watts is generally sufficient in many organic syntheses carried out at the laboratory scale.

The semiconductor generator can reproduce similar temperature rise conditions in a liquid sample compared with the magnetron generator using remarkably lower microwave input power. Moreover, the semiconductor generator is an optimal microwave emitter in frequency-dependent reactions (processes) because the frequency can easily be set in the semiconductor generator. A demonstration model using these advantages is displayed in Figure 1.8 that shows the inner chamber of a commercial microwave cooking oven equipped with two GaN semiconductor generators (maximum power, 250 W). Each semiconductor generator is connected to directional antennas inside the oven (top and bottom positions). The microwave radiation from the semiconductor transmitter consists of a wave of regular frequency, and so it is possible to synthesize by the phase control of the microwaves emitted from the two antennas depicted in the oven. Accordingly, selective heating of food can easily be achieved in such microwave cooking ovens. For instance, if two different foodstuffs are placed on a dish (meat and potatoes), it is possible to selectively heat only the desired foodstuff (e.g., meat). Moreover, even though the two foodstuffs may have different dielectric

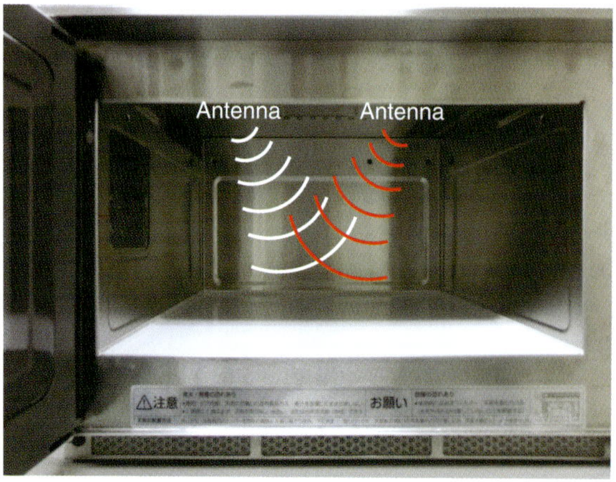

Figure 1.8 Photograph of the inner chamber of a microwave apparatus with microwaves emitted from the two semiconductor generators (GaN; maximum power at 250 W).

loss factors, it is possible to heat both uniformly. Such a microwave oven can also be used for chemical reactions and materials processing.

1.8 Chemical Reactions Using the Characteristics of Microwave Heating

Materials synthesis possessing nanoscale mesoscopic structures, for example, SnO_2/TiO_2, using SnO_2 (high microwave heating efficiency) and TiO_2 (low microwave heating efficiency) has been reported by Takizawa and coworkers [30] using microwave selective heating (Figure 1.9). Such heating by the microwaves produces a supersaturated solid solution, after which the nanoscale mesoscopic structures are generated in the microphase.

When titanium metal is heated under an air atmosphere, TiO_2 is formed by oxidation, whereas titanium nitride (TiN) is generally prepared by a sputtering method, which is expensive and requires equipment of a complicated organization and not least coating of an uneven sample presents some difficulties. On the other hand, when microwave heating is used, it is possible to synthesize TiN easily from a titanium metal substrate under air conditions at 800 °C for 10 min (Figure 1.10) [31]. TiN possesses excellent wear resistance, corrosion resistance, and thermal conductivities. Mechanistic details of the preparation of TiN have not been elucidated.

Microwave calcination equipment for the fabrication of nanometallic inks is displayed in Figure 1.11a. The lower melting point of the metal nanoparticles compared with bulk metal makes the former useful in new electronic wiring, among other applications. Since differential heating can be carried out in the atmosphere

Figure 1.9 Transmission electron microscopy (TEM) image of SnO_2/TiO_2 nanoscale mesoscopic structures. Photograph provided by courtesy of Prof. H. Takizawa, Tohoku University, Japan.

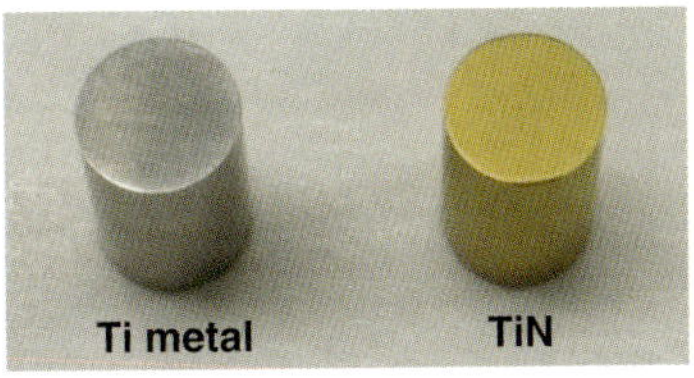

Figure 1.10 Pictures of naked Ti metal and TiN after microwave irradiation with a commercial microwave oven. Photograph provided by courtesy of Prof. H. Takizawa, Tohoku University, Japan.

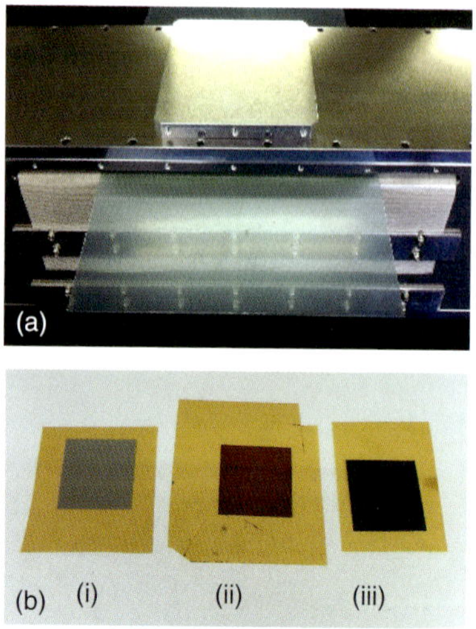

Figure 1.11 (a) Microwave nanometallic ink continuous processor system and (b) microwave processing of (i) nano-Ag ink, (ii) nano-Cu ink, and (iii) nano-CuO ink on a PET plastic base. Photographs provided by courtesy of Fuji Electronic Industrial Co. Ltd, Japan.

under microwave irradiation, such plastics as polyethylene terephthalate (PET) can form the base of the electronic wiring. When the pattern on the plate is written with a paste of either silver or gold, sparks will result if microwave irradiation was to be carried out in a regular microwave oven. However, a significant advantage of the microwaves' magnetic field heating is that no sparks occur when heating the metal nanoparticles. Low-temperature calcination of an indium tin oxide (ITO) film or a ZnO film, and recovery of a rare earth (among others) are possible with such a methodology.

An additional advantage of the microwaves' magnetic field heating method is on the coating of a PET plastic base to form Ag, Cu, and CuO nanoparticle ink (Figure 1.11b). Generally, oxidation of the surface of copper occurs when heated by a conventional heating method in an open atmosphere. However, with microwave magnetic field heating, even if heating was carried out in the presence of air, the copper surface is not oxidized because oxidation would drastically change the electrical properties of the ink. Clearly the microwave method that avoids formation of an oxide film is an effective methodology in such materials processing.

A relation between oxygen and the microwaves also seems appropriate in solution chemistry. To the extent that some reactions are oxygen sensitive, degassing of dissolved oxygen in water and in 2-propanol solvent was examined by microwave heating [32]. Degassing dissolved oxygen from a water sample is more efficient when the solution is exposed to microwave irradiation (heating) than when heated by conventional means (oil bath) under otherwise identical temperature conditions. The efficacy of the microwaves was greater at the lower frequency of 915 MHz than with the more commonly used frequency of 2.45 GHz as a result of the greater penetration depth of the 915 MHz microwaves into the reactor. Degassing was carried out by exposing the water sample to 2.45 GHz microwaves with the sample located within the waveguide at positions rich in either the electric field or magnetic field densities ($T = 80\,°C$). Microwaves

Figure 1.12 Photograph of the solutions before (a) and after (b–d) the synthesis of cyclohexanol from cyclohexanone in the presence of 2-propanol and Wilkinson's catalyst on removing dissolved oxygen by the pipette degassing method after a 10-min heating period: (b) 915 MHz, (c) 2.45 GHz, and (d) oil bath. Reproduced from Ref. [32]. Copyright 2014 by Elsevier B.V.

were again more effective than conventional heating (38% vs. 30%) in degassing dissolved oxygen in 2-propanol solvent. Figure 1.12 illustrates a photograph of the reaction solutions degassed with argon purging the solution with a Pasteur pipette after heating for a 10-min period [32]. The initial light yellow solution (Figure 1.12a) changed to a brownish color after this time period with the color intensity differing according to the heating method used. The intensity of the brownish color increased in the order: 915 MHz (Figure 1.12b) $<$ 2.45 GHz (Figure 1.12c) $\ll$ oil bath (Figure 1.12d). The solution heated with the oil bath displayed the most intense color change (Figure 1.12d), suggesting some oxidative decomposition of Wilkinson's catalyst present in the synthesis of cyclohexanol from cyclohexanone in 2-propanol solvent. Similar observations of color change were made when the initial solution that had been purged with oxygen gas was heated. Evidently, degassing the oxygen-rich 2-propanol solution was slow compared to the inactivation of Wilkinson's catalyst. With regard to the effects of microwave irradiation versus conventional heating on product yields in the conversion of cyclohexanone to cyclohexanol in 2-propanol media and in the presence of Wilkinson's catalyst, $ClRh(PH_3)_3$, little difference was found in product yields irrespective of whether the reaction was activated by microwave irradiation or by conventional heating from an oil bath (81–86% after 10 min). However, argon purging the solution with a Pasteur pipette did reveal some changes in product yields: about 13% by conventional heating, 37% by 2.45 GHz microwaves, and 41% with 915 MHz microwaves also after 10 min. Wilkinson's catalyst appeared more stable under microwave irradiation than under conventional heating.

Organic syntheses have been carried out using some of the features of microwave heating. The efficiency of microwave heating changes according to the dielectric loss of a substance. Organic syntheses that use this characteristic have been reported by Raner [33]. After commencement of microwave heating a two-phase water/chloroform system (1 : 1 by volume; 100 ml) for 40 s, the temperatures of the aqueous and organic phases were 105 and 48 °C, respectively, owing to differences in the dielectric properties of the solvents. A sizable differential could be maintained for several minutes before cooling. Comparable conditions would be difficult to obtain by traditional heating methods. Differential heating is particularly advantageous in carrying out Hofmann elimination reactions (Figure 1.13) [33]. A mixture of *N*-[2-(4′-ethoxybenzoyl)ethyl]-*N*,*N*,*N*-trimethylammonium iodide, water, and chloroform was heated with stirring for 1 min at 110 °C (temperature of the aqueous phase). During the reaction, the product 4′-ethoxyphenylvinyl ketone was extracted and diluted into the poorly microwave-absorbing but cooler organic phase. The mixture was then subjected to rapid post-reaction cooling; the ketonic monomer was obtained in 97% yield. Selective heating can occur in a heterogeneous catalyzed reaction.

A study that considered microwave frequency effect has also been reported. For example, the usefulness of the 5.8 GHz microwaves was demonstrated by carrying out the organic solvent-free synthesis of 2-allylphenol through the Claisen rearrangement process, as well as the synthesis of the $C_{12}-C_2-C_{12}$ Gemini surfactant in ethanol solvent to demonstrate the usefulness of the 915 MHz microwave frequency [15].

Microwave chemistry has been a subject that has generated much excitement of late. Prompt and high-quality syntheses can be achieved using the microwave technique. Some of the fundamental research in microwave radiation has been developed by industry. The next stage will involve the improvement of efficient chemical syntheses as might be required for possible industrial applications in the preparation of significant quantities of high-value chemicals. For this to occur, however, it will necessitate a combination of microwave chemistry with highly coherent microwaves as established in the communication field. In this regard, such coherence of electromagnetic waves is typically found in lasers. The amplification by interference of a highly coherent wave promises to enhance the efficiency of microwave chemistry accompanied with high energy saving. However, it is very difficult for a chemist to understand, let alone master microwave radiation.

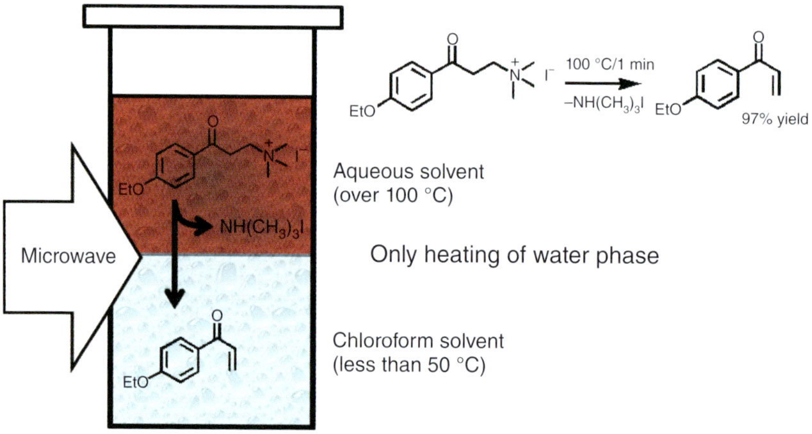

Figure 1.13 Hofmann elimination reaction taking place by selective heating. Reproduced from Ref. [33]. Copyright 1995 by the American Chemical Society.

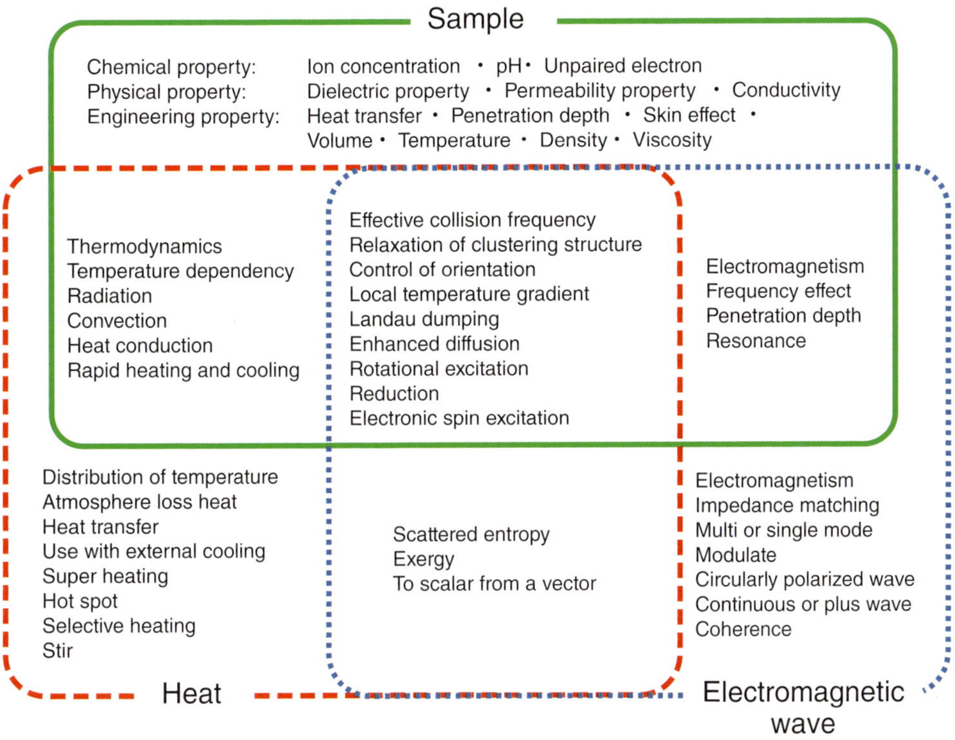

Figure 1.14 Factors of the samples, heat, and electromagnetic waves (microwaves) in the fields of chemical reactions and materials processing; relationships between each factor are also shown. Reproduced from Ref. [34]. Copyright 2013 by S. Horikoshi.

Properties of the sample (i), heating the sample (ii), and the electromagnetic waves (iii) must always be controlled. Various factors impact microwave chemistry. The interconnections between these factors are summarized in Figure 1.14 [34].

1.9
Microwave Frequency Effect in Chemical Syntheses

Microwaves are electromagnetic waves that span the frequency range 0.3–300 GHz so that there is no need to adhere to the commonly used 2.45 GHz frequency. When examining frequency effects, two barriers have obstructed the researcher's curiosity. In the first, the microwave device that can generate different microwave frequencies have tended to be very expensive compared to devices that operate at the common frequency of 2.45 GHz. Recently, however, chemical reaction equipment that can generate different frequencies has become available at costs otherwise similar to those available at the 2.45 GHz frequency. The second barrier has been with the industrial, scientific, and medical (ISM) bands that have been reserved internationally for the use of radio frequencies employed for ISM purposes other than communications [35]. Examples of applications in these bands include radio frequency process heating, microwave heating source, and medical diathermy machines. The powerful emissions of these devices can create electromagnetic interference and disrupt radio communications that use the same frequency. Accordingly, these devices have been limited only to certain frequency bands. The ISM bands are defined by the International Telecommunication Union – radio communications sector (ITU-R). The globally used frequencies are 2.45, 5.8, and 24 GHz; other frequencies are established by each country (Table 1.2) [14].

For several years, the microwave (MW) frequency response and impact on chemical reactions have been examined in only but a few cases. Research reports regarding frequency effects in microwave-assisted organic syntheses have tended to be rather scarce. In the field of inorganic chemistry, Möller and Linn [36] used 5.8 GHz microwaves to calcine ceramics, and Takizawa and coworkers [37] utilized 28 GHz microwaves for the rapid synthesis of $(In_{0.67}Fe_{0.33})_2O_3$, whereas 5.5 GHz microwaves were used [38] in the rapid syntheses of cryptomelane-type manganese oxides for catalysis. In the field of organic chemistry, Gedye and Wei [39] used a variable-frequency microwave oven to effect the Knoevenagel reaction of acetophenone with ethyl cyanoacetate in the presence of piperidine (no solvent; microwave frequency, 8.1 GHz); with anisole as the solvent, the reaction was carried out at 12.2 GHz. Séguin and coworkers [40] reported a new IR reactor cell built with a microwave cavity operating at a frequency of 5.8 GHz; this equipment allowed efficient and time-resolved heating of the catalyst. The authors also showed that the 5.8 GHz microwave frequency is the optimal frequency for catalyst heating. Microwave-induced oligomerization of methane

Table 1.2 Frequency allocation for industrial, medical, and scientific (ISM) purposes in the range 0.4–25 GHz.[a]

Frequency (GHz)	Tolerance level (GHz)	Area permitted
0.43392	0.43305–0.43479	Austria, Netherlands, Portugal, West Germany, Switzerland, Africa
0.896	0.886–0.906	United Kingdom
0.915	0.902–0.928	North and South America
2.375	2.325–2.425	Albania, Bulgaria, Hungary, Romania Czechoslovakia, USSR
2.450	2.400–2.500	Worldwide except where 2375 MHz is used
3.390	3.1866–3.5934	The Netherlands
5.800	5.725–5.875	Worldwide
6.780	6.765–6.795	The Netherlands
24.125	24.000–24.250	Worldwide

a) Reproduced from Ref. [14]. Copyright 2012 by Wiley-VCH Verlag GmbH.

with nickel, iron, and activated carbon catalysts has been examined relative to microwave frequency and other parameters by Conde and Suib [41]. When frequency was increased, the activity of the catalysts was enhanced. Watanabe and coworkers have proposed microwave solvolysis as a pretreatment of wood biomass using 5.8 GHz microwaves [42].

Both the chemical structure of the substrates and the frequency of the microwave radiation have been shown to be important parameters in the synthesis of 3,6-diphenyl-4-*n*-butylpyridazine through a Diels–Alder process [43], in the syntheses of [bmim]BF_4 ionic liquid [44], in the synthesis of cyclohexanol with Wilkinson's catalyst [32], in the synthesis of C_{12}–C_2–C_{12} Gemini surfactant [15], in preparing 4-methylbiphenyl using the Suzuki–Miyaura coupling reaction [14], in nanoparticle synthesis [45], in photocatalytic wastewater treatments [46], and in the enzymatic proteolysis of a peptide [47]. Despite the preponderance of worldwide microwave ovens (including commercial systems used by research chemists) that operate at 2.45 GHz, the use of other microwave frequencies should not and must not be disregarded in furthering the development of microwave chemistry and the possible scale-up of microwave-assisted organic syntheses. For instance, using a single-mode applicator, microwaves at the 915 MHz frequency (wavelength: 32.79 cm) can permit irradiation of a reactor of larger size because of the inherent larger size of the waveguide (sectional area of the waveguides: $24.77 \times 12.38\ cm^2$) compared to the 2.45 GHz (wavelength: 12.24 cm) waveguide (sectional area of the waveguides: $10.92 \times 5.46\ cm^2$) [15]. Furthermore, heating large quantities of alcohols is remarkably effective at the lower frequency compared to the higher microwave frequencies, namely 915 MHz and 2.45 GHz.

At the industrial scale, it is well known that the extent of energy conversion (more than 80%) and the lifetime (more than 10 000 h) of a 915 MHz magnetron are two attractive features of this frequency than those of the 2.45 GHz magnetron (70% and 8000 h). Conditions for which 915 MHz was more suitable for making large-sized equipment have been met. On the other hand, heating a nonpolar solvent with 5.8 GHz microwaves is particularly effective owing to the higher densities of the electric and magnetic fields. The wavelength (sectional areas of the waveguides: $4.0 \times 2.0\ cm^2$) of the 5.8 GHz (wavelength: 7.46 cm) is most suitable for heating a small sample because it is short compared with 2.45 GHz (Figure 1.5). It is difficult from the point of view of controlling the temperature to heat a small quantity with microwaves [48]. The domestic multi-mode technique provides a field pattern with areas of high and low field strengths, such that the temperature of small samples (e.g., biological materials) can vary drastically between different positions of the samples, which could lead to formation of localized hot spots and cold spots. For example, the microwave apparatus proposed for enzymatic reactions at the microliter scale can achieve strict temperature control when using a 5.8 GHz microwave system [47] that incorporates a semiconductor generator. Therefore, power-saved microwaves (microwatt level) can stably irradiate small-scale biomaterial samples (Figure 1.15) [47].

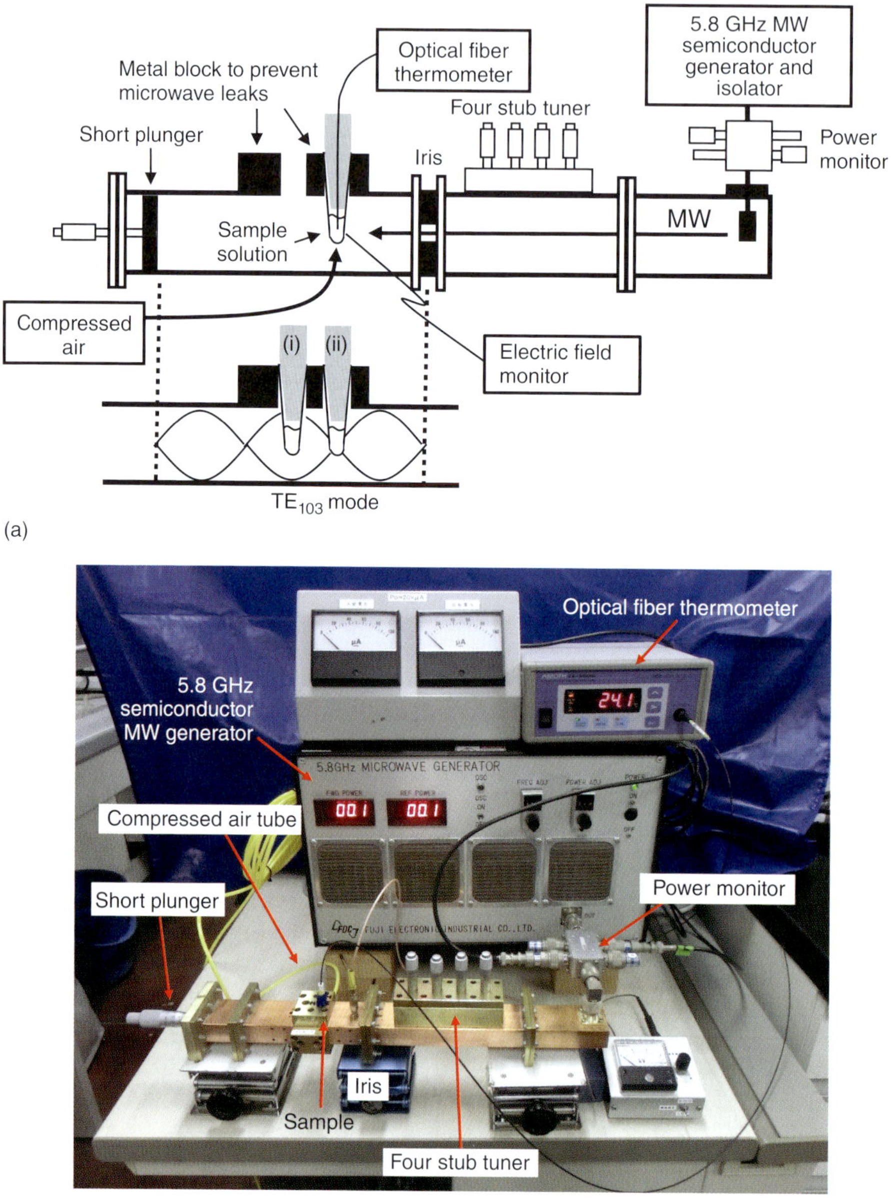

(a)

(b)

Figure 1.15 (a) Details of the experimental setup and positioning of the samples in the single-mode 5.8 GHz microwave resonator; (i) maximal position of the electric field (*E* field) density and (ii) maximal position of the magnetic field (*H* field) density. (b) Photograph of the single-mode microwave resonator and the 5.8 GHz semiconductor microwave generator; the photograph also shows the actual position of the sample at the *H* field maximum. Reproduced from Ref. [47]. Copyright 2015 by Elsevier B.V.

1.10 Summary

The fields of microwave chemistry and material processing have grown immensely since the 1990s, and considerations from a physicochemical point of view are advancing the microwave effect. Studies of chemical processes using the features of microwave heating have also become popular. Choosing the optimal frequency of the microwaves for each situation can promote microwave chemistry further. Moreover, the use of several microwave frequencies could also prove beneficial to the chemical industry in microwave organic chemistry, as it may open up new applications of microwave-assisted organic syntheses, such as solvent-free reactions and supported reactions, as well as the preparation and reactions of inorganic solids, particularly where solvents cannot be heated by the most commonly used 2.45 GHz microwaves.

References

1. Horikoshi, S. and Serpone, N. (2013) in *Microwaves in Nanoparticle Synthesis: Fundamentals and Applications*, Chapter 3, (eds S. Horikoshi and N. Serpone), Wiley-VCH Verlag GmbH, Weinheim, pp. 39–54.
2. Spencer, P.L. (1950) Method of treating foodstuff. US Patent 2,495,429, Jan. 24, 1950.
3. McArthur, E.D. (1993) Electric Heating Apparatus, US Patent 1,900,573, March 7, 1933.
4. Cherbański, R. and Molga, E. (2008) Intensification of desorption processes by use of microwaves – An overview of possible applications and industrial perspectives. *Chem. Eng. Process.*, **1**, 48–58.
5. Eskicioglu, C., Terzian, N., Kennedy, K.J., Droste, R.L., and Hamoda, M. (2007) Athermal microwave effects for enhancing digestibility of waste activated sludge. *Water Res.*, **41**, 2457–2466.
6. Agilent Technologies (2006) Agilent Basics of Measuring the Dielectric Properties of Materials, Application Note, June 26, 2006; see *http://www3.imperial.ac.uk/pls/portallive/docs/1/11949698.PDF* (accessed March 2015).
7. Metaxas, A.C. and Meredith, R.J. (1983) *Industrial Microwave Heating*, Peter Peregrinus, London.
8. Peng, Z., Hwang, J.-Y., and Andriese, M. (2012) Magnetic loss in microwave heating. *Appl. Phys. Express*, **5**, 027304-1-3.
9. Horikoshi, S., Sumi, T., and Serpone, N. (2012) Unusual effect of the magnetic field component of the microwave radiation on aqueous electrolyte solutions. *J. Microwave Power Electromagn. Energy*, **46**, 215–228.
10. Horikoshi, S. and Serpone, N. (2011) Microwave frequency effect(s) in organic chemistry. *Mini-Rev. Org. Chem.*, **8**, 299–305.
11. Whittaker, A.G. and Mingos, D.M.P. (1992) Microwave-assisted solid-state reactions involving metal powders. *J. Chem. Soc., Dalton Trans.*, **18**, 2751–2752.
12. Bossavit, A. (1998) *Computational Electromagnetism*, Academic Press, Boston, MA.
13. (a) Roy, R., Agrawal, D., Cheng, J., and Gedevanishvil, S. (1999) Full sintering of powdered-metal bodies in a microwave field. *Nature*, **399**, 668; (b) Cheng, J., Roy, R., and Agrawal, D. (2001) Experimental proof of major role of magnetic field losses in microwave heating of metal and metallic composites. *J. Mater. Sci. Lett.*, **20**, 1561–1563; (c) Cheng, J., Roy, R., and Agrawal, D. (2002) Radically different effects on materials by separated microwave electric and

magnetic fields. *Mater. Res. Innovations*, **5**, 170–177.
14. Horikoshi, S. and Serpone, N. (2012) in *Microwaves in Organic Synthesis*, Chapter 9, 3rd edn (eds A. de la Hoz and A. Loupy), Wiley-VCH Verlag GmbH, Weinheim, vol. 1, pp. 377–423.
15. Horikoshi, S., Matsuzaki, S., Mitani, T., and Serpone, N. (2012) Microwave frequency effects on dielectric properties of some common solvents and on microwave-assisted syntheses: 2-allylphenol and the C_{12}-C_2-C_{12} gemini surfactant. *Radiat. Phys. Chem.*, **81**, 1885–1895.
16. Horikoshi, S. and Serpone, N. (2013) in *Microwaves in Nanoparticle Synthesis: Fundamentals and Applications*, Chapter 2 (eds S. Horikoshi and N. Serpone), Wiley-VCH Verlag GmbH, Weinheim, pp. 25–37.
17. Shimomura, K., Miyazaki, T., Taniguchi, N., and Tutiya, A. (1972) Study of ceramics processing. *Annu. Rep. RIKEN Jpn.*, **48**, 11–19.
18. Kingston, H.M. and Haswell, S.J. (eds) (1997) *Microwave-Enhanced Chemistry*, American Chemical Society, Washington, DC.
19. Sutton, W.H. (1989) Microwave processing of ceramic materials. *Am. Ceram. Soc. Bull.*, **68**, 376–386.
20. See e.g. (a) National Research Council (1994) *Materials Research Advisory Board, Microwave Processing of Materials*, Publication NMAB-473, National Academy Press, Washington, DC; (b) Folz, D.C., Booske, J.H., Clark, D.E., and Gerling, J.F. (eds) (2003) *Microwave and Radio Frequency Applications, Proceedings from the Third World Congress on Microwave and RF Processing*, American Ceramic Society, Westerville, OH.
21. Takizawa, H. (2012) Microwave non-equilibrium chemistry in an inorganic chemistry and material chemistry. *Inst. Elect. Eng. Jpn.*, **132**, 17–19.
22. Wan, J.K.S. (1993) Microwaves and chemistry: the catalysis of an exciting marriage. *Res. Chem. Intermed.*, **2**, 147–158.
23. Wan, J.K.S. and Ioffe, M.S. (1994) Surface heating and energy transfer in pulsed microwave catalytic systems: a microwave-induced acoustic study. *Res. Chem. Intermed.*, **20**, 115–132.
24. Gedye, R., Smith, F., Westaway, K., Ali, H., Baldisera, L., Laberge, L., and Rousell, J.R. (1986) The use microwave ovens for rapid organic synthesis. *Tetrahedron Lett.*, **27**, 279–282.
25. Giguere, R.J., Bray, T.L., Duncan, S.M., and Majetich, G. (1986) Application of commercial microwave ovens to organic synthesis. *Tetrahedron Lett.*, **27**, 4945–4948.
26. Mannhold, R., Kubinyi, H., and Folkers, G. (2006) in *Microwaves in Organic and Medicinal Chemistry*, Chapter 1 (eds C.O. Kappe and A. Stadler), Wiley-VCH Verlag GmbH, Weinheim, pp. 4–5.
27. Namboodiri, V.V. and Varma, V.S. (2001) Microwave-accelerated Suzuki cross-coupling reaction in polyethylene glycol (PEG). *Green Chem.*, **3**, 146–148.
28. See e.g. de la Hoz, A. and Loupy, A. (eds) (2012) *Microwaves in Organic Synthesis*, 3rd edn, Wiley-VCH Verlag GmbH, Weinheim.
29. Horikoshi, S., Osawa, A., Abe, M., and Serpone, N. (2011) On the generation of hot-spots by microwave electric and magnetic fields and their impact on a microwave-assisted heterogeneous reaction in the presence of metallic Pd nanoparticles on an activated carbon support. *J. Phys. Chem. C*, **115**, 23030–23035.
30. Takizawa, H., Hagiya, A., Aoyagi, T., and Hayashi, Y. (2008) Preparation of mesoscopic TiO_2-SnO_2 composite grains by spinodal decomposition under 28 GHz microwave irradiation. *Chem. Lett.*, **37**, 714–715.
31. Takizawa, H. (2009) Inorganic material processing. *J. Jpn. Soc. Colour Mater.*, **82**, 56–60.
32. Horikoshi, S., Matsuzaki, S., Sakamoto, S., and Serpone, N. (2014) Efficient degassing of dissolved oxygen in aqueous media by microwave irradiation and the effect of microwaves on a reaction catalyzed by Wilkinson's catalyst. *Radiat. Phys. Chem.*, **97**, 48–55.
33. Raner, K.D., Strauss, C.R., Trainor, R.W., and Thorn, J.S. (1995) A new microwave

reactor for batchwise organic synthesis. *J. Org. Chem.*, **60**, 2456–2460.
34. Horikoshi, S., Shinohara, N., Takizawa, H., and Fukushima, J. (2013) *Microwave Chemistry*, Sankyo Publishing Co. Ltd.
35. Technical report by Metaxas, A.C. see *http://www.pueschner.com/downloads/MicrowaveHeating.pdf#search='40680ISM band 3390'* (accessed March 2015).
36. Möller, M. and Linn, H. (2006) New microwave frequency 5.8 GHz for industrial applications. *Key Eng. Mater.*, **264–268**, 735–739.
37. Takizawa, H., Uheda, K., and Endo, T. (2000) Rapid formation and growth of bixbyite-type $(In_{0.67}Fe_{0.33})_2O_3$ by 28-GHz microwave irradiation. *J. Am. Ceram. Soc.*, **83**, 2321–2323.
38. Malinger, A.K., Ding, Y.-S., Sithambaram, S., Espinal, L., Gomez, S., and Suib, S.L. (2006) Microwave frequency effects on synthesis of cryptomelane-type manganese oxide and catalytic activity of cryptomelane precursor. *J. Catal.*, **239**, 290–298.
39. Gedye, R.N. and Wei, J.B. (1998) Rate enhancement of organic reactions by microwaves at atmospheric pressure. *Can. J. Chem.*, **76**, 525–532.
40. Séguin, E., Thomas, S., Bazin, P., Bond, G., Henriques, C., and Thibault-Starzyk, F. (2009) Infrared and microwaves at 5.8 GHz in a catalytic reactor. *Phys. Chem. Chem. Phys.*, **11**, 1697–1701.
41. Conde, L.D. and Suib, S.L. (2003) Catalyst nature and frequency effects on the oligomerization of methane via microwave heating. *J. Phys. Chem. B*, **107**, 3663–3670.
42. see e.g. Mitani, T., Suzuki, H., Oyadomari, M., Shinohara, N., Watanabe, T., Tsumiya, T., and Segoh, H. (2008) Study of Microwave Irradiation Systems for Pretreatment of Woody Biomass toward Bio-Ethanol Production. Technical Report, The Institute of Electronics, Information and Communication Engineers (IEICE), SPS2008-18.
43. Horikoshi, S., Iida, S., Kajitani, M., Sato, S., and Serpone, N. (2008) Chemical reactions with a novel 5.8-GHz microwave apparatus. 1. Characterization of properties of common solvents and application in a Diels–Alder organic synthesis. *Org. Process Res. Dev.*, **12**, 257–263.
44. Horikoshi, S., Hamamura, T., Kajitani, M., Yoshizawa-Fujita, M., and Serpone, N. (2008) Green chemistry with a novel 5.8-GHz microwave apparatus; Prompt one-pot solvent-free synthesis of a major ionic liquid: the 1-butyl-3-methylimidazolium tetrafluoroborate system. *Org. Process Res. Dev.*, **12**, 1089–1093.
45. Horikoshi, S., Abe, H., Sumi, T., Torigoe, K., Sakai, H., Serpone, N., and Abe, M. (2011) Access to small size distributions of nanoparticles by microwave-assisted synthesis. Formation of Ag nanoparticles in aqueous carboxymethylcellulose solutions in batch and continuous-flow reactors. *Nanoscale*, **3**, 1697–1702.
46. Horikoshi, S., Sakai, F., Kajitani, M., Abe, M., and Serpone, N. (2009) Microwave frequency effects on the photoactivity of TiO_2: dielectric properties and the degradation of 4-chlorophenol, bisphenol A and methylene blue. *Chem. Phys. Lett.*, **470**, 304–307.
47. Horikoshi, S., Nakamura, T., Kawaguchi, M., and Serpone, N. (2015) Enzymatic proteolysis of peptide bonds by a metallo-endoproteinase under precise temperature control with 5.8-GHz microwave radiation. *J. Mol. Catal. B: Enzym.*, **116**, 52–59.
48. Fermér, C., Nilsson, P., and Larhed, M. (2003) Microwave-assisted high-speed PCR. *Eur. J. Pharm. Sci.*, **18**, 129–132.

Part I
Fundamentals

Microwaves in Catalysis: Methodology and Applications, First Edition.
Edited by Satoshi Horikoshi and Nick Serpone.

2
Loss Mechanisms and Microwave-Specific Effects in Heterogeneous Catalysis

A.E. Stiegman

2.1
Introduction

The application of microwaves to chemical process has increased dramatically in recent years, particularly in the area of organic synthesis, where the reported observation of large accelerations in reaction rate enhancements has been the motivating factor. Studies to date in the area of microwave-enhanced organic chemistry have been well reviewed in a number of articles and books [1–7]. In spite of intense activity in the study of microwave-driven homogeneous reactions, considerable controversy still exists over the degree to which microwave radiation can exert a specific enhancement effect on the rate of these reactions. Conversely, the use of microwaves to drive heterogeneous catalyzed reaction systems is much less controversial as clearly defined mechanisms exist by which microwaves can enhance chemical reactions on solid surfaces [8–12].

The most obvious advantage that microwave radiation affords in driving a heterogeneously catalyzed reaction is the ability to selectively heat the catalyst. Many industrial processes utilizing heterogeneous catalysts are high-temperature processes wherein both components of the reaction (i.e., catalyst and medium) are heated to the temperature required for the reaction to occur.

Microwave heating can, under appropriately designed conditions, selectively heat the catalyst to the temperature required for substrate activation, allowing the medium to remain at a substantially lower temperature. A potential advantage, shown schematically in Figure 2.1, is that rapid activation of the substrate occurs at the hot surface of the catalysts, which also imparts kinetic energy that rapidly ejects the product from the hot surface into the cooler medium, thereby impeding further reaction [9, 13, 14]. In addition, it has also been reported that if this set of conditions is optimized (i.e., very hot catalyst and cool surroundings), then product selectivity can favor the kinetic product over the thermodynamic one. This latter observation is especially true with a gas–solid reaction [15, 16]. The selective heating process can be complex, however, depending on the actual system. Figure 2.1 shows possible heat transfer pathways in a microwave-driven heterogeneously catalyzed system. The microwaves heat the catalyst through direct

Microwaves in Catalysis: Methodology and Applications, First Edition.
Edited by Satoshi Horikoshi and Nick Serpone.

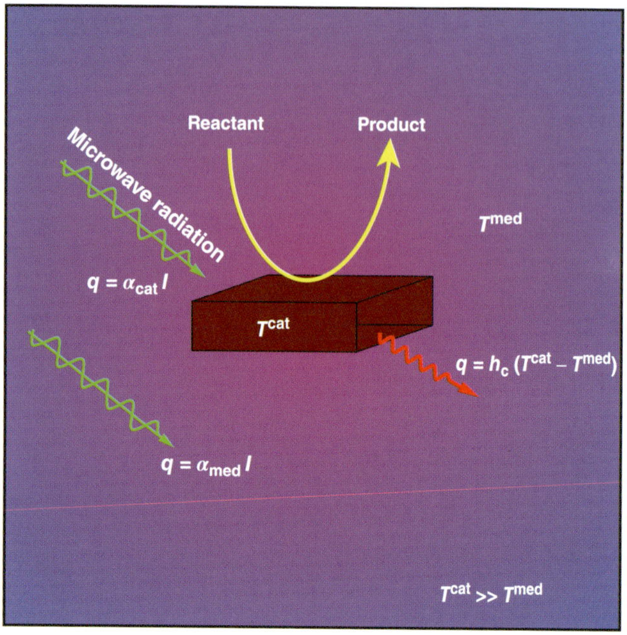

Figure 2.1 Energy transfer processes in the heating of a heterogeneous catalyst by microwave radiation.

absorption of the radiation, the magnitude of which is dictated by the radiation intensity (I) and the absorption cross-section of the catalyst material (α_{cat}). If the reaction takes place in a liquid medium, the solution can also heat through direct absorption of radiation, which will also be dependent on the absorption cross-section (α_{med}) and the intensity of the radiation. In addition, the solution will also be heated indirectly by convective heat transport from the solid catalyst to the medium (h_c). Clearly, in a highly absorbing medium, there may be little temperature difference between the catalyst and the solution ($T^{med} \approx T^{cat}$). As such, *if temperature is the only relevant factor in driving the reaction*, then for highly absorbing solvent systems, there may be little difference between convective and microwave heating. In a gas–solid reaction, however, the opposite is true, and $T^{cat} \gg T^{med}$ as there can be no heating of the gaseous medium ($\alpha_{med} = 0$), and convective heat flow from the catalyst to the surroundings (h_c) will be significantly less than in a condensed phase reaction. This means the gas–solid reactions will exhibit a high degree of selective heating under microwave irradiation and are more likely to provide an energy savings when compared to convective heating.

It is important to note, however, that while selective heating of the solid catalyst is a well-established, advantageous microwave effect in heterogeneous reactions, other in-depth studies, including recent ones from our laboratory, have demonstrated that other microwave-specific processes can dramatically and propitiously alter catalytic activity. In considering other microwave-specific effects, it is important to understand the fundamental physical processes by which microwaves interact with various catalyst materials.

2.2 Heterogeneous Catalyst Systems

There are a number of different classes of heterogeneous catalyst materials, each of which will have different microwave absorption processes. We mention three significant ones here, noting, however, that this breakdown is far from exhaustive [17–19]:

1) *Solid oxide catalysts:* This includes any bulk oxide used to catalyze a reaction at its surface. Typical examples are the simple binary oxides such as SiO_2, Al_2O_3, TiO_2, and ZrO_2, and ternary oxides such as spinels and perovskites. Porous silicate and alumino silicate materials such as zeolites and templated mesoporous sieves also fall into this category.
2) *Metals:* Metal surfaces, particularly late metals such as Ni, Cu, and Ag, catalyze many commercial processes.
3) *Support catalysts:* The supported catalyst is primarily the category defined by an oxide support with an active site deposited on the surface that performs all or part of the catalytic function. The active site can be an isolated transition metal ion or complex or a metal particle.

All of these systems will exhibit different microwave absorption processes that originate from different loss processes. As such, some will exhibit significant microwave-specific rate enhancements. To understand this, it is necessary to understand what those loss processes are and how they affect chemical reactions on the surface of the catalyst.

2.3 Physics of Microwave Absorption

The frequency range for the microwave region is typically designated as occurring from 0.3 to 300 GHz (1 mm–1 m) [20–22]. Within this very broad frequency range, microwave power devices are required to operate within one of the industrial, scientific, or medical (ISM) radio bands to prevent interference with broadcast and communications bands. Microwave ovens utilize a frequency in the S-band of 2.45 GHz (~122 mm). Because of these constraints, any heterogeneous catalyst that we expect to heat using microwaves must have an absorption at 2.45 GHz. Therefore, unlike conventional photochemistry, where the frequency of the applied radiation is matched to the absorption bands of the materials, in microwave-driven processes, the material absorption must match the set frequency of microwave heating devices. Importantly, such limitation would not apply to an appropriately shielded device for specialized applications.

At the frequencies used in heating, the absorption of radiation results in a *relaxation* process in the material, not a *resonance* process. It is the dynamics of these relaxation processes, which are characterized by the functional form of the relaxation time as measured by dielectric spectroscopy that helps define the type of relaxation process occurring in the materials. For a lucid and comprehensive discussion of the dynamics of dielectric relaxation, the reader is referred to a recent review by Richert [23]. What is important to note is that the result of these relaxation processes leads to the generation of heat, with the energy of the photons being insufficient to create an excited state through resonant absorption that can result in a chemical transformation (i.e., bond-making and bond-breaking processes).

Electromagnetic field-induced relaxation is an energy loss process that involves complex interactions of both the E and H fields with the magnetic and dielectric parameters of the material. The amount of heat transferred will depend on the magnitude of the loss, and the final temperature attained will depend on the heat and the thermal properties of the material (i.e., heat capacity and thermal conductivity) [20].

Dielectric loss processes are those associated with interactions of the electric field (E) of the microwave radiation with the material. The propagation of electromagnetic radiation through the material and the associated loss processes are treated using the complex valued form of the permittivity (i.e., dielectric constant), which is given in Eq. (2.1). In Eq. (2.1), the dielectric permittivity, ε, is frequency dependent and is a function of the real (ε') and imaginary (ε'') permittivity. The real part of the permittivity, ε, is the storage component (which approaches the static dielectric constant in the limit of low frequencies) and the imaginary part, ε'', is the dielectric loss. With conducting solids, loss processes due to conduction also occur. Equation (2.2) reflects this loss mechanism where σ is the conductivity and ε_0 is the free space permittivity:

$$\varepsilon(\omega) = \varepsilon'(\omega) - i\varepsilon''(\omega) \tag{2.1}$$

$$\varepsilon(\omega) = \varepsilon'(\omega) - i\left(\varepsilon''(\omega) + \frac{\sigma}{\omega\varepsilon_0}\right) \tag{2.2}$$

A useful measure of the magnitude of the loss process, which dictates the amount of heating that will take place, is given by the loss tangent (Eq. (2.3)), which is the ratio of the imaginary-to-real permittivity:

$$\tan(\delta_E) = \frac{\varepsilon''}{\varepsilon'} \tag{2.3}$$

Loss processes that lead to heating of solid materials can also occur through interactions with the magnetic field. Such loss processes are described in terms of the real and imaginary permeability (μ), in a manner analogous to the dielectric loss (Eq. (2.4)), with the magnitude of the loss being given by a magnetic loss tangent Eq. (2.5):

$$\mu(\omega) = \mu'(\omega) - \mu''(\omega) \tag{2.4}$$

$$\tan(\delta_H) = \frac{\mu''}{\mu'} \tag{2.5}$$

In general, permittivity loss is more often treated than permeability loss as it is the dominant loss process in microwave heating of many materials.

Insertion of the complex permittivity and permeability into the Maxwell equations yields solutions of the form given in Eqs. (2.6) and (2.7). The negative exponential term is the damping factor, which attenuates the electromagnetic field as it penetrates a material. The attenuation factor, α, is a function of the real and imaginary permittivity (Eq. (2.6)) and permeability (Eq. (2.7)). If the medium is a loss at a certain frequency, then the value of ε'' and μ'' will be large and the propagation of the electromagnetic wave will be strongly attenuated.

$$E = |E|e^{-\alpha(\varepsilon)z}e^{i(\omega t-\beta z)} \tag{2.6}$$

$$H = |H|e^{-\alpha(\mu)z}e^{i(\omega t-\beta z)} \tag{2.7}$$

For materials with dielectric loss only (i.e., $\mu'' = 0$), the attenuation factor is given by Eq. (2.8) [22, 24].

$$\alpha = \omega\left(\frac{\mu_0\mu'\varepsilon'\varepsilon_0}{2}\right)^{1/2}\left[\left(1+\left(\frac{\varepsilon'}{\varepsilon_0}\right)^2\right)^{1/2} - 1\right]^{1/2} \tag{2.8}$$

$$D_{\mathrm{p}} = \frac{1}{1\alpha} \tag{2.9}$$

A useful concept in describing microwave heating is the penetration's depth, which is the depth in the materials at which the field strength drops to $1/e$ of its initial value (Eq. (2.9)). When the materials exhibit both dielectric and magnetic loss processes, the total attenuation factor, α, will be a function of ε'' and μ'' [25].

Obviously, for a solid material to be a viable microwave catalyst, significant loss processes must exist at 2.45 GHz for selective heating to occur. For solids that are known to catalyze a particular reaction, the act of selectively heating the solid catalyst may well result in an enhanced reaction rate. A significant issue in the field of microwave catalysis (and in microwave chemistry in general) is whether other rate enhancement processes are also present.

2.4 Microwave Loss Processes in Solids

2.4.1 Dielectric Loss

Heating through dielectric loss is by far the most discussed process in chemical applications of microwave heating. In liquid phase molecular reaction systems, microwave heating occurs primarily through dielectric loss due to electric field-induced orientation and relaxation processes of the molecular dipole as described by Debye [26]. For solid materials, the mechanisms of dielectric loss differ but involve a specific interaction between the solid and the oscillating electric field of the microwave radiation.

There are a number of loss mechanisms that lead to heating through interaction of the E field of the applied microwave radiation with the material. In practice, dielectric loss processes are characterized by their dielectric relaxation dynamics. As described by Jonscher, based on their dielectric responses, loss processes in solids can be divided into two generic processes: "dipolar" and "charge carrier" [27]. Dipolar processes leave no residual polarization upon relaxation while charge-migration processes leave a finite polarizing charge in the system after relaxation. From the standpoint of catalysis, it is important to consider what particular loss processes are operational at the frequencies at which microwave heating is carried out and, further, whether the specific, molecular scale mechanism associated with the possible loss processes can affect the reaction process.

2.4.2 Charge Carrier Processes

2.4.2.1 Conduction Loss

In the low-frequency regime when microwave heating takes place, losses can occur due to electrical conduction. In fact, there is a general correlation between the measured conductivity and the heating rate under microwave absorption at 2.45 GHz over a very broad range of materials [28, 29]. A caveat, however, is that for some of the listed materials, other loss processes, including permeability losses, will also be operative and may be the dominant factor in microwave heating. The conductive contribution to loss is given in Eq. (2.2), where the complex part of the dielectric constant contains the ratio between the conductivity (σ) and the frequency (ω) of the radiation.

The mechanism by which heating occurs is a manifestation of Joule heating when electrical conduction in the materials is induced by the oscillating field. Specifically, when the electromagnetic wave couples with the mobile charges (electrons) within the material, the charges oscillate with the applied field creating an alternating current. The energy loss is due to the inherent resistance of the materials and will be described by Eq. (2.10), where the differential represents the energy (U) loss rate, and R and i are the resistance and current, respectively:

$$\frac{dU}{dt} = Ri \tag{2.10}$$

A notable exception to the correlation between microwave heating and conductivity among various materials is that of pure metals, which, despite their high conductivity, show only modest microwave-induced heating. The origin of this phenomenon arises from the fact that the induced electric current in very good conductors generates a magnetic field that is much stronger than the electric field and lags in phases by almost 45°. The generation of the strong magnetic field and the current it induces causes an exponential damping of the applied electromagnetic field as it enters the materials. In a strong conductor, the penetration depth is referred to as the *skin depth* in metals and is given by Eq. (2.11), where c is the speed of light and the other symbols have the meanings defined previously:

$$\delta \approx \frac{c}{\sqrt{2\pi\mu_0\mu'\omega\sigma}} \tag{2.11}$$

Given that metal surfaces are important catalysts in a number of processes, the skin effect suggests that microwave-driven reactions on metal surfaces are unlikely [19]. In fact, however, metal powders can heat efficiently depending on the metal. This is due to several factors. One is that the skin depth for many important catalytic metals (such as Pt, Pd, and Fe) is in the order of ~3 μm so that powders and small particles will readily heat [30]. In addition, recent studies have found that in highly conductive metal, the loss processes responsible for heating are due to permeability losses owing to the magnetic field of the radiation [31, 32].

2.4.2.2 Space–Charge Recombination

From the standpoint of catalysis, one of the most important dielectric loss mechanisms is space–charge recombination, in which the electric field coerces the movement of charge (electrons or ions), giving a charge-separated state that recombines (Figure 2.2). Charge recombination (relaxation) is inhibited by resistance and other barriers so that recombination of the charges is out of phase with the applied radiation, giving rise to energy loss and consequent heating. The process is shown schematically for charges distributed across a surface where the relaxation time, τ, affects the magnitude of the loss tangent.

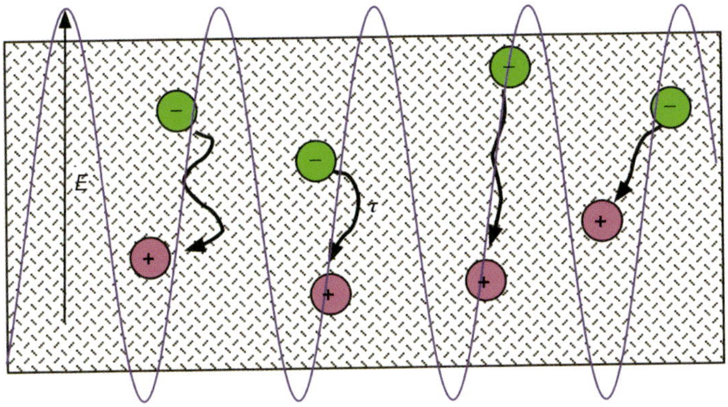

Figure 2.2 Space–charge (Maxwell–Wagner) recombination dielectric loss process in a solid.

This basic description comprises a number of processes that may occur depending on the material. Interfacial or Maxwell–Wagner polarization is the buildup of charges at the interface of heterogeneous dielectrics where a conducting particle is placed in a nonconducting medium. In this process, charges in the conductor, driven by the alternating electric field, become trapped in surface sites and defects and, as described earlier, their recombination leads to heating. Loss processes due to interfacial polarization contribute to the heating of, for example, carbon materials and high permittivity metal oxides [33, 34].

2.4.2.3 Dipolar Loss

Dipolar loss processes in solids occur through interactions of the radiation groups possessing a dipole moment. The archetypal description of this type of loss process, in which radiation couples with an assembly of noninteracting dipoles, was given by Debye [26], who theorized that dipolar molecules in a condensed medium experience coercive rotational orientation of the dipole in the direction of the electric field. The loss process occurs due to "frictional" interactions with the medium so that the rotational motion will be out of phase with the oscillations of the radiation.

In solids, the dipole moments that couple with the radiation exist as localized sites within or on the surface of the solid that can originate from a number of sources depending on the materials. In a bulk solid, vacancy, interstitial, or point defect sites within the crystal lattice can create a localized internal dipole that can couple with the radiation [35]. Insulating solid oxide materials such as alumina, silica, and other metal oxides, which are often used as catalyst supports, are poor microwave absorbers unless they have a significant amount of defect sites or impurities.

More significantly, at least for catalytic consideration, are dipolar surface sites. The surface of a solid material will necessarily have surface sites composed of dangling atoms or groups. For a metal oxide, for example, this can be terminal oxide, a hydroxide group, or a surface vacancy where oxygen was removed. These sites will selectively couple with the radiation, giving rise to dipolar loss processes, which will result in localized heating. This is important as this type of selective heating process can create hot spots on the surface, localized around a specific group (e.g., dangling oxygen), as shown schematically in Figure 2.3.

This can potentially affect catalysis in two ways. The presence of hot spots formed from these types of interactions can *appear* to affect the reaction rate simply because the average temperature of the surface is greater than its measured value. In fact, hot spot formation from this and other sources (i.e., arcing) has been demonstrated for several reaction systems [16, 36, 37]. Other ramifications of this loss process for heterogeneous catalysis occur if the surface groups are reactive and contribute to the mechanism of the reaction. Selective acceleration of a primary reaction step, such as an oxygen transfer process (Figure 2.4) to yield as oxidized substrate, can potentially result in significant rate enhancement. This could be viewed as microwave enhancement of the Mars van Krevelen

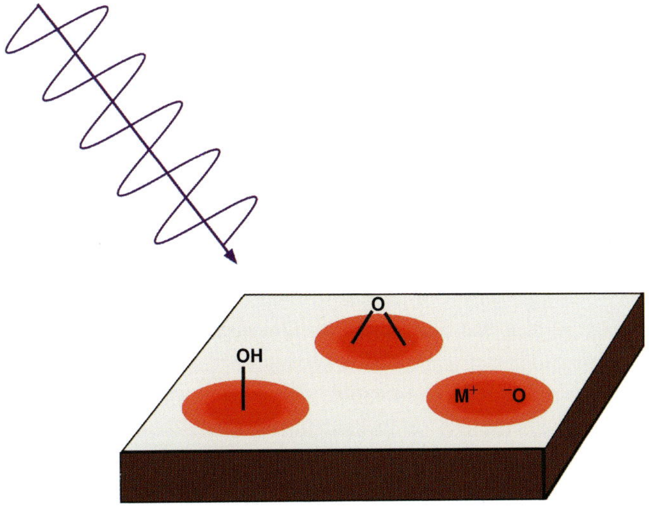

Figure 2.3 Surface sites that can couple to the incident microwave radiation and result in localized heating.

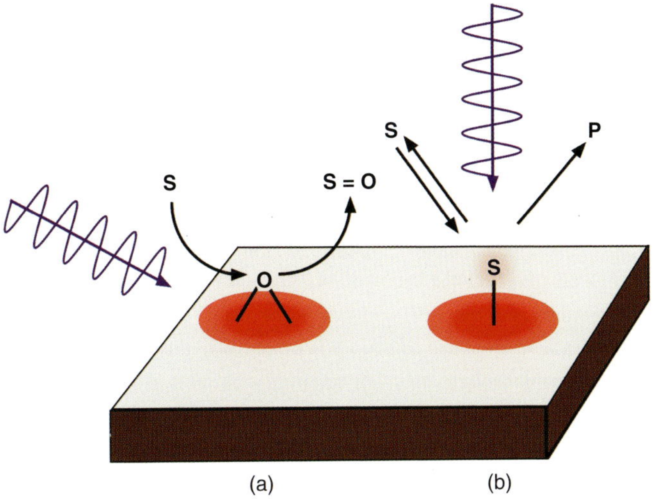

Figure 2.4 Dipolar loss processes that can result in microwave-specific chemical reactivity at the surface through (a) accelerated atom transfer to a substrate (enhanced Mars van Krevelen mechanism) and (b) selective coupling to an adsorbed substrate.

mechanism, which would represent a particular type of microwave-specific effect on a catalytic process [17].

A selective dipolar heating process also occurs for molecules adsorbed at the surface. Any molecule, even nonpolar ones, will have a dipole moment when adsorbed on the surface since they must necessarily lose their center of symmetry. The selective heating of these adsorbed sites can change the adsorption–desorption process but more importantly can also accelerate chemical reactions by orders of magnitude [38–40].

2.4.3 Magnetic Loss Processes

In magnetic materials, loss processes can occur through interactions of the magnetic moment of the materials with the magnetic field component of the radiation. Such loss processes generate heat with the magnitude of the loss being measured by the real (μ') and imaginary (μ'') component of the permeability and the associated loss tangent (Eqs. (2.4) and (2.5)). There are a number of mechanisms associated with magnetic loss processes that ultimately depend on the magnetization of the materials. In general, no heating occurs for diluted paramagnetic ions; the loss process occurs through interactions of the H field with the magnetic moment and is the largest for material with aligned spins (i.e., ferromagnets). These processes were tabulated by Newnham *et al.* [28] from the classic text by Chikazumi and Charap [41]. Among the various processes listed are Eddy currents, where induction currents are established in a conducting material by the oscillating magnetic field, which causes resistive heating of the solid. Many of the loss processes involve a radiation field with a magnetic domain, including hysteresis loss from the irreversible displacement of domain walls and domain wall oscillations.

From the standpoint of heterogeneous catalysis, these magnetic loss processes can be very efficient at selectively heating the substrate. It is not clear, however, whether magnetic loss processes, in and of themselves, will contribute to microwave-specific rate enhancement. Interestingly, one of the properties of microwave instrumentation is that, by the use of specific waveguides, it is possible to separate spatially the electric and magnetic field components of the standing wave of the radiation. Because of this, it is possible to irradiate materials with fields that are predominately electric (E) or magnetic (H). The application of distinct fields (E or H) of the microwave radiation was developed for material applications such as sintering by Roy and Agrawal [31, 42–44]. Their studies provide a comparison of the heating properties of various materials that allows the primary mechanism for heating to be discerned. For example, rapid H field heating occurs with metal particles such as iron and cobalt powders while oxides such as alumina and zinc heat primarily dramatically with the E field. This is consistent with the dominance of magnetic losses such as Eddy current heating in the metals and conduction and dipolar dialectic loss processes dominating in the oxides. In terms of catalysis, the effect of E and H fields on the Pd-catalyzed Suzuki–Miyaura coupling reaction for the synthesis of 4-methylbipyridine was investigated by Horikoshi. It was found that the yield dramatically increased when H field irradiation was used. This was attributed to the avoidance of hot spots, which are prevalent under E-field irradiation and, in this reaction, lead to degradation of the product [45].

2.5
Loss Processes and Microwave-Specific Catalysis: Lessons from Gas–Carbon Reactions

A well-defined example of how a loss mechanism, specifically interfacial dipolar loss processes, can affect a chemical reaction comes from our investigations of gas–carbon reactions [46, 47]. These reactions were selected because carbon is an extremely good microwave absorber and, since they are gas–solid reactions, the only microwave loss processes will be at the carbon. The reactions are ones that comprise the carbon gasification processes related to the generation of synthesis gas ($CO + H_2$) from carbon sources such as coal or biomass. The principle reactions, all of which are in equilibria, are listed in Eqs. (2.12)–(2.15):

$$\text{Steam–carbon} \qquad C + H_2O \rightleftharpoons CO + H_2 \qquad \Delta H = +131\ \text{kJ mol}^{-1} \tag{2.12}$$

$$\text{Boudouard} \qquad C + CO_2 \rightleftharpoons 2CO \qquad \Delta H = +172\ \text{kJ mol}^{-1} \tag{2.13}$$

$$\text{Water–gas shift} \qquad CO + H_2O \rightleftharpoons CO_2 + 2H_2 \qquad \Delta H = -41\ \text{kJ mol}^{-1} \tag{2.14}$$

$$\text{Methanation} \qquad C + 2\,H_2 \rightleftharpoons CH_4 \qquad \Delta H = -75\ \text{kJ mol}^{-1} \tag{2.15}$$

The primary reaction for energy production is the steam–carbon reaction (Eq. (2.12)), which produces synthesis gas (CO and H_2). In conjunction with the water–gas shift reaction (Eq. (2.14)), the steam–carbon reaction is used to produce hydrogen, or, alternatively, the CO and H_2 can produce hydrocarbons via the Fischer–Tropsch process. The Boudouard reaction, named after its discoverer, is the reaction of CO_2 with carbon to produce CO. Both the steam–carbon (Eq. (2.12)) and the Boudouard reactions (Eq. (2.13)) are highly endothermic with positive entropies, so very high temperatures are required to drive the reaction to the right.

For the Boudouard reaction (Eq. (2.13)), comparative kinetic studies of thermal and microwave-driven reactions were carried out as a function of temperature where it was found that the apparent activation energy under microwave conditions decreased dramatically to 38.5 kJ mol^{-1} compared to 118.4 kJ mol^{-1} under convective heating [47]. If the microwave were simply heating the carbon to the reaction temperature and exerting no other effect on the reaction process, we would not expect to see any difference in the activation energy. This suggests the existence of a microwave-specific effect for the forward rate of the reaction. Because all of the other reactions that make up the steam–carbon system (Eqs. (2.12)–(2.15)) are equilibria, changes in the position of the equilibria will necessarily mean a change in the thermodynamics of the process.

Table 2.1 Thermodynamic parameters for thermal and microwave gas–carbon reactions.

Reaction	Thermal[a)]		Microwave	
	ΔH (kJ mol^{-1})	ΔS (J mol^{-1})	ΔH (kJ mol^{-1})	ΔS (J mol^{-1})
Steam–carbon	144.2	158.1	15.2	29.5
Boudouard	180.2	191.2	27.0	43.1
Water–gas shift	−36.0	−58.1	−11.4	−14.1
Carbon–hydrogen	−79.7	−90.4	−9.1	−32.2

a) ΔH and ΔS are intermediate values between the two temperature extremes of the experiment: 764 and 1265 K as calculated from thermodynamic table.

The equilibrium constants, as a function of temperature, for all of the steam–carbon reactions, were measured under microwave irradiation. From these equilibrium constants, the free energy (ΔG) was calculated at each temperature using the standard relationship ($\Delta G = -RT\ln(K_p)$). The free energy for the thermal reaction, over the same range of temperatures, was calculated from the thermodynamic tables. What was observed was that, comparatively, there was a dramatic shift of the equilibrium to the right for the endothermic reactions (Eqs. (2.12) and (2.13)) and to the left for the exothermic reactions (Eqs. (2.14) and (2.15)) under microwave conditions. From the equilibrium constants as a function of temperature, the enthalpy and entropy of each reaction under microwave conditions was determined (Table 2.1). For the steam–carbon reaction, the enthalpy (ΔH) was found to be 15.2 kJ mol^{-1} compared to 144.2 kJ mol^{-1} under thermal conditions. Similarly, ΔH of the Boudouard reaction was found to be 27.0 kJ mol^{-1} under microwave heating compared to 180.2 kJ mol^{-1} under conventional thermal heating. We also observed a similar reduction in the entropy (Boudouard: $\Delta S = 191.2$ J mol^{-1} thermal and 43.1 MW; steam–carbon: $\Delta S = 158.1$ J mol^{-1} thermal and 29.5 MW). The thermodynamic parameters found for all the reactions in the steam–carbon system are given in Table 2.1.

2.5.1 Thermochemical Considerations

Obviously, since thermodynamic functions are state functions, if the microwave were merely heating the carbon, then there should be no change in ΔH or ΔS; in other words, it would not matter which path is used to heat the carbon, as only its final temperature is important. As such, the observed microwave effect on the thermodynamic properties of the reaction must be due to changes in the relative thermochemical kinetics of the forward and reverse reactions, which define the position of equilibrium and which are related to the equilibrium constant based on Eqs. (2.6)–(2.8) [48]:

$$K_p = e^{\frac{-\Delta G}{RT}} = \frac{k_1}{k_{-1}} = \frac{a_1 e^{\frac{-G_1^*}{RT}}}{a_{-1} e^{\frac{-G_{-1}^*}{RT}}} = \frac{a_1 e^{\frac{-(H_1^* - T\Delta S_1^*)}{RT}}}{a_{-1} e^{\frac{-(H_{-1}^* - T\Delta S_{-1}^*)}{RT}}} \tag{2.16}$$

$$\Delta H^\circ = (H_1^* - H_{-1}^*) \tag{2.17}$$

$$\Delta S^\circ = (S_1^* - S_{-1}^*) \tag{2.18}$$

Since the gas phase species do not absorb microwaves, these thermodynamic changes must necessarily arise from changes in the primary reactivity of the carbon surface, or the reactivity of species adsorbed on the surface, through dipolar loss processes.

One of the possible origins of a microwave effect that changes the relative rates of the forward and reverse reactions is through the formation of hot spots on the surface. Hot spot formation is common in microwave-driven reactions and is frequently invoked to explain microwave effects in catalysis. In a classic study by Mingos on the endothermic dehydrogenation of H_2S on a supported Pt catalyst, a similar shift in equilibrium to right was observed [49, 50]. In the case of the gas–carbon reactions, a careful consideration of the thermodynamics allows us to rule out hot spots as a likely mechanism. In particular, the standard expression for the enthalpy of the conventional thermal reaction in terms of the enthalpies of formation of the reactants and products is given in Eq. (2.9), where ΔH_T^{o} is the standard enthalpy of formation corrected for a specific reaction temperature, T, and n is the stoichiometric factor for each product and reactant. What is important to note is that the only species in the gas–carbon reactions that absorbs microwaves is carbon, and it appears only as a reactant in Eqs. (2.12), (2.13), and (2.15). We note here that the water–gas shift reaction (Eq. (2.14)) is not an independent reaction but can be written as a function of the steam–carbon and the Boudouard reaction; as such, carbon does not appear explicitly in the equation. In Eq. (2.20), the contribution to the standard enthalpy of the reaction due to microwave loss processes is shown explicitly as an effective enthalpy ($\Delta H_{\text{microwave}}$):

$$\Delta H_{\text{thermal}} = \left[\sum_i n_i \left(\Delta H_T^{\text{o}}\right)\right]_{\text{product}} - \left[\left[\sum_i n_i \left(\Delta H_T^{\text{o}}\right)\right]_{\text{other reactants}} + (\Delta H_T^{\text{o}})_{\text{carbon}}\right]_{\text{reactants}} \tag{2.19}$$

$$\Delta H_{\text{microwave}} = \left[\sum_i n_i \left(\Delta H_T^{\text{o}}\right)\right]_{\text{product}} - \left[\left[\sum_i n_i \left(\Delta H_T^{\text{o}}\right)\right]_{\text{other reactants}} + (\Delta H_T^{\text{o}})_{\text{carbon}} + (\Delta H_{\text{microwave}})_{\text{carbon}}\right]_{\text{reactants}} \tag{2.20}$$

As can be seen in Table 2.1, the enthalpy of the reaction shifts relative to the thermal reaction exothermically for Eqs. (2.12) and (2.13) and endothermically for Eq. (2.15). This means that the effective enthalpy due to the loss process is positive for some reactions and negative for others (Eqs. (2.12) and (2.13)). However, if the microwave effect is due to hot spot formation, then it means that the actual temperature of the surface is higher than what is measured. This would necessarily mean that the enthalpy of the carbon would increase, and, since it only appears as a reactant, all of the reaction would have to show an endothermic shift. This was not observed.

We hypothesize that there are two dielectric loss mechanisms that can alter the equilibrium position by changing the forward and reverse reaction rates. The changes would be dependent on the specific mechanisms of the reaction and, as such, result in some reactions undergoing an endothermic shift and others an exothermic shift. As discussed earlier, one of the primary dielectric loss processes is a space–charge mechanism (Maxwell–Wagner) where electron–hole pairs are generated and can be trapped at defect sites at the surface [51–53]. The electron–hole pairs are reactive and can potentially accelerate the oxidation of the surface by a substrate (CO_2 and H_2O for the Boudouard and steam–carbon reactions) (Figure 2.5a). The strength of this mechanism is that it is consistent with the known physics of microwave heating of carbon. It also would accelerate the oxidation of the surface, thereby shifting the equilibrium to the right, as was observed.

The second place where microwave radiation can accelerate the forward reaction is by selectively heating the surface oxides that form from the initial oxidation event (Figure 2.5b). This is known to be the rate-determining step for the steam–carbon and Boudouard reactions, and, if they were accelerated by selective coupling of the microwave radiation, it is likely they would accelerate the forward reaction.

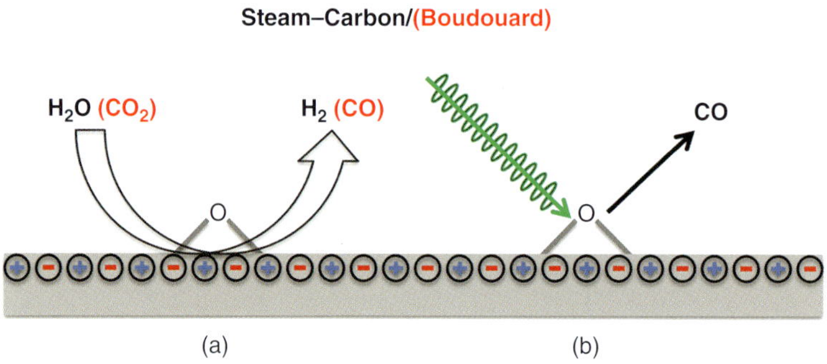

Figure 2.5 Proposed mechanisms for microwave enhancement of gas–carbon reactions: (a) microwave-induced space charges (electron–hole pairs) react with a substrate and (b) the dipoles of the surface oxides couple to the microwave and are rapidly ejected from the surface.

2.6 Final Comments on Microwave-Specific Effects in Heterogeneous Catalysis

In both homogeneous solution reactions and in reactions that are heterogeneously catalyzed, the hypothesis of microwave-specific effects is that the reaction rates under microwave irradiation can differ significantly from what is obtained through conventional convective heating. This has been a topic of great controversy in the organic chemistry community, where there is much prevailing opinion that incident microwave radiation can only heat the solution, and, while that can happen very rapidly, in the end there are no other additional effects possible due to the radiation [54]. Although several recent publications have begun to show this to be a failed hypothesis, the presence of microwave-specific effects in homogeneous reactions is frequently difficult to discern [55–58]. In heterogeneously catalyzed reactions, the concept of microwave effects is much less controversial since the ability of a microwave to selectively heat the catalyst is not disputed. The presence of other microwave-specific effects at the catalyst surface may not necessarily be oblivious but is also less controversial, based on some of the loss mechanisms outlined here.

Much discussion of microwave-specific rate enhancement has involved the concept of thermal versus nonthermal processes. As the name implies, thermal processes proceed through the heat generated by the microwave while nonthermal processes proceed through other interactions of the radiation that propel the reaction by means of processes that are not inherently thermal. Although there is little rationale for it in the synthetic organic community, microwave-specific effects have become synonymous with nonthermal effects [54].

In heterogeneous catalysis, the distinction between thermal and nonthermal processes is also made, though, quite correctly, it is not often used to discredit claims of a microwave effect. In our view, when discussing microwave effects in heterogeneous catalysis, the distinction between thermal and nonthermal effects is of dubious merit. As outlined in this discussion, there are a large number of different loss processes, occurring through different mechanisms, which result in the deposition of heat in the catalyst. Microwave effects can arise from some of these processes based on the specifics of these mechanisms and how they affect reaction steps at the surface. In some cases, the generation of heat may be almost incidental to the mechanism of microwave enhancement. It is also the case that there are nonresonant, nonrelaxation processes that can affect a reaction. In this regard, we point to the study by Horikoshi of the effects of microwave on the photo-assisted reaction of TiO_2 [59].

Acknowledgments

We thank Professor Ranko Richert for helpful discussions. Funding for this work was provided by the National Science Foundation under grant CHE-1112046 and by the Catalysis Science Initiative of the U.S. Department of Energy, Basic Energy Sciences (DE-FG02-03ER15467).

References

1. Kappe, C.O. (2004) Controlled microwave heating in modern organic synthesis. *Angew. Chem. Int. Ed.*, **43** (46), 6250–6284.
2. Kappe, C.O. (2008) Microwave dielectric heating in synthetic organic chemistry. *Chem. Soc. Rev.*, **37** (6), 1127–1139.
3. Kappe, C.O., Dallinger, D., and Murphree, S. (2009) *Practical Microwave Synthesis for Organic Chemists: Strategies, Instruments, and Protocols*, Wiley-VCH Verlag GmbH, Weinheim.
4. Hajek, M. (2002) in *Microwaves in Organic Synthesis* (ed A. Loupy), Wiley-VCH Verlag GmbH, Weinheim, pp. 345–374.
5. Loupy, A. (ed) (2006) *Microwaves in Organic Synthesis*, 1st edn, Vol. 1, Wiley-VCH Verlag GmbH, Weinheim, p. 532.
6. Leadbeater, N.E. (ed) (2011) *Microwave Heating as a Tool for Sustainable Chemistry*, CRC Press, Boca Raton, FL.
7. Horikoshi, S. and Serpone, N. (2011) Features of microwaves in synthetic organic chemistry. *Mini-Rev. Org. Chem.*, **8** (3), 281.
8. Bond, G., Moyes, R.B., and Whan, D.A. (1993) Recent applications of microwave-heating in catalysis. *Catal. Today*, **17** (3), 427–437.
9. Desai, B. and Kappe, C.O. (2004) *Immobilized Catalysts*, Springer-Verlag, Berlin, pp. 177–208.
10. Durka, T., Van Gerven, T., and Stankiewicz, A. (2009) Microwaves in heterogeneous gas-phase catalysis: experimental and numerical approaches. *Chem. Eng. Technol.*, **32** (9), 1301–1312.
11. Horikoshi, S. and Serpone, N. (2014) Role of microwaves in heterogeneous catalytic systems. *Catal. Sci. Technol.*, **4** (5), 1197–1210.
12. Zhang, X.L. and Hayward, D.O. (2006) Applications of microwave dielectric heating in environment-related heterogeneous gas-phase catalytic systems. *Inorg. Chim. Acta*, **359** (11), 3421–3433.
13. Chemat, F., Esveld, D.C., Poux, M., and Di-Martino, J.L. (1998) The role of selective heating in the microwave activation of heterogeneous catalysis reactions using a continuous microwave reactor. *J. Microwave Power Electromagn. Energy*, **33** (2), 88–94.
14. Bogdal, D., Lukasiewicz, M., Pielichowski, J., Miciak, A., and Bednarz, S. (2003) Microwave-assisted oxidation of alcohols using Magtrieve (TM). *Tetrahedron*, **59** (5), 649–653.
15. Gopalakrishnan, S., Munch, J., Herrmann, R., and Schwieger, W. (2006) Effects of microwave radiation on one-step oxidation of benzene to phenol with nitrous oxide over Fe-ZSM-5 catalyst. *Chem. Eng. J.*, **120** (1–2), 99–105.
16. Zhang, X.L., Hayward, D.O., and Mingos, D.M.P. (2003) Effects of microwave dielectric heating on heterogeneous catalysis. *Catal. Lett.*, **88** (1–2), 33–38.
17. Thomas, J.M. and Thomas, W.J. (1997) *Principles and Practice of Heterogeneous Catalysis*, 1st edn, VCH Publishers, New York.
18. Hodnett, B.K. (2000) *Heterogeneous Catalytic Oxidation*, 1st edn, John Wiley & Sons, Ltd, Chichester.
19. Somorjai, G.A. (1994) *Introduction to Surface Chemistry and Catalysis*, John Wiley & Sons, Inc, New York.
20. Hill, J.M. and Marchant, T.R. (1996) Modelling microwave heating. *Appl. Math. Modell.*, **20** (1), 3–15.
21. Jackson, J.D. (1975) *Classical Electrodynamics*, John Wiley & Sons, Inc., New York.
22. Metaxas, A.C. and Meredith, R.J. (1983) in *Industrial Microwave Heating*, IET Power and Energy Series (eds A.T. Johns, G. Ratcliff, and J.R. Platts), The Institution of Engineering and Technology, Herts, pp. 78–81.
23. Richert, R. (2014) *Advances in Chemical Physics*, John Wiley & Sons, Inc., pp. 101–195.

24. von Hippel, A. (1954) *Dielectrics and Waves*, John Wiley & Sons, Inc., New York.
25. Peng, Z.W., Hwang, J.Y., Mouris, J., Hutcheon, R., and Huang, X.D. (2010) Microwave penetration depth in materials with non-zero magnetic susceptibility. *ISIJ Int.*, **50** (11), 1590–1596.
26. Debye, P. (1929) *Polar Molecules*, 1st edn, Chemical Catalog Company, New York.
27. Jonscher, A.K. (1999) Dielectric relaxation in solids. *J. Phys. D: Appl. Phys.*, **32** (14), R57–R70.
28. Newnham, R.E., Jang, S.J., Xu, M., and Jones, F. (1991) in *Microwaves: Theory and Application in Materials Processing* (eds D.E. Clark, F.D. Gac, and W.H. Sutton), American Ceramic Society, Westerville, OH, pp. 51–67.
29. Stein, D.F. and Edgar, R.H. (1994) *Microwave Processing of Materials*, National Academy Press, Washington, DC.
30. Crane, C.A., Pantoya, M.L., Weeks, B.L., and Saed, M. (2014) The effects of particle size on microwave heating of metal and metal oxide powders. *Powder Technol.*, **256**, 113–117.
31. Cheng, J., Roy, R., and Agrawal, D. (2002) Radically different effects on materials by separated microwave electric and magnetic fields. *Mater. Res. Innovations*, **5** (3–4), 170–177.
32. Porch, A., Slocombe, D., and Edwards, P.P. (2013) Microwave absorption in powders of small conducting particles for heating applications. *Phys. Chem. Chem. Phys.*, **15** (8), 2757–2763.
33. Hotta, M., Hayashi, M., Lanagan, M.T., Agrawal, D.K., and Nagata, K. (2011) Complex permittivity of graphite, carbon black and coal powders in the ranges of X-band frequencies (8.2 to 12.4 GHz) and between 1 and 10 GHz. *ISIJ Int.*, **51** (11), 1766–1772.
34. Masud, G., Ghosh, A., and Chaudhri, B.K. (2011) Maxwell Wagner polarization and high dielectric permittivity of charge disproportionate system La0.5Ba0.5FeO3 perovskite. *AIP Conf. Proc.*, **1349** (1), 847–848.
35. Rosenberg, H.H. (1988) *The Solid State*, Oxford University Press, Oxford.
36. Zhang, X.L., Hayward, D.O., and Mingos, D.M.P. (1999) Apparent equilibrium shifts and hot-spot formation for catalytic reactions induced by microwave dielectric heating. *Chem. Commun.*, (11), 975–976.
37. Horikoshi, S., Kamata, M., Mitani, T., and Serpone, N. (2014) Control of microwave-generated hot spots. 6. Generation of hot spots in dispersed catalyst particulates and factors that affect catalyzed organic syntheses in heterogeneous media. *Ind. Eng. Chem. Res.*, **53** (39), 14941–14947.
38. Conner, W.C. and Tompsett, G.A. (2008) How could and do microwaves influence chemistry at interfaces? *J. Phys. Chem. B*, **112** (7), 2110–2118.
39. Vallee, S.J. and Conner, W.C. (2006) Microwaves and sorption on oxides: a surface temperature investigation. *J. Phys. Chem. B*, **110** (31), 15459–15470.
40. Vallee, S.J. and Conner, W.C. (2008) Effects of microwaves and microwave frequency on the selectivity of sorption for binary mixtures on oxides. *J. Phys. Chem. C*, **112** (39), 15483–15489.
41. Chikazumi, S. and Charap, S.H. (1964) in *Physics of Magnetism*, Wiley Series on the Science and Technology of Materials (eds J.H. Holloman, J.E. Burke, B. Chalmers, R.L. Sproull, and A.V. Tobolsky), John Wiley & Sons, Inc., New York, pp. 537–582.
42. Cheng, J.P., Roy, R., and Agrawal, D. (2001) Experimental proof of major role of magnetic field losses in microwave heating of metal and metallic composites. *J. Mater. Sci. Lett.*, **20** (17), 1561–1563.
43. Roy, R., Fang, Y., Cheng, J.P., and Agrawal, D.K. (2005) Decrystallizing solid crystalline titania, without melting, using microwave magnetic fields. *J. Am. Ceram. Soc.*, **88** (6), 1640–1642.
44. Roy, R., Peelamedu, R., Hurtt, L., Cheng, J.P., and Agrawal, D. (2002) Definitive experimental evidence for Microwave Effects: radically new effects of separated *E* and *H* fields, such as decrystallization of oxides in seconds. *Mater. Res. Innovations*, **6** (3), 128–140.
45. Horikoshi, S., Osawa, A., Abe, M., and Serpone, N. (2011) On the generation

of hot spots by microwave electric and magnetic fields and their impact on a microwave-assisted heterogeneous reaction in the presence of metallic Pd nanoparticles on an activated carbon support. *J. Phys. Chem. C*, **115** (46), 23030–23035.

46. Ferrari, A., Hunt, J., Lita, A., Ashley, B., and Stiegman, A.E. (2014) Microwave-specific effects on the equilibrium constants and thermodynamics of the steam–carbon and related reactions. *J. Phys. Chem. C*, **118** (18), 9346–9356.

47. Hunt, J., Ferrari, A., Lita, A., Crosswhite, M., Ashley, B., and Stiegman, A.E. (2013) Microwave-specific enhancement of the carbon–carbon dioxide (Boudouard) reaction. *J. Phys. Chem. C*, **117** (51), 26871–26880.

48. Benson, S.W. (1976) *Thermochemical Kinetics*, 2nd edn, John Wiley & Sons, Inc., New York.

49. Zhang, X.L., Hayward, D.O., and Mingos, D.M.P. (2001) Microwave dielectric heating behavior of supported MoS_2 and Pt catalysts. *Ind. Eng. Chem. Res.*, **40** (13), 2810–2817.

50. Zhang, X.L., Hayward, D.O., and Mingos, D.M.P. (2002) Dielectric properties of MoS2 and Pt catalysts: effects of temperature and microwave frequency. *Catal. Lett.*, **84** (3–4), 225–233.

51. Atwater, J.E. and Wheeler, R.R. (2003) Complex permittivities and dielectric relaxation of granular activated carbons at microwave frequencies between 0.2 and 26 GHz. *Carbon*, **41** (9), 1801–1807.

52. Atwater, J.E. and Wheeler, R.R. (2004) Microwave permittivity and dielectric relaxation of a high surface area activated carbon. *Appl. Phys. A: Mater. Sci. Process.*, **79** (1), 125–129.

53. Atwater, K.E. and Wheeler, R.R. (2004) Temperature dependent complex permittivities of graphitized carbon blacks at microwave frequencies between 0.2 and 26 GHz. *J. Mater. Sci.*, **39** (1), 151–157.

54. Kappe, C.O., Pieber, B., and Dallinger, D. (2013) Microwave effects in organic synthesis: myth or reality? *Angew. Chem. Int. Ed.*, **52** (4), 1088–1094.

55. Chen, P.-K., Rosana, M.R., Dudley, G.B., and Stiegman, A.E. (2014) Parameters affecting the microwave-specific acceleration of a chemical reaction. *J. Org. Chem.*, **79** (16), 7425–7436.

56. Dudley, G.B., Stiegman, A.E., and Rosana, M.R. (2013) Correspondence on microwave effects in organic synthesis. *Angew. Chem. Int. Ed.*, **52** (31), 7918–7923.

57. Rosana, M.R., Hunt, J., Ferrari, A., Southworth, T.A., Tao, Y., Stiegman, A.E., and Dudley, G.B. (2014) Microwave-specific acceleration of a Friedel–Crafts reaction: evidence for selective heating in homogeneous solution. *J. Org. Chem.*, **79** (16), 7437–7450.

58. Rosana, M.R., Tao, Y.C., Stiegman, A.E., and Dudley, G.B. (2012) On the rational design of microwave-actuated organic reactions. *Chem. Sci.*, **3** (4), 1240–1244.

59. Horikoshi, S. and Serpone, N. (2014) On the influence of the microwaves' thermal and non-thermal effects in titania photoassisted reactions. *Catal. Today*, **224**, 225–235.

3
Transport Phenomena and Thermal Property under Microwave Irradiation

Yusuke Asakuma

3.1
Introduction

Over the last two decades, microwave (MW) technology has been explored extensively in materials science and chemistry. Earlier studies have suggested that the enhanced rates of aqueous reaction were caused by the nonthermal effects of microwave [1–3]. Consequently, there are a number of experimental results in different aqueous reactions [4, 5]. Special effect caused by microwave irradiation such as higher yields and selective reactions have been extensively reported [6–8]. On the other hand, nonthermal effects of microwave irradiation on water are not well understood in spite of reported effects on aqueous reactions. On the contrary, special microwave effect has been reported as thermal effect [9]. In order to understand nonthermal effect of microwave, transport phenomena and thermal property under the irradiation are essential. However, microwave leakage to the outside of the reactor must be cared for *in situ* observation. Accordingly, it is necessary to establish noncontact observation tool from the outside of the reactor.

Microwave irradiation is well known for high efficient heating of liquids such as water and alcohol (ethylene glycol), which has higher dielectric constant and loss factors. This chapter focuses on interesting behavior obtained through *in situ* observation in microwave reactor. In particular, characteristics caused by nanobubbles under microwave irradiation are interesting for materials development and process engineering. This chapter introduces experimental investigations on bubble size, the convection behavior, and the surface tension profile for water and ethylene glycol through various *in situ* observation tools. Finally, we hope to clarify nonthermal effect and to gain new insights into enhanced microwave effects on aqueous processes.

Microwaves in Catalysis: Methodology and Applications, First Edition.
Edited by Satoshi Horikoshi and Nick Serpone.

3.2 Bubble Formation

Microwave acceleration has been incorporated into the synthesis of ceramic powders, polymers, and composites in the field of hydrothermal synthesis, as well as emulsion polymerization for fine particles with a monodisperse size distribution [10, 11]. Recently, microwave technology has been applied to industrial processes for material characterization to save energy and time. However, the mechanism of monodisperse fine particle formation under microwave irradiation remains unclear. However, it is difficult for commercial nanoparticle measuring apparatuses to adjust the light source power and the angle between the light axis and detector probe. Accordingly, handmade observation tools for nanoparticle measurement were installed within the reactor.

The movement of particles smaller than a few micrometers is governed by Brownian motion, which causes flow around the particles. The force acting upon Brownian particles is measured using a technique called dynamic light scattering (DLS) [12], which provides the diffusion coefficient and particle size. By introducing an *in situ* observation technique for nanoparticles or bubbles in a microwave reactor, particle movement and bubble formation in liquid are clarified during and after microwave irradiation.

Figure 3.1 shows an overview of the microwave reactor (Micro Denshi Co., Ltd, model). A suspension of monodisperse polystyrene latex (PSL; 100 nm, Duke Scientific Corp.) particles is used as the standard. The PSL particle suspension is added to water or ethylene glycol, and the low-density suspension is poured into an optical quartz cell (10 mm × 10 mm × 40 mm; 4 ml). The cell is set on a sample holder made of polyether ether ketone (PEEK) resin in the center of the waveguide tube in a microwave reactor. The position and angle of the holder

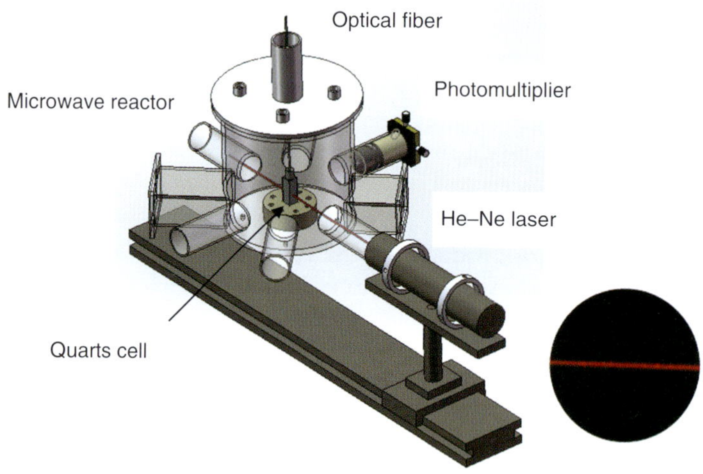

Figure 3.1 Microwave reactor with DLS.

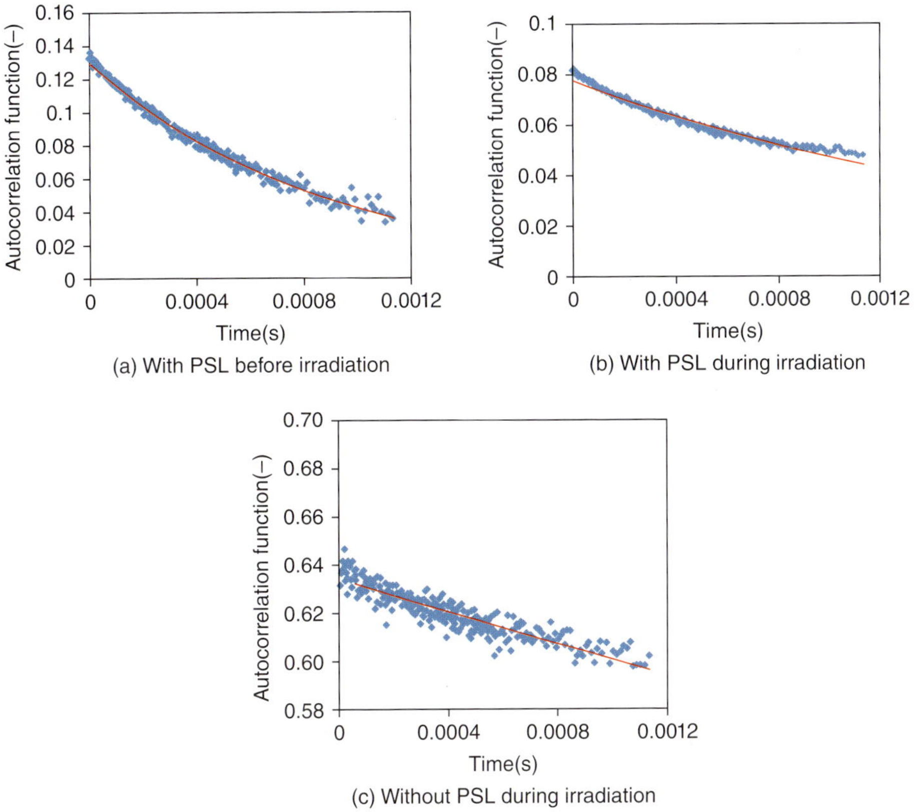

Figure 3.2 Time-averaged autocorrelation function (a) with PSL before irradiation, (b) with PSL during irradiation, and (c) without PSL during irradiation.

are adjusted against the pathway of a He–Ne laser and scattered light detector. When the laser irradiates a sample containing nanoparticles or nanobubbles, light is scattered as shown in Figure 3.1, and the number of scattered photons is counted by a photomultiplier (Hamamatsu Photonics Co., Ltd, model). The time-averaged autocorrelation function of photon, which is shown in Figure 3.2, leads to diffusion coefficient and size for nanoparticle or nanobubble. A fluorescent temperature probe (Anritsu Meter Co., Ltd, FL-2000; Optical fiber: FS100-M) measures the temperature throughout the experiment. Dynamic changes in particle or bubble size during or after microwave irradiation were measured every 10–15 s.

Figure 3.2 shows examples of the time-averaged autocorrelation function (a) with PSL before irradiation, (b) with PSL during irradiation, and (c) without PSL during irradiation in the case of water. The autocorrelation function for the PSL suspension was clear for particle size estimation even under microwave irradiation. For example, the correlation coefficients of the least-squares method for plots and lines in Figure 3.2a, b were almost 1 and about 0.9, respectively. The coefficient became a little worse under irradiation, indicating that the size distribution during irradiation slightly widened, despite the size distribution before irradiation being monodisperse. On the other hand, bubble formation around particles can be considered as a reason for size changes obtained from different correlation functions before, during, and after microwave irradiation. If the dispersion medium does not include nanoparticles, scattering light from nanobubbles is detected during the irradiation as shown in Figure 3.2c. The autocorrelation function of pure water during microwave irradiation did not show good exponential correlation, indicating that precise size measurement is difficult because of wide size distribution or large size. Accordingly, heterogeneous nucleation of bubbles occurred on the quartz wall in the case of no particles in the water. In measurement under microwave heating, nanoparticles played an important role in more stable nucleation of bubble formation around the particles because the sharper size distribution was maintained by the addition of fine particles. On the other hand, since scattering light of bubbles was observed under microwave irradiation in the case of no particles, it has found that nanobubbles in pure water were formed by microwave irradiation.

Figure 3.3 shows size and temperature profiles in water and ethylene glycol (EG) during and after microwave irradiation for irradiation time until 90 °C. Closed and open symbols indicate sizes during and after microwave irradiation, respectively, and lines indicate the temperature profile. The bubble size in water increased during microwave irradiation. Temperature increased with irradiation time and dropped gradually after irradiation. Accordingly, the bubble size was a maximum around the time the microwave was turned off. When the size slowly returned to the initial size, the microwave irradiation effect indirectly continued for sometime as nanobubble formed by water molecule rotation even at temperature below

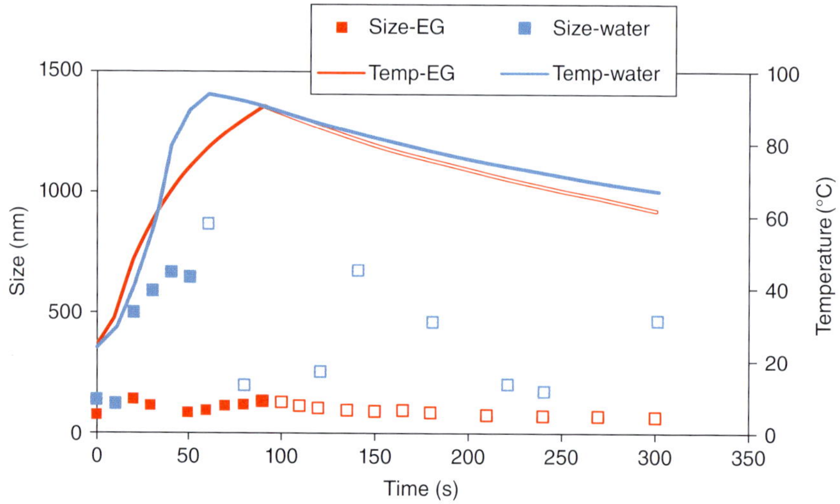

Figure 3.3 Size and temperature profiles for water and ethylene glycol during and after microwave irradiation.

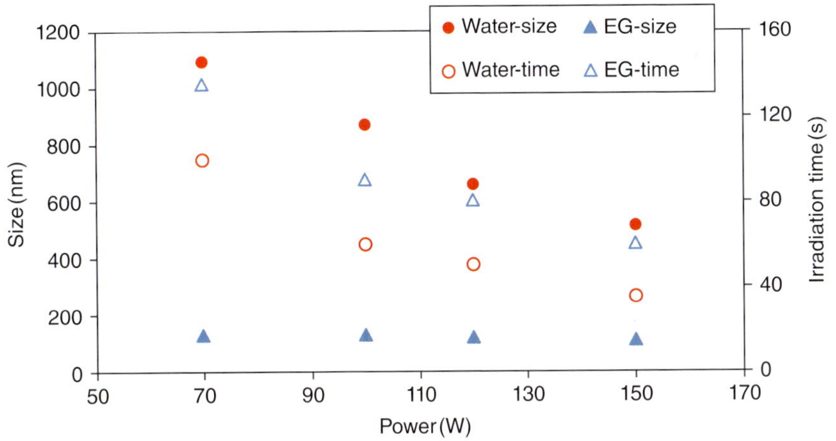

Figure 3.4 Maximum size and irradiation time versus irradiation power during and after microwave irradiation (90 °C).

boiling point. However, in the case of ethylene glycol, bubble size did not become larger so much even at high temperature. The maximum sizes of water and ethylene glycol are plotted against irradiation power in Figure 3.4. Longer irradiation time is more effective for attaining maximum bubble size in water because it takes much time for lower power to reach designated temperature. However, in the case of ethylene glycol, maximum bubble sizes caused by irradiation are not so large although the irradiation time is longer than that of water.

Usually, nanobubble has higher energy because the pressure and temperature are too high theoretically. Moreover, free radicals, which are important for chemical reaction, are generated after the collapse due to the self-pressurizing effect of nanobubbles (bubble contraction) [13]. However, bubble size profiles in water and ethylene glycol could not be explained by irradiation time only. Accordingly, the vapor–liquid phase equilibrium for vaporization and the capability for radical production must be cared when considering nonthermal effect of microwave irradiation.

3.3 Convection

The convection behavior is important in chemical process because it affects heat and mass transfer too much. Usually, the natural convection and Marangoni surface flow are caused by temperature gradient. Moreover, the convection caused by electric and magnetic distributions is interesting and must be predicted for microwave application in aqueous system.

The convection within the pendant droplet, during and after microwave, was monitored with Particle Imaging Velocimetry (PIV) as shown in Figure 3.5. PIV apparatus consists of camera (Sigma Koki Co., Ltd, Model SK-TC202USB-AT) and

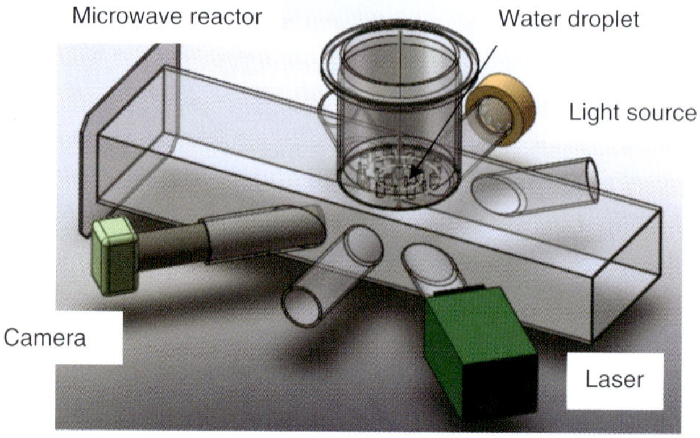

Figure 3.5 Microwave reactor with PIV system.

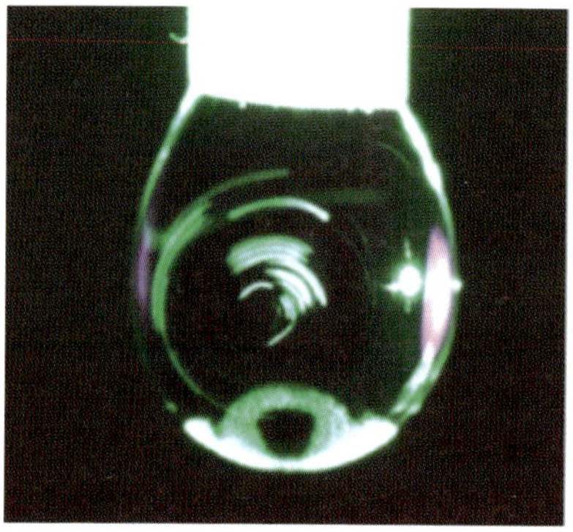

Figure 3.6 Convection observation of droplet by PIV system.

green laser (solid state laser 532 nm, 50 mW, Kato Koken Co., Ltd). Tracer particle (40 μm, Nylon) was seeded to visualize flow. Droplet with tracer particles was set through Teflon pipe with inside diameter of 1 mm. Laser is introduced to the central axis of the droplet and movie is captured by camera, which is located at 90° angle from the direction of laser line. For example, convection in water droplet is observed under microwave irradiation as shown in Figure 3.6. Angular velocity of circular flow for water and ethylene glycol droplet is measured when irradiation time is 30 s.

Figure 3.7 shows the angular velocity of circular flow inside water during microwave irradiation. For first 5 s, the convection is unstable. After that, the circulation flow becomes more stable gradually. Angular velocity is plotted for the different power. The velocity of lower microwave power (150 W) is lower than that of the higher power. However, the velocity might be saturated in the case of higher power (300 and 600 W).

Although the velocity inside ethylene glycol droplet was measured during microwave irradiation, the stable circulation flow was not observed consistently.

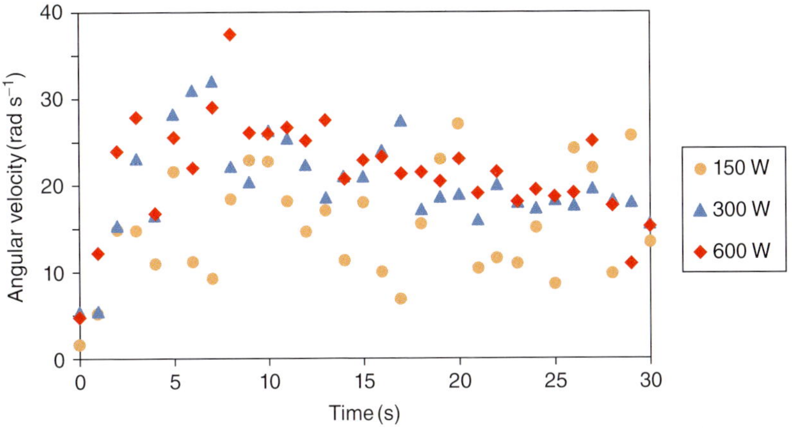

Figure 3.7 Angular velocity of circular flow for water droplet.

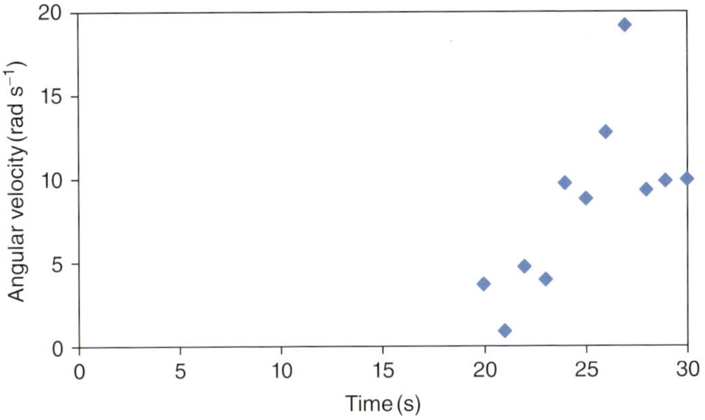

Figure 3.8 Angular velocity of circular flow for ethylene glycol droplet.

In most cases, the convection did not develop perfectly. Hence, effective observation was obtained only ~10% of the time. Figure 3.8 shows an example of the angular velocity. For first 20 s, the convection was not observed. After that, the circulation flow becomes more stable gradually. The circulation flow continued for about 5 s immediately until microwave is turned off.

On the other hand, the stable circulation flow was observed by microwave irradiation every time in the case of water droplet. The observed flow rate in water droplet was higher than that of ethylene glycol and developed easily. The velocity difference between water and ethylene glycol can be explained by their penetration depths. The depth of ethylene glycol (3 mm) is almost of same size with droplet, higher microwave absorption limited near the surface. However, the depth of water (12 mm) is much wider than droplet diameter, and microwave is strongly absorbed into the whole droplet. Accordingly, the convection behaviors for the velocity and the stabilization between water and ethylene glycol should be attributed to different parameters for microwave absorption, such as dielectric constant and penetration depth.

3.4 Surface Tension

Application of microwave irradiation for chemical processes, such as emulsification and polymerization, has been reported in the literature [7, 8, 14]. Surface tension or interfacial tension is an important property for the industrial process. On the other hand, surface tension is determined by molecular arrangement within the interfacial zone, ~1 nm. We introduce influence of microwave on surface tension of water and ethylene glycol droplet [15, 16].

The pendant drop was formed at the tip of polytetrafluoroethylene (Teflon), with inside diameter of 1 mm and a wall thickness of 0.5 mm. The tube was placed inside the reactor, which is microwave-proof. After the droplet formation, an optic fiber was inserted inside the droplet (Figure 3.9). Light source was used from one side of the reactor. A digital camera (Sigma Koki Co., Ltd, Model SK-TC202USB-AT) was employed from the opposite side to capture the shape of the droplet during and after microwave irradiation.

Temperature inside the droplet both during and after microwave irradiation was transferred to the computer through a digital controller (Anritsu meter Co., Ltd; Model: FL-2000). Images of the droplet were recorded by a digital. The edge profiles of the droplet were analyzed by Axisymmetric Drop Shape Analysis (ADSA) to calculate surface tension [17]. Examples of raw image and fitting results are shown in Figure 3.10.

Figure 3.11a shows surface tension of water droplet as a function of time. During microwave irradiation, the temperature increased as expected. Generally, the

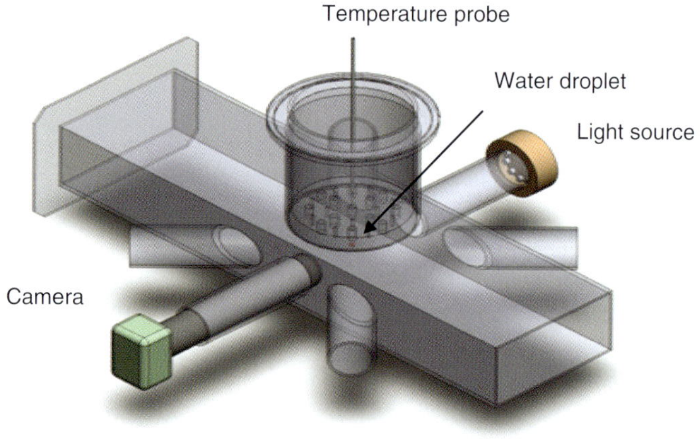

Figure 3.9 Microwave reactor for surface tension measurement.

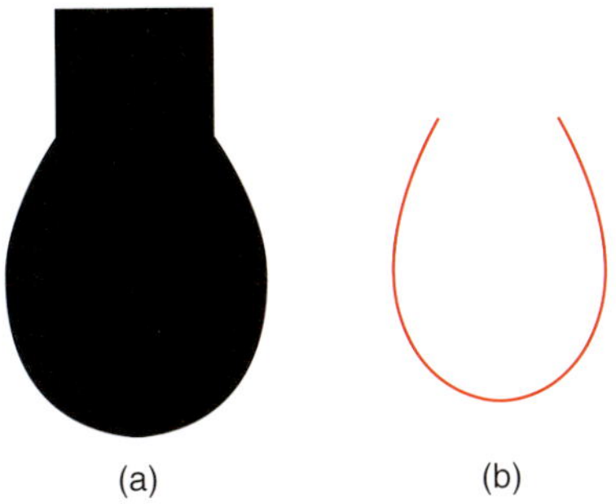

Figure 3.10 Drop profile: (a) raw image and (b) edge for surface tension measurement.

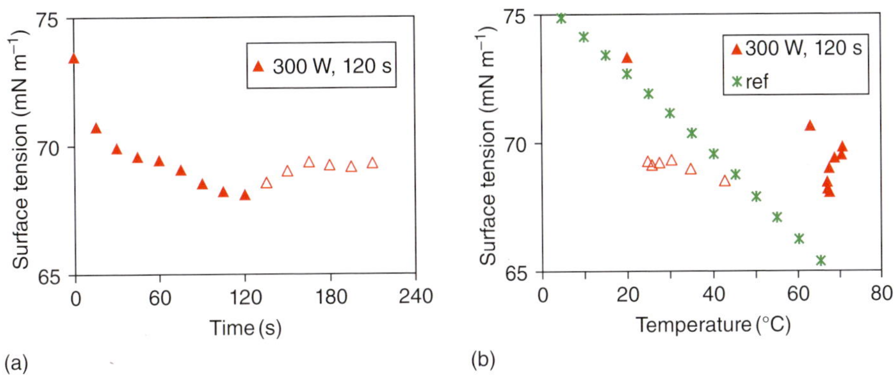

Figure 3.11 Surface tension of water droplet as a function of time and temperature (solid symbols: with MW; unfilled symbols: without MW) for 300 W, 120 s. (a) Time and (b) temperature.

temperature reached a plateau within about 20 s and remained at that temperature until the microwave irradiation was turned off at 120 s. At that time, the droplet gradually returned to its original temperature. Once irradiation had ceased, the surface tension gradually increased, but unlike the temperature, it did not return to the original value.

Surface tension of water droplet was plotted against temperature as shown in Figure 3.11b. The results indicated that the surface tension reduction is not due to thermal effect solely. The data after microwave irradiation clearly demonstrated the nonthermal effect on surface tension because these plots were below the reference values [18].

EG droplet was also investigated using the same procedure. The surface tension of ethylene glycol (Figure 3.12) was reduced during microwave heating as expected. However, the surface tension quickly recovered during the cooling period. The surface tension plot was close to the reported data in the literature [19]. Hence, it can be concluded that the surface tension reduction of ethylene glycol was purely due to thermal effect of microwave. The sharp contrast between Figures 3.11 and 3.12 highlighted that the "memory" effect is a unique property of water.

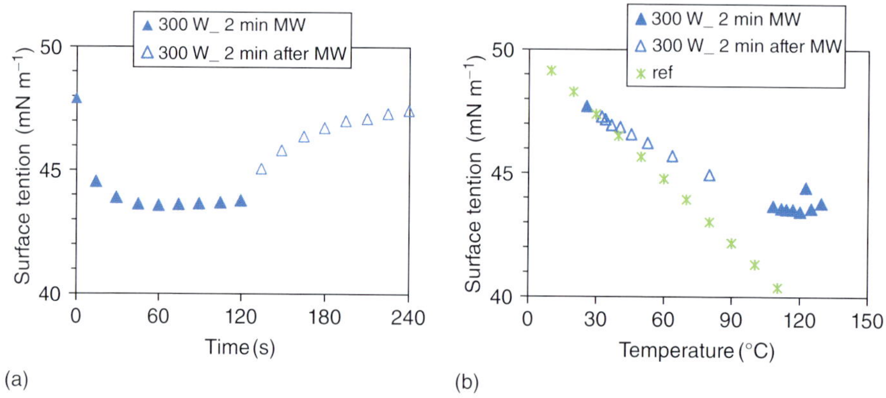

Figure 3.12 Surface tension of ethylene glycol droplet as a function of time and temperature (solid symbols: with MW; unfilled symbols: without MW) for 300 W, 120 s. (a) time and (b) temperature.

A recent study has indicated that water surface tension is determined by the H-bond network between the interfacial water molecules. The H-bond network at the interface depends on the dynamics of H-bond forming and breaking [20], which is directly related to rotational motion of the water molecules by microwave irradiation. Figure 3.11 shows that water retains the effects by microwave irradiation. The mechanism for a reduction in surface tension might be caused by the change of water molecule network. The change will continue as nanobubble or cluster structure of water molecule because it is theoretically more stable. The effects observed in this study are for small droplets with a diameter of only a few millimeters, which is well below the limit of microwave penetration depth [14].

3.5 Discussion of Nonthermal Effect for Nanobubble Formation

Many researchers are thinking about nonthermal effect by microwave irradiation for special behaviors such as fast reaction rate. According to the results of nanobubble formation and surface tension reduction by microwave irradiation, energy and free radical released by the collapse of nanobubbles or water cluster [13] might be an origin of reason for special microwave effect. For example, when metal nanoparticles are added in the solution as a heterogeneous catalyst, fine bubbles are produced around the particle, which plays an important role in more stable bubble nucleation. In particular, if catalyst particle has magnetic property, more bubbles with higher energy are produced on the surface by the irradiation. This behavior should be related with nonequilibrium local heating [21] or superheating [22]. Moreover, the mechanism for faster reaction with homogeneous catalyst under the microwave irradiation will be explained by nanobubble formation in aqueous solution as well.

During microwave heating, local temperature gradients and internal evaporation can promote transport of mobile components (liquids and gases) and energy. As compared with conventional heating, microwaves can generally provide faster and selective heating. Moreover, energy stored as nanobubbles might have many potential benefits for high yield and quality. Further nanoscale information through *in situ* observation under microwave as shown in this chapter will cause optimum process control and new process applications of microwave.

References

1. Birdwhistell, K.R., Schulz, B.E., and Dizon, P.M. (2012) Rapid synthesis of Group VI carbonyl complexes by coupling borohydride catalysis and microwave heating. *Inorg. Chem. Commun.*, **26**, 69–71.
2. Laurent, R., Laporterie, A., Dubac, J., Berlan, J., Lefeuvre, S., and Audhuy, M. (1992) Specific activation by microwaves: myth or reality? *J. Org. Chem.*, **57** (26), 7099–7102.
3. Goncalo, P., Roussel, C., Marie Melot, J., and Vebrel, J. (1999) Contribution of microwaves in organic synthesis: statement of a methodology for the microwave-induced preparation of benzofuran-2(3H)-one and its comparison with classical heating. *J. Chem. Soc., Perkin Trans.*, **2** (10), 2111–2115.
4. Kuhnert, N. (2002) Microwave-assisted reactions in organic synthesis—are there any nonthermal microwave effects? *Angew. Chem. Int. Ed.*, **4** (11), 1863–1866.
5. Kappe, C.O. (2008) Microwave dielectric heating in synthetic organic chemistry. *Chem. Soc. Rev.*, **37** (6), 1127–1139.
6. Loupy, A., Monteux, D., Petit, A., Aizpurua, J.M., Domínguez, E., and Palomo, C. (1996) Towards the rehabilitation of the Leuckart reductive amination reaction using microwave technology. *Tetrahedron Lett.*, **37** (45), 8177–8180.
7. Bao, J. and Zhang, A. (2004) Poly(methyl methacrylate) nanoparticles prepared through microwave emulsion polymerization. *J. Appl. Polym. Sci.*, **93** (6), 2815–2820.
8. Hoogenboom, R. and Schubert, U.S. (2007) Microwave-assisted polymer synthesis: recent developments in a rapidly expanding field of research. *Macromol. Rapid Commun.*, **28** (4), 368–386.
9. Herrero, M.A., Kremsner, J.M., and Kappe, C.O. (2008) Nonthermal microwave effects revisited: on the importance of internal temperature monitoring and agitation in microwave chemistry. *J. Org. Chem.*, **73**, 36–47.
10. Hu, S., Lu, C., Wang, W., Ding, M., Ni, Y., and Xu, Z. (2013) Synthesis of monodisperse erbium aluminum garnet (EAG) nanoparticles via a microwave method. *J. Rare Earths*, **31**, 490–496.
11. Ma, L., Yang, L.-L., Wang, Y.-G., Zhou, X.-P., and Xu, X.-Y. (2013) Microwave-assisted preparation of nearly monodisperse flower-like CaF_2 microspheres. *Ceram. Int.*, **39**, 5973–5977.
12. Asakuma, Y. and Miura, M. (2014) Effect of microwave radiation on diffusion behavior of anti-solvent during crystallization. *J. Cryst. Growth*, **402**, 32–36.
13. Miichi, T., Ihara, S., Satoh, S., and Yamabe, C. (2000) Spectroscopic measurements of discharges inside bubbles in water. *Vacuum*, **59**, 236–243.
14. Nüchter, M., Ondruschka, B., Bonrath, W., and Gum, A. (2004) Microwave assisted synthesis – a critical technology overview. *Green Chem.*, **6**, 128–141.
15. Parmar, H., Kanazawa, Y., Asada, M., Asakuma, Y., Phan, C., Pareek, V., and Evans, G. (2014) Influence of microwave on water surface tension. *Langmuir*, **30**, 9875–9879.
16. Asakuma, Y., Kanazawa, Y., Parmar, H., Phan, C., Pareek, V., and Evans, G. (2014) Surface tension profiles under various microwave radiation modes. *J. Energy Power Eng.*, **8**, 585–588.

17. Zuo, Y.Y., Ding, M., Bateni, A., Hoorfar, M., and Neumann, A.W. (2004) Improvement of interfacial tension measurement using a captive bubble in conjunction with axisymmetric drop shape analysis (ADSA). *Colloids Surf., A*, **250** (1–3), 233–246.
18. Gittens, G.J. (1969) Variation of surface tension of water with temperature. *J. Colloid Interface Sci.*, **30** (3), 406–412.
19. Wohlfarth, C. (2008) *Supplement to IV/16*, Springer, pp. 421–422.
20. Park, S. and Fayer, M.D. (2007) Hydrogen bond dynamics in aqueous NaBr solutions. *Proc. Natl. Acad. Sci. U.S.A.*, **104** (43), 16731–16738.
21. Tsukahara, Y., Higashi, A., Yamauchi, T., Nakamura, T., Yasuda, M., Baba, A., and Wada, Y. (2010) In situ observation of nonequilibrium local heating as an origin of special effect of microwave on chemistry. *J. Phys. Chem. C*, **114** (19), 8965–8970.
22. Stuerga, D. and Gaillard, P. (1996) Microwave heating as a new way to induce localized enhancements of reaction rate. Non-isothermal and heterogeneous kinetics. *Tetrahedron*, **52**, 5505–5510.

4
Managing Microwave-Induced Hot Spots in Heterogeneous Catalytic Systems

Satoshi Horikoshi and Nick Serpone

4.1
What Are Hot Spots?

The term "*hot spot* " in the field of microwave chemistry and materials processing refers to an inhomogeneous dissipation of microwave energy through selective heating of different parts of a material that causes the generation of temperature gradients within various domains of the material or sample [1]. Where high-temperature microdomains (so-called hot spots) are formed when the material is microwave-irradiated, they must undergo fast heat transfer processes for the material to come into thermal equilibrium. However, where conditions are such that the rate of heat transfer is slow, formation of steady-state hot spots occurs that, in principle, can enhance the rate of a chemical process occurring within or near those hot spots; however, formation of hot spots can also prove deleterious to the chemical process.

4.2
Microwaves in Heterogeneous Catalysis

Acceleration of a large number of organic syntheses is no longer in doubt when these are carried out under microwave radiation, which provides the thermal energy necessary to drive the reactions. In this regard, different types of examples of reactions have been summarized in a technical book edited by de la Hoz and Loupy [2]. The possibility of scale-up of some reactions has also been examined [3]. However, such processes must also be examined in more detail using different approaches. One such process might be the microwave-assisted organic synthesis of a product using a solid catalyst that is also of interest and meaningful to Green Chemistry [4].

When microwave selective heating is desired, it is possible to carry out organic syntheses using some of the features of the microwave radiation. For example, when microwave nonabsorbing substrates, such as a nonpolar solvent or a gas, are involved in a catalyzed chemical process, then only the solid catalyst may

Microwaves in Catalysis: Methodology and Applications, First Edition.
Edited by Satoshi Horikoshi and Nick Serpone.

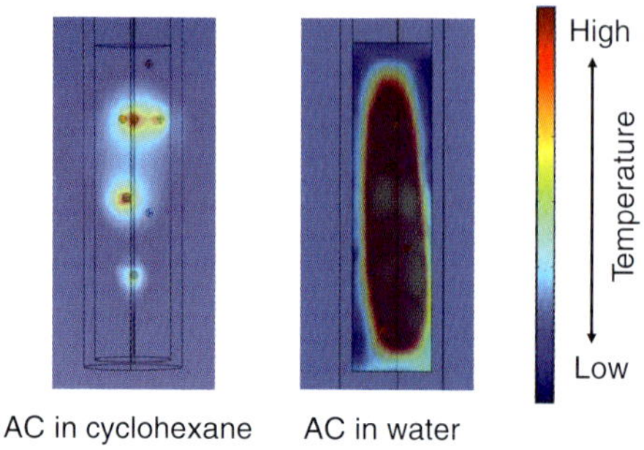

Figure 4.1 Simulated temperature distribution in water and cyclohexane solvents in the presence of Pt/activated carbon (AC) catalyst particles subjected to microwave radiation. The simulation was carried out using an RF module and the COMSOL Multiphysics software Version 4.3a. Reproduced from Horikoshi and Serpone [5]. Copyright 2014 by the Royal Society of Chemistry.

be selectively heated. Figure 4.1 illustrates the results of a simulation experiment of the temperature distributions in water and in cyclohexane (a poor microwave absorber) in the presence of an activated carbon (AC) catalyst support (a strong microwave absorber) exposed to microwave radiation [5]. The simulations were performed using an radio frequency (RF) module and the COMSOL Multiphysics software Version 4.3a. The activated carbon was selectively heated in cyclohexane by Joule loss heating, while selective heating of activated carbon did not occur in aqueous media as water is a strong microwave absorber. The chemical reaction typically occurs at the catalyst particle surface so that when only the dispersed catalyst particles are selectively heated by the microwaves not only does the reaction take place but significant energy conservation can also be achieved.

Relevant to the present discussion, a study by Matsuzawa *et al.* ([6]; see also Chapter 7) examined the synthesis of 4-phenyltoluene by the Suzuki–Miyaura coupling reaction using both heterogeneous and homogeneous Pd-based catalysts. The heterogeneous catalyzed process was enhanced by the microwaves, while the same reaction in homogeneous media was only slightly enhanced under similar conditions. Moreover, when carried out in a heterogeneous solvent-free system, the reaction was only slightly enhanced by the microwaves. Another example is the synthesis of the monoglycerylcetyldimethylammonium chloride (MGCA) surfactant from the 3-chloro-1,2-propanediol (CP) and *N*,*N*-dimethylhexadecylamine (DMHA) precursors in 2-propanol solvent and under solvent-free conditions [7] to expose the possible existence of specific effects in microwave-assisted reactions. Results showed no specific microwave effect(s) when the synthesis was carried out in homogeneous 2-propanol media under otherwise identical temperature conditions to those of conventional heating with an oil bath. However, when the reaction was performed under heterogeneous solvent-free conditions with microwave dielectric heating and oil-bath heating at 130 °C for a 30-min heating period, a specific microwave effect(s) was evidenced by differences in product yields (62% by microwave heating vs 47% by conventional heating). Hence, a heterogeneous system provides optimal conditions to delineate the microwave effect(s) in chemical reactions. By contrast, specific microwave effects have tended to be very difficult to ascertain in chemical syntheses carried out in homogeneous media.

The above notwithstanding, however, though a heterogeneous catalyst system is ideal for performing a catalyzed reaction by the microwave-assisted method, the nature of the solvent medium remains nonetheless an important factor to contend with. For instance, if a polar solvent were used in carrying out the Suzuki–Miyaura cross-coupling reaction in the presence of the palladium/activated carbon (Pd/AC) catalyst in heterogeneous media, the advantages of using microwaves are diminished as the microwaves will heat nonselectively the whole of the bulk solution. Despite this, however, Arvela and Leadbeater [8] have witnessed significant enhancement effects of the microwaves using a novel methodology of microwave heating and simultaneously cooling externally in the synthesis of biphenyl compounds from the Suzuki–Miyaura cross-coupling reaction, which led to significant product yields under such experimental conditions.

Studies reported so far have demonstrated that selective heating of a solid catalyst in a heterogeneous reaction is a typical occurrence when using microwaves. The combination of the catalyst and catalyst support, which is a strong microwave absorber, and a nonpolar solvent satisfy this condition. A fixed bed reactor is also an effective component of a reaction system. The temperature distribution peculiar to microwave irradiation can enhance adsorption and desorption (contact efficiency) of the substrate onto the catalyst surface and can also lead to energy saving during the heating stage(s). However, the heat developed on the catalyst can escape to the atmosphere through the reactor walls. To avoid such heat losses, insulation of the reaction container is usually used, with such heat insulating materials as fiberglass wrapped around the external walls of the reactor to prevent escape of the heat to the surroundings. Unfortunately, such materials attenuate the microwave radiation reaching the reacting substrates in the reactor. Accordingly, novel reactor designs have been developed and their advantages have been demonstrated [9].

4.3 Microwave-Induced Formation of Hot Spots in Heterogeneous Catalysis

4.3.1 Hot Spot Phenomenon

Several researchers have attempted to elucidate the existence of hot spots in chemicals subjected to microwave radiation. Indeed, heating a solid catalyst by microwave radiation frequently generates hot spots as a result of differential heating. For example, an initial report by Mingos and coworkers [10] showed significant apparent shifts in the equilibrium constant in the microwave-assisted gas phase decomposition of H_2S catalyzed by metal sulfides on a γ-Al_2O_3 support (solid/gas system). The authors attributed the shifts to development of hot spots with dimensions of 90–1000 μm in the catalytic beds. In solid/liquid systems, hot spots can reach temperatures 100–200 K above the temperature of the remaining bulk and cause a possible reorganization of the catalyst under microwave irradiation conditions. In this regard, Tsukahara *et al.* [11] observed the occurrence of nonequilibrium local heating of dimethyl sulfoxide (DMSO) molecules in the proximity of Co and Fe particles as a result of a faster input of microwave power than heat loss induced by temperature gradients established between the high-temperature microdomains and the surrounding low-temperature domains. Such nonequilibrium local heating led to enhanced dechlorination of 2-chloroethylbenzene and 4-phenylbutylchloride. In that instance, the reaction temperature of the microwave system was about 55 K greater than the bulk solution temperature (473 K) under microwave irradiation. The occurrence and the existence of hot spots induced by microwave radiation on materials have often been described by indirect suppositions, but have seldom been observed visually. For instance, Gutmann *et al.* [12] have shown that formation of an organomagnesium Grignard reagent could be both activated and deactivated by the microwaves' electric field (*E*-field), and attributed the activation event as originating either from a "cleansing effect" of the Mg surface, or from the formation of more reactive spherical Mg particles owing to mild electrostatic discharges (arcing, hot spots) between individual Mg turnings. However, neither the cause of the generation of the hot spots nor their features were examined [12].

Accordingly, it is relevant to examine and explain the mechanism of generation as well as the minimization or suppression of hot spots. Events occurring during the synthesis of 4-methylbiphenyl with the Suzuki–Miyaura coupling reaction (Eq. (4.1)) under microwave irradiation investigated by Horikoshi *et al.* [13] using a high-speed camera provide the basis to explain the occurrence of hot spots. Figure 4.2 displays photographs recorded in real time, which show that the Pd/AC catalyst particulates immersed in toluene solvent were selectively heated by the microwaves' *E*-field radiation:

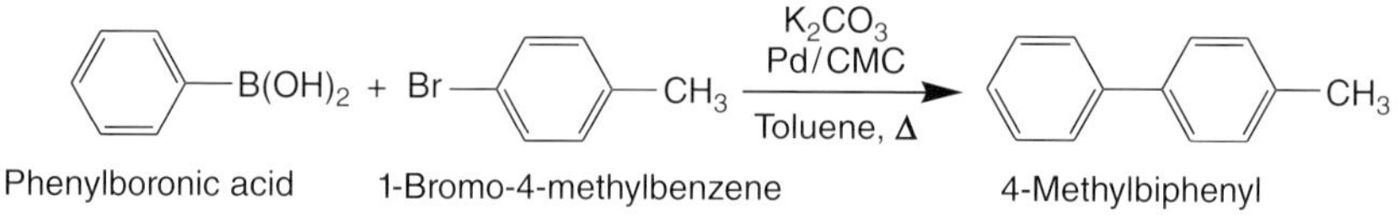

(4.1)

In the case considered, as soon as microwave irradiation was initiated, bubbles could be seen immediately formed on the activated carbon surface after only 3 s (Figure 4.2a), owing to differential heating of the activated carbon particles. Moreover, also observed was the aggregation of the activated carbon particulates as though they had been magnetized by the microwave radiation. An orange-colored light was observed from the aggregated activated carbon particles after 16 s (see

Figure 4.2 High-speed camera photograph of the electrical arc discharge occurring on the activated carbon in toluene solvent under the microwaves' *E*-field irradiation: (a) after 3 s of microwave irradiation, (b) after 16 s of irradiation, (c) under continuous irradiation, and (d) emitted light reaching maximal intensity. Reproduced from Horikoshi *et al.* [13]. Copyright 2013 by Elsevier B.V.

the lower left side in Figure 4.2b). Subsequently, the light emanating from the activated carbon surface increased as did its intensity, together with a greater number of bubbles formed around the emitted light (Figure 4.2c). The emitted light became even more intense on the activated carbon surface under exposure to continued microwave radiation. Figure 4.2d shows the light from the hot spots reaching maximum intensity, after which the luminescence from the activated carbon surface disappeared, and the aggregated activated carbon particles were re-dispersed.

Chemical yields of 4-methylbiphenyl produced in toluene solvent from the Suzuki–Miyaura coupling reaction (Eq. (4.1)) under microwave *E*-field and *H*-field irradiation together with those from the more conventional oil-bath heating method are reported in Figure 4.3. Under *E*-field conditions, the yield was about 6% after a 60-min irradiation period at an input microwave power of 70 W, whereas the corresponding yield under *H*-field irradiation conditions was 34% (also after 60 min). With the oil-bath heating method employed for comparison with the microwave method, the chemical yield of 4-methylbiphenyl was 21% for a heating time of 60 min. It was noted that when hot spots form, the efficiency of the synthesis of 4-methylbiphenyl tended to decrease [13, 14]. On the other hand, when no hot spots form, as occurred under *H*-field irradiation, the synthesis efficiency of 4-methylbiphenyl was significantly greater by nearly sixfold (Figure 4.3). Accordingly, it was relevant to investigate the influence that hot spots may have

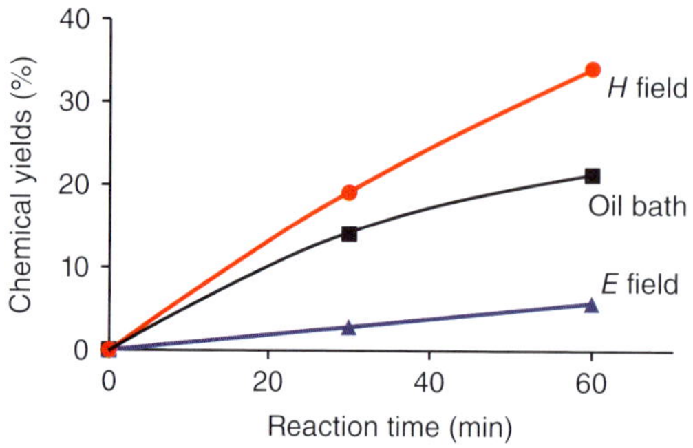

Figure 4.3 Product yields of 4-methylbiphenyl in toluene solvent under irradiation with the microwaves' *H*-field and *E*-field components and oil-bath heating. Reproduced from Horikoshi *et al.* [13]. Copyright 2013 by Elsevier B.V.

on possible occurrence of side reactions other than the principal Suzuki–Miyaura cross-coupling reaction. With an electrical power input of the microwaves fixed at 70 W, generation of hot spots occurred with *E*-field irradiation, but none under *H*-field irradiation. This also led to a closer examination of the Suzuki–Miyaura cross-coupling reaction and side reactions that might include the formation of biphenyl through a Suzuki–Miyaura homocoupling process and formation of 4,4′-dimethylbiphenyl through an Ullmann-type coupling (Eq. (4.2)) [15] under hot spot and no hot spot formation conditions:

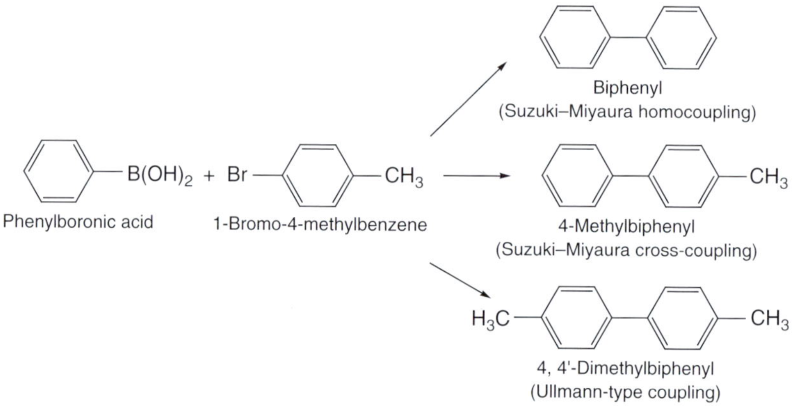

(4.2)

When hot spots do occur, the cause for the decrease of the yields in a chemical synthesis could be attributed to a decline in catalyst activity. In the present case, formation of hot spots on the catalyst surface did decrease the yield of 4-methylbiphenyl, yet the amount of by-product(s) generated increased. To understand this assertion, Figure 4.4 illustrates scanning electron microscope (SEM) images of the Pd/AC surface before and after the formation of hot spots [14]. The Pd catalyst particles are dispersed rather uniformly on the AC surface prior to

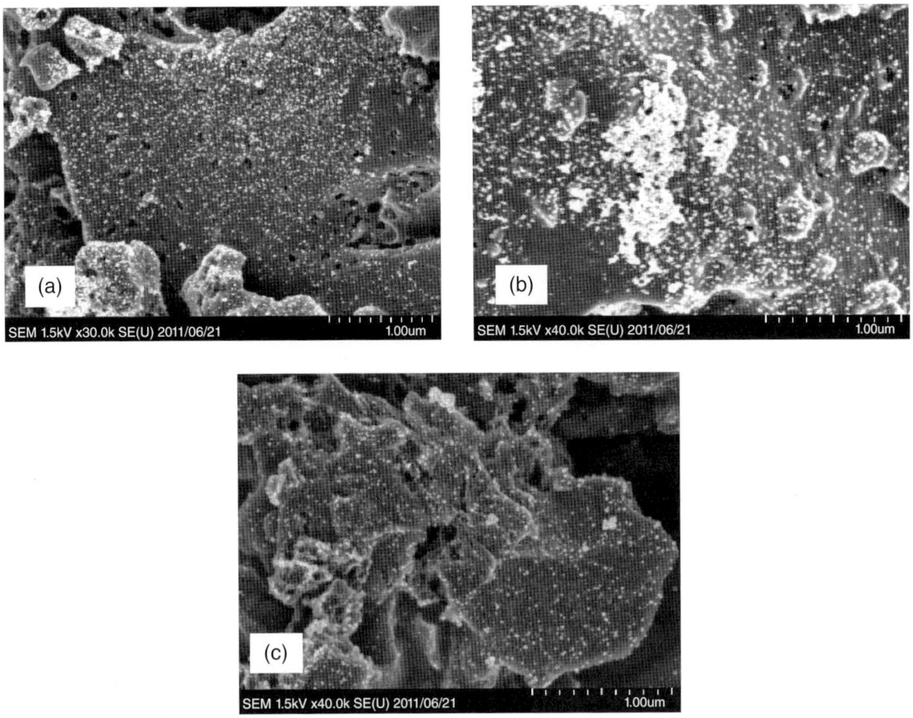

Figure 4.4 SEM images of Pd/AC catalyst surface: (a) 0 min irradiation, (b) after 30 min of microwave irradiation after formation of hot spots, and (c) under conditions of no formation of hot spots. Reproduced from Horikoshi *et al.* [14]. Copyright 2011 by the American Chemical Society.

microwave irradiation (Figure 4.4a). However, large Pd aggregates formed after 30 min of microwave irradiation of the Pd/AC catalyst under *E*-field heating conditions (Figure 4.4b). By contrast, no aggregation of Pd particles occurred and no hot spots formed under conditions when heating was carried out by the microwaves' magnetic field (Figure 4.4c). Evidently, aggregation of Pd particles and generation of hot spots (electric discharge) on the catalyst surface are events that occurred concurrently under *E*-field heating conditions. Accordingly, the lower chemical yields of 4-methylbiphenyl (12% vs 22% after 2 h of microwave irradiation) were attributed, in part, to the formation of less reactive larger Pd aggregates as one of several related factors that affect yields; for example, (i) a lower number of reactive sites on the Pd particles through annihilation by the Joule heating effect, (ii) possible obstruction of reactive sites on the Pd particles by a small layer of carbonaceous residue that might have formed by dielectric heating of the AC support under the microwaves' *E* field, and lastly (iii) to an excessive number of hot spots formed on the Pd/AC catalyst surface. It was deduced that the decrease in efficiency was due to aggregation of the Pd catalyst because of the hot spots on the activated carbon surface [14].

4.3.2
Mechanism(s) of Formation of Hot Spots

To further understand the phenomenon of hot spots, the distribution and the relationship of the microwaves' E-field intensity (2.45 GHz) within the spatial volume between the activated carbon particles were examined by a simulation technique using an RF module and the COMSOL Multiphysics software Version 4.3a [13]. The distances between the activated carbon particles (size, 0.65 mm diameter) in toluene were set up at intervals of 0, 0.001, and 0.1 mm. Figure 4.5a-i shows that

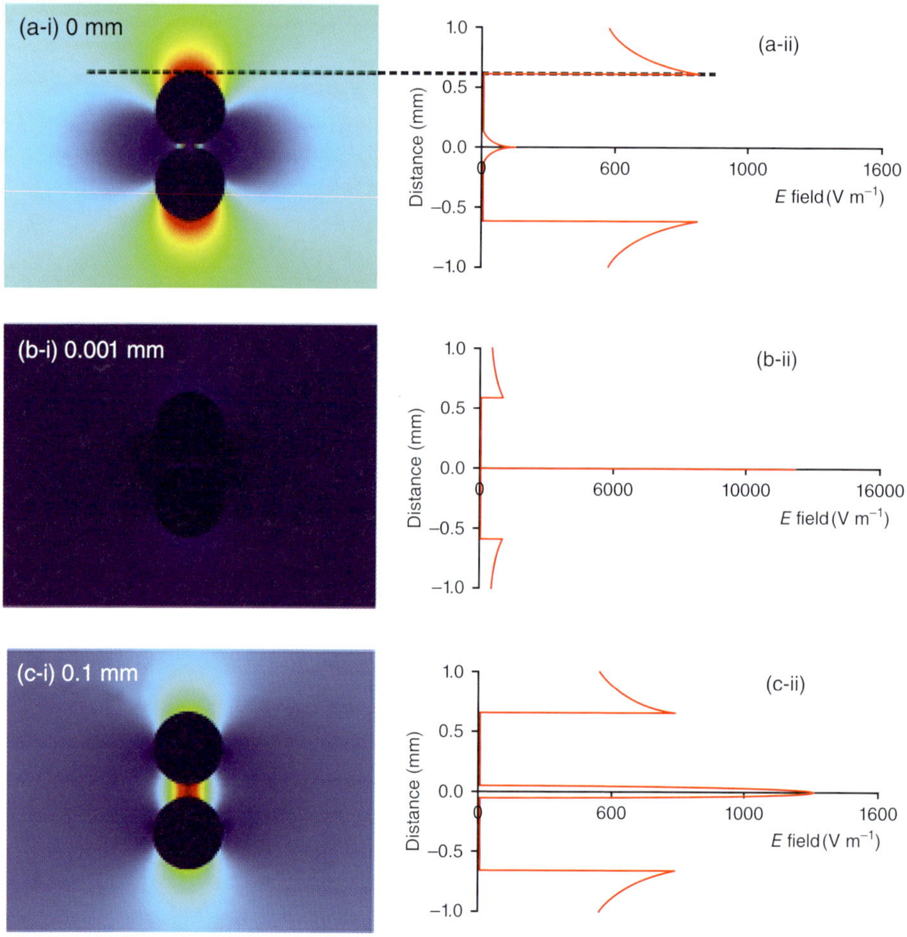

Figure 4.5 Distribution of the microwaves' E fields (2.45 GHz) around two activated carbon particles in toluene solvent simulated with the COMSOL Multiphysics software Version 4.3a. Colored images (a-i), (b-i), and (c-i) display the E-field intensity in two activated carbon particles in toluene together with the magnitude of the gap between the particles, whereas (a-ii), (b-ii), and (c-ii) graphics show the magnitude of the E field (V m^{-1}) within this gap. Reproduced from Horikoshi *et al.* [13]. Copyright 2013 by Elsevier B.V.

at a distance of 0 mm (particles in contact with each other), the maximal electric field distribution was seen at both ends of the particles and decayed at longer distances from the particles; the electric field intensity between the two particles was $813\,V\,m^{-1}$ (Figure 4.5a-ii). On the other hand, when the distance between the activated carbon particles was set at 0.001 mm (Figure 4.5b-i), the field was concentrated entirely within the gap that separated the activated carbon particles; the maximum E-field intensity was $11\,838\,V\,m^{-1}$ (Figure 4.5b-ii). Compared with the activated carbon particles that were glued to each other (0 mm), the change in the electric field intensity was nearly 15 times greater if the particles were ever so slightly separated by a mere 0.001 mm, not to mention the location of the electric field observed in the volume between the particles. When the distance between two activated carbon particles was extended to 0.1 mm (Figure 4.5c-i), the maximal E-field intensity decreased down to $1253\,V\,m^{-1}$ (Figure 4.5c-ii).

Clearly, hot spots are generated whenever the particles are separated by very small distances; however, when the distance was 0 mm no hot spots formed, at least none were observed. Such a phenomenon recalls the *surface enhanced plasmon resonance* (SEPR) phenomenon that occurs between metal nanoparticles, wherein the electric field of light is concentrated between nanoparticles within a narrow gap [16]. To confirm experimentally the phenomenon predicted from the simulations with respect to the effect of distances between particles and to the generation of hot spots, activated carbon particles were pasted on a glass plate at various distances using an adhesive made of polyvinyl acetate resin. Hot spots were easily generated at distances between the activated carbon particles comparable to those used in the simulation (Figure 4.6). The distance between particles is therefore an important parameter in the formation of hot spots. Nonetheless, generation of hot spots can be suppressed in a reaction taking place in a fixed bed reactor under flow-through conditions.

4.3.3
Particle Aggregation by Polarization of Activated Carbon Particulates

Being able to observe visually a catalyst particle inside a reactor during an organic synthesis catalyzed by metal catalyst deposits on AC support particles revealed that the particles moved in a regular manner, which changed by whether the magnetic field or the electric field was used whenever microwave radiation is a part of the chemical process. Moreover, the catalyst particles were mutually connected end-to-end along the particles' major axis. As an example, it is useful to consider the nature of the connection of Pd deposits on the AC support in the Suzuki–Miyaura cross-coupling reaction as illustrated in Figure 4.7. The type of connection between particles depends on either the electric field (E field) or the magnetic field (H field) inside the microwave resonator. The direction of the E-field and H-field radiation typically changes according to the position of the sample in the waveguide; the latter can be arranged either vertically (Figure 4.7a-i) or horizontally (Figure 4.7b-i), while the reactor is perpendicularly suspended from the upper part of the waveguide [11]. Following the addition of activated

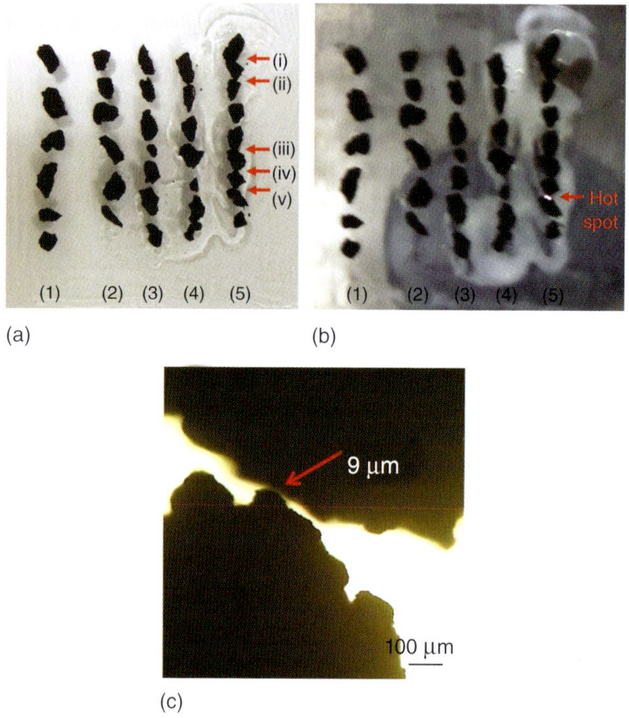

Figure 4.6 Observation of hot spots generated in activated carbon particles fixed at various intervals on a glass plate: (a) photograph of the initial sample; (b) formation of hot spots between the AC particles under microwave irradiation; and (c) observed distance between the particles in the fourth column with an optical microscope. Reproduced from Horikoshi *et al.* [13]. Copyright 2013 by Elsevier B.V.

carbon particles (no Pd), phenylboronic acid, 1-bromo-4-methylbenzene, and K_2CO_3 to the toluene solvent, the suspension was then irradiated continually by either the microwaves' *E*-field or *H*-field components. The microwave input power level for the *E*-field irradiation was 70 W, while the power level for the *E*- and *H*-field irradiation ranged from 90 to 120 W. No hot spots were generated at the lower power level. Inside the resonator waveguide, the *E* field is in a transverse direction, whereas the *H* field is in the direction of the length of the waveguide. As such, the *E* field irradiated the reactor positioned parallel to the field, while the *H* field irradiated the reactor positioned perpendicularly to the field.

Connected particles of activated carbon under *E*-field irradiation are illustrated in Figure 4.7a-ii, which shows that many of the AC particulates are connected horizontally, in addition to being stacked. By contrast, under *H*-field irradiation, the AC particles are connected in a circular manner (Figure 4.7a-iii) that takes the form of the inner walls of the reactor. When placed in a conical reactor and irradiated by the *H* field of the microwaves, the AC particles formed a small ring in accordance with the aspect of the inner reactor wall (Figure 4.7a-iv).

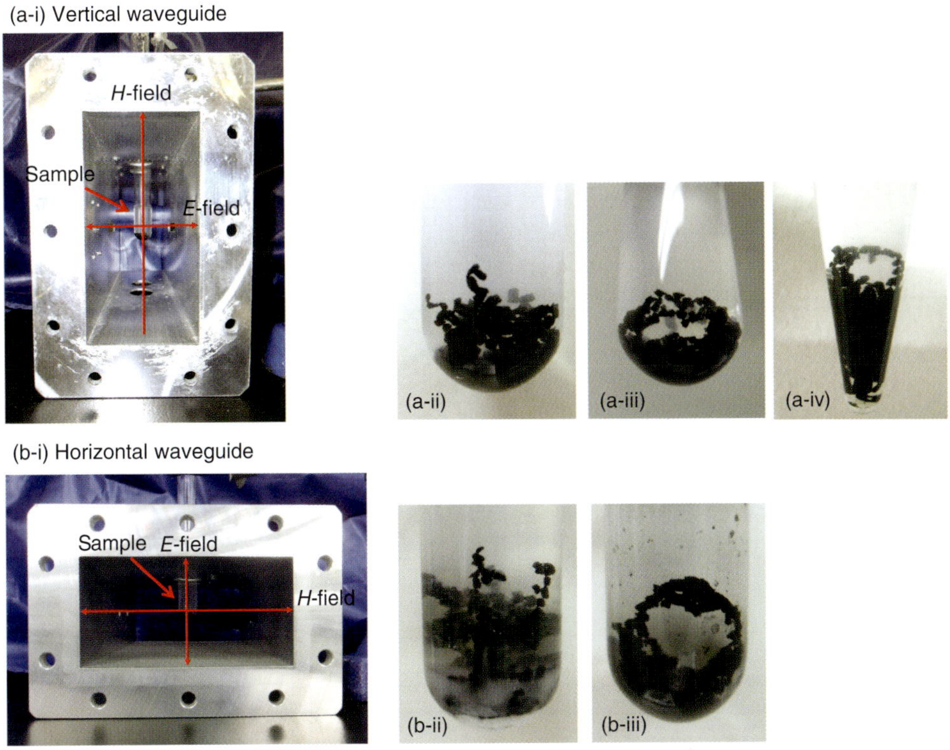

Figure 4.7 Photograph of the reactor introduced into (a-i) the vertically positioned waveguide and (b-i) horizontally positioned waveguide. Appearance of the Pd/AC catalyst in solution under *E* field (a-ii), *H* field (a-iii), and *H* field (a-iv; conical reactor) in the vertically positioned waveguide, and under the *E* field (b-ii) and *H* field (b-iii) in the horizontally positioned waveguide. Note that the photographs of a-ii and a-iv were taken in a slant position. Reproduced from Horikoshi *et al.* [13]. Copyright 2013 by Elsevier B.V.

The waveguide was also arranged horizontally, in which the reactor was suspended from the upper part and then irradiated with both *E*-field and *H*-field microwaves. Having arranged the waveguide horizontally, the directions of the *E*-field and *H*-field components are now the reverse (Figure 4.7b-i). Under *E*-field irradiation, the activated carbon particulates were connected horizontally in the direction of the *E*-field radiation (Figure 4.7b-ii). On the other hand, under *H*-field irradiation, the rings formed horizontally (Figure 4.7a-iii).

The directions of the *E* and *H* fields to the activated carbon particles distributed in solution in the reactor are depicted in Figure 4.8. The *E* field irradiated horizontally the activated carbon particles. Therefore, polarization arises on both sides of the surface of an activated carbon particle (Figure 4.8a). For this reason, an activated carbon particle is electrostatically connected horizontally to another activated carbon particle (Figure 4.7a-ii). However, since neither the reaction solution nor the activated carbon is magnetic, there should have been

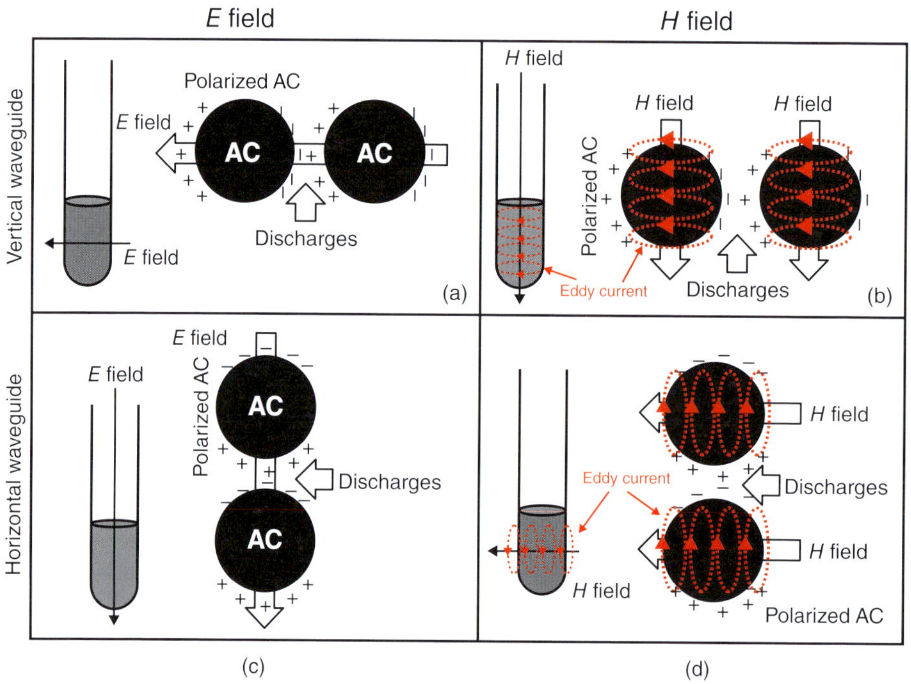

Figure 4.8 Illustration of the *E* and *H* fields in the reactor and the AC particles in a different location of the waveguides, and image of the polarization on the activated carbon surface. (a) Vertical waveguide and *E*-field condition; (b) vertical waveguide and *H*-field condition; (c) horizontal waveguide and *E*-field condition; and (d) horizontal waveguide and *H*-field condition. Note that the circular red-dotted line denotes the electric field produced by the magnetic field. Reproduced from Horikoshi *et al.* [13]. Copyright 2013 by Elsevier B.V.

no influence of the *H* field on the activated carbon particulates. However, the vertically aligned *H* field generates circular Eddy currents according to Ampere's right-handed screw rule relative to the reactor (Figure 4.8b). Polarization thus arises on the right and left of an activated carbon particle by this *E* field, which led the activated carbon particles to be connected to each other. In accordance with the aspect of the inner reactor wall, the activated carbon particles combined to form a circle (Figure 4.7a-iii,a-iv). Since aggregation of the activated carbon particles by a magnetic field is influenced by Eddy currents, irradiation required a stronger microwave input power level (see later).

Under the conditions of Figure 4.7b-i, since the vertical *E* field parallels the reactor, polarization is produced on both the upper and lower sides of the activated carbon particle surface (Figure 4.8c) so that the particles are connected vertically (Figure 4.7b-ii). Under *H*-field irradiation, an Eddy current is generated (Figure 4.8d) so that, in this case, connection of activated carbon particles form a vertical circle as illustrated in Figure 4.7b-iii. Polarization occurred through the microwave's *E* field, and connection of activated carbon particles took place by electrostatic attraction. Moreover, connection of particles also takes place through formation of Eddy currents under *H*-field irradiation. As a consequence, at a certain optimal distance, a hot spot is formed in the process in which particles approach each other by electrostatic attraction.

4.3.4
Control of the Occurrence of Hot Spots

As inferred earlier, the occurrence of hot spots on a solid catalyst surface can also be deleterious to microwave-assisted catalyzed reactions. Accordingly, studies have been undertaken to further clarify their generation and their influence on chemical reactions, and to find means on how they could be controlled. When aiming at the industrialization of a chemical process that involves microwave radiation, this hot spot phenomenon peculiar to microwaves needs to be minimized if not entirely suppressed, which can be achieved using only the microwaves' magnetic field [14].

In addition, generation of hot spots can also be avoided by using Pt catalyst particles deposited on carbon microcoil (CMC) supports (Figure 4.9) [17], which indeed proved effective catalyst supports in microwave organic chemistry, even though the available surface area was significantly smaller than that of activated carbon particles. The CMCs also proved to be better microwave absorbers than the ACs and thus optimal for the selective heating of metal catalyst deposits.

In addition, to the extent that toluene solvent in the process examined could be heated with 5.8-GHz microwaves (a frequency change from the usual 2.45-GHz microwaves), the temperature gradient between the catalyst and the solution bulk decreased such that there was a remarkable drop in the generation of hot spots [3].

Chemical reactions can also be carried out with microwave dielectric heating accompanied by the more conventional heating method, with the benefit that the applied microwave power can be lowered. For instance, using an internal/external

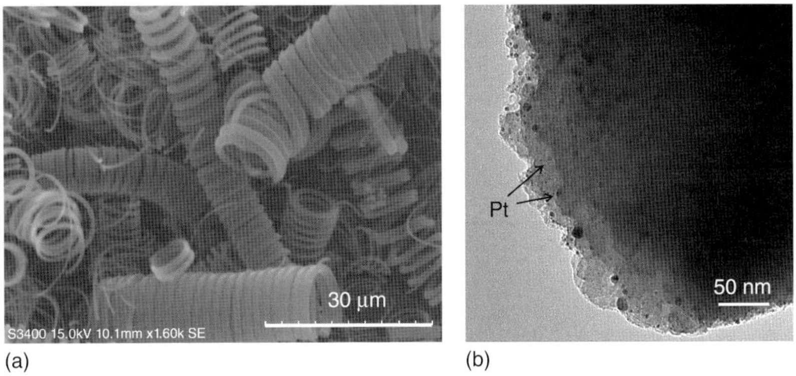

Figure 4.9 (a) Low-resolution SEM image of Pt/CMCs and (b) transmission electron microscope (TEM) image of Pt-deposits on CMCs at greater resolution. Reproduced from Horikoshi *et al.* [17]. Copyright 2012 by Elsevier B.V.

hybrid dielectric heating methodology, in which indirect microwave irradiation is achieved with the usage of microwave absorbers as heating sources (MAHSs; external heating) and direct heating of the Pd/AC catalyst particles with the microwaves (internal heating), led to a significant enhancement of the chemical yields of 4-methylbiphenyl from the Suzuki–Miyaura coupling reaction in toluene solvent [18]. An important result of this methodology is that the generation of hot spots can be controlled by the distribution of the microwave energy and by the low-temperature gradient. Another advantage of the methodology is that a high temperature can be maintained with significant energy saving.

The occurrence of hot spots in a heterogeneous catalyzed process can also be controlled largely by the experimental conditions used [19], for instance (i) by the size of the reactor, (ii) by the extent to which the solid catalyst particles (e.g., Pd/AC) are dispersed within the reaction volume through changes in stirring rates – that is, their dispersiveness, (iii) by irradiation with pulsed versus continuous microwaves, (iv) by the microwaves' input power levels, and (v) when irradiating with microwave radiation in the presence of a standing wave or in its

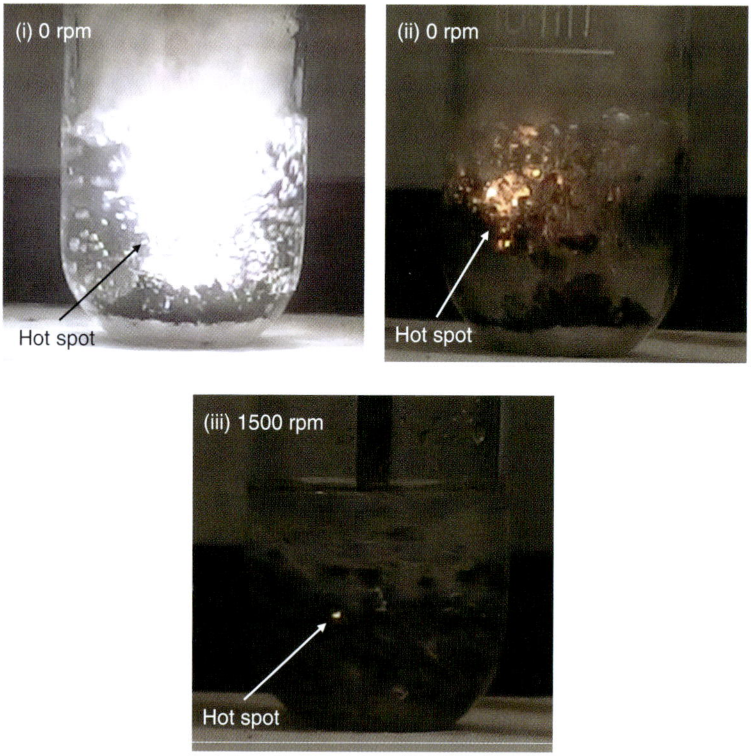

Figure 4.10 High-speed camera photograph of the electrical arc discharge occurring on the Pd/AC catalyst surface during the reaction in the 25-mm wide tube reactor under 550-W microwave irradiation and at a stirring rate of (a) 0 rpm, (b) 0 rpm, and (c) 1500 rpm. Reproduced from Horikoshi *et al.* [19]. Copyright 2014 by the American Chemical Society.

absence (nonstanding wave conditions). Increasing the rate of stirring of the reactants tends to minimize the occurrence of hot spots as demonstrated in Figure 4.10 that displays the generation of hot spots in the heterogeneous sample at 0 rpm (no mixing) and under conditions of significant mixing (stirring at 1500 rpm) [19].

The hot spots that formed under nonstirring conditions showed up as intense white sparkles (Figure 4.10a). With the turbulence created by the microwave heating, a large number of orange-like colored hot spots were observed over a wide area (Figure 4.10b). By contrast, no hot spots displaying the whitish intense sparkles were observed at 500 rpm stirring; only the orange-colored hot spots were seen (not shown in the figure). The number and frequency of formation of the orange-colored hot spots decreased considerably on increasing the stirring rate to 1500 rpm (Figure 4.10c); in fact, there was a hint of only a small hot spot.

In summary, we have shown that hot spots do occur, under certain conditions, at the catalyst particle surface, and their role in heterogeneous catalysis has been described, together with the possible mechanism(s) through which such hot spots are formed. Means to minimize or otherwise suppress the occurrence of hot spots during the organic syntheses have also been described.

References

1. See for example, Wikipedia. Microwave Chemistry, *http://en.wikipedia.org/wiki/Microwave_chemistry* (accessed 08 May 2015).
2. de la Hoz, A. and Loupy, A. (eds) (2012) *Microwaves in Organic Synthesis*, 3rd edn, Wiley-VCH Verlag GmbH, Weinheim.
3. Horikoshi, S. and Serpone, N. (2012) in *Microwaves in Organic Synthesis*, 3rd edn (eds A. de la Hoz and A. Loupy), Wiley-VCH Verlag GmbH, Weinheim, pp. 377–423.
4. Chemat-Djenni, Z., Hamada, B., and Chemat, F. (2007) Atmospheric pressure microwave assisted heterogeneous catalytic reactions. *Molecules*, **12**, 1399–1409.
5. Horikoshi, S. and Serpone, N. (2014) Role of microwaves in heterogeneous catalytic systems. *Catal. Sci. Technol.*, **4**, 1197–1210.
6. Matsuzawa, M., Togashi, S., and Hasebe, S. (2012) Investigation of microwave effects on transition metal catalyzed reaction using an isothermal reaction. *J. Chem. Eng. Jpn.*, **45**, 429–435.
7. Horikoshi, S., Fukui, M., Tsuchiya, K., Abe, M., and Serpone, N. (2010) Microwave specific effects in organic synthesis: a proposed model from the solvent-free synthesis of monoglycerylcetyl- dimethylammonium chloride. *Chem. Phys. Lett.*, **491**, 244–247.
8. Arvela, R.K. and Leadbeater, N.E. (2005) Suzuki coupling of aryl chlorides with phenylboronic acid in water using microwave heating with simultaneous cooling. *Org. Lett.*, 7, 2101–2104.
9. Horikoshi, S., Osawa, A., Suttisawat, Y., Abe, M., and Serpone, N. (2010) A novel Dewar-like reactor for maintaining constant heat and enhancing product yields during microwave-assisted organic syntheses. *Org. Process Res. Dev.*, **14**, 1461–1464.
10. Zhang, S., Hayward, D.O., and Mingos, D.M.P. (1999) Apparent equilibrium shifts and hot-spot formation for catalytic reactions induced by microwave dielectric heating. *Chem. Commun.*, 975–976.
11. Tsukahara, Y., Higashi, A., Yamauchi, T., Nakamura, T., Yasuda, N., Baba, A., and Wada, Y. (2010) *In situ* observation of nonequilibrium local heating as an origin of special effect of microwave on chemistry. *J. Phys. Chem. C*, **114**, 8965–8970.

12. Gutmann, B., Schwan, A.M., Reichart, B., Gspan, C., Hofer, F., and Kappe, C.O. (2011) Activation and deactivation of a chemical transformation by an electromagnetic field: evidence for specific microwave effects in the formation of Grignard reagents. *Angew. Chem. Int. Ed.*, **50**, 4636–4640.
13. Horikoshi, S., Osawa, A., Sakamoto, S., and Serpone, N. (2013) Control of microwave-generated hot spots. Part V. Mechanisms of hot-spot generation and aggregation of catalyst in a microwave-assisted reaction in toluene catalyzed by Pd-loaded AC particulates. *Appl. Catal., A*, **460–461**, 52–60.
14. Horikoshi, S., Osawa, A., Abe, M., and Serpone, N. (2011) On the generation of hot spots by microwave electric and magnetic fields and their impact on a microwave-assisted heterogeneous reaction in the presence of metallic Pd nanoparticles on an activated carbon support. *J. Phys. Chem. C*, **115**, 23030–23035.
15. Cravotto, G., Beggiato, M., Penoni, A., Palmisano, G., Tollari, S., Lévêques, J.-M., and Bonrath, W. (2005) High-intensity ultrasound and microwave, alone or combined, promote Pd/C-catalyzed aryl–aryl couplings. *Tetrahedron Lett.*, **6**, 2267–2271.
16. McMahon, J.M., Li, S., Ausman, L.K., and Schatz, G.C. (2012) Modeling the effect of small gaps in surface-enhanced Raman spectroscopy. *J. Phys. Chem. C*, **116**, 1627–1637.
17. Horikoshi, S., Suttisawat, Y., Osawa, A., Takayama, C., Chen, X., Yang, S., Sakai, H., Abe, M., and Serpone, N. (2012) Organic syntheses by microwave selective heating of novel metal/CMC catalysts – The Suzuki–Miyaura coupling reaction in toluene and the dehydrogenation of tetralin in solvent-free media. *J. Catal.*, **289**, 266–271.
18. Horikoshi, S., Osawa, A., Sakamoto, S., and Serpone, N. (2013) Control of microwave-generated hot spots. Part IV. Control of hot spots on a heterogeneous microwave-absorber catalyst surface by a hybrid internal/external heating method. *Chem. Eng. Process.*, **69**, 52–56.
19. Horikoshi, S., Kamata, M., Mitani, T., and Serpone, N. (2014) Control of Microwave-Generated Hot Spots. 6. Generation of hot spots in dispersed catalyst particulates and factors that affect catalyzed organic syntheses in heterogeneous media. *Ind. Eng. Chem. Res.*, **53**, 14941–14947.

Part II
Applications – Preparation of Heterogeneous Catalysts

Microwaves in Catalysis: Methodology and Applications, First Edition.
Edited by Satoshi Horikoshi and Nick Serpone.

5
Preparation of Heterogeneous Catalysts by a Microwave Selective Heating Method

Satoshi Horikoshi and Nick Serpone

5.1
Introduction

Studies on the microwave-assisted synthesis of heterogeneous catalysts (solid catalysts) originated in the 1980s with the microwave sintering of ceramics of interest mainly to the American Ceramics Society [1, 2], while the synthesis of metallic photocatalyst nanoparticles were investigated extensively in the 1990s [3]. These studies provided the basis for today's catalyst synthesis research in microwave chemistry. Metallic catalysts are found deposited on the surface of carbonaceous materials and can be found in the literature using such keywords as: catalyst synthesis, microwave heating, zeolites [4], and photocatalysts [5]. A nonexhaustive introduction into the study of microwave-assisted syntheses of catalysts/carbonaceous materials is hereby given as exemplified by the synthesis of TiO_2-based catalysts and others using selective microwave heating and most often carbon as the support in one form or other (e.g., activated carbon, carbon nanotubes, graphene), and where available the activity of such catalysts.

5.2
Synthesis of Metal Catalysts on Carbonaceous Material Supports

Carbonaceous materials have been investigated widely as functional materials for various applications that include electrodes for batteries and fuel cells, adsorbents for water purification, and as catalyst supports [6–10]. With the continued development of activated carbon (AC) technologies, new applications of microwave heating in preparing carbon supports have witnessed successful and continued interests [11]. Microwave-induced chemical reactions have been investigated toward a large number of applications such as, for example, in the pyrolysis of high volatile bituminous coal with nitrogen and coal gasification in water vapor. Char – a solid material that remains after light gases (e.g., coal gas) and tar have been released from a carbonaceous material during the initial stage of combustion known as *carbonization, charring*, or *pyrolysis* – is a fairly good absorber of

Microwaves in Catalysis: Methodology and Applications, First Edition.
Edited by Satoshi Horikoshi and Nick Serpone.

microwave energy [12, 13] that has led to applying a certain amount of microwave energy for char synthesis to be established with minimal reaction conditions.

Microwave ovens are generally smaller than conventional ovens. When compared with conventional electrical heating [14], microwave heating is a more efficient and attractive way of removing oxygenated functionalities from carbon surfaces. Microwave treatments produce comparable changes in the textural and chemical properties of activated carbonaceous materials, although over a far shorter time period. Microwave activation of AC produces a product that performs better than standard activated carbon [15]. Accordingly, microwaves are now being used in various technological and scientific domains where heating dielectric materials is required.

Palladium nanoparticles deposited on carbon nanotubes (CNTs) find applications that range from hydrogen storage to catalysis [16–18]. The one-step linker-free synthesis of hybrid materials that consist of palladium nanoparticles and multiwall carbon nanotubes (Pd-NP/MWCNTs) has been described, a method that uses microwave radiation for the effective decomposition of the tris(dibenzylideneacetone)dipalladium(0), $Pd_2(dba)_3$, complex in the presence of MWCNTs. High loadings of Pd nanoparticles (up to 40 wt%) having sizes between 3 and 5 nm could be deposited on the surface of MWCNTs within a very short time of only 2 min [19].

The microwave-assisted method to synthesize graphene-supported metallic catalysts has become a widely documented method [20, 21]. Although graphene-supported metal catalysts have high surface areas and stabilities, only a few studies have examined their application toward organic reactions, whereas greater attention has been focused toward applications in fuel cells, hydrogen storage, sensors, and solar cells [22, 23]. To reduce a solution of graphene from being agglomerated and restacked to form graphite through van der Waals attractive forces owing to the absence of O groups, some efforts has been expended on synthesizing graphene-supported Pt nanostructures by the microwave heating method [24, 25]. Aggregation of graphene sheets reduces the high surface area of graphene and so limits the effective dispersion of Pt nanoparticles being deposited onto graphene sheets [26]. Microwave-assisted syntheses have also been reported on the preparation of graphene-supported Pt/Pd nanostructures, in which graphene oxide (or graphitic oxide, GO), and Pt precursors (K_2PtCl_4) or Pd precursors ($PdCl_2$) were reduced using ascorbic acid in the presence of hexadecyltrimethyl ammonium bromide through microwave irradiation [27]. Poly(diallydimethyl ammonium chloride) is effective in modifying Pd/Pt nanoparticles through the functionalization of graphene oxide in ethylene glycol (EG) [28]. Other reports have described the deposition of Pt and Pd/Pt nanocrystals on exfoliated few layers of graphene by solvothermal reactions [29]. In addition, the preparation and the deposition of Pd/Pt nanoparticles in cubic form on graphene have been carried out in *N*,*N*-dimethylformamide solvent [30].

Microwaves can be used to synthesize graphitic carbon particles at temperatures lower than a conventional method, the most common synthesis route being chemical vapor deposition at temperatures around 800 °C in the presence of a transition metal catalyst [31]. Microwave irradiation of polyethylene glycol at temperatures between 160 and 220 °C for 40 min in the absence of a catalyst yields chains of graphitic carbon particles.

Galletti and coworkers [32] synthesized ruthenium nanocatalysts using a microwave-assisted solvothermal method that were deposited on a γ-Al_2O_3 catalyst support. The resulting material could hydrogenate phenol to cyclohexanone. A feature of this method is the rapid synthesis of Ru nanoparticles with a small diameter and a narrow size distribution. Similarly, Glaspell and coworkers [33] reported a simple method for the microwave-assisted synthesis and characterization of Au and Pd nanoparticle catalysts supported on CeO_2 nanoparticles for CO oxidation; catalysts so produced displayed high activity and stability. The method offers extremely short reaction times and produces high yields of highly efficient pure nanoparticle catalysts.

5.3 Photocatalysts

Through the years, advanced oxidation processes (AOPs) have been shown to be effective in the photooxidative disposal of various volatile pollutants in the gas phase and pollutants in aqueous media [34]. The most widely adopted AOPs include photodegradation in the presence of TiO_2, the Fenton, and photo-Fenton processes, together with ultrasonication and ozonation (O_3). AOPs rely on the generation of reactive-free radicals, especially the hydroxyl and hydroperoxyl radicals ($OH^{\cdot}$, $HOO^{\cdot}$), and the superoxide radical anion ($O_2^{-\cdot}$) to effect photooxidations. Disposal of environmental pollutants through semiconductor photocatalysis has attracted extensive interest over the last few decades. Among various semiconductors, TiO_2 has been the leading photocatalyst owing to its good photoactivity, high chemical stability, low cost, and nontoxicity [35]. The field of heterogeneous photocatalysis has expanded rapidly within the last four decades, having undergone various developments especially in relation to energy and the environment. It has been defined as the acceleration of photoreactions in the presence of a catalyst.

Inertness to the chemical environment and its long-term photostability has made TiO_2 an important material in many practical applications and in commercial products ranging from drugs to foods, cosmetics to catalysts, paints to pharmaceuticals, and sunscreens to solar cells in which TiO_2 has been used as a desiccant, as a brightener, or as a reactive mediator [36].

The two most significant and potential applications of photocatalysis have been in solar water splitting and the purification of air and water that contain low concentrations of pollutants. The multidisciplinary nature of the field has also increased significantly and now includes semiconductor physics, surface science, photochemistry, physical chemistry, materials science, and chemical engineering [37]. Recent years have witnessed increased interest in applying semiconductor-based photocatalysis toward environmental clean-up, drinking water treatment, and industrial and health applications [38]. The catalyst itself is unchanged during the process and no consumable chemicals are required, which results in considerable savings and simpler operation of the equipment involved. The influence of the environment or ecosystem on the photocatalysts must also be considered because most photocatalysts are in the form of nanoparticulates. To avoid such issues, the photocatalyst nanoparticles are immobilized on some suitable catalyst support. In this regard, significant progress has been made in recent years in heterogeneous photocatalysis on immobilizing titanium dioxide particles on such support materials as glass, silica beads, polymers, and zeolites [39], as the re-use of dispersed TiO_2 nanoparticles can cause undue problems of filtration. Accordingly, it is important to fix TiO_2 particles for possible recycling after the photocatalytic events.

Some reports have described how to immobilize photocatalysts on the support surface by the microwave heating method. For instance, Hu *et al.* [40] reported a microwave-assisted synthesis of graphene/ZnS nanocomposites to be used as photocatalysts in aqueous media. In this experiment, graphitic oxide and zinc acetate dihydrate $\{Zn(CH_3COO)_2{\cdot}2H_2O\}$ were reduced simultaneously with thioacetamide during the microwave irradiation process. Compared to other *in situ* approaches, the microwave-assisted synthesis method allows for selective activation of the target precursor to initiate nucleation, thereby generating smaller and more uniform particles. Unfortunately, only scant studies have been carried out on the immobilization of photocatalysts on catalyst supports by the microwave method. Accordingly, additional studies will be required in the future to find more performing photoactive immobilized catalytic materials.

Immobilization of TiO_2 nanoparticles on AC nanoparticulates has drawn great attention owing to the high adsorption capability of activated carbon that can enrich an organic substrate close to the catalyst, thereby promoting pollutant transfer from the support onto the photocatalyst TiO_2 and potentially increasing photocatalytic efficiency. The synergistic effect of adsorption of TiO_2 particulates by AC can have beneficial consequences in the photo-induced transformation of several types of organic pollutants, such as the defluorination of pentafluorobenzoic acid [41] and the photodegradation of a dye [42] in aqueous media, as well as the photodecomposition of the volatile organic compound (VOC) isopropanol (IPA) in air [43] in the presence of dispersed TiO_2/AC particulates. However, the hydrophilic characteristics of TiO_2 particles do not facilitate adsorption on a hydrophobic-activated carbon surface. TiO_2 must first be coated if it is to be fixed onto the AC surface, following which the surface area-to-volume of activated carbon diminishes.

Finally, nanoparticles have important consequences on human health and on ecological systems [44]. In this regard, the U.S. Food and Drug Administration (FDA) permits up to 1% TiO_2 as an inactive ingredient in food products. While there are no known health effects associated with the use of TiO_2, a recent study found that 3- to 6-year-old children are the most affected group of people that consume TiO_2 particles from food products. A regulatory framework for the use of TiO_2 in food products has yet to be firmly established in many countries, especially in developing nations. Many new properties of TiO_2 have been explored during the past few years [45].

5.3.1 Preparation of TiO_2/AC Particles

This section focuses on the preparation of TiO_2 particles supported on AC particulates using the microwave (MW) heating method, which is then compared to the more traditional hydrothermal (oil-bath) heating method. Details of the procedure are taken from some of our recent studies [46].

To prepare TiO_2/AC particulates, an aqueous titanium oxysulfate solution (0.125 M; 20 ml) and AC (1 g; diameter, 0.65 mm) were introduced into an Anton Paar high-pressure Pyrex cylindrical reactor (30 ml), following which the reactor was subjected to microwave irradiation using an Anton Paar Monowave 300 microwave apparatus under stirring conditions (400 rpm). Determination of the temperature distribution in a reactor can pose frequent problems when using microwave heating [47]. Accordingly, the temperature distribution in the sample solution was measured using both a ruby fiber optic sensor located at the center of the solution and a radiation thermometer near the external walls of the reactor. The difference in temperature of the solution at the two locations was less than 2 °C, indicating a nearly uniform temperature distribution throughout the suspension. Soon after the microwave heating step, the sample of TiO_2/AC particles was filtered and washed repeatedly with methanol and water, and then dried at 500 °C overnight in an electric furnace. For comparison, conventional heating with an oil-bath was also used in the synthesis of TiO_2/AC particles under otherwise identical temperature conditions achieved by soaking the cylindrical reactor in the oil-bath preheated to 170–190 °C.

The initial transparent solution became cloudy at 70 °C, turning to a white dispersion of TiO_2 particles generated at 80 °C; at 90 °C the white color of the suspension was enhanced even further. The TiO_2 particles formed a thin film coating on the activated carbon surface at 70 °C (Figure 5.1a; [46]) as confirmed by SEM–EDX techniques (Figure 5.2; [46]), whereas TiO_2 particles formed on the AC surface at 80 °C (Figure 5.1b). Scanning electron microscopic (SEM) photographs revealed an average particle size of TiO_2 on the AC surface of 426 nm at 80 °C; at 90 °C the average particle diameter was 415 nm (Figure 5.1c). At a reaction temperature of 120 °C (note that since the reactor used in this experiment was of a closed type, temperatures greater than 100 °C can be achieved by changing the microwave input power) TiO_2 particles adsorbed on the AC surface as displayed in Figure 5.1d, which shows small (357 nm) and larger intermingled particles on the AC surface.

A comparison between the composite TiO_2/AC particulates prepared by the microwave heating method and the more conventional oil-bath heating method at 80 °C is made in the SEM images of Figure 5.1e (oil-bath method) and Figure 5.1b (MW method). Although the same heating rate and the same reaction temperature were used, TiO_2 formed on the AC surface was different from that of the microwave method (Figure 5.1b), with oil-bath heating only a TiO_2 thin film formed on the AC support at 70 °C, and no changes occurred even when the temperature reached 90 °C by the latter heating method (Figure 5.1f). Similar findings were reported for the oil-bath heating method when the TiO_2 precursor

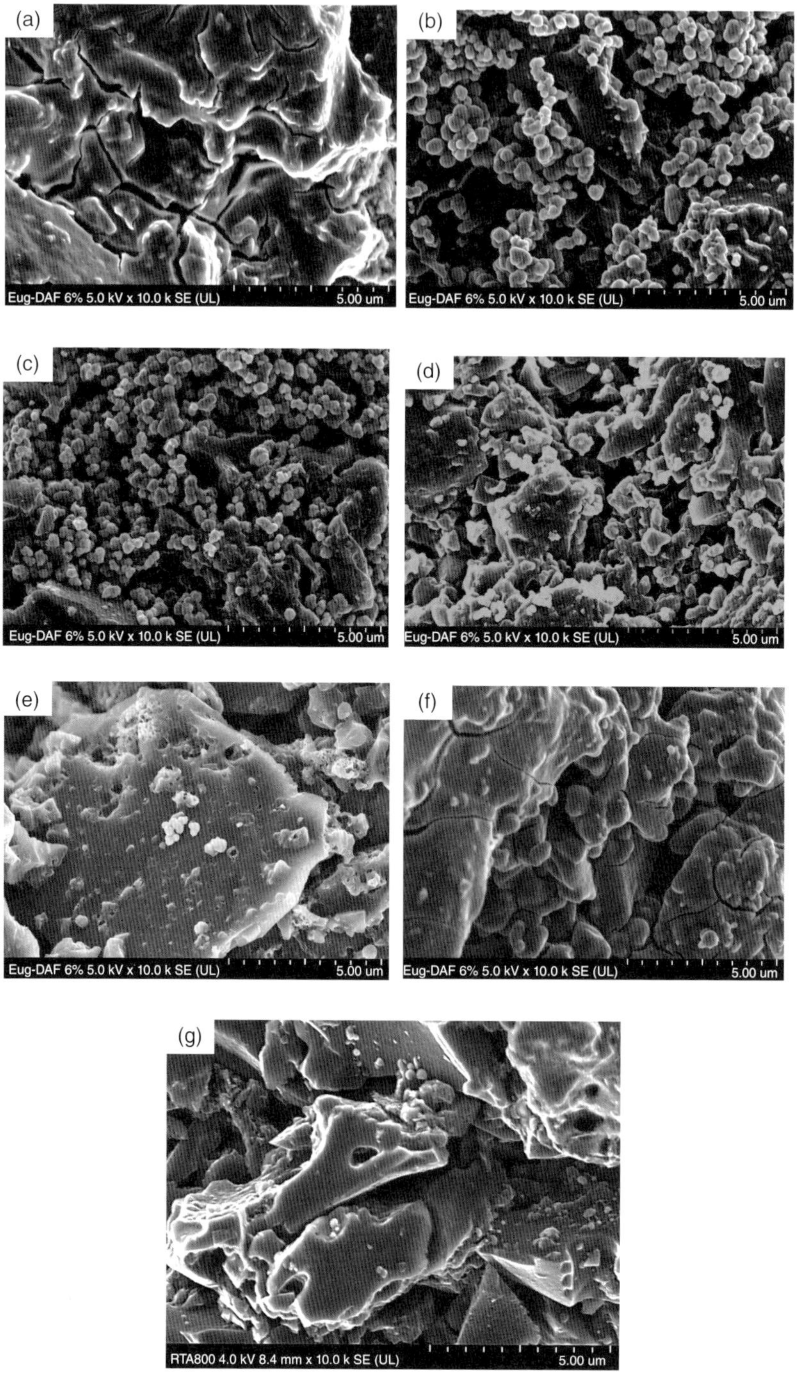
(a)
Eug-DAF 6% 5.0 kV x 10.0 k SE (UL) 5.00 um
(b)
Eug-DAF 6% 5.0 kV x 10.0 k SE (UL) 5.00 um
(c)
Eug-DAF 6% 5.0 kV x 10.0 k SE (UL) 5.00 um
(d)
Eug-DAF 6% 5.0 kV x 10.0 k SE (UL) 5.00 um
(e)
Eug-DAF 6% 5.0 kV x 10.0 k SE (UL) 5.00 um
(f)
Eug-DAF 6% 5.0 kV x 10.0 k SE (UL) 5.00 um
(g)
RTA800 4.0 kV 8.4 mm x 10.0 k SE (UL) 5.00 um

Figure 5.1 Scanning electron microscopic images of TiO_2/activated carbon particulates under various synthesis conditions: reaction temperatures were (a) 70 °C, (b) 80 °C, (c) 90 °C, and (d) 120 °C under microwave heating conditions; using the oil-bath heating method the reaction temperatures were (e) 80 °C and (f) 90 °C; (g) naked activated carbon as the control. Reproduced from Ref. [46]. Copyright 2013 by Elsevier B.V.

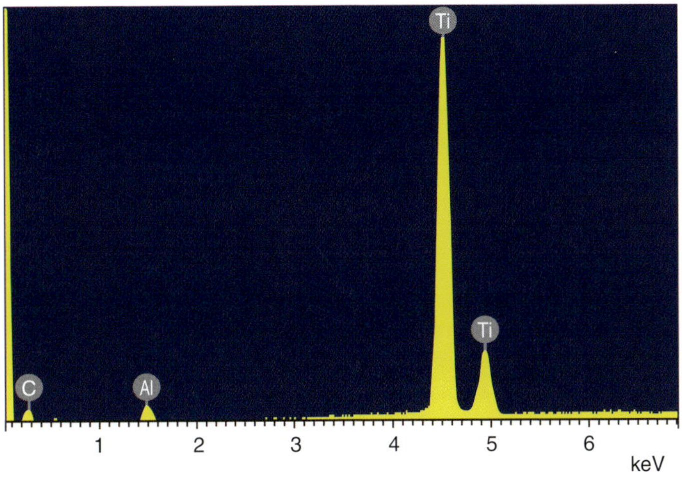

Figure 5.2 SEM–EDX pattern of TiO_2 coating on the activated carbon particulates upon heating to a temperature of 70 °C by the microwave method. Reproduced from Ref. [46]. Copyright 2013 by Elsevier B.V.

was a titanium(IV) isopropoxide solution [48]. No growth of TiO_2 particles on activated carbon by the microwave method had been hitherto reported. No doubt the formation mechanisms of TiO_2 on the AC surface between the microwave and the oil-bath heating methods must have differed as evidenced by the SEM results and as described below.

Figure 5.3 displays a plot of the monotonic decrease in TiO_2 particle size on the AC surface with increasing reaction temperatures [46]. Evidently, the TiO_2/AC

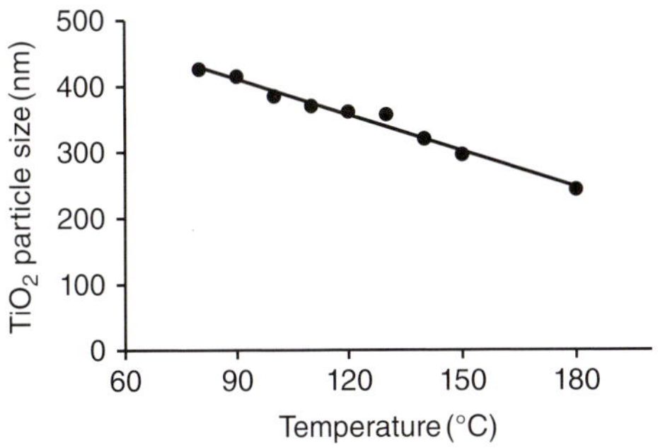

Figure 5.3 Plot illustrating the decrease in TiO_2 particle size on the AC surface at various reaction temperatures. Reproduced from Ref. [46]. Copyright 2013 by Elsevier B.V.

composite particles from the synthesis at the higher temperatures tended to have lower specific surface areas. However, at temperatures greater than 100 °C, the TiO_2 particles tended not to be adsorbed on some sites of the AC surface. Results suggested an optimal reaction temperature of 90 °C.

5.3.2
Proposed Mechanism of Formation of TiO_2/AC Particles

A most important characteristic feature of heating a solvent medium by microwave irradiation is the relative dielectric loss (ε_r'') factor [49], which represents the quantity of input microwave energy lost to the sample by being dissipated as heat. This factor is a useful index of the generation of heat because of the interaction of the solvent with the microwave radiation field. Dielectric losses at a microwave frequency of 2.45 GHz can be determined using an Agilent Technologies HP-85070B Network Analyzer and an Agilent dielectric high-temperature probe (up to ~200 °C) at various temperatures (at 5 °C intervals) using a conventional plate heater. Temperatures of the solutions can be measured with an optical fiber thermometer. As an example, the dielectric losses (ε_r'') recorded for AC particles dispersed in an aqueous titanium oxysulfate solution at various temperatures are reported in Figure 5.4 [46]. For comparison, dielectric losses for pure water are also displayed that indicate a general decrease with an increase in temperature.

The initial dielectric loss of the AC aqueous dispersion ($\varepsilon_r'' = 33.4$) at near-ambient temperature decreased slightly up to about 60 °C, after which it increased with an increase in temperature owing to some microscopic chemical changes occurring in the TiO_2 precursor (titanium oxysulfate). This contrast is likely due to efficient direct microwave heating of the activated carbon in the dispersion [50]. Accordingly, even though the AC particles were present in a high dielectric loss solvent such as water, selective heating of the AC particles nonetheless occurred

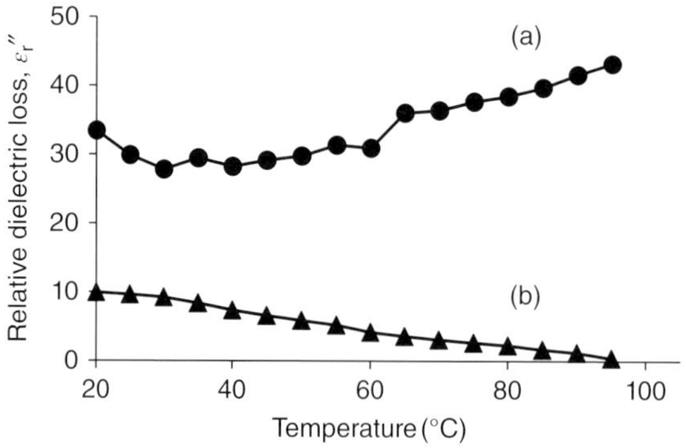

Figure 5.4 Relative dielectric losses (ε_r'') at the microwave frequency of 2.45 GHz and at various temperatures (a) for the aqueous dispersion of activated carbon and titanium oxysulfate and (b) for pure water. Reproduced from Ref. [46]. Copyright 2013 by Elsevier B.V.

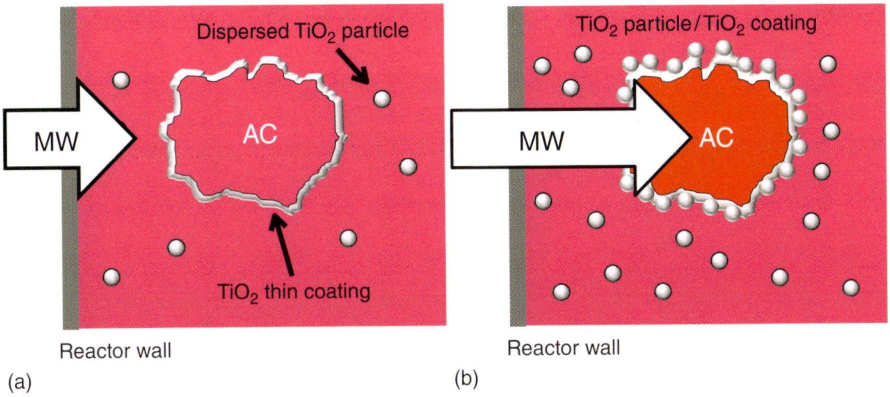

Figure 5.5 Cartoon illustrating the growth mechanism of TiO_2/AC particles produced by the microwave heating method at (a) a temperature of 70 °C and (b) at temperatures greater than 80 °C. Reproduced from Ref. [46]. Copyright 2013 by Elsevier B.V.

under microwave irradiation [51] and provided the necessary energy for the formation and growth of the TiO_2 particles.

The proposed growth mechanism of TiO_2 under microwave heating at 70 °C and temperatures greater than 80 °C conditions is illustrated in the cartoons of Figure 5.5 [46]. Both the dispersed TiO_2 particles in solution and the TiO_2 thin coating on the AC surface formed under efficient microwave heating at 70 °C (Figure 5.5a). Apparently, the growth of the TiO_2 thin film took place through heat conduction from the solution. At temperatures greater than 80 °C, the heating efficiency of the water by microwave heating decreased, whereas microwave direct heating of the AC particles increased (Figure 5.5b). Evidently, direct heating of the AC particles produced TiO_2 particles on the AC surface, together with smaller TiO_2 particles formed at the higher reaction temperatures. In the case of the oil-bath heating method, formation of TiO_2 on the AC surface occurred only through heat conduction from the heated oil outside the reactor. The growth mechanism was independent of temperature changes.

5.3.3 Photoactivity of MW-Prepared TiO_2/AC Composite Particles in the Degradation of Isopropanol

Approximately 94% of IPA was either chemisorbed and/or physisorbed after 40 min under dark conditions on the TiO_2/AC catalyst surface produced by microwave and oil-bath heating methods, even though the specific surface areas differed [46]. The extent of adsorption of IPA showed no further changes after 40 min, following which UV-light irradiation of the TiO_2/AC/IPA system led to the photodecomposition of the volatile IPA via first-order kinetics (12.4×10^{-3} min^{-1}), whereas with the TiO_2/AC particles produced by the oil-bath method the photodegradation kinetics were nearly sixfold slower

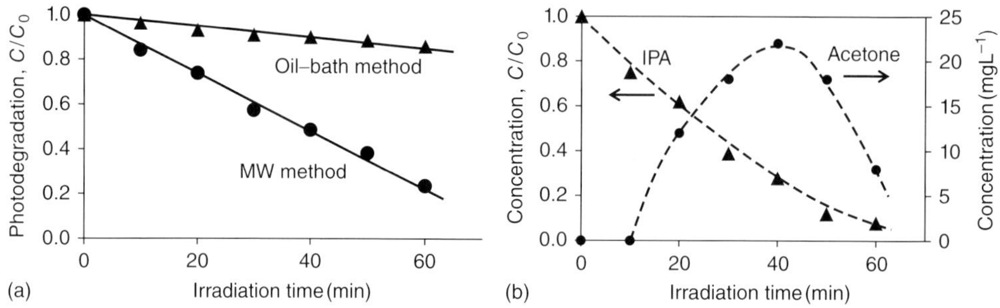

Figure 5.6 (a) Photodecomposition kinetics of isopropanol with the microwave-synthesized and oil-bath-synthesized TiO_2/AC particles. (b) Photodegradation dynamics of isopropanol and formation and degradation of the intermediate product acetone in the presence of TiO_2 alone (no AC support). Reproduced from Ref. [46]. Copyright 2013 by Elsevier B.V.

(2.2×10^{-3} min^{-1}) – Figure 5.6a [46]. The greater degradation efficiency for the microwave-prepared TiO_2/AC composite was likely due to the direct formation of TiO_2 particles on the AC surface. Figure 5.6 also compares the photodegradation of IPA with and without the presence of the AC support.

Acetone and CO_2 gas are produced as intermediate and final products, respectively, in the photodecomposition of IPA in the presence of TiO_2 only (no AC support) – see Figure 5.6b [46]. With the TiO_2 particles produced by the microwave method, the acetone intermediate formed after 20 min of UV irradiation. However, with the TiO_2/AC composites, the amount of acetone formed could not be quantified because it adsorbed on the TiO_2/AC particulates and underwent further degradation to carbon dioxide as evidenced by a control experiment carried out with dried TiO_2 particles (100 mg) and 300 mg l^{-1} (ppm) of IPA. In this control experiment, approximately 270 mg l^{-1} (90%) of IPA was chemically and/or physically adsorbed on the TiO_2 surface after 20 min in the dark; no further adsorption occurred after this time. There was no difference in the initial adsorption under dark conditions between the particulates produced by the microwave method and those from the oil-bath method, suggesting that the photodegradation of IPA occurred through the mediation of the TiO_2 particles formed on the AC surface [46].

5.4 Microwave-Assisted Syntheses of Catalytic Materials for Fuel Cell Applications

Fuel cells convert the chemical energy of a fuel into electrical energy (direct current) through a chemical reaction with O_2 or with some other oxidizing agent. Typically, H_2 produced from steam reforming of CH_4 is most commonly used; however, hydrocarbons (e.g., methane) or such alcohols as methanol can also be used directly (Figure 5.7) [52]. Though fuel cells produce electricity from the *burning* of such fuels, they differ from batteries in that a continuous supply of the

(a)

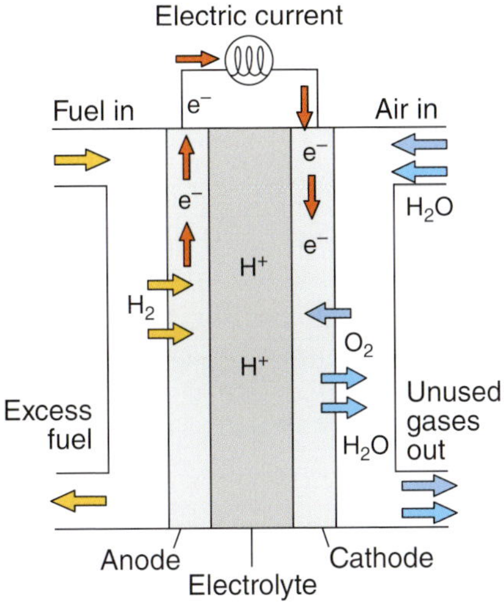

(b)

Figure 5.7 (a) Model of a direct methanol fuel cell; the actual fuel cell stack is the layered cube shape in the center of the image. (b) Scheme of a proton-conducting fuel cell.

fuel and the oxidizing agent are a necessity. And although the first fuel cell was reported in the early 1800s, it was only over a century later that fuel cells saw their first applications with the coming of space exploration.

Presently, fuel cells are used in various modes of transport, such as forklifts, automobiles, buses, boats, motorcycles, and submarines, as well as providing primary and backup power in industrial and residential buildings [52]. Different types of fuel cells are available that consist mainly of an anode, a cathode, and an electrolyte that permits charges to migrate between the two compartments of the fuel cell. They are usually classified by the type of electrolyte used and by the start-up time. It must be noted that the *burning* of the fuel typically requires a catalyst, particularly for hydrogen as the fuel where platinum particulates are the catalytic entities often deposited on some appropriate support.

5.4.1
Microwave-Assisted Synthesis of Pt/C Catalyst Particulates for a H_2 Fuel Cell

The most efficient metal catalyst for fuel cell applications is platinum, which is the only catalyst active for the oxidation of hydrogen, methanol, and reduction of oxygen at low temperature. Carbon has proven a very good catalyst support for Pt as it improves growth, structure, and dispersion of the Pt particulates that in turn enhances the catalytic properties and stability of electrocatalysts such as platinum [53]. In this regard, Chen and coworkers [54] reported a simple procedure that makes use of microwaves to prepare Pt^0 nanoparticles supported on carbon

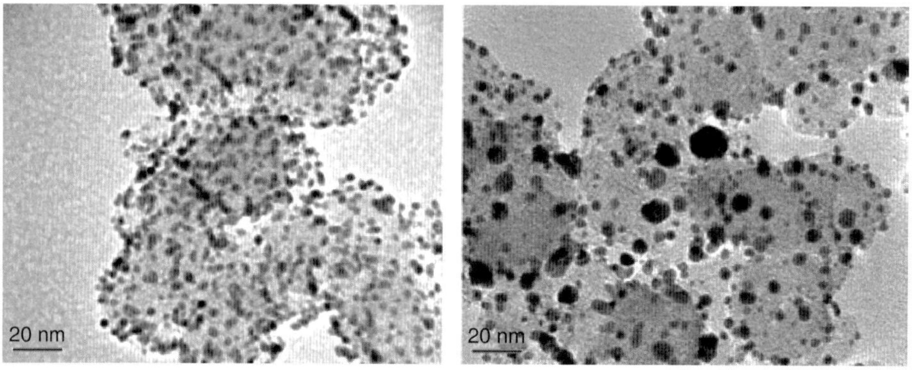

(a) (b)

Figure 5.8 TEM images of (a) the microwave-assisted synthesis of Pt nanoparticles supported on Vulcan carbon XC-72 and of the (b) commercially available E-TEK Pt/C catalyst (nominal Pt loading 20 wt%). Reproduced from Chen *et al.* [54]. Copyright 2002 by the Royal Society of Chemistry.

particulates yielding Pt/C nanocomposites consisting of 10, 15, and 20 wt% Pt wherein spherical Pt nanoparticles (diameter 3.5–4.0 nm) with a narrow particle size distribution were uniformly dispersed on the carbon support (TEM images in Figure 5.8). Tests revealed greater electrochemical activity of the Pt/C catalyst nanocomposites for the oxidation of liquid methanol at ambient temperatures than commercially available catalysts [54].

The pulsed microwave-assisted polyol synthesis method is a useful method to deposit Pt particulates onto multiwalled CNTs to be used as electrocatalysts for proton exchange membrane fuel cells [55] – see, for example, Figure 5.7b. The method is effective in growing Pt nanoparticles on the CNT surface with smaller sizes (2–3 nm) and with good uniformity. Also examined was the effect of the different microwave pulse factors ($MP = t_{on}/t_{off}$) on the size and morphology of Pt nanoparticles over the CNT support. It appears that for a small MP factor, the relaxation time (t_{off}) is significant compared to the irradiation time (t_{on}).

An optimal combination of irradiation time and relaxation time tends to induce a rapid growth of nuclei in the reaction mixture, as a consequence of which Pt particles with smaller size and good uniformity can be attained [55]. On the other hand, if the relaxation time is less than the irradiation time by a large amount (i.e., $t_{on} > t_{off}$), the reaction temperature rises rapidly and the Pt^{4+} ions do not have sufficient time to create new nucleation sites such that most of the metallic ions would then adsorb over the previously deposited Pt particles. This phenomenon is known as the *gradual island growth* , which results in Pt nanoparticles on the CNT support with bigger size and poor uniformity, and forms Pt aggregates. The viability of the pulsed microwave-assisted deposition method as a simple and effective method was shown for the commercial feasibility in fuel cell application.

5.4.2 Preparation of Nanocatalysts for a Methanol Fuel Cell

A recent 2011 study by Amin and coworkers [56] reported on the preparation of Pt/C and Pt/Co_3O_4/C catalysts and also examined their catalytic activity for the oxidation of methanol through cyclic voltammetric and chronoamperometric techniques. The Pt nanoparticles were dispersed by two different methods: (i) by an impregnation method and (ii) by a microwave-assisted method with Vulcan carbon and Co_3O_4/Vulcan carbon as the supports for the platinum particulates. Apparently, catalyst nanocomposites prepared by the microwave method exhibited a lower catalytic activity than the nanocomposites prepared by the impregnation method; the Pt/Co_3O_4/C composite catalysts displayed better electrocatalytic activity for the oxidation of methanol than did the Pt/C catalyst. Moreover, significant electrocatalytic activities and good stabilities of the catalytic entities were attributed to a synergistic effect between the Pt and the Co_3O_4, which was also responsible for greater dispersion of the Pt nanoparticles on the support surface support. Both the impregnation and the microwave preparation methods produced a uniform dispersion of small particles as was evidenced by the TEM technique (see Figures 5.9 and 5.10).

The synthesis of Pt/Co_3O_4 nanoparticles supported on C (Vulcan XC-72R) was carried out in two steps [56]. In the first step, an appropriate amount of $Co(NO_3)_2{\cdot}6H_2O$ and carbon were mixed together in double distilled water (loading Co_3O_4, 5 wt%; pH 10) after which the mixture was stirred constantly for 3 h, then filtered, washed several times with double distilled water, dried at 80 °C for 6 h, and then calcined at 400 °C for 3 h in air to yield Co_3O_4/C. In the second step, the desired quantity of prepared powdered support and a Pt precursor (i.e., H_2PtCl_6) were ultrasonically mixed in double distilled water (pH 11) for 1 h at 70 °C following which a sodium borohydride solution was added to the mixture (molar ratio of Pt to $NaBH_4$, 1 : 70). The process was completed either by stirring

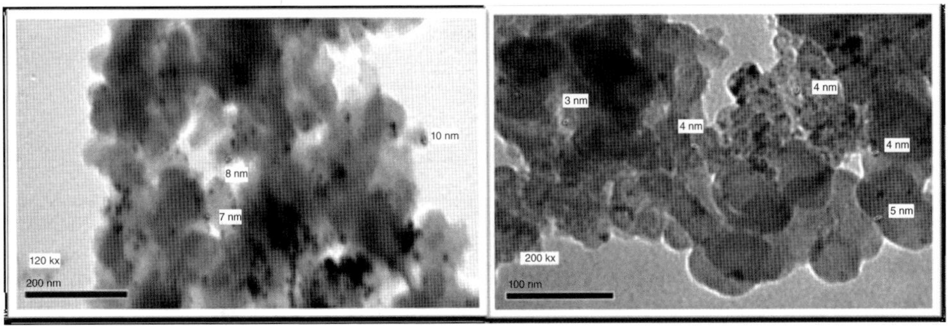

(a) (b)

Figure 5.9 TEM micrographs of Pt/C nanocomposites obtained by the (a) impregnation method and by the (b) microwave synthesis method. Reproduced from Amin *et al.* [56]. Copyright 2011 by ESG, Belgrade, Serbia.

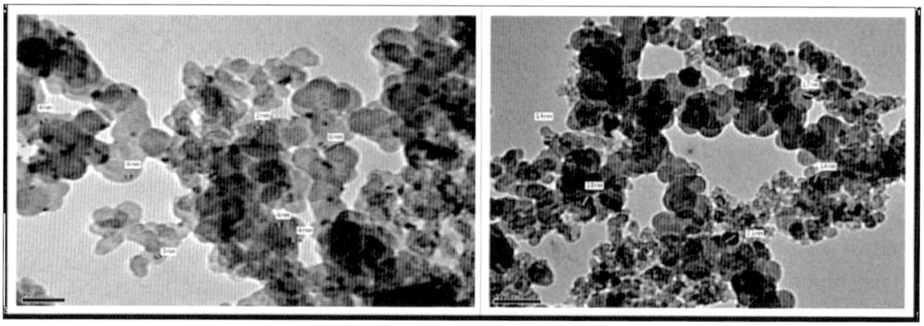
(a) (b)

Figure 5.10 TEM micrographs of $Pt/Co_3O_4/C$ obtained by the (a) impregnation method and by the (b) microwave synthesis method. Reproduced from Amin *et al.* [56]. Copyright 2011 by ESG, Belgrade, Serbia.

the mixture for 3 h in air (impregnation method) or by placing the mixture in a household microwave oven (50 GHz, 1400 W) for 5 min (20 s ON and 10 s OFF). The resulting samples were filtered, washed several times with double distilled water, and dried in an oven at 80 °C for 6 h (Pt loading, 25 wt%).

At about the same time, Guo and coworkers [57] reported the preparation of a catalyst for methanol oxidation in a fuel cell that consisted of Pt particles supported on carbon aerogel (Pt/CA) using a microwave-assisted polyol process. High-resolution transmission electron microscopy (HRTEM) and X-ray diffraction (XRD) results revealed a uniform dispersion of spherical Pt nanoparticles (diameter, 2.5–3.0 nm), while cyclic voltammetric and chronoamperometric techniques showed the Pt/carbon aerogel catalyst to display greater electrocatalytic activity and electrochemical stability for methanol oxidation at ambient temperature than did a commercial Pt/C catalyst sample with the same Pt loading (40 wt%). A mechanism for methanol oxidation in acid aqueous media was also suggested by the authors (Eqs. (5.1)–(5.3)) [57]:

$$CH_3OH_{ads} + Pt \rightarrow PtCO_{ads} + 4H^+ + 4e^- \quad (5.1)$$

$$H_2O + Pt \rightarrow PtOH_{ads} + H^+ + e^- \quad (5.2)$$

$$PtOH_{ads} + PtCO_{ads} \rightarrow Pt + Pt + CO_2 + H^+ + e^- \quad (5.3)$$

The synthesis of the platinum/carbon aerogel catalyst involved first the preparation of the organic aerogel by a sol–gel process using resorcinol and formaldehyde mixtures dissolved in water with Na_2CO_3 as the catalyst; the carbon aerogel was then obtained by carbonizing the dried organic aerogel in a quartz tube furnace at 1050 °C for 4 h under an inert N_2 atmosphere [57]. Subsequently, the Pt/CA catalysts were prepared from an EG solution of $H_2PtCl_6{\cdot}6H_2O$, followed by 30 min of ultrasonic agitation and then by dropwise addition of a KOH solution. The carbon aerogel powder was uniformly dispersed in the final solution by ultrasonic agitation, after which it was placed in a microwave oven (2.45 GHz, 800 W) and heated for 180 s (6 cycles of 20 s ON and 10 s OFF). After filtration, the residue was washed with deionized water and dried in a vacuum oven at 80 °C for 12 h.

5.4.3 Effects of pH on Pt Particle Size and Electrocatalytic Activity of Pt/CNTs for Methanol Electro-oxidation

The electrocatalytic activity of Pt and Pt-based alloy particles is known to depend on the type of carbon supports. In this regard, Li and coworkers [58] prepared Pt nanoparticles with different mean sizes by microwave heating EG solutions of H_2PtCl_6 at different pH's in the presence of CNTs as supports. The latter were prepared by catalytic chemical vapor deposition using Ni_4MoMg_4 nanoparticles as the catalyst for CNTs growth; the carbon source was methane. Transmission electron microscopy showed that Pt particles were smaller and more uniform when the synthesis pH increased from 3.4 to 9.2; the mean particle size was 5.8, 5.2, 3.4, and 2.7 nm when the synthesis pH was 3.6, 5.8, 7.4, and 9.2, respectively (Figure 5.11).

Clearly, the pH was an important factor that influenced the particle size so that the size of the Pt particles could be selected by a simple adjustment of the synthesis

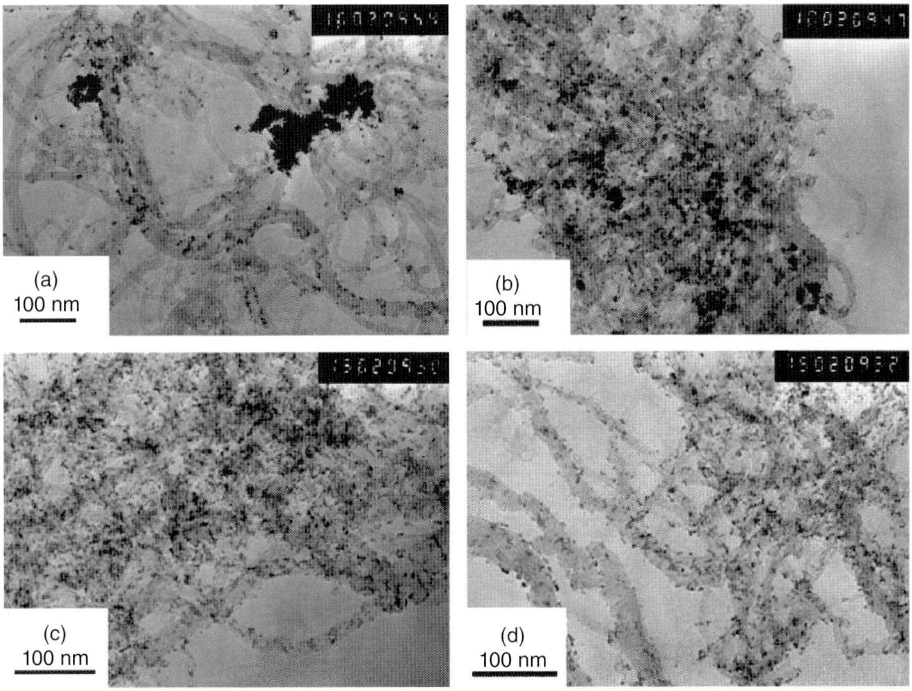

Figure 5.11 TEM images of microwave-synthesized Pt/CNTs from ethylene glycol solutions of H_2PtCl_6 at different pH's in the presence of CNTs: (a) pH 3.6; (b) pH 5.8; (c) pH 7.4, and (d) pH 9.2. Reproduced from Li *et al.* [58]. Copyright 2005 by Elsevier Ltd.

solution pH [58]. In acidic media the Pt nanoparticles were apparently agglomerated and not well dispersed on CNTs, and the particles' size ranged from 2.0 to 9.0 nm. By contrast, in alkaline media, the Pt nanoparticles were hardly agglomerated and were well dispersed on the surface of CNTs. The majority of the Pt particles varied from 2.5 to 4.5 nm for pH 7.4 and from 1.5 to 3.5 nm for pH 9.2, indicating that the increase in the synthesis solution pH could improve the size uniformity and dispersion of Pt nanoparticles on the surface of the CNTs. Moreover, electrochemical measurements showed that the Pt/CNTs catalyst prepared from the synthesis solution pH of 7.4 exhibited better performances for methanol electrooxidation than other samples [58].

5.5 Other Catalysts Prepared by Microwave-Related Procedures

In their article of 2003, Shi and Hwang reviewed some microwave-assisted wet chemical syntheses of various catalytic materials in which they described the advantages, the significance, and steps needed toward possible industrialization [59]. Steps were identified and discussed to bring the microwave-assisted syntheses of catalysts to industrialization, namely (i) identify and evaluate the microwave nonthermal effect, (ii) examine the microwave effects at different synthesis stages, and (iii) examine the temperature influence on the microwave effects.

In this context, the influence of microwave radiation on the properties and structure of the Fe_2O_3/SO_4^{2-} solid superacid catalyst was examined by Ning and coworkers [60] who noted that compared to the traditional method of preparing such superacids, the noncrystalline Fe_2O_3/SO_4^{2-} superacid prepared in a microwave field displayed the highest magnetic susceptibility and catalytic activity toward esterification reactions.

The characterization and catalytic properties of nickel aluminate nanoparticles, prepared by both a conventional combustion method (CCM) and a microwave combustion method (MCM) using nickel nitrate and aluminum nitrate as precursors in aqueous media, were reported by Ragupathi and coworkers [61]. The preparation involved the *Opuntia dillenii haw* plant extract, which simplified the process, provided an alternative process for a simple and an economical synthesis that required no surfactant, and led to a fast method for the synthesis of $NiAl_2O_4$ nanoparticles. The MCM produced $NiAl_2O_4$ with uniform size and well-defined shape with crystallinity and whose optical property was determined by diffuse reflectance spectroscopy. The $NiAl_2O_4$ prepared by the MCM possessed a higher surface area, lower crystallite size than the $NiAl_2O_4$ nanoparticles prepared by the CCM, which in turn led to improved performance toward the selective oxidation of benzyl alcohol to benzaldehyde.

Microwave hydrothermal processing has been used by Kumada *et al.* [62] to prepare two bismuth oxides, ABi_2O_6 (A = Mg, Zn) with the unusual oxidation state of Bi^{5+} that were previously prepared under limited hydrothermal conditions only. The authors discovered that microwave hydrothermal processing was far more advantageous in kinetics than conventional hydrothermal processing for soft-chemical reactions and for dissolution–precipitation reactions.

More interestingly, microwave irradiation can lead to morphology-controlled growth of nano- and microstructures of ZnO. Various basic ZnO structures were thus obtained by Cho and coworkers [63] that included nanorods, nanocandles, nanoneedles, nanodisks, nanonuts, microstars, microUFOs, and microballs (Figure 5.12) prepared by a fast, simple, and reproducible method that required no template, catalyst, or surfactant at relatively low temperatures (90 °C) using low-power microwave-assisted heating (~50 W) and a subsequent aging process. Such results were obtained by changing the precursor chemicals, the capping agents, and the aging times. Even more complex ZnO structures, including ZnO bulky stars, cakes, and jellyfishes, were constructed by microwave irradiation of a mixture of the as-prepared basic ZnO structures. Growth mechanisms for the shape-selective ZnO synthesis were proposed based on these results; they are displayed in Figure 5.13.

The industrial relevance of the use of microwaves in the preparation of catalysts is best exemplified by a 2013 US patent from El-Shall and coworkers [64] who produced graphene using microwave irradiation of solution phase graphitic oxide in the presence of a chemical reducing agent, or alternatively by subjecting the graphitic oxide to the effects of microwave plasma that needed no chemical reducing agent. The latter method was environmentally advantageous, cost effective, and encompassed the simultaneous reduction of solution or solid phase graphitic oxide together with a variety of metal (e.g., Pd, Co, Au, Ag, Cu, Pt, Ni, Fe, Mn, Cr, V, Ti, and Sc) precursors that resulted in the dispersion of metallic nanoparticles supported on the large surface area of the thermally stable two-dimensional graphene sheets. The graphene-supported metal nanoparticles were profitably used as catalysts toward various Suzuki-type reactions involving aryl bromides and were free of contaminants from residual solvent or reducing agent.

The microwave method has also been used [65] to synthesize solids with the saponite structure with Mg^{2+} or Ni^{2+} as octahedral cations and Fe^{3+} in the tetrahedral sheet; the formulae of the resulting solids were calculated from the element chemical analysis data resulting in SNiFe: $\{[Si_{7.136}Fe_{0.864}][Ni_{5.813}][Na_{0.810}]O_{20}(OH)_4{\cdot}13H_2O\}$ and SMgFe $\{[Si_{7.200}Fe_{0.800}][Mg_{5.820}][Na_{0.7020}]O_{20}(OH)_4{\cdot}9H_2O\}$. A special type of silico-alumino-phosphate with opal structure (Opal-SAPO) has been prepared under microwave irradiation conditions using tetraethyl orthosilicate, triethanolamine, aluminum isopropoxide, and phosphoric acid (85%) [66]. The catalytic behavior of the Opal-SAPO on regeneration of methylethylketone, benzaldehyde, and cyclohexanone from the corresponding semicarbazone derivatives was examined with results indicating that the OH groups on the Opal-SAPO surface provided the Bronsted acidity to the substrates. The first step in the regeneration of aldehydes or ketones from semicarbazone derivatives involved the chemisorption of semicarbazones on the Bronsted acid sites. Cleavage of the carbon–nitrogen bond was catalyzed by H^+ ions followed by hydrolysis with small amounts of water; no reaction occurred without the presence of

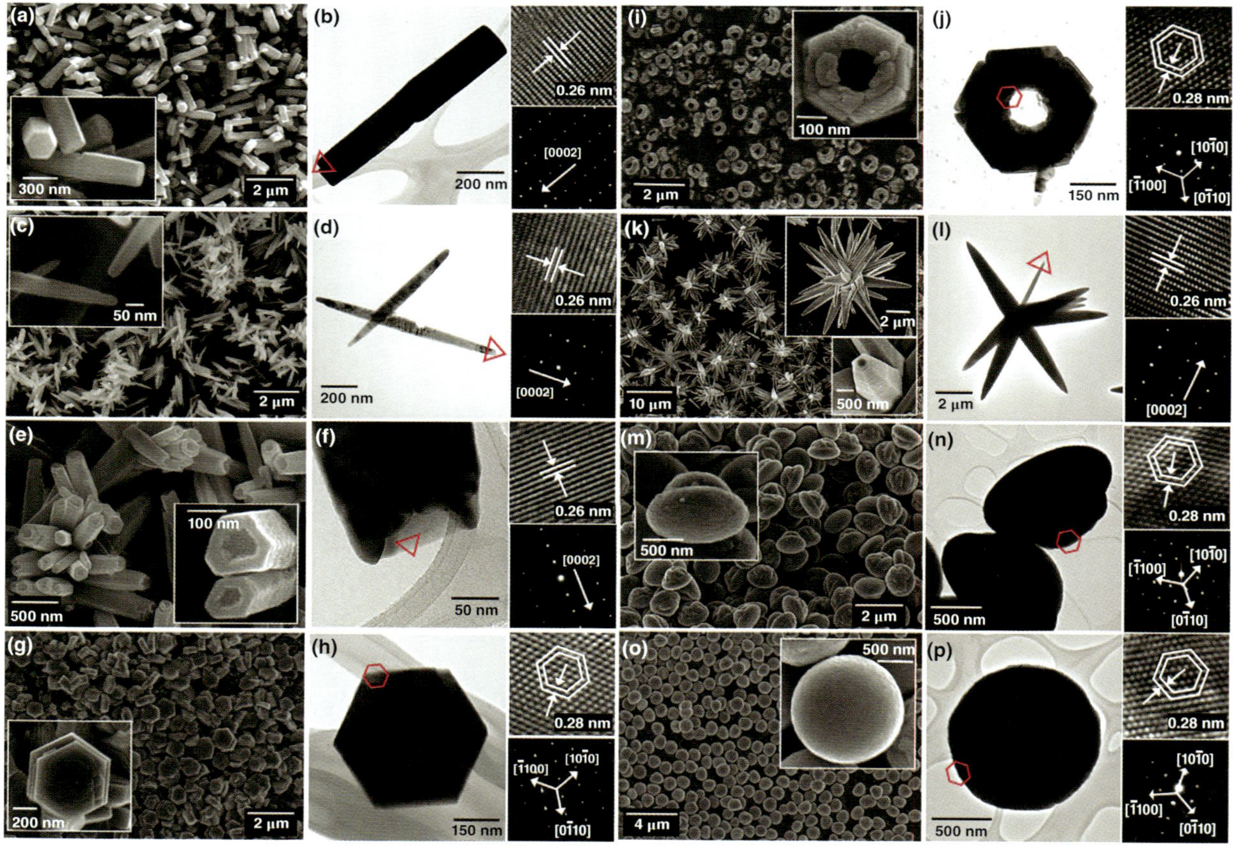

Figure 5.12 SEM (left) and TEM (right) images of the basic ZnO structures synthesized by microwave irradiation: (a, b) nanorods, (c, d) nanoneedles, (e, f) nanocandles, (g, h) nanodisks, (i, j) nanonuts, (k, l) microstars, (m, n) microUFOs, and (o, p) microballs. Reproduced from Cho *et al.* [63]. Copyright 2008 by the American Chemical Society.

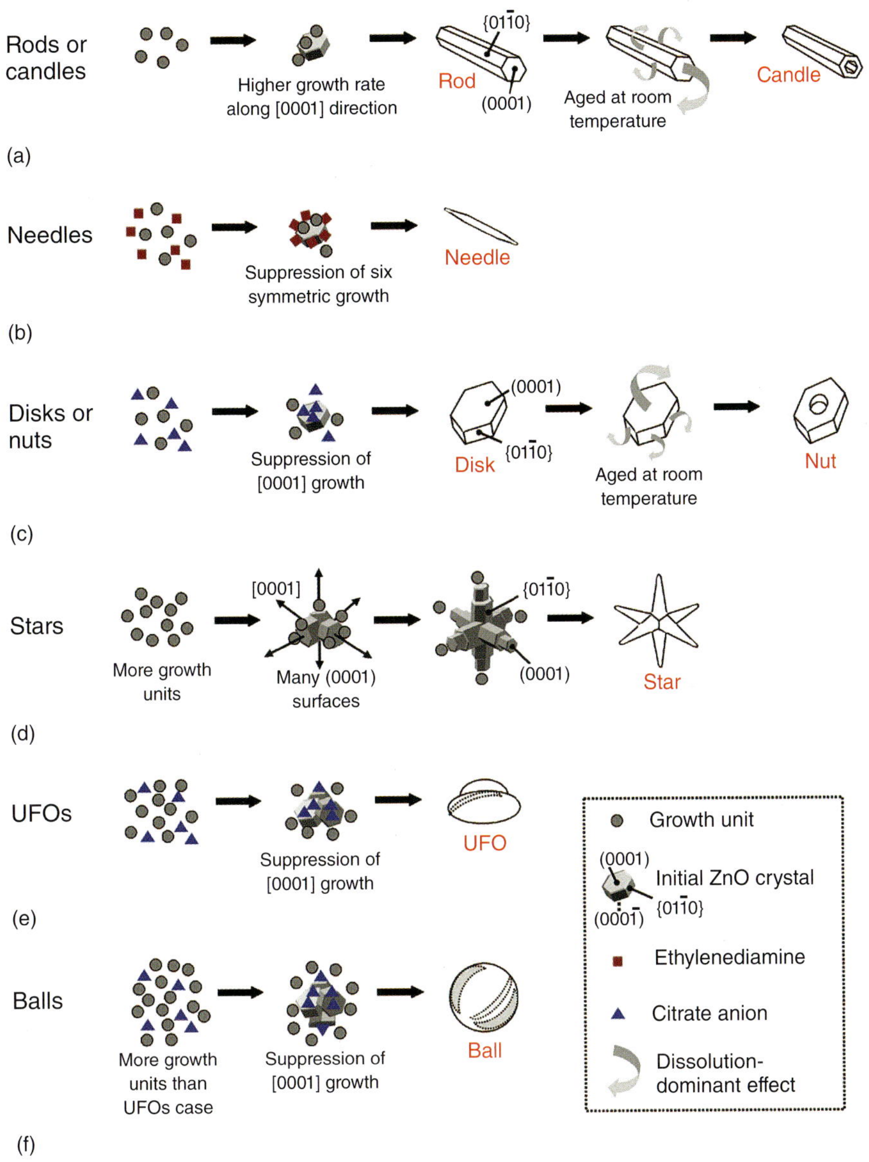

Figure 5.13 Schematic diagram of the proposed formation processes of some basic ZnO structures: (a) nanorods or nanocandles, (b) nanoneedles, (c) nanodisks or nanonuts, (d) microstars, (e) microUFOs, and (f) microballs. Reproduced from Cho *et al.* [63]. Copyright 2008 by the American Chemical Society.

Figure 5.14 Scanning electronic microscopic spectra of (a) $(NH_4)_3PW_{12}O_{40}$, (b) $Cs_3PW_{12}O_{40}$, (c) $Ag_3PW_{12}O_{40}$, and (d) $Cu_3(PW_{12}O_{40})_2$. Reproduced from Ref. [67]. Copyright 2007 by Elsevier B.V.

water. The reactivity of Opal-SAPO toward the regeneration of carbonyls was methylethylketone > benzaldehyde > cyclohexanone [66].

Photocatalysts of the type $M_3PW_{12}O_{40}$ ($M = NH_4^+$, Cs^+, Ag^+, Cu^{2+}), prepared by Deng *et al.* [67] using a microwave radiation solid-phase synthesis method, possessed an average particle size of about 15–80 nm (SEM and XRD techniques) – see Figure 5.14. Diffuse reflectance UV–vis spectra showed that the absorption band at 260 nm was shifted to 270–350 nm and became broad in comparison with those of phosphotungstic acid $H_3PW_{12}O_{40}$ (Figure 5.15). Results from the photocatalytic degradation of formaldehyde demonstrated that the shift of the UV–vis absorption band was proportionally related to the activity of the photocatalyst. The $Cu_3(PW_{12}O_{40})_2$ catalyst that displayed the maximal red shift to 350 nm of its UV–vis absorption band was the most efficient catalyst in the photodegradative conversion of formaldehyde, and possessed a network structure for better shape-selective catalysis and for some photocatalytic activity under visible

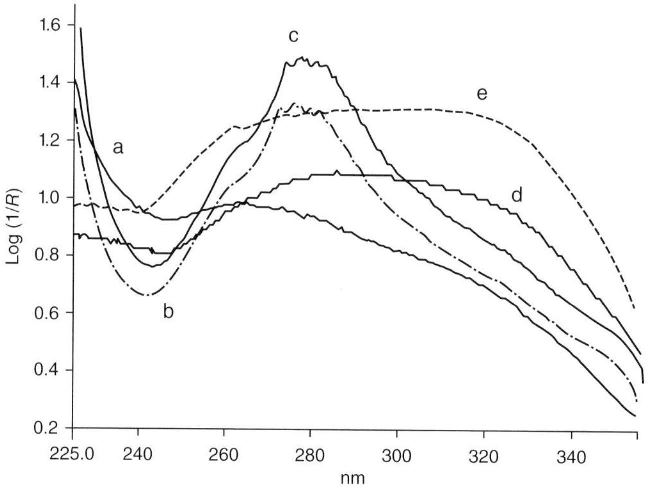

Figure 5.15 Diffuse reflectance UV–vis spectra of (a) $H_3PW_{12}O_{40}$, (b) $(NH_4)_3PW_{12}O_{40}$, (c) $Cs_3PW_{12}O_{40}$, (d) $Ag_3PW_{12}O_{40}$, and (e) $Cu_3(PW_{12}O_{40})_2$. Reproduced from Ref. [67]. Copyright 2007 by Elsevier B.V.

light. The $M_3PW_{12}O_{40}$ catalysts prepared by the microwave method were stable and easily regenerated.

Preoxidized activated carbons impregnated with iron-based nanoparticles have been prepared in a single step under hydrothermal conditions with microwave radiation [68]. The hydrothermal treatment provided led to the formation of fine particles that could easily be impregnated deep into the porous support by means of water. Their efficiency for the removal of As(V) from water was compared with the pure preoxidized activated carbon and iron oxide nanoparticles impregnated without microwave radiation. As such, microwave radiation provided much faster iron impregnation on the active carbon (AC) surface. At the first stage of microwave radiation, iron oxide impregnation was low; however, after 6 min the iron oxide nanoparticles (100 nm size) started to cover the surface uniformly. Further treatment with microwaves increased the size of the particles and the amount of surface coverage. Moreover, the microwave hydrothermal treatment led to relatively higher iron oxide loadings within 10 min. XRD characterization revealed that at the first stage of radiation, iron deposited in the form of β-FeOOH, which ultimately turned into Fe_2O_3. Although radiation increased the surface area of the material during the first stages, at the last stage the surface area increased no further because of complete surface coverage. Laboratory experiments showed that adsorption of arsenic on the iron oxide was strongly dependent on pH, adsorbent dose, and As(V) concentration, with the percent removal of As(V) increasing with decreasing pH of the solution so as to obtain an effective arsenate removal; adsorption experiments required pH between 3 and 5 for the adsorbent materials.

A novel microwave-assisted synthesis of Ni/C nanocomposites was reported by Gunawan and coworkers [69], which was achieved within a few minutes starting from nickel salts and a renewable high-content carbon source, tannin. The carbon precursor was *Quebracho tannin*, a renewable-resource material obtained from the hot-water extraction of *Schinopsis lorentzii* and *Schinopsis balansae* that are indigenous to Argentina and Paraguay. The process involved a simultaneous carbonization of the carbon precursor as well as the reduction of nickel ions to elemental nickel nanoparticles in an ambient atmosphere. The technique provided a fast, easy, and economical way to produce nickel/carbon nanocomposites that required neither hydrogen nor an inert gas during the transformation. The resulting nanocomposites possessed a high surface area and were said to be suitable as high-efficiency catalysts. SEM images (Figure 5.16) showed that the solution dispersion method gave a better dispersion of nickel nanoparticles than the solid-state dispersion method. The surface area of the nanocomposites was high, indicating a highly porous nanocomposite material.

Supported Ni–B catalysts have been synthesized by a silver-catalyzed electrodeless plating method in a water bath (W) and in an EG bath under microwave irradiation [70] and used in the hydrogenation of acetophenone. Results showed that nickel loading on Ni–B/MgO catalyst was aided by microwave irradiation. The increment of nickel loading resulted in growing particle size of Ni–B particles in Ni–B/MgO(W). By contrast, the particle size of Ni–B particle in Ni–B/MgO(EG)

Figure 5.16 Scanning electron microscopic images of the Ni/C nanocomposites prepared by solution dispersion at (a) 5000× magnification; (b) 10 000× magnification; (c) 25 000× magnification; and (d) 75 000× magnification. Reproduced from Ref. [69]. Copyright 2011 by the Taylor & Francis Publisher, London, UK.

catalyst changed somewhat under microwave irradiation. The best preparation temperature under microwave irradiation was 328 K. Results from the hydrogenation of acetophenone showed that the supported Ni–B catalysts exhibited excellent phenyl-1-ethanol (PE) selectivity, and that the Ni–B/MgO(EG) showed better catalytic activity and selectivity. The catalytic properties of Ni–B/MgO(EG) could be promoted by increasing microwave irradiation temperature and time. The hydrogenation of acetophenone was particularly important because of the extensive industrial application of its possible reaction products, namely phenyl-1-ethanol (PE) and cyclohexyl-1-ethanol (CHE) (Figure 5.17) [70]. The selectivity of PE and CHE depended on the electronic properties of catalyst, in which active metals rich in electrons prefer to adsorb C=O groups rather than aromatic rings, resulting in a higher PE selectivity. Further hydrogenolysis of PE to styrene can occur with further hydrogenation leading to the formation of ethylbenzene (EB) with increasing partial pressures of hydrogen or hydrogenation time so that hydrogenation should be carried out at low hydrogen pressure and in short time to avoid formation of the latter two species.

Titanium dioxide, iron-doped TiO_2, and sulfur-doped TiO_2 photocatalysts were reported by Esquivel and coworkers [71] and synthesized using a microwave-assisted method. The prepared samples were tested toward the photodegradation

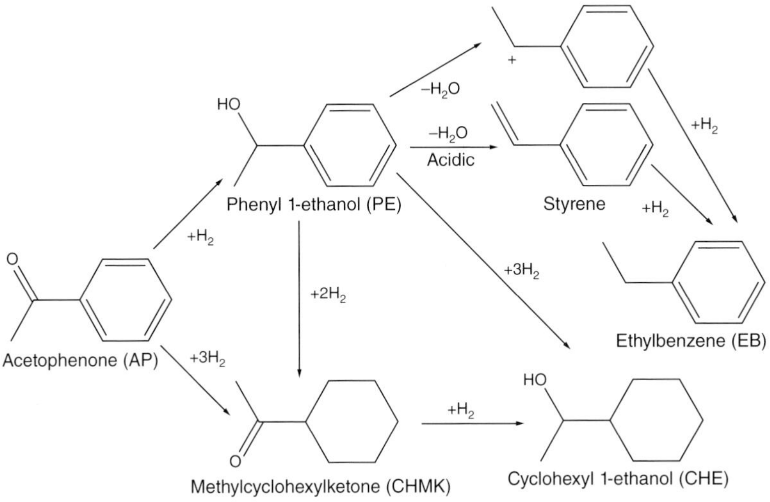

Figure 5.17 Scheme illustrating the possible pathway in the hydrogenation of acetophenone, which upon further hydrogenation yields styrene and ethylbenzene. Reproduced from Ref. [70]. Copyright 2008 by Elsevier B.V.

of methyl red dye under UV irradiation at ambient temperature. Changes in temperature and duration of the microwave synthesis caused some changes in the textural and crystalline structure of the TiO_2's that influenced their photocatalytic activity. Best photocatalyst was the (0.1 wt%) S-doped TiO_2 prepared by microwave irradiation at 215 °C for 60 min and subsequently calcined at 550 °C. The authors ascribed the enhancement of color removal (49%) of the dye over the latter photocatalyst to the formation of the anatase phase and enhancement of sample acidity with respect to undoped TiO_2 [71]. Calcination at 700 °C led to a drastic drop in activity because of formation of large rutile TiO_2 crystals. An increase in the photoactivity with an increase in S content strongly suggested that the positive effect of sulfur on the photoactivity of TiO_2 was due to its ability to trap electrons. For the most active S-doped TiO_2 sample, the authors inferred that the S species decreased electron–hole recombination, and contrary to TiO_2 doping with S^{2-} ions, the microwave synthesis was not effective for the preparation of effective Fe-doped TiO_2 photocatalysts for the targeted reaction.

Mixing mechanically the zeolite HZSM-5 (Si/Al = 25) and an appropriate amount of ammonium heptamolybdate ($(NH_4)_6Mo_7O_{24}{\cdot}4H_2O$) yielded mixtures that were subsequently calcined in a microwave oven for some time, after which the solid samples were pressed, crushed, and sieved to separate the catalyst grains in the size range 20–35 mesh for subsequent use in the aromatization of methane [72]. For comparison, the same Mo catalyst loading was prepared by an impregnation method. Results indicated that the catalysts prepared by microwave heating performed better than the catalysts prepared by the impregnation method. XRD analyses showed that the Mo species (20–30 nm) were highly dispersed on the outer surface of the CNTs (from the aromatization of CH_4), as also evidenced by TEM for the catalyst prepared by the microwave heating method. The actual active species during the reaction was Mo_2C.

Silver nanoparticles (AgNPs) have been synthesized in aqueous media using a microwave-assisted synthetic route with hexamine as the reducing agent and the biopolymer pectin (extracted from citrus peels and apple pomace under mildly acidic conditions) as the stabilizer [73]. Transmission electron microscopic images indicated that the nanoparticles espoused a spherical architecture with average diameter of about 19 nm. Reduction of 4-nitrophenol to 4-aminophenol by $NaBH_4$ in aqueous media was selected as the model reaction with which to investigate the catalytic activity of the pectin-stabilized AgNP, which did in fact display very good catalytic activity.

Crosswhite and coworkers [74] prepared a series of nanoscale magnetic spinel oxides of the composition MCr_2O_4 (M = Cu, Co, Fe) to serve as heterogeneous catalysts with good microwave absorption properties. Their activity as oxidation catalysts in aqueous methanol media were investigated under microwave irradiation, which led to rapid conversion of methanol to formaldehyde, directly generating aqueous formalin solutions. The Cu, Fe, and Co chromite spinel catalyzed reaction occurred under relatively mild conditions (1 atm O_2, 60 °C), with irradiation times of 80 min converting methanol to formaldehyde (Figure 5.18) [74].

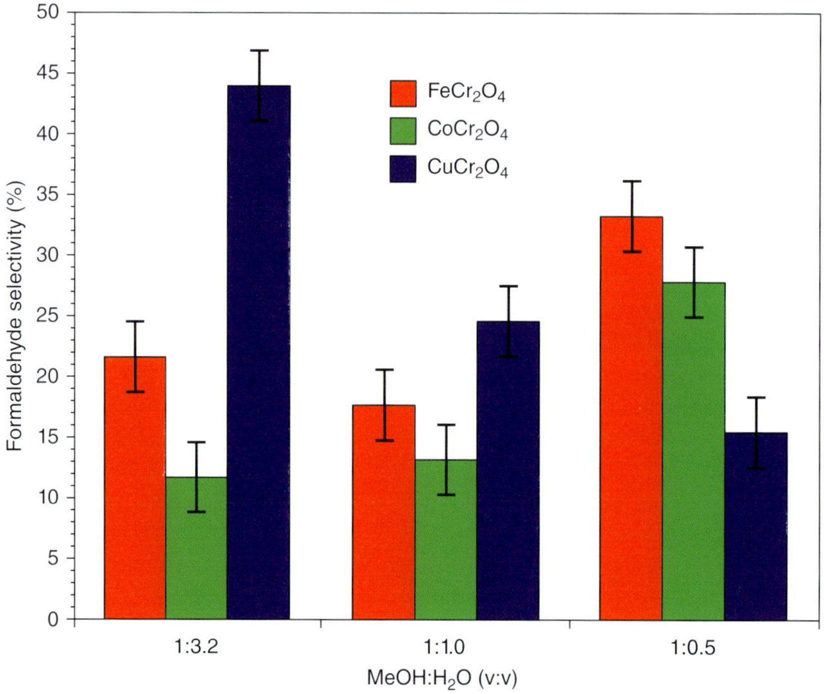

Figure 5.18 Conversion of methanol to formaldehyde after 80 min at 60 °C under 1 atm O_2 for the three spinel catalysts under varying $MeOH:H_2O$ ratios. Reproduced from Ref. [74]. Copyright 2013 by the American Chemical Society.

Reaction run under identical conditions of concentration, time, and temperature using traditional convective heating yielded dramatically lower percent conversions for the Cu and Co spinels; no observable thermal products were obtained from the Fe spinels.

An investigation into the influence of microwave radiation on the thermostability of chromia-alumina catalysts used in the dehydrogenation of low molecular weight paraffins (a mixture of hydrocarbons mainly of the alkane series) has been reported by Karimov *et al.* [75]. The authors discovered that the use of microwave radiation in the preparation of the catalysts led to an increase in their thermostability through the stabilization of active centers during the drying of the catalysts in a microwave radiation field. The thermostability of the catalysts correlated with the temperature and the duration of calcination.

5.6
Concluding Remarks

This chapter has provided a nonexhaustive introduction into the study of microwave-assisted syntheses of catalysts/carbonaceous materials as exemplified by the synthesis of TiO_2-based catalysts using selective microwave heating and activated carbon as the support, and where available their activity. In addition, some examples have been described that concerned mostly the preparations of catalysts *per se*, while others also provided some clues as to their application – for example, fuel cells. The reader may wish to consult another book that presents additional information into the microwave-assisted preparative methods of nanoparticulates [76].

References

1. Kingston, H.M. and Haswell, S.J. (eds) (1997) *Microwave-Enhanced Chemistry*, American Chemical Society, Washington, DC.
2. Kitchen, H.J., Vallance, S.R., Kennedy, J.L., Tapia-Ruiz, N., Carassiti, L., Harrison, A., Whittaker, A.G., Drysdale, T.D., Kingman, S.W., and Gregory, D.H. (2014) Modern microwave methods in solid-state inorganic materials. Chemistry: from fundamentals to manufacturing. *Chem. Rev.*, **114**, 1170–1206.
3. Horikoshi, S. and Serpone, N. (eds) (2013) *Microwaves in Nanoparticle Synthesis – Fundamentals and Applications*, Wiley-VCH Verlag GmbH, Weinheim.
4. Li, Y. and Yang, W. (2008) Microwave synthesis of zeolite membranes: a review. *J. Membr. Sci.*, **316**, 3–17.
5. Wilson, G.J., Will, G.D., Frost, R.L., and Montgomery, S.A. (2002) Efficient microwave hydrothermal preparation of nanocrystalline anatase TiO_2 colloids. *J. Mater. Chem.*, **12**, 1787–1791.
6. Hsieh, C.-T., Chen, W.-Y., Chen, I.L., and Roy, A.K. (2012) Deposition and activity stability of Pt–Co catalysts on carbon nanotube-based electrodes prepared by microwave-assisted synthesis. *J. Power Sources*, **199**, 94–102.
7. Kalpana, D., Cho, S.H., Lee, S.B., Lee, Y.S., and Misra, R. (2009) Recycled waste paper—A new source of raw material for electric double-layer capacitors. *J. Power Sources*, **190**, 587–591.
8. Wang, J., Chen, M., Wang, C., Wang, J., and Zheng, J. (2011) Preparation of mesoporous carbons from

amphiphilic carbonaceous material for high-performance electric double-layer capacitors. *J. Power Sources*, **196**, 550–558.
9. Yang, C.-M., Kim, Y.-J., Endo, M., Kanoh, H., Yudasaka, M., Iijima, S., and Kaneko, K. (2007) Nanowindow-regulated specific capacitance of supercapacitor electrodes of single-wall carbon nanohorns. *J. Am. Chem. Soc.*, **129**, 20–21.
10. Yuan, Y., Smith, J.A., Goenaga, G., Liu, D.-J., Luo, Z., and Liu, J. (2011) Platinum decorated aligned carbon nanotubes: electrocatalyst for improved performance of proton exchange membrane fuel cells. *J. Power Sources*, **196**, 6160–6167.
11. See e.g.Yuen, F.K. and Hameed, B.H. (2009) Recent developments in the preparation and regeneration of activated carbons by microwaves. *Adv. Colloid Interface Sci.*, **149**, 19–27.
12. Kong, Y. and Cha, C.Y. (1996) Reduction of NO_X adsorbed on char with microwave energy. *Carbon*, **34**, 1035–1040.
13. Djebabra, D., Dessaux, O., and Goudmand, P. (1991) Coal gasification by microwave plasma in water vapor. *Fuel*, **70**, 1473–1475.
14. Menéndez, J.A., Menéndez, E.M., Iglesias, M.J., García, A., and Pis, J.J. (1999) Modification of the surface chemistry of active carbons by means of microwave-induced treatments. *Carbon*, **37**, 1115–1121.
15. van Wyk, E.J., Bradshaw, S.M., and de Swardt, J.B. (1998) The dependence of microwave regeneration of activated carbon on time and temperature. *J. Microwave Power Electromagn. Energy*, **33**, 151–157.
16. Sayago, I., Terrado, E., Lafuente, E., Horrillo, M.C., Maser, W.K., and Benito, A.M. (2005) Hydrogen sensors based on carbon nanotubes thin films. *Synth. Met.*, **148**, 15–19.
17. Ansón, A., Lafuente, E., Urriolabeitia, E.P., Navarro, R., Benito, A.M., Maser, W.K., and Martínez, M.T. (2006) Hydrogen capacity of palladium-loaded carbon materials. *J. Phys. Chem. B*, **110**, 6643–6648.
18. Ansón, A., Lafuente, E., Urriolabeitia, E.P., Navarro, R., Benito, A.M., Maser, W.K., and Martínez, M.T. (2007) Preparation of palladium loaded carbon nanotubes and activated carbons for hydrogen sorption. *J. Alloys Compd.*, **436**, 294–297.
19. Cano, M., Benito, A., Maser, W.K., and Urriolabeitia, E.P. (2011) One-step microwave synthesis of palladium–carbon nanotube hybrids with improved catalytic performance. *Carbon*, **49**, 652–658.
20. Linyuan, C., Yanlan, L., Baohua, Z., and Lehui, L. (2010) In situ controllable growth of Prussian blue nanocubes on reduced graphene oxide: facile synthesis and their application as enhanced nanoelectrocatalyst for H_2O_2 reduction. *ACS Appl. Mater. Interfaces*, **2**, 2339–2346.
21. Subrahmanyam, K.S., Arun, K.M., Swapan, K.P., and Rao, C.N.R. (2010) A study of graphene decorated with metal nanoparticles. *Chem. Phys. Lett.*, **497**, 70–75.
22. Chang, H., Wang, G., Yang, A., Tao, X., Liu, X., Shen, Y., and Zheng, Z. (2010) A transparent, flexible, low- temperature, and solution-processible graphene composite electrode. *Adv. Funct. Mater.*, **20**, 2893–2902.
23. Lotya, M., Hernandez, Y., King, P.J., Smith, R.J., Nicolosi, V., Karlsson, L.S., Blighe, F.M., De, S., Wang, Z., McGovern, I.T., Duesberg, G.S., and Coleman, J.N. (2009) Liquid phase production of graphene by exfoliation of graphite in surfactant/water solutions. *J. Am. Chem. Soc.*, **131**, 3611–3620.
24. Xina, Y., Liu, J.-G., Zhou, Y., Liu, W., Gao, J., Xie, Y., Yin, Y., and Zou, Z. (2011) Preparation and characterization of Pt supported on graphene with enhanced electrocatalytic activity in fuel cell. *J. Power Sources*, **196**, 1012–1018.
25. Jasuja, K., Linn, J., Melton, S., and Berry, V. (2010) Microwave-reduced uncapped metal nanoparticles on graphene: tuning catalytic, electrical, and Raman properties. *J. Phys. Chem. Lett.*, **1**, 1853–1860.
26. Li, F., Guo, Y., Liu, Y., Qiu, H., Sun, X., Wang, W., Liu, Y., and Gao, J. (2013) Fabrication of Pt–Cu/RGO hybrids and their electrochemical performance for

the oxidation of methanol and formic acid in acid media. *Carbon*, **64**, 11–19.
27. Zhang, H., Xu, X., Gu, P., Li, C., Wu, P., and Cai, C. (2011) Microwave-assisted synthesis of graphene-supported Pd_1Pt_3 nanostructures and their electrocatalytic activity for methanol oxidation. *Electrochim. Acta*, **56**, 7064–7077.
28. Toshima, N., Harada, M., Yonezawa, T., Kushihashi, K., and Asakura, K. (1991) Structural analysis of polymer-protected palladium/platinum bimetallic clusters as dispersed catalysts by using extended X-ray absorption fine structure spectroscopy. *J. Phys. Chem.*, **95**, 7448–7453.
29. Qian, W., Hao, R., Zhou, J., Eastman, M., Manhat, B.A., Sun, Q., Goforth, A.M., and Jiao, J. (2013) Exfoliated graphene-supported Pt and Pt-based alloys as electrocatalysis for direct methanol fuel cells. *Carbon*, **52**, 595–604.
30. Chen, X., Cai, Z., Chen, X., and Oyama, M. (2014) Synthesis of bimetallic Pt/Pd nanocubes on graphene with *N,N*-dimethylformamide and their direct use for methanol electrocatalytic oxidation. *Carbon*, **66**, 387–394.
31. Gogotsi, Y., Libera, J., Koster van Groos, A.J., and Yoshimura, M. (2001) Hydrothermal synthesis of carbon nanotubes. Proceedings Joint 6th International Symposium on Hydrothermal Reactions (ISHR) & 4th International Conference on Solvothermal Reactions (ICSTR), 2001, pp. 350–355.
32. Galletti, A.M.R., Antonetti, C., Longo, I., Capannelli, G., and Venezia, A.M. (2008) A novel microwave assisted process for the synthesis of nanostructured ruthenium catalysts active in the hydrogenation of phenol to cyclohexanone. *Appl. Catal. A: Gen.*, **350**, 46–52.
33. Glaspell, G., Fuoco, L., and El-Shall, M.S. (2005) Microwave synthesis of supported Au and Pd nanoparticle catalysts for CO oxidation. *J. Phys. Chem. B*, **109**, 17350–17355.
34. Rao, D.G., Senthilkumar, R., Byrne, J.A., and Feroz, S. (eds) (2012) *Wastewater Treatment: Advanced Processes and Technologies*, CRC Press, New York.
35. See e.g.Hoffmann, M.R., Martin, S.T., Choi, W.Y., and Bahnemann, D.W. (1995) Environmental applications of semiconductor photocatalysis. *Chem. Rev.*, **95**, 69–96.
36. Kamat, P. (2012) TiO_2 nanostructures: recent physical chemistry advances. *J. Phys. Chem. Lett.*, **116**, 11849–11851.
37. Teoh, W.Y., Amal, R., and Scott, J. (2012) Progress in heterogenours photocatalysis: from classical radical chemistry to engineering nanomaterials and solar reactors. *J. Phys. Chem. Lett.*, **3**, 629–639.
38. Daghrir, R., Drogui, P., and Robert, D. (2013) Modified TiO_2 for environmental photocatalytic applications: a review. *Ind. Eng. Chem. Res.*, **52**, 3581–3599.
39. Shan, A.Y., Ghazi, T.I.M., and Rashid, S.A. (2010) Immobilization of titanium dioxide onto supporting materials in heterogeneous photocatalysis: a review. *Appl. Catal. A: Gen.*, **389**, 1–8.
40. Hu, H., Wang, X., Liu, F., Wang, J., and Xu, C. (2011) Rapid microwave-assisted synthesis of graphene nanosheets–zinc sulfide nanocomposites: optical and photocatalytic properties. *Synth. Met.*, **161**, 404–410.
41. Ravichandran, L., Selvam, K., and Swaminathan, M. (2010) Highly efficient activated carbon loaded TiO_2 for photo defluoridation of pentafluorobenzoic acid. *J. Mol. Catal. A: Chem.*, **317**, 89–96.
42. Gao, B., Yap, P.S., Lim, T.M., and Lim, T.-T. (2011) Adsorption-photocatalytic degradation of Acid Red 88 by supported TiO_2: effect of activated carbon support and aqueous anions. *Chem. Eng. J.*, **171**, 1098–1107.
43. Matos, J., García-López, E., Palmisano, L., García, A., and Marcì, G. (2010) Influence of activated carbon in TiO_2 and ZnO mediated photo-assisted degradation of 2-propanol in gas–solid regime. *Appl. Catal., B*, **99**, 170–180.
44. Horie, M., Kato, H., Fujita, K., Endoh, S., and Iwahashi, H. (2012) *In vitro* evaluation of cellular response induced by manufactured nanoparticles. *Chem. Res. Toxicol.*, **25**, 605–619.
45. Weir, A., Westerhoff, P., Fabricus, L., Hristovski, K., and von Goetz, N. (2012) Titanium dioxide nanoparticles in food

and personal care products. *Environ. Sci. Technol.*, **46**, 2242–2250.

46. Horikoshi, S., Sakamoto, S., and Serpone, N. (2013) Formation and efficacy of TiO_2/AC composites prepared under microwave irradiation in the photoinduced transformation of the 2-propanol VOC pollutant in air. *Appl. Catal., B*, **140–141**, 646–651.
47. Gutmann, B., Schwan, A.M., Reichart, B., Gspan, C., Hofer, F., and Kappe, C.O. (2011) Activation and deactivation of a chemical transformation by an electromagnetic field: evidence for specific microwave effects in the formation of Grignard reagents. *Angew. Chem. Int. Ed.*, **50**, 4636–4640.
48. e.g.Nishikawa, N., Murayama, M., and Kamiya, K. (2006) Adsorption and photocatalytic characteristics of activated carbon made from TiO_2-coated woody waste. *Mokuzai Gakaishi*, **52**, 50–54.
49. Horikoshi, S., Matsuzaki, S., Mitani, T., and Serpone, N. (2012) Microwave frequency effects on dielectric properties of some common solvents and on microwave-assisted syntheses: 2-Allylphenol and the $C_{12}-C_2-C_{12}$ Gemini surfactant. *Radiat. Phys. Chem.*, **81**, 1885–1895.
50. Mingos, D.M.P. and Baghurst, D.R. (1997) Applications of Microwave Dielectric Heating Effects to Synthetic Problems in Chemistry, in *Microwave-Enhanced Chemistry: Fundamentals, Sample Preparation and Applications*, Chapter 1 (eds H.M. Kingston and S.J. Haswell), American Chemical Society, Washington, DC, pp. 3–53.
51. Horikoshi, S., Osawa, A., Abe, M., and Serpone, N. (2011) On the generation of hot-spots by microwave electric and magnetic fields and their impact on a microwave-assisted heterogeneous reaction in the presence of metallic Pd nanoparticles on an activated carbon support. *J. Phys. Chem. C*, **115**, 23030–23035.
52. See e.g. Wikipedia Fuel Cell, *http://en.wikipedia.org/wiki/Fuel_cell* (accessed 07 May 2015).
53. Shams, M.M., Zamiri, B., and Fatoorehchi, H. (2010) Microwave-assisted synthesis of fuel cell catalyst. Proceedings of the 14th International Electronic Conference on Synthetic Organic Chemistry (ECSOC-14), Basel, Switzerland, November 1–30, 2010, paper No. C002, *http://www.usc.es/congresos/ecsoc* (accessed 07 May 2015).
54. Chen, W.X., Lee, J.Y., and Liu, Z. (2002) Microwave-assisted synthesis of carbon supported Pt nanoparticles for fuel cell applications. *Chem. Commun.*, 2588–2589.
55. Roy, A.K. and Hsieh, C.-T. (2013) Pulse microwave-assisted synthesis of Pt nanoparticles onto carbon nanotubes as electrocatalysts for proton exchange membrane fuel cells. *Electrochim. Acta*, **87**, 63–72.
56. Amin, R.S., Elzatahry, A.A., El-Khatib, K.M., and Elsayed Youssef, M. (2011) Nanocatalysts prepared by microwave and impregnation methods for fuel cell application. *Int. J. Electrochem. Sci.*, **6**, 4572–4580.
57. Guo, Z., Zhu, H., Zhang, X., Wang, F., Guo, Y., and Wei, Y. (2011) Microwave-assisted synthesis of high-loading, highly dispersed Pt/carbon aerogel catalyst for direct methanol fuel cell. *Bull. Mater. Sci.*, **34**, 577–581.
58. Li, X., Chen, W.-X., Zhao, J., Xing, W., and Xu, Z.-D. (2005) Microwave polyol synthesis of Pt/CNTs catalysts: effects of pH on particle size and electrocatalytic activity for methanol electrooxidization. *Carbon*, **43**, 2168–2174.
59. Shi, S. and Hwang, J.-Y. (2003) Microwave-assisted wet chemical synthesis: advantages, significance, and steps to industrialization. *J. Miner. Mater. Charact. Eng.*, **2**, 101–110.
60. Ning, Y., Wang, J., and Hong, P. (1996) Influence of microwave radiation on properties and structure of Fe_2O_3/SO_4^{2-} solid superacid catalyst. *J. Mater. Sci. Technol.*, **12**, 307–311.
61. Ragupathi, C. *et al.* (2014) Preparation, characterization and catalytic properties of nickel aluminate nanoparticles: a comparison between conventional and microwave method. *J. Saudi Chem. Soc.* doi: 10.1016/j.jscs.2014.01.006
62. Kumada, N., Kinomura, N., and Komarneni, S. (1998) Microwave hydrothermal synthesis of ABi_2O_6

(A = Mg, Zn). *Mater. Res. Bull.*, **33**, 1411–1414.

63. Cho, S., Jung, S.-H., and Lee, K.-H. (2008) Morphology-controlled growth of ZnO nanostructures using microwave irradiation: from basic to complex structures. *J. Phys. Chem. C*, **112**, 12769–12776.
64. El-Shall, M.S., Abdelsayed, V., Hassan, H.M.A., Rahman, A.E.I., Khder, S., Abouzeid, K.M., Dai, Q., Afshani, P., Gupton, F., Siamaki, A.R., Abdullah, Z., Alothman, M., and Alkhathlan, H.Z. (2013) Production of graphene and nanoparticle catalysts supported on graphene using microwave radiation. US Patent US2013/0211106 A1, Aug. 15, 2013.
65. Trujillano, R., Rico, E., Vicente, M.A., Rives, V., Bergalaouis, L., Chaabene, S.B., and Ghorbel, A.D. (2009) Microwave-assisted synthesis of Fe^{3+} saponites. Characterization by X-ray diffraction and FT-IR spectroscopy. *Rev. Soc. Esp. Miner.*, **11**, 189–190.
66. Derikvand, Z. and Farzaneh, F. (2003) Synthesis, characterization and catalytic behavior of special type of silico-alumino-phosphate with Opal structure. *J. Sci. (Iran)*, **14**, 235–238.
67. Deng, Q., Zhou, W., Li, X., Peng, Z., Jiang, S., Yue, M., and Cai, T. (2007) Microwave radiation solid-phase synthesis of phosphotungstate nanoparticle catalysts and photocatalytic degradation of formaldehyde. *J. Mol. Catal. A*, **262**, 149–155.
68. Yurum, Y., Yurum, A., Kocabas, Z.O., and Semiat, R. (2013) Synthesis and characterization of iron- impregnated pre-oxidized activated carbon prepared by microwave radiation for As(V) removal from water, *Geophys. Res. Abst.*, **15** (2013) EGU2013-6865, EGU General Assembly 2013.
69. Gunawan, G., Bourdo, S., Saini, V., Biris, A.S., and Viswanathan, T. (2011) Novel microwave-assisted synthesis of nickel/carbon (Ni/C) nanocomposite with tannin as the carbon source. *J. Wood Chem. Technol.*, **31**, 345–356.
70. Wu, Z.-J., Ge, S.-H., Zhang, M.-H., Li, W., and Tao, K.-Y. (2008) Synthesis of a supported nickel boride catalyst under microwave irradiation. *Catal. Commun.*, **9**, 1432–1438.
71. Esquivel, K., Nava, R., Zamudio-Méndez, A., Vega González, M., Jaime-Acuna, O.E., Escobar- Alarcón, L., Peralta-Hernández, J.M., Pawelec, B., and Fierro, J.L.G. (2013) Microwave-assisted synthesis of (S)Fe/TiO_2 systems: effects of synthesis conditions and dopant concentration on photoactivity. *Appl. Catal., B*, **140–141**, 213–224.
72. Qi, S., Yang, B., and Lang, L. (2004) Surface analysis of Mo/HZSM5 catalyst prepared by microwave heating for non-oxidative aromatization of methane. *Prepr. Pap. - Am. Chem. Soc., Div. Fuel Chem.*, **49**, 186–187.
73. Siby, J. and Beena, M. (2014) Synthesis of silver nanoparticles by microwave irradiation and investigation of their catalytic activity. *Res. J. Recent Sci.*, **3**, 185–191.
74. Crosswhite, M., Hunt, J., Southworth, T., Serniak, K., Ferrari, A., and Stiegman, A.E. (2013) Development of magnetic nanoparticles as microwave-specific catalysts for the rapid, low-temperature synthesis of formalin solutions. *ACS Catal.*, **3**, 1318–1323.
75. Karimov, O.K., Daminev, R.R., and Kasyanova, L.Z. (2013) Increase in thermo-stability of chromia-alumina catalysts used in dehydrogenation of paraffines by means SHF radiation. *Middle-East J. Sci. Res.*, **18**, 127–130.
76. Horikoshi, S. and Serpone, N. (eds) (2013) *Microwaves in Nanoparticle Synthesis*, Wiley-VCH Verlag GmbH, Weinheim.

Part III
Applications – Microwave Flow Systems and Microwave Methods Coupled to Other Techniques

Microwaves in Catalysis: Methodology and Applications, First Edition.
Edited by Satoshi Horikoshi and Nick Serpone.

6
Microwaves in Cu-Catalyzed Organic Synthesis in Batch and Flow Mode

Faysal Benaskar, Narendra Patil, Volker Rebrov, Jaap Schouten, and Volker Hessel

6.1
Introduction

Recent years have witnessed microwave (MW) radiation become an increasingly attractive and novel energy source to conduct synthetic organic chemistry to perform high-speed synthesis with excellent process operability and product yields [1–8]. In addition, the combination of microwave heating and flow chemistry has recently emerged as a highly attractive opportunity in novel processing, which has enabled a variety of tools to scale up and increase productivity for a large number of chemical reactions [9–13]. The separate development of both flow-through reactors, such as micro- and milli-reactors, and the field of microwave-assisted chemistry has provided a relatively facile method to join both efforts into a so-called micro-assisted (microreactors and microwaves) process [14–20]. Kappe and coworkers have recently presented an overview on the "Microwave-to-Flow" paradigm, where they demonstrated extremely high temperature and pressure windows to conduct organic reactions enabled by the combination of mesoscale reactors and microwave radiation [13, 21]. In addition, Organ *et al.* have considerably contributed to the current developments in microwave-assisted capillary-type flow reactors for metal-catalyzed organic reactions [22–24]. More examples on dedicated flow systems in combination with microwave heating have been reported for large-scale synthesis and industrial applications [25]. The pharmaceutical company AstraZeneca studied various examples of microwave-assisted organic reactions for flow synthesis of pharmaceuticals, yielding productivities up to $6\,l\,h^{-1}$ [26]. Earlier, Moseley *et al.* developed automated microwave stop-flow reactors that provided competitive productivities as compared to batch-scale reactions [27]. However, the current successes in combining microwave and flow systems have mainly been achieved in multimode microwave cavities, which generally suffer from microwave field inhomogeneities and nonuniform temperature distribution due to a limited microwave penetration depth. So far, only a few reports on highly focused single-mode microwave energy have been published. Recently, Patil *et al.* have introduced a new scale-up method for single-mode microwave-assisted flow

Microwaves in Catalysis: Methodology and Applications, First Edition.
Edited by Satoshi Horikoshi and Nick Serpone.

processing where modular scale-up was reached by implementing single-mode microwave cavities in series [28]. In this way, a packed-bed reactor gave a conversion of 99% with the highest production rate of 170 kg_{prod}/kg_{cat} h for esterification of acetic acid and ethanol catalyzed by ion-exchange resin in 18 single mode cavities. A similar approach was carried out for a multicomponent reaction of benzaldehyde, piperidine, and phenylacetylene catalyzed by a thin Cu film in a wall-coated tubular reactor that provided 99% conversion with the highest production rate of 7740 kg_{prod}/kg_{cat} h in 28 cavities. The authors introduced a novel way of combining flow reactors via a modular scale-up approach, with implementation of a multicavity single-mode microwave assembly [28].

Recent progress in the field of microprocessing provided many remarkable routes to replace typical batch processes with flow processes at various scales [29, 30]. In particular, microstructured reactors are currently regarded as a separate class of chemical reactors with their specific characteristics, such as very low time constants for heat and mass transfer, and well-tuned residence times [31–33]. However, for Cu-catalyzed liquid-phase organic reactions in the fine-chemicals industry, different criteria exist to beneficially exploit microwave and microprocess technologies. Of particular interest are the use of solids that make flow processing in liquid-phase systems complex, and that can be circumvented by pretreating the reactants [34], or else coating the catalyst onto predesigned supports such as coated or packed-bed milli- or micro-tubular reactors [35]. Nevertheless, milli- and micro-process technologies present a unique way to operate at high pressures and temperatures as a result of the smaller reaction volumes [36–41], and thus provide safe conditions for less risky or complex investments. Accordingly, this chapter will focus on recent developments of heterogeneous Cu-catalysis used in microwave-assisted chemical reactions, which in some cases have also proven of great benefit in flow-type reactors.

6.2 Microwave-Assisted Copper Catalysis for Organic Syntheses in Batch Processes

6.2.1 Bulk and Nano-structured Metals in a Microwave Field

6.2.1.1 Interaction of Bulk Metal with Microwaves

Currently, it is well accepted that, because of the so-called skin effect, metallic powders above a certain size preferably reflect microwaves, while a fine metallic powder is able to absorb microwave radiation and efficiently convert the microwave energy into heating [42]. This effect can best be described according to the penetration depth theory, where the small penetration depth of the electromagnetic field into the metals limits the temperature rise of a bulk metal and allows effective MW heating only for metal particles or thin metal films [43, 44]. Typically, iron-based powders show the highest heating rates and temperatures as a result of both Eddy current loss (in an alternating magnetic field) and magnetic reversal loss (in an alternating electric field). Indeed, diamagnetic metals (e.g., Sn and Cu) are rather efficiently heated as compared to paramagnetic metals (e.g., Ti). On the other hand, noble metals show very little heating owing to the absence of an oxidic layer, which causes the Eddy current flow for heating [45, 46]. It has also been reported that selective heating is significantly affected by the metal's particle size and the microwave frequency. The effect of particle sizes on heating was found to be optimal in the range 50–100 nm, and would be highly amplified if the metal particle were embedded in a nonmagnetic matrix, such as SiO_2, TiO_2, and Al_2O_3 [47, 48]. These matrices are typically found in heterogeneous catalysts, where metal particles are supported on these matrices [49]. The efficient coupling of metal powders with microwave radiation generally leads to high heating rates and does not necessarily cause major arcing issues, provided the heat transfer to the reaction medium is sufficiently fast. This property can be used selectively in solid-state reactions, where the heat supply to the solid is eventually transferred to the reaction process. For example, the synthesis of metal chalcogenides and metal cluster compounds has previously demonstrated to be highly effective, especially since the reaction products appeared to couple efficiently with microwave radiation [50]. For large particles (>10 μm), the phenomenon of arcing has also been regarded as an important issue that should be addressed, especially when operating with flammable solvents. In the case of copper, Whittaker *et al.* showed that the total amount of Cu powder in a solvent significantly influences microwave absorption and arcing formation [51]. They correlated the frequency of arcing mainly to interaction of particles, which increased at increasing Cu loading in the reaction system. This effect was also observed by Benaskar *et al.* [52] in the microwave-assisted Cu-catalyzed Ullmann C–O coupling reaction where it was observed that the amount of sparks formed in a dimethylacetamide solvent was linearly related to the Cu-catalyst loading (see Table 6.1).

6.2.1.2 Metallic Catalyst Particle Size and Shape Effect on Microwave Heating

The heating efficiency of metal particle supported on a support matrix has long been a point of discussion with regard to a theoretical approach. Several studies have been done to investigate whether a significantly high metal particle temperature can be attained within a nonabsorbing catalyst support matrix, resulting from the energy dissipation to the metal particle environment [53, 54].

Table 6.1 Number of sparks formed in the Cu-catalyzed Ullmann-type C–O coupling reaction using dimethylacetamide (DMA) as the solvent.

Catalyst loading (mg)	Number of sparks (min^{-1})
0	0
20	2
40	5
80	8
100	12

However, from the heterogeneous catalysis point of view, the local hot spots, where only the catalyst particle is heated and not the support matrix, lead to highly efficient chemical conversion owing to heating-up from the catalyst and quenching-down from matrix and (nonpolar) solvent. This selective heating has been demonstrated by Benaskar and coworkers for a Cu nanoparticles-based catalyst supported on a TiO_2 mesoporous film, which were coated onto glass beads and applied as a fixed-bed reactor (Figure 6.1) [14]. These authors demonstrated that the alternating segmentation of the catalyst-coated and noncoated sections led to significant temperature rise at the catalyst section and temperature stagnation at the inert segments (Figure 6.2). These authors also showed that the use of Cu-absorbing catalyst nanoparticles supported on poorly absorbing TiO_2 in a nonabsorbing *p*-xylene flow could still reach 120 °C, which is a typical reaction temperature for some organic reaction. Also, the use of micron-sized copper particles has shown great effect on the heat-up rates and final temperatures when irradiated with microwaves.

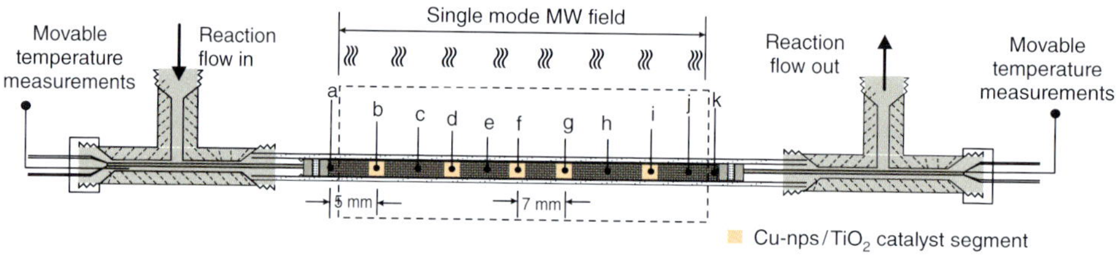

Figure 6.1 Experimental setup for local temperature measurements at local Cu-catalyst-deposited section.

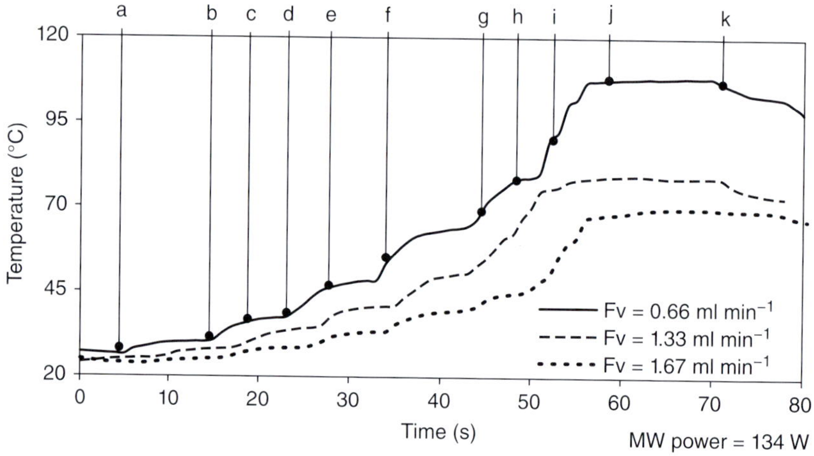

Figure 6.2 Microwave temperature profiles obtained at various axial positions from the multisegmented $Cu/TiO_2/SiO_2$ micro fixed-bed reactor using *p*-xylene as nonabsorbing solvent.

Ma *et al.* found that the effect on heating is strongly dependent on conductivity, permittivity, and permeability of the copper powder [55]. Rybakov *et al.* developed a model that described the microwave absorption capacity of electrical conductive materials under various microwave irradiation conditions. The theoretic model showed good agreement with their experimental observation of microwave irradiation regarding both volumetric and localized temperature rise [56]. Mishra *et al.* were able to describe an electromagnetic thermal model and experimentally validate the temperature profile of Sn, Cu, and W alloys under multimode microwave heating [57]. Mondal *et al.* reported that the thermal profile of electrically conductive copper is affected by the particle size and the porosity (resulting from pressing the powder) under 2.45 GHz multimode microwave irradiation [58]. They studied copper powders with mean particle sizes varying from 6 to 383 μm with initial porosities from 24% to 44% and showed very effective coupling with microwaves attaining temperatures above 1000 °C. They further observed a very significant dependency of the temperature rate on both the particle sizes and porosity. In the case of particle sizes, there was a significant effect on both the initial temperature increase and final temperature, whereas the variation of porosity (after pressing the particles) appeared to affect only the time required to reach the final temperature, which was similar for all the porosities. Table 6.2 shows the time required to reach the final temperatures and the recorded final temperatures as a function of particle sizes. These authors also explained the effect of particle size on the heating rate and on the final temperature as resulting from the difference in the so-called skin depth and the change of surface area as a function particle size. The effect of porosity (or green density) on the heating rate is summarized in Table 6.3 as the time required to reach the final temperature, which in all cases appeared to be 1000 °C. To obtain the various densities, Cu powder (particle size of 18 μm) was compacted via a compaction press from 90 to 350 MPa to attain theoretical densities ranging from 56% to 76%.

6.2.1.3 Polymetallic Systems in Microwave Chemistry

The use of polymetallic systems has shown great benefits in recent decades from the point of view of both chemical activity and physical stability. Using a polymetallic catalyst in microwave-assisted chemistry has also recently shown benefit with respect to selective heating [14]. Lee *et al.* used a suspension of copper/iron bimetallic nanoparticles in combination with microwave heating

Table 6.2 Time and final temperature obtained as function of the copper particle sizes.

Particle size (μm)	Time required to reach T_{final} (min)	Final temperature (°C)
6	20	1003
12	25	958
18	29	929
63	30	912
383	37	809

Table 6.3 Time and final temperature obtained as function of the copper particle sizes.

As pressed densities (%)[a]	Time required to reach T_{final} (min)[b]
56	5
65	8
71	12
76	13

a) The porosity was measured as the inverse of density, thus 44%, 35%, 29%, and 24%.
b) Final temperature for all experiments was 1000 °C.

to enhance the chlorobenzene decomposition in aqueous solutions [59]. They showed that the degree of nanoparticles in the aqueous suspension played a major role in MW absorption that consequently led to increased chlorobenzene decomposition. Their hypothesis was that microwave irradiation significantly enhanced the catalytic effect of Cu^0 in chlorobenzene removal, whereas the Fe^0 played a major role as a heating source. Engels *et al.* reported on highly oxidative stable Cu nanoparticles catalyst using hydrazine hydrate, sodium borohydride, and sodium hypophosphite to synthesize copper nanoparticles with different morphologies and mean particle sizes via a chemical reduction method [60]. This synthesis method for highly stable copper bimetallic nanoparticle catalyst has also shown great benefit in other applications, which also demonstrated to be applicable for microwave-assisted chemistry [61].

6.2.2 Microwave-Assisted Copper Catalysis for Chemical Synthesis

6.2.2.1 Bulk Copper Particles for Catalysis and Microwave Interaction

The use of micron-sized copper powders and particles has shown great benefit in microwave-assisted catalyzed reaction. Although it is generally assumed that micron-sized particles possess less surface for catalytic reactions, as compared to nanosized particles, the interaction of microwaves with micron-sized powders provides beneficial effects compared to well-dispersed nanoslurries [57]. A powdered copper catalyst was also used in the microwave-assisted Ullmann ether synthesis using a variety of chloroheterocycles with different phenols (Scheme 6.1). D'Angelo *et al.* reported a significant yield increase in considerably shorter reaction times when using microwaves instead of conventional

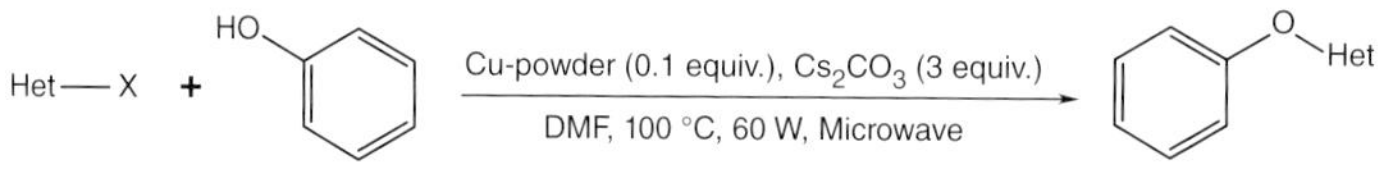

Scheme 6.1 The Ullmann-type C–O coupling from 4-chlorpyridine·HCl and phenol using microwaves and Cu nanoparticles.

heating [62]. They proposed a facile method for the aryl ether synthesis from chloropyridines, chloroquinolines, and chlorobenzothiazoles using micron-sized Cu powders in conjunction with microwaves.

6.2.2.2 Microwave-Assisted Copper-Catalyzed Bond Formation Reactions

Copper-mediated bond formation reactions represent important transformation processes that have recently been the object of an intensive focus to develop a wide range of substrates and products. Recently, Zhao *et al.* reported an efficient and practical protocol toward the synthesis of 5,6-dihydroindolo[1,2-*a*]quinoxaline derivatives via the microwave-assisted CuI-catalyzed intramolecular *N*-arylation mechanism [63]. Their method showed a major improvement to afford tetracyclic products with good to excellent yields (83–97%) in short reaction times (45–60 min) using microwave irradiation (Scheme 6.2).

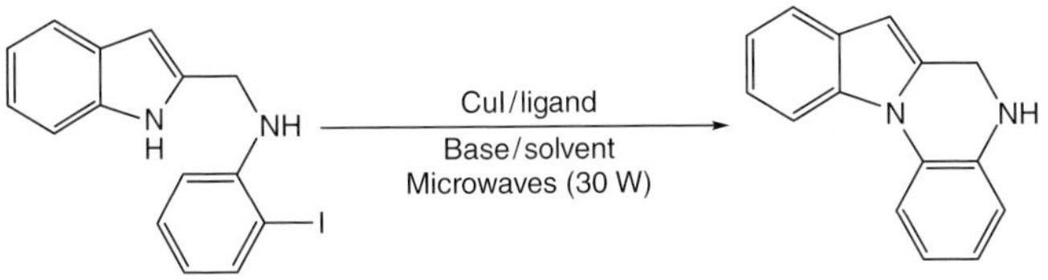

Scheme 6.2 Microwave-assisted synthesis of 5,6-dihydroindolo[1,2-*a*]quinoxaline using a CuI–ligand based catalyst.

Benaskar *et al.* reported the first example of poly(*N*-vinylpyrrolidone)-protected copper nanoparticles used in the microwave-assisted Ullmann ether synthesis [64]. They compared the activity with various micron-sized Cu powders, Cu wires, and Cu salts under both microwave and oil-bath heating (Scheme 6.3).

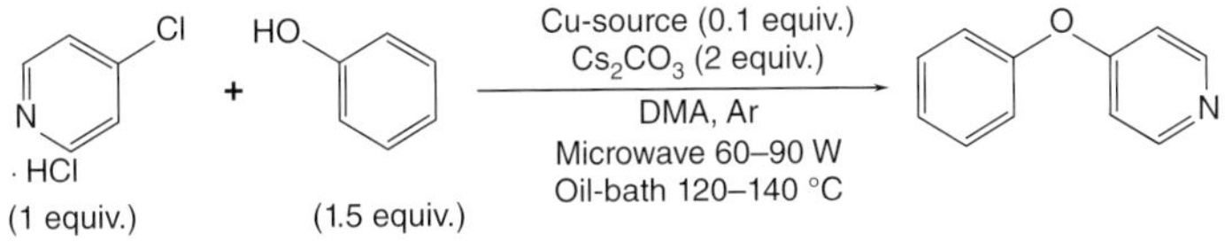

Scheme 6.3 The Ullmann-type C–O coupling from 4-chlorpyridine·HCl and phenol using microwaves and Cu nanoparticles.

Nanoscopic Cu catalysts provided very good yields in the microwave-assisted Ullmann ether synthesis of 4-phenoxypyridine. This was enabled using oxidation-resistant nanoparticle catalysts. However, microwires proved to be excellent catalysts in oil-bath heated systems [64]. Liu *et al.* reported a microwave-assisted copper-catalyzed coupling reaction of halopyridines using various nitrogen nucleophiles under solvent-free and ligand-free conditions yielding the corresponding coupling products in moderate-to-high yields (Scheme 6.4) [65]. In their investigation toward solvent-free and ligand-free microwave-assisted coupling reaction, the authors screened various bases and copper sources using the model reaction of the coupling of 3-iodopyridine with pyrazole.

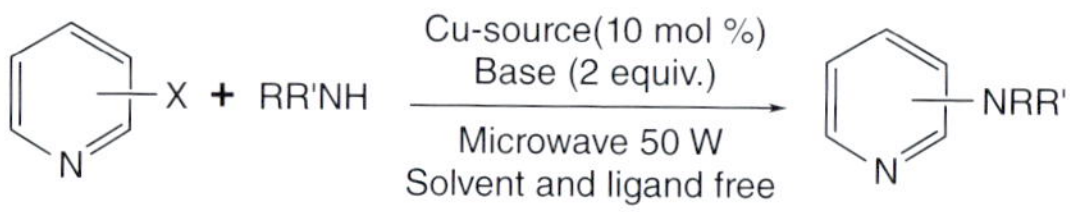

Scheme 6.4 Microwave-assisted solvent-free and ligand-free copper-catalyzed cross-coupling between halopyridines and nitrogen nucleophiles.

In line with previously reported work, the use of Cu_2O as a catalyst and Cs_2CO_3 as a base appeared to be most favorable with respect to product yield [66]. It was concluded that the heteroarylated product would not form if either the base or the copper were missing, suggesting that the mechanism does not follow the S_NAr mechanism [67]. Ke *et al.* reported on the simple and efficient synthesis protocol of a variety of alkylated phenols via a microwave-assisted copper-catalyzed hydroxylation of aryl halides in an aqueous environment (Scheme 6.5); yields were obtained in moderate-to-excellent yields of up to 95% [68]. This protocol lays an important basis to synthesize 2,3-dihydroxy-1,4-naphthoquinone, well known for its antiproliferation effect.

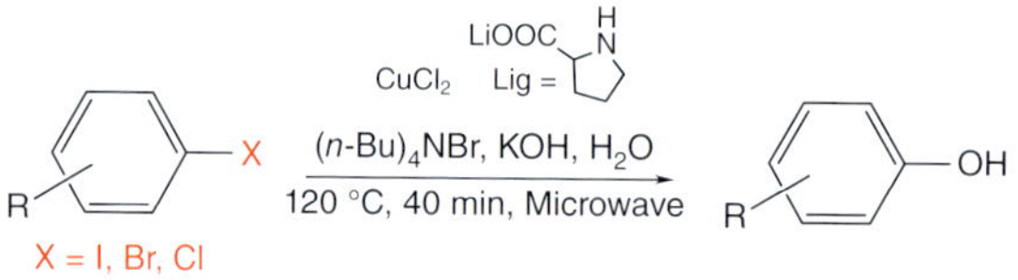

Scheme 6.5 Microwave-assisted copper-catalyzed hydroxylation of aryl halides to a variety of alkylated phenols.

Earlier He *et al.* developed an operationally simple and efficient method for the copper-catalyzed coupling of a variety of aryl bromides and iodides with phenols using microwave heating [69]. However, the excess of phenol, expensive aryl iodides, and bromides, in combination with the relatively high reaction temperatures, make this protocol highly inefficient (Scheme 6.6).

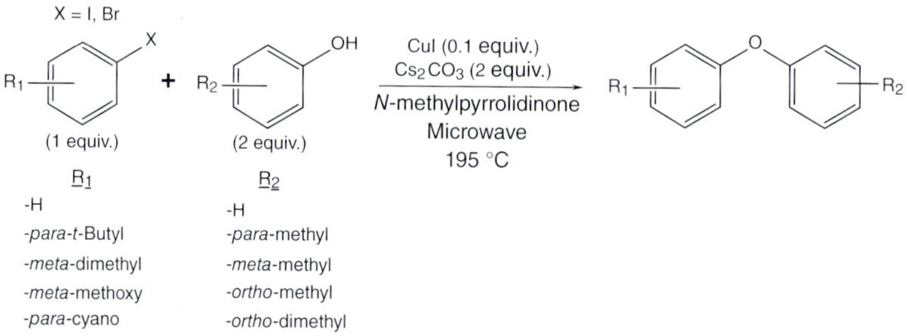

Scheme 6.6 Arylation of various substituted phenols with various aryl halides.

A catalyst-free microwave-assisted diaryl ether synthesis taking place at high reaction temperatures was reported previously [70]. The authors claimed to have prepared diaryl ethers by direct coupling of phenols with electron-deficient aryl halides, obtaining excellent yields under microwave irradiation at about 190 °C within 10 min. Enantioselective production of *N*-aryl-α-amino acids was reported via a microwave-assisted copper-catalyzed *N*-arylation of various α-amino acids and aryl halides [71]. A highly efficient catalytic microwave system was demonstrated that led, in most cases, to much higher product yields in less than 30 min compared to a conventionally heated reaction (Scheme 6.7).

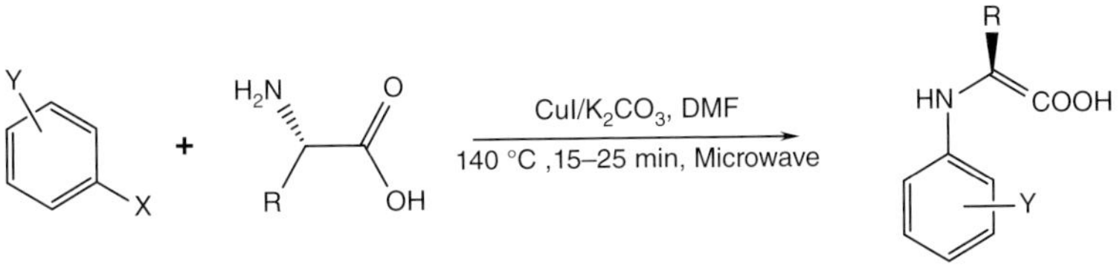

Scheme 6.7 Production of *N*-aryl-α-amino acids via a microwave-assisted copper-catalyzed *N*-arylation of various α-amino acids and aryl halides.

Wu *et al.* developed an experimental method for the *N*-arylation of various N–H containing heteroarenes with (*S*)-[1-(3-bromophenyl)-ethyl]-ethylamine using microwave irradiation [72]. Although their method appeared to be reproducible, the reaction time and temperature makes this reaction still not efficient for large-scale production (Scheme 6.8).

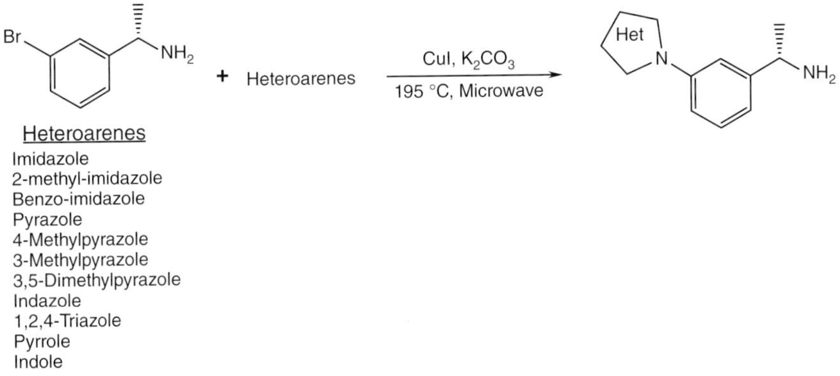

Scheme 6.8 Microwave-assisted copper-catalyzed *N*-arylation of N–H heteroarenes with various heteroarenes.

Kale *et al.* developed a new catalyst system for efficient synthesis of 1,2,3-triazoles in aqueous media with excellent yields via a multicomponent reaction using microwave heating (Scheme 6.9) [73]. The main advantage of microwave heating was the dramatic reduction of the reaction time, from 90 min by conventional heating to 5 min under microwave heating. Their methodology

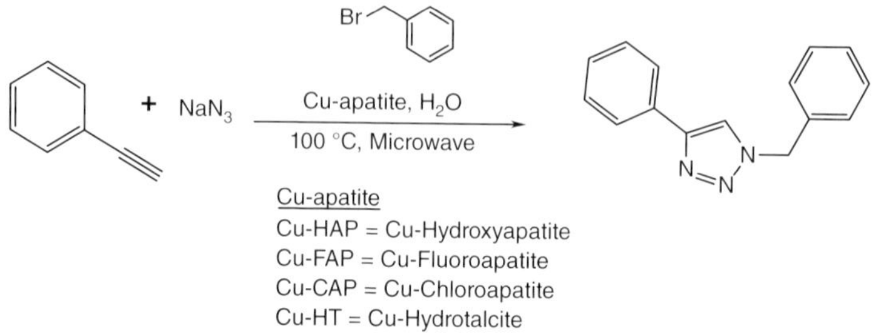

Scheme 6.9 Microwave-assisted multicomponent reaction of alkyne, halide, and sodium azide catalyzed by copper apatite as heterogeneous base and catalyst in water.

was applicable to a wide range of alkynes and halides; various Cu apatite-based heterogeneous catalysts were proposed. Among the fluoro-, chloro-, and hydroxyl-apatite, the latter demonstrated the highest activity, whereas the hydrotalcite showed the least activity.

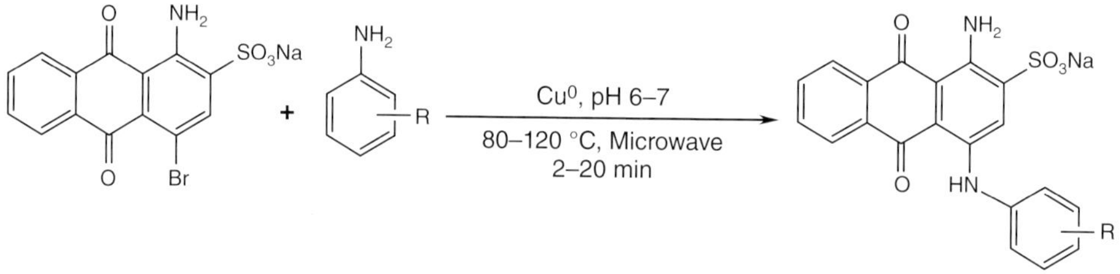

Scheme 6.10 Microwave-assisted Cu(0)-catalyzed Ullmann coupling toward the production of anilinoanthraquinones.

Baqi *et al.* reported the synthesis of a range of anilinoanthraquinones using a novel, Cu(0)-catalyzed, microwave-assisted Ullmann C–N coupling reaction of bromaminic acid with a range of aniline derivatives in a slightly acidic phosphate solution (Scheme 6.10) [74]. The synthesis of unprecedented compounds using this methodology could find industrial applications in the preparation of dyes. Good-to-excellent yields were obtained within 2–20 min at 80–120 °C and 40–100 W microwave power [74]. This methodology provided a novel way to access anilinoanthraquinones, thereby enabling the synthesis of previously inaccessible compounds. Also investigated were various copper sources; highest activity was obtained from metallic copper. In addition, for all 26 substituted anilines that were studied, microwave heating resulted in much higher yields and shorter reaction times compared to conventional heating.

6.2.3 Supported Cu-Based Catalyst for Sustainable Catalysis in Microwave Field

6.2.3.1 Microwave Activation and Synthesis of Cu-Based Heterogeneous Catalysts

In addition to the application of Cu catalyst in microwave heating, examples of microwave-assisted Cu catalyst have also been reported by Lin and coworkers who found a significant improvement on the methanol reforming reaction activity using a microwave-treated Cu/ZnO catalyst [75]. The activity and stability of the catalyst was confirmed using transmission microscopy and X-ray diffraction, as well as photoelectron and absorption spectroscopic analyses. Apparently, the enhanced activity and stability may be the result of defect formation and strong metal support interactions after microwave irradiation.

Efficient microwave calcination of Cu-based hydrotalcite, which enhanced the crystallinity of the oxidic and spinel phases, was reported by Cross *et al.* [76] who used an integrated feedback system, wherein the catalyst temperature was maintained constant and the microwave power was varied. Under microwave irradiation, they also observed crystal phases that would otherwise only be observed at much higher temperatures in the case of conventional heating. The catalyst demonstrated much higher activity in the base-catalyzed transesterification of glyceryl tributyrate with methanol. The authors attributed the increased activity to the increased concentration and strength of basic sites and to the occurrence of surface defects. It was postulated that this was caused by the release of water during dehydroxylation, as a result of localized temperature variation under microwave irradiation (Scheme 6.11).

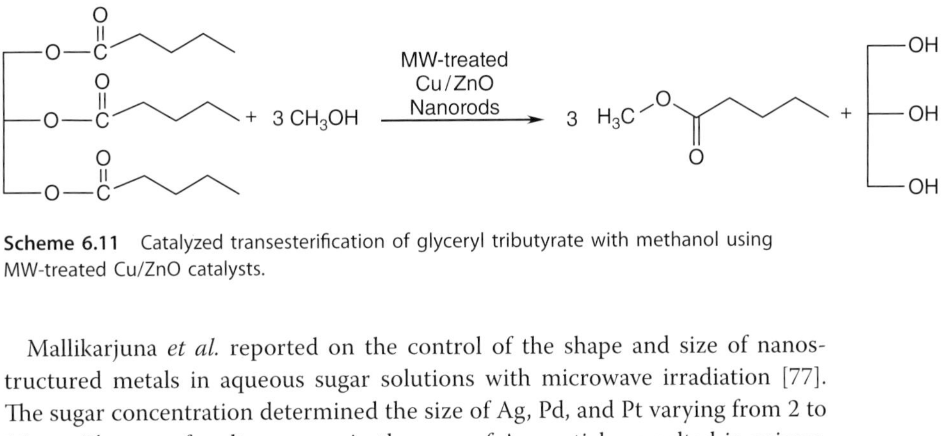

Scheme 6.11 Catalyzed transesterification of glyceryl tributyrate with methanol using MW-treated Cu/ZnO catalysts.

Mallikarjuna *et al.* reported on the control of the shape and size of nanostructured metals in aqueous sugar solutions with microwave irradiation [77]. The sugar concentration determined the size of Ag, Pd, and Pt varying from 2 to 15 nm. The use of maltose sugar in the case of Au particles resulted in prisms, cubes, and hexagonal nanostructured shapes, whereas spherical nanostructures resulted from α-D-glucose and sucrose. A catalytic oxidation method was proposed using pyrrole monomers to synthesize catalytically active nanocomposites from the nanocrystals.

6.2.3.2 **Cu-Supported Catalyst Systems for C–O, C–C, C–S, and C–N Coupling Reactions**

A new route of supported Cu catalyst on TiO_2 film was recently introduced for the C–O coupling reactions for the preparation of Cu and CuZn nanoparticles supported on nonporous TiO_2 and mesoporous TiO_2 films as supports [35]. Thin films of TiO_2 were also deposited on SiO_2 and TiO_2 beads as substrates for activity and stability testing. The catalytic activity of Cu nanoparticulate-based catalysts has been examined in the Ullmann C–O coupling reaction of 4-phenoxypyridine and potassium phenolate [34]. A mesoporous TiO_2 support consisting of 20 nm pores appeared to provide the highest product yield at a catalyst loading of 2 wt% Cu/meso-TiO_2 films with considerably high stability against leaching (Scheme 6.12).

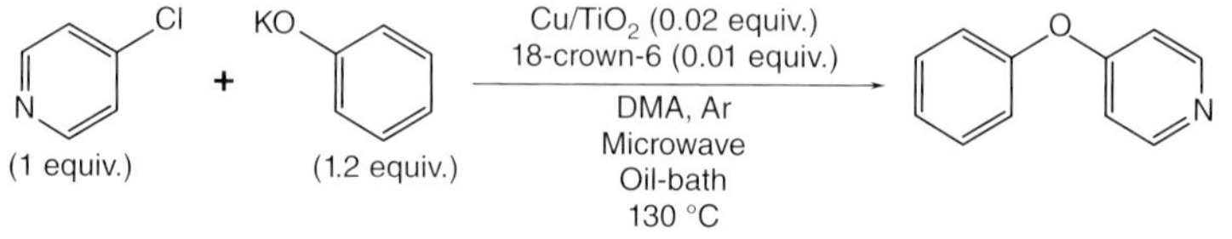

Scheme 6.12 The Ullmann-type C–O coupling from 4-chlorpyridine·HCl and phenol using microwaves and Cu nanoparticles.

6.3 Microwave-Assisted Copper Catalysis for Organic Syntheses in Flow Processes

6.3.1 Microwave-Assisted Catalyzed Organic Synthesis in Flow Processes

Various types of microwave-assisted flow reactors for catalyzed reactions have been reported in the last few decades mainly to translate the small-scale efforts into large scales. An important drawback accompanying scale-up of batch reactors originates from the restricted penetration depth of microwaves, which are typically in the order of a few centimeters. On the other hand, operating a heterogeneously catalyzed reaction in flow mode is typically hindered by operational issues, such as clogging, viscosity, pump capacity, and residence times [16]. Therefore, intensive research has been carried out in the combination of reactor design with catalyst immobilization techniques for flow processing, allowing maximal interaction between reagents and a catalyst and avoiding the very serious issue of pipe clogging, for example, in microreactors. In particular, metal-catalyzed bond formation reactions have shown great strides in recent research with enormous advantages for the academic and industrial environment [78]. Microwave heating finds its particular competitiveness with conventional techniques in transition metal-catalyzed product, providing added value by reduced reaction times and product selectivity [79]. Currently, several tailor-built microwave systems have solved a variety of safety and reproducibility concerns, enabling promising efficiencies in terms of production scale. However, a significant step toward industrial plant scales remains an intensive topic of process development for which the challenges are currently being dealt with [32].

6.3.1.1 Microwave Heating in Homogeneously Catalyzed Processes

Various pharmaceutically relevant reactions for scale-out purposes in a so-called automated stop-flow microwave reactor have been reported by Moseley *et al.*

[27] who described the Claisen rearrangement and benzofuran formation, the Heck reaction (Scheme 6.13), hydrolysis of *S*-thiocarbamates, and the combined hydrolysis and alkylation reaction. Productivities between 50 and 250 g day^{-1} were reported at standard conditions, while at intensified operating conditions productivities up to 1.5 kg day^{-1} were achieved.

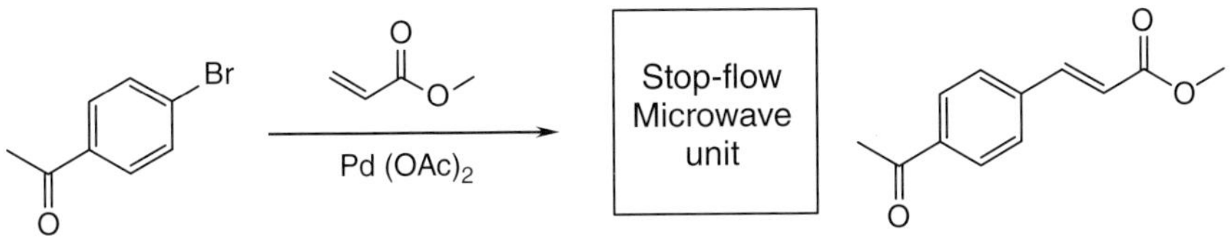

Scheme 6.13 The Heck reaction used in the stop-flow microwave reaction.

Jachuck *et al.* have developed an isothermal reactor where microwaves heat a water stream indirectly connected to a reaction stream, leading to a heat-exchange system [80]. This microwave flow reactor was used for the oxidation reaction of benzyl alcohol by $Fe(NO_3)_3$ (Scheme 6.14), with a reported conversion of 75% of the starting material to benzaldehyde within a residence time of 17 s and a microwave power of 40 W.

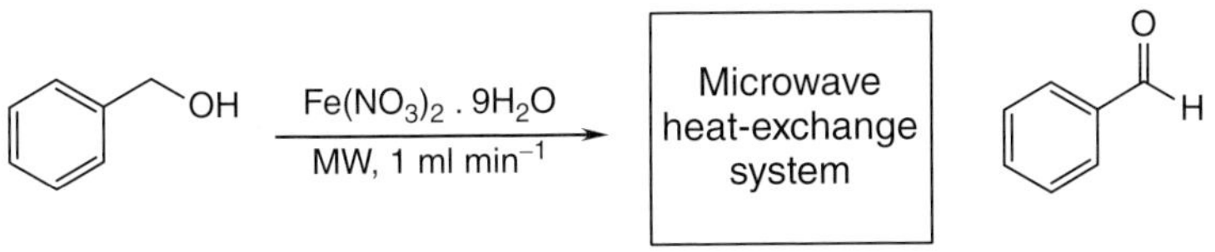

Scheme 6.14 Microwave-assisted flow synthesis of benzaldehyde under isothermal conditions.

Savin and coworkers reported a microwave continuous-flow system in the [3+2] cycloaddition of dimethyl acetylenedicarboxylate with benzyl azide in toluene as the reaction medium (Scheme 6.15) [81]. The authors initially optimized the reaction conditions in batch mode, after which the optimal protocol was used in a continuous-flow mode using a commercially available Kevlar-enforced Teflon coil placed in a dedicated single-mode microwave flow reactor.

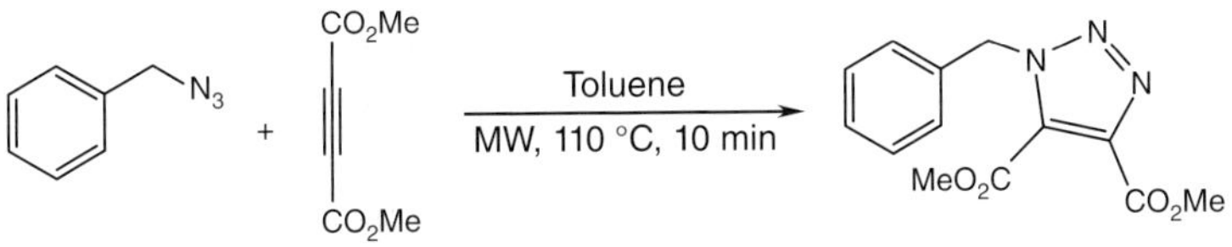

Scheme 6.15 Cycloaddition reactions under continuous-flow and microwave conditions.

6.3.1.2 Microwave Energy Efficiency and Uniformity in Catalyzed Flow Processes

One of the important aspects on which a new technology is generally evaluated is the energy efficiency. Microwaves cannot escape from such an evaluation. An important topic of discussion should, however, be the efficiency of microwaves in terms of heating and not the efficiency of a magnetron in transforming the supplied power into microwaves. This is because the application of microwaves in chemical syntheses focuses mainly on the heat generated, either volumetrically or locally, in a certain part of the system, for example, on the catalyst surface. Additionally, it is of high importance that the efficiency evaluation should clearly describe the details of the procedure. In their work, Patil *et al.* defined it as the ratio of the amount of energy absorbed by the sample to the set point of the microwave power input [82]. They proposed to take into account the global energy loss to the surrounding, while calculating the enthalpy change of the sample. The aim was to account for all the energy that has been transformed from microwaves into heat. This study [82] also demonstrated that the natural choice of chemists to go for a multimode-type microwave cavity for scale-up studies scored poorly not only on energy efficiency but also on energy uniformity (Figure 6.3). On the other hand, a

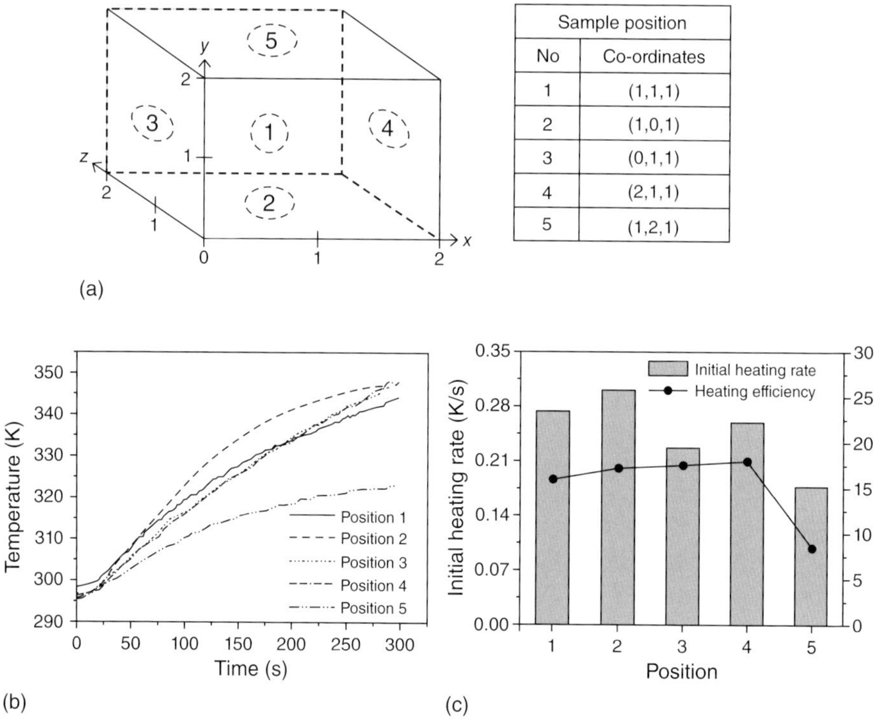

Sample position	
No	Co-ordinates
1	(1,1,1)
2	(1,0,1)
3	(0,1,1)
4	(2,1,1)
5	(1,2,1)

Figure 6.3 Microwave heating pattern in a multimode-type microwave cavity. Effect of sample position (a) on heating rates (b) and heating efficiency (c). Reproduced from Ref. [82].

monomode-type cavity, although limited in terms of volume, gives not only higher heating efficiency but also greater uniformity (Figure 6.4).

The efficiency of microwave heating, has also been raised as an important issue for scale-up by several other research groups [83–85]. The efficiency of microwave heating is better for the monomode-type microwaves than for the conventional multimode-type, relatively large, microwave cavities. In the case of monomode cavities under stop-flow conditions, the heating efficiency is highest at the load diameter equal to and above the half wavelength of the electromagnetic field in the liquid (Figure 6.5) [86]. On the other hand, under continuous-flow conditions, the heating efficiency increases monotonously with the load diameter (Figure 6.6).

For the case of monomode cavities, the high energy intensity of the focused microwaves, however, appeared to be a difficulty for controlled operation, especially when a strong microwave absorbing component was present [2, 87]. Some

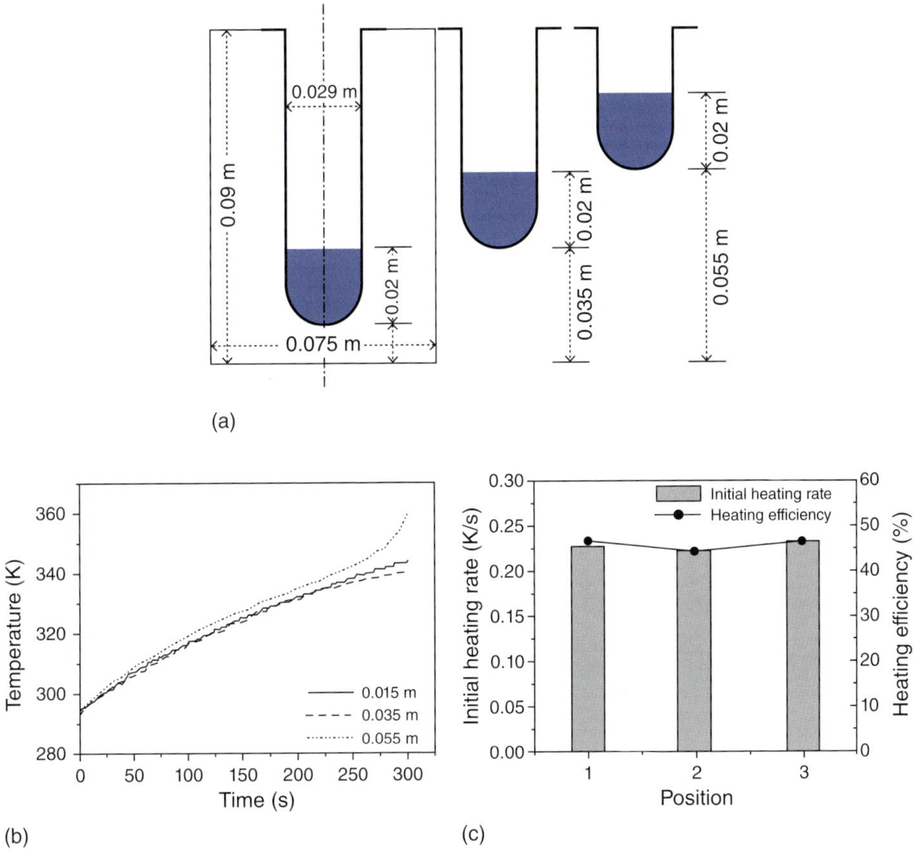

Figure 6.4 Microwave heating pattern in a monomode-type microwave cavity. Effect of sample position on heating rates and heating efficiency. Reproduced from Ref. [82].

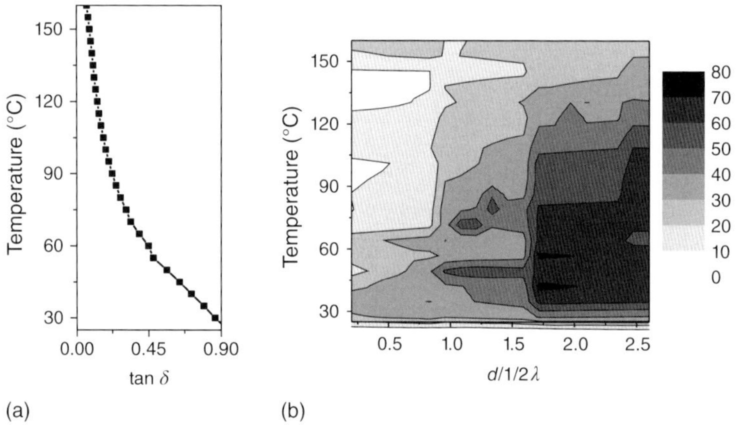

Figure 6.5 (a) Loss tangent and (b) heating efficiency as a function of temperature and $d/{}^{1}/_{2}\lambda$ (ratio of load diameter over half wavelength) for ethylene glycol at stop-flow conditions. The loss tangent monotonously decreases with temperature from 0.94 to 0.05. Reproduced from Ref. [86].

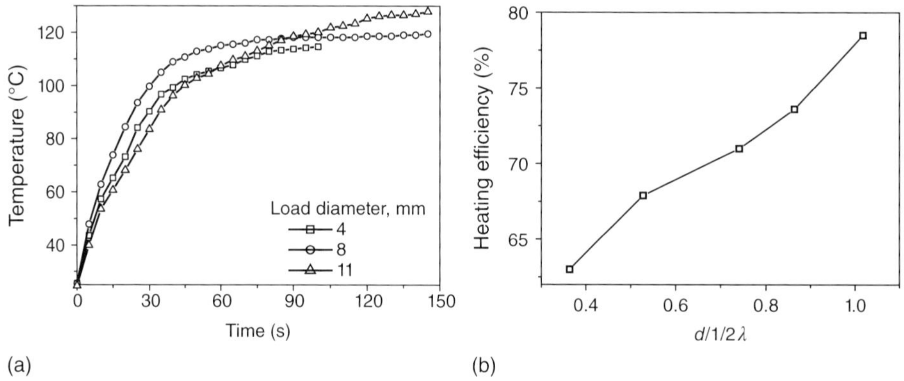

Figure 6.6 (a) Temperature increase as a function of time and (b) heating efficiency dependence on $d/{}^{1}/_{2}\lambda$ (ratio of load diameter over half wavelength) in case of ethylene glycol under continuous-flow conditions. Flow rate: 15 ml min^{-1}. Microwave irradiation at 2.45 GHz at a constant applied power of 100 W. Reproduced from Ref. [86].

of the proposed flow reactor systems suggested the use of a dead load to extract the excess microwave energy so as to avoid runaways [88, 89]. A beneficial use of the extracted excess energy has, however, not been discussed.

Accurate temperature measurements are crucial for a good control in a microwave integrated reactor system. Several articles have described indirect methods of temperature measurements, for example, infrared detectors [83, 89, 90]. Such indirect methods resulted in measuring the temperatures of the reactor wall instead of the reaction medium, which subsequently led either to alleged microwave effects or to poor understanding of the temperature distribution. However, temperature measurements can be done more accurately using a fiber-optic sensor [91, 92]. When inserted in the reaction mixture, these sensors measure the temperature directly, are microwave transparent, and do not disturb the microwave field [93].

Even after using fiber-optic sensors to measure temperature, Patil *et al.* found a hot spot in a reactor-heat exchanger flow system [2]. Although experimentally demonstrated, the selective nature of microwave heating and the interaction of system components with the oscillating electric field of the microwaves are, in most of the cases, not explicitly understood [18–20, 22, 94]. Most reaction mixtures contain polar and nonpolar components, some of which interact very strongly with microwaves while others are completely microwave transparent. Additionally, many chemical reactors require some internals, for example, mixers, distributors, measurement ports (i.e., temperature sensors), for various functions during operation. In general, these internals influence the mixing/velocity profiles in the (batch/continuous) reactors and, consequently, can interfere in the microwave heating process. Therefore, in the early stage of process design, understanding of the system interactions (in a geometrical as well as in a fluid-dynamic sense) with microwaves and their influence on the microwave heating process is necessary to allow a robust process control and operation. Particularly, in the case of flow reactors, the velocity profiles can strongly influence both processes, that is, heat transport and microwave absorption/interaction.

The modeling efforts to understand the detailed mechanism of microwave-assisted operation, in combination with convective heat transport, are largely applied in the food and drying industry [95, 96]. In contrast, qualitative insight into the mechanisms behind microwave heating in reactive systems is rather limited [97–100]. Most of the efforts are used either to show the existence of a hot spot or to provide design guidelines for microwave applicators. Datta *et al.* have nicely demonstrated the possibility of combining the heat transfer and interactions of the oscillating electric field with the applied load, which they broadly referred to as *microwave combination heating* [101, 102]. They proposed an iterative procedure to reach the steady-state solution. In this iterative procedure, the temperature change simulated by heat transfer, which consequently changed the dielectric properties of the load, was used to quantify interactions with the applied microwave field.

Following this approach, Patil *et al.* demonstrated that the temperature increase in the highly microwave-absorbing reaction mixture was two to four times higher in the almost stagnant regions, for example, reactor walls, compared to the bulk liquid (Figure 6.7) [103]. Additionally, the stagnant layer formation caused by insertion of a fiber-optic probe from the inlet (i.e., from the direction of the flow) resulted in higher temperatures (Figure 6.8). They established that the temperature profile depends more on the reaction mixture velocity profile than on the electric field intensity, especially in systems where the focused microwave field is applied over a small tubular reactor. This study [103] also showed that fully coupled simulation of microwave field and nonisothermal flow

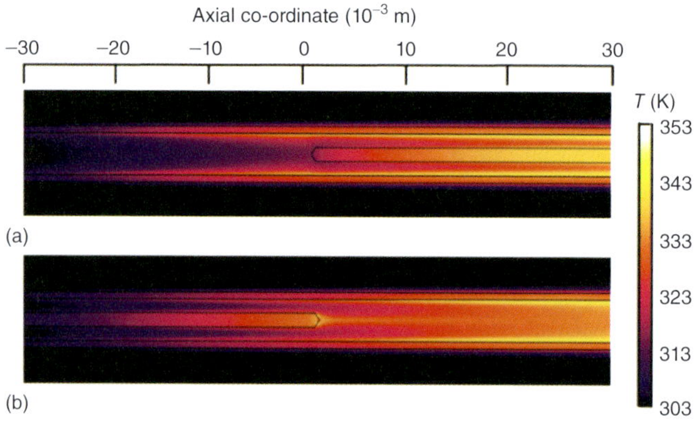

Figure 6.7 Influence of the direction of the probe insertion on the temperature profiles obtained by modeling, (a) probe inserted from the outlet and (b) probe inserted from the inlet. See Figure 6.1 for dimensions of the reactor assembly. Reproduced from Ref. [103].

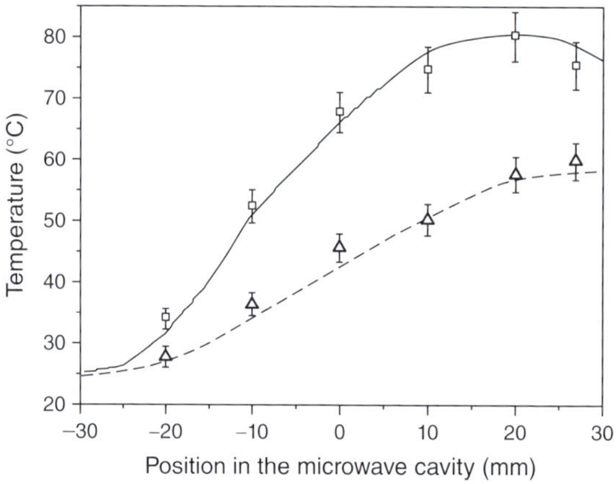

Figure 6.8 Influence of the direction of the probe insertion on the temperature profiles obtained by modeling (lines) and experiments (data points). Solid line and squares: probe inserted from inlet, dotted line and triangles: probe inserted from outlet. Reproduced from Ref. [103].

is feasible, and is highly recommended to understand the thermal performance of microwave-heated flow reactors.

The topic of energy efficiency and uniformity of microwaves also raises the discussion on economic viability of such systems on an industrial scale. Recently, Benaskar *et al.* reported an extensive cost study that revealed the major cost factors in the use of microwaves, microprocessing, or a combination of both [78]. The authors performed this technoeconomic analysis using two liquid-phase model reactions, namely the heterogeneously catalyzed Ullmann C–O cross-coupling reaction (Scheme 6.16) and the homogeneous acid-catalyzed synthesis of aspirin (Scheme 6.17).

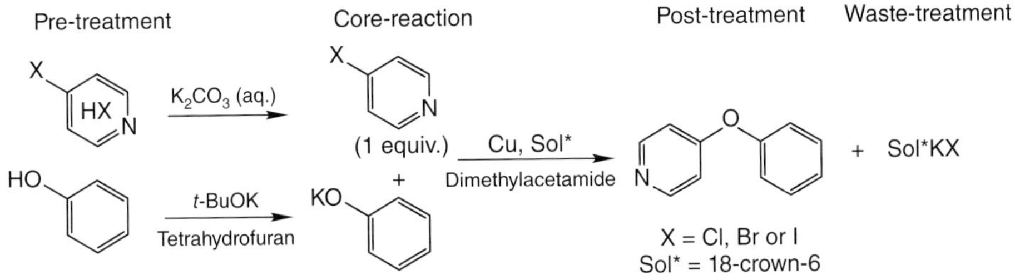

Scheme 6.16 Liquid-phase heterogeneously Cu-catalyzed Ullmann C–O cross-coupling reaction.

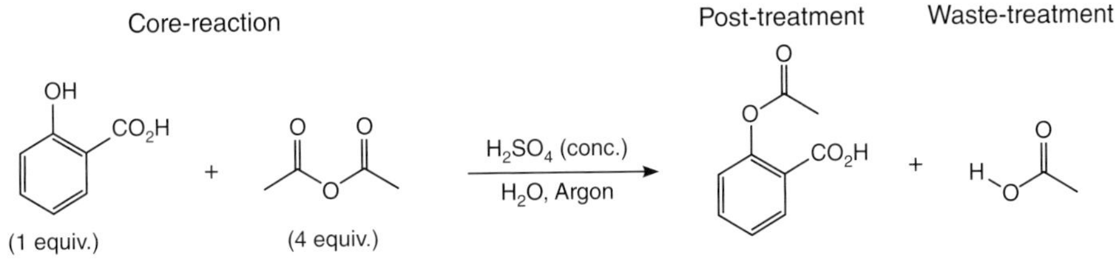

Scheme 6.17 Liquid-phase homogeneously acid-catalyzed synthesis of aspirin.

In their study, the authors compared the production of 4-phenoxypyridine (as precursor of Vancomycin) with the synthesis of aspirin [78]. The operating costs in the Ullmann synthesis were mainly related to upstream processing (reactant excess, pretreatment, and catalyst synthesis). The costs of the synthesis of aspirin, on the other hand, were mainly related to a downstream-based process (workup, waste treatment). The impact of an integrated microwave heating and microprocessing system on profitability was evaluated by comparing the operating costs and chemical activity of the two processes at equal capital costs. Various existing microwave technologies were compared with conventional pilot-scale heating techniques, both in batch and continuous operations. This cost evaluation led to the conclusion that in the case of the Ullmann synthesis, the CAPEX (capital expenditure) was negligible compared to the OPEX (operational expenditure), whereas the aspirin synthesis showed a significant CAPEX contribution of about 40%. In the former case, the type of catalyst strongly determined the cost-effectiveness of a continuously operated Ullmann process owing to the critical chemical activity. In general, the energy contribution to the overall cost was negligible, although a single-mode microwave system showed higher energy efficiency. In terms of cost, the best-case scenario for a combined microwave heating and microflow process was attained using a micropacked-bed reactor in comparison to a wall-coated microreactor.

6.3.2
Structured Catalyst in Microwave-Assisted Flow Processing for Organic Reactions

6.3.2.1 Thin-Film Flow Reactors for Organic Syntheses

Thin-film catalyst-coated capillary reactors have been reported for a variety of cross-coupling and ring-closing metathesis reactions, utilizing metal catalysts with microwave heating resulting in excellent conversions. The same setup was used to conduct the catalyst-free Wittig reaction and some nucleophilic aromatic substitution reactions. The Suzuki–Miyaura reaction showed significant yield in the Pd-coated thin-film capillary reactor [18]. Comer and Organ developed a capillary-based flow system to perform small-scale organic synthesis using microwaves as the energy source that they optimized with regard to reaction concentration and power input [18]. An efficient segmented system that included injection, mixing, and reaction zones avoiding laminar flow constrains in traditional microreactor systems was demonstrated.

Shore and Organ reported a microwave-assisted Pd-coated capillary reactor that was used for the Diels–Alder cycloaddition of various reagents, who observed a dramatic rate acceleration, and significantly shortened reaction times attributed to both selective heating and catalytic activity of the metal film [104]. It was postulated that the observed high yield originated from the reduced residence time, which avoided the formation of by-products (Scheme 6.18).

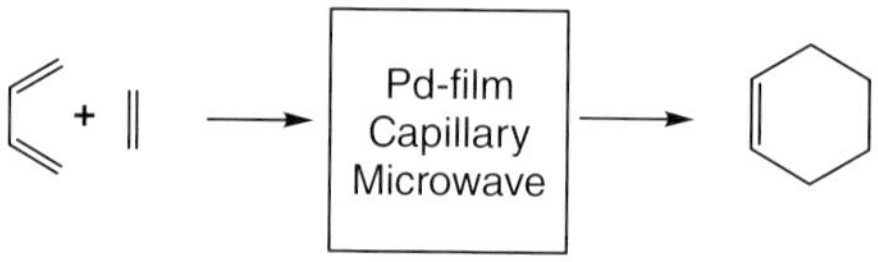

Scheme 6.18 The Diels–Alder cycloaddition reaction used in Pd-coated MW capillary.

He *et al.* used microwave energy to provide heat locally to a heterogeneous Pd-supported catalyst onto alumina situated inside a microreactor [20]. They used a 10–15 nm gold film on the reactor wall surrounding the catalyst-bed to provide heat to the Suzuki–Miyaura cross-coupling reaction, which occurred by highly selective interaction with microwaves.

Benaskar and coworkers recently introduced a novel Cu/ZnO catalyst coated as a thin film onto a tubular glass reactor [52]. After reactor surface treatment, ZnO nanowires were grown onto the reactor glass surface by circulating an equimolar aqueous solution of hexamethylenetetramine (HMTA) and $Zn(NO_3)_2$, after which it was followed by a drying step [52]. The synthesized copper nanoparticles suspended in methanol were deposited by flowing the suspension in the reactor, followed by drying and calcination at 350 °C (Scheme 6.19). The wall-coated milli-reactor was applied in the Cu-catalyzed Ullmann-type C–O coupling reaction of potassium phenolate and 4-chloropyridine toward the formation of 4-phenoxypyridine, which was shown to be highly effective at low Cu loadings and microwave powers. The use of a wall-coated Cu/ZnO catalyst showed particularly high activity at low flow rate of reactants, as a result of higher contact time

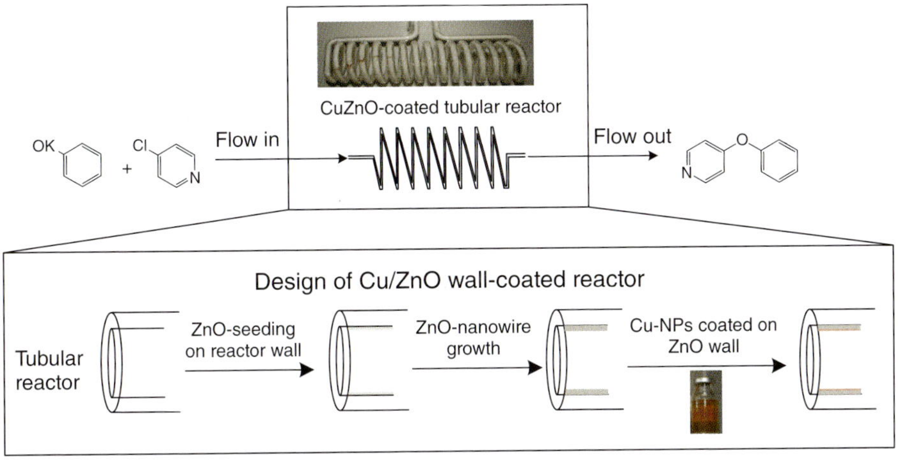

Scheme 6.19 Catalyst coating procedure for the wall-coated Cu/ZnO catalytic reactor used in the Ullmann-type C–O coupling reaction toward 4-phenoxypyridine.

between the reactants and the catalyst surface. Catalyst deactivation in such a system was mainly attributed to Cu oxidation and coke formation, which in either cases resulted in leaching of the Cu catalyst at longer reaction times. Yields up to 60% were reported using the Cu/ZnO wall-coated reactor for two consecutive runs without there being any significant activity drop.

Utilization of metal coating for enhanced microwave absorption was also demonstrated by some research groups [20, 94], who demonstrated that the temperatures achievable are dependent on the microwave power, the flow rate, and the magnitude of the dielectric loss (ε'') at the microwave frequency. However, even for strong absorbing solvents such as N-Methylformamide (NMF), the overall heating efficiency was very low, typically less than 1% for an 800 μm diameter capillary. It was proposed that controlled heating of a reaction can be achieved using selective microwave absorption by a gold film placed on the external surface of a capillary reactor. Such a method showed at least a 10-fold increase in efficiency, thus enabling a reduction in reaction time as well as enhancements in product yields.

6.3.2.2 **Structured Fixed-Bed Reactors for Flow Synthesis**

Another example of combined flow reactors with microwave heating has been reported by Benaskar *et al.* in a so-called μ^2-process, which was successfully applied in the Ullmann-type C–O coupling of potassium phenolate and 4-chloropyridine (Scheme 6.20) [14]. The authors demonstrated highly selective microwave absorption in a micro-fixed-bed reactor (μ-FBR) using a supported Cu nanocatalyst, which resulted in a significant activity enhancement in comparison to an oil-bath-heated process (Scheme 6.21). The demonstrated yields up to 80% were reported using a multisegmented μ-FBR without noticeable activity loss; the μ-FBR was packed with glass beads coated with Cu/TiO_2 and $CuZn/TiO_2$ catalysts.

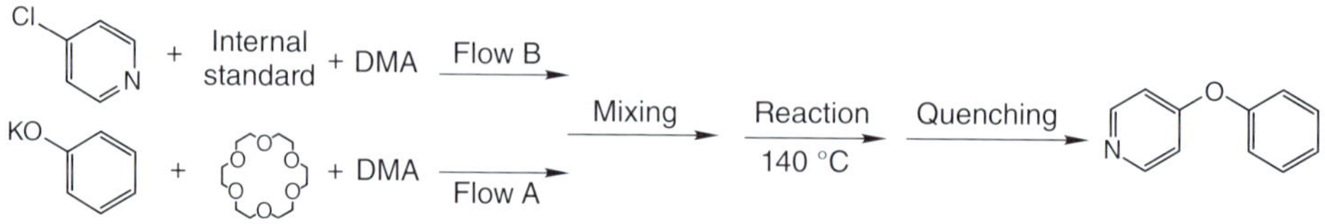

Scheme 6.20 Cu-catalyzed Ullmann etherification of a 4-chloropyridine and potassium phenolate flow in DMA to 4-phenoxypyridine.

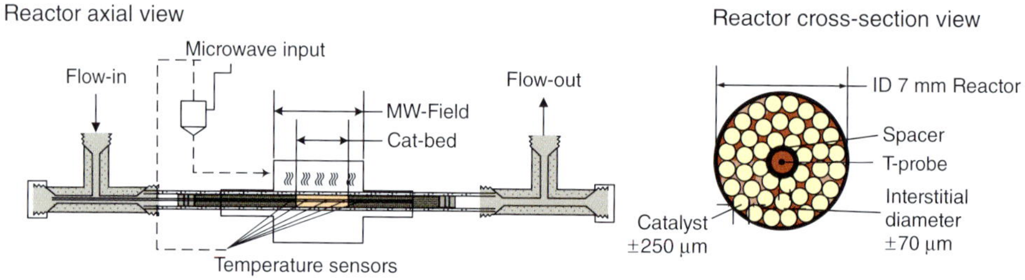

Scheme 6.21 Micro-fixed-bed reactor setup using in-line microwave and temperature controlling applied in the Cu/TiO_2 catalzyed C–O coupling reaction.

In their study, Benaskar and coworkers reported an elegant way to measure and control the reactor temperature along the axial direction of the μ-FBR by using a mobile fiber-optic probe along the catalyst-bed [14]. This system provided highly accurate temperature control in an isothermal reactor at 20 W. To calibrate the microwave power to the temperature-controlling probes, they first adjusted the MW field density in the cavity with only a solvent flow, before conducting the reaction mixture through the reactor. This calibration method enabled a highly accurate microwave input, since the changing reaction mixture composition (and therefore the dielectric properties) during the reaction was corrected for by adjustment of the waveguide stub tuners. In a final step, heating and control of selective microwave absorption by metallic Cu nanoparticles were investigated in a μ-FBR coated with Cu/TiO_2 and $CuZn/TiO_2$ for further fine-tuning of the microwave energy input. The Cu and CuZn catalyst nanoparticles were synthesized following a sol–gel deposition and impregnation method that yielded Cu loadings of about 1 wt%. Adding Zn to the Cu nanoparticles demonstrated an increased catalyst activity as a result of increased Cu^0 oxidation stability. They reported one of the first examples to combine microprocess and single-mode microwave technology in the concept of novel-process-windows for organic syntheses.

Other possible uses of fixed-bed reactor technology could also be to understand the systems where the reaction media is highly polar and the catalyst itself is inert. In this case, the use of microwaves would then be as a volumetric heating source. One such example is the work of Patil *et al.* who used an integrated heat exchanger reactor system to demonstrate efficient and controlled flow processing of highly microwave absorbing reaction media [2]. Esterification of ethanol and acetic acid was chosen as a model reaction (Scheme 6.22). Being a microwave transparent

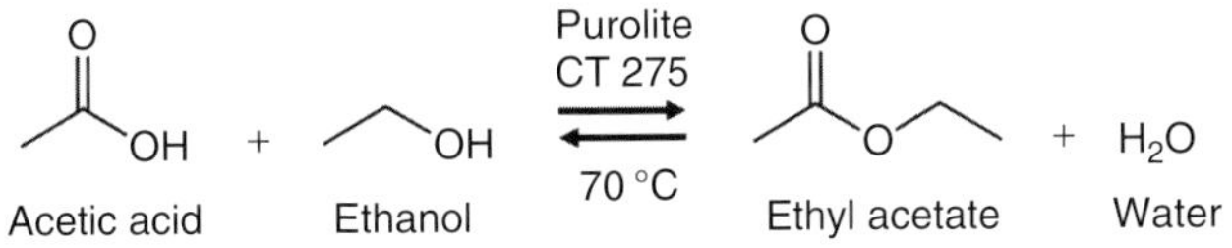

Scheme 6.22 Ester formation from acetic acid and ethanol.

solvent, toluene was used as the coolant. The authors established that integration of generated (microwave) heat leads to good prediction of reactor lengths.

6.3.2.3 Scale-Up of Microwave-Assisted Flow Processes

The scale-up of microwave-assisted organic syntheses can be classified in terms of the mode of operation, that is, (semi-)batch or continuous [12, 26, 27, 105–115]. In most cases, the choice is based on the benefits and limitations associated with the respective mode of operation. Batch scale-up literature mostly reported scaling small reaction volumes of 5 ml conducted in single-mode type of microwave cavities to 1 dm^3 multimode microwave ovens [94]. Most of the studies evaluated the limitations related to the volumetric scale-up using standard microwave reactors available commercially [110]. However, a few studies looked into the possibility of designing a microwave setup that satisfied the requirements of large batch processes assuring homogeneous heating [105, 106, 109]. Similarly, studies on continuous operation evaluated the design of commercial continuous-flow microwave reactors [12, 112, 113] and micro-reactors [26], together with process-specific microwave-integrated reactor setups [114].

Realization of microwave-assisted continuous organic synthesis (MACOS) on a commercial scale, however, requires a proper design of microwave-integrated tubular reactors [2, 86]. Patil *et al.* presented the following important functionalities that are a must in a microwave setup for development of microwave-integrated reactor systems: (i) a predictable electric field pattern, (ii) a tunable cavity, (iii) reflected power measurements, and (iv) unitized scale-up [103]. Considering the importance of these functionalities, an alternative microwave setup has been designed (Figure 6.9) [82]. This novel monomode microwave setup not only allows for proper formulation of the complete energy balances on an individual cavity, but also reduces the grid to applicator losses by utilizing only one magnetron to supply the four individual cavities. The functionality for this numbering-up approach for monomode microwave cavities was further demonstrated by optimizing reactor performance at a cavity level and then scaling out by transient operation through cavities in series [28].

The theoretically predicted number of cavities was validated by experiments where the steady-state conversion and production rate were shown to increase from the first cavity to the eighth cavity in series (Figure 6.10) [28]. This study clearly established the feasibility of modular scale-up of production capacity in flow reactors by implementing microwave cavities in series.

A similar numbering-up approach can also be applied to reactors. In their work on MACOS, Organ *et al.* discussed the possibility of parallelization of milli-reactor tubes [19, 22]. Although this work did not focus on scale-up, it

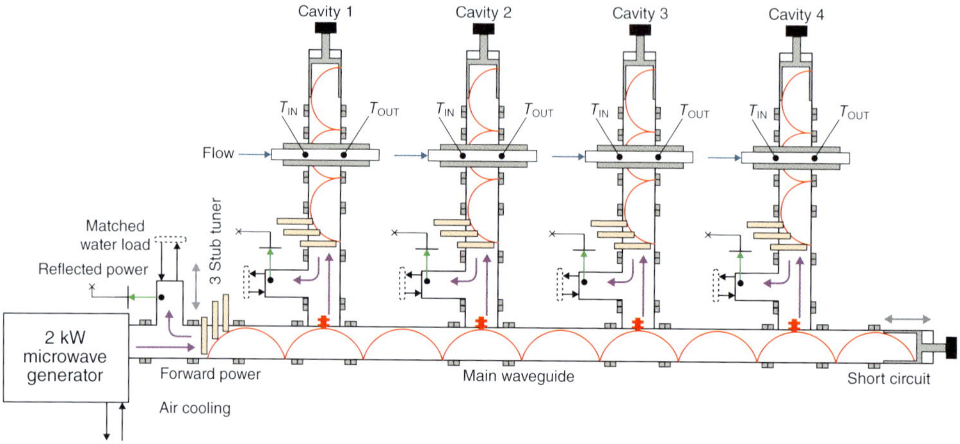

Figure 6.9 Schematic view of a dedicated microwave setup (with labeling of all the parts) designed for continuous-flow fine-chemicals synthesis and multiphase flow handling. Red lines designate field pattern. Arrows signify flow of energy (purple), signals (green), liquids (blue), and movement of stub tuner, short circuit (gray). Design: TU/e; manufacturer: Fricke und Mallah GmbH, Germany. Reproduced from Ref. [82].

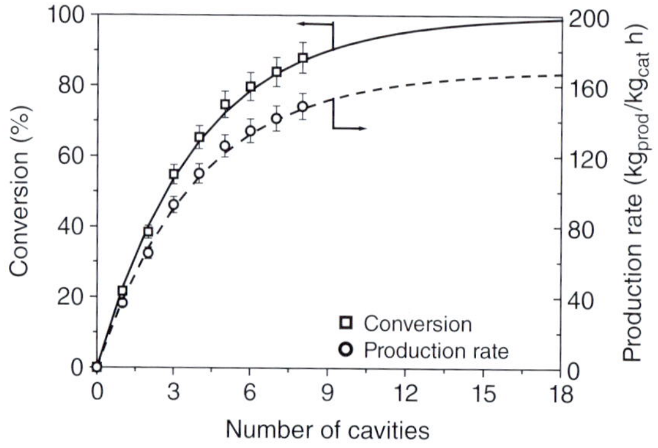

Figure 6.10 Experimental validation of the predicted conversion of acetic acid and the production rate of the ethyl acetate for the esterification reaction in the packed-bed reactor as a function of the number of cavities in series. Symbols: experimental results, lines: theoretical predictions. Reproduced from Ref. [28].

addressed the possibility of library (organic) synthesis in multiple parallel tubes. Patil *et al.*, on the other hand, explored parallelization as one of the ways to scale-up MACOS [15]. The use of a multi-tubular milli-reactor/heat exchanger system (MTMR) has been demonstrated to achieve a commercially viable capacity of 1 kg day^{-1} (Figure 6.11) [28]. The MTMR system was successfully tested for the Cu-catalyzed production of 1,3-diphenyl-2-propynylpiperidine starting from benzaldehyde, piperidine, and phenylacetylene (Scheme 6.23).

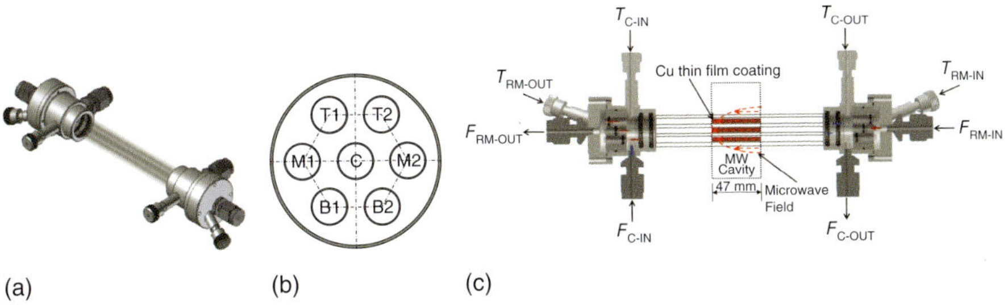

Figure 6.11 Schematic of a multi-tubular milli-reactor/heat exchanger (MTMR) setup. (a) Complete assembly. (b) Tube distribution. (c) Schematic of the MTMR assembly with coated milli-reactor tubes in the microwave cavity. Reproduced from Ref. [28].

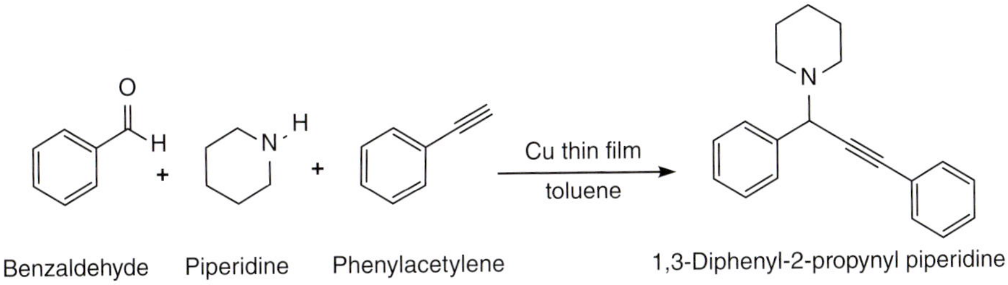

Scheme 6.23 Multicomponent reaction of benzaldehyde, piperidine, and phenylacetylene to produce 1,3-diphenyl-2-propynyl piperidine.

Cu thin films (350 ± 40 nm) were deposited on the inner wall of the reactor tube for uniform and efficient microwave absorption by all the tubes in the MTMR assembly [28]. The temperature of the copper surface in the reactor tubes was found to be at least 100 K higher than the bulk temperature of the reaction mixture. However, the oxidation and leaching of the Cu coatings led to a decreased activity during the course of operation (Figure 6.12). Cooling the outer surface of

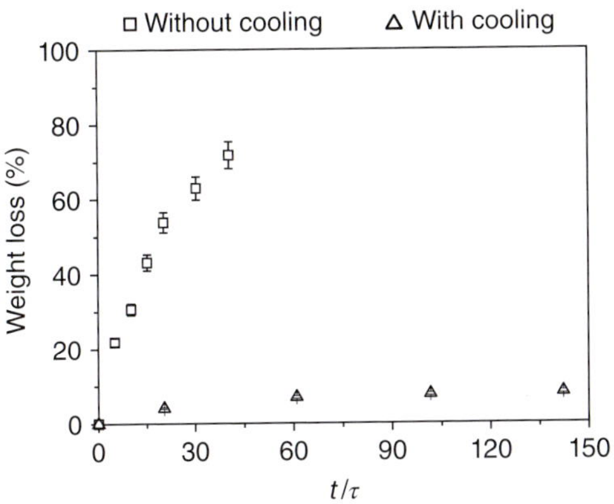

Figure 6.12 Weight-loss profile of Cu thin film over time (normalized with residence time, τ) measured by Inductively Coupled Plasma (ICP) analysis. Reproduced from Ref. [28].

the tubes by a counter-current flow of a microwave transparent coolant (toluene) reduced the copper leaching, which was attributed to uncontrolled and excessive heat build-up. The obtained throughput factor of 5.6 was, however, lower than the expected factor of 6 as a result of the flow distribution. Although this work provides an insight into the possibility of engineering the microwave-integrated Cu-coated reactor systems and scale up by parallelization, it also brings forth the challenge of copper being a highly oxidizing metal.

The above-discussed scale-up with each of the investigated numbering-up approaches has proven successful. However, as an overall conclusion, it is worthwhile noting that application of microwaves as a process intensification tool, especially in the case of organic synthesis, is more attractive for liquid–solid reactions, where the solid is the selectively (microwave) heated catalyst. Targeting direct and selective heating of the catalytically active surface, that is, the locus of the reaction, results in elevated reaction temperatures and, therefore, in high reaction rates with limited bulk liquid heating.

6.4 Concluding Remarks

There is no doubt that Cu-catalysis has made great strides in recent years in both catalyst development and application in organic syntheses. Supported and immobilized nanocatalyst systems are currently under extensive development that could provide special advantages for continuous-flow processes. Replacement of conventional heating techniques with microwave heating adds a special value when selective microwave absorption can preferably be done at catalytic sites, leaving the remainder of the reaction mixture and the catalyst support unheated. As such, direct heating "on-the-spot" is a highly efficient way of conducting chemistry for selective chemical reactions.

References

1. Öhrngren, P., Fardost, A., Russo, F., Schanche, J.-S., Fagrell, M., and Larhed, M. (2012) *Org. Process Res. Dev.*, **5**, 1053–1063.
2. Patil, N.G., Hermans, A.I.G., Benaskar, F., Meuldijk, J., Hulshof, L.A., Hessel, V., Schouten, J.C., and Rebrov, E.V. (2012) *AIChE J.*, **58**, 3144–3155.
3. Gallert, T., Hahn, M., Sellin, M., Schmöger, C., Stolle, A., Ondruschka, B., Keller, T.F., and Jandt, K.D. (2011) *ChemSusChem*, **4**, 1654–1661.
4. Kappe, C.O. (2010) *ChemSusChem*, **3**, 1085.
5. Dressen, M.H.C.L., van de Kruijs, B.H.P., Meuldijk, J., Vekemans, J.A.J.M., and Hulshof, L.A. (2010) *Org. Process Res. Dev.*, **14**, 351–361.
6. de la Hoz, A. (2011) *ChemSusChem*, **4**, 666.
7. Schmink, J. (2010) *Microwave Heating as a Tool for Sustainable Chemistry*, CRC Press, pp. 1–24.
8. Kappe, C.O. (2004) *Angew. Chem. Int. Ed.*, **43**, 6250–6284.

9. Moseley, J.D. (2009) *Chim. Oggi*, **27**, 6–10.
10. Schön, U., Messinger, J., Eichner, S., and Kirschning, A. (2008) *Tetrahedron Lett.*, **49**, 3204–3207.
11. Benaskar, F., Hessel, V., Krtschil, U., Löb, P., and Stark, A. (2009) *Org. Process Res. Dev.*, **13**, 970–982.
12. Moseley, J.D., Lenden, P., Lockwood, M., Ruda, K., Sherlock, J.-P., Thomson, A.D., and Gilday, J.P. (2007) *Org. Process Res. Dev.*, **12**, 30–40.
13. Glasnov, T.N. and Kappe, C.O. (2007) *Macromol. Rapid Commun.*, **28**, 395–410.
14. Benaskar, F., Patil, N.G., Rebrov, E.V., Ben-Abdelmoumen, A., Meuldijk, J., Hulshof, L.A., Hessel, V., and Schouten, J.C. (2013) *ChemSusChem*, **6**, 353–366.
15. Patil, N.G., Benaskar, F., Rebrov, E.V., Meuldijk, J., Hessel, V., Hulshof, L.A., and Schouten, J.C. (2012) *Ind. Eng. Chem. Res.*, **51**(44), 14344–14354.
16. Matsuzawa, M., Togashi, S., and Hasebe, S. (2011) *J. Therm. Sci. Technol.*, **6**, 69–79.
17. Togashi, S., Miyamoto, T., Asano, Y., and Endo, Y. (2009) *J. Chem. Eng. Jpn.*, **42**, 512–519.
18. Comer, E. and Organ, M.G. (2005) *J. Am. Chem. Soc.*, **127**, 8160–8167.
19. Comer, E. and Organ, M.G. (2005) *Chem. Eur. J.*, **11**, 7223–7227.
20. He, P., Haswell, S.J., and Fletcher, P.D.I. (2004) *Lab Chip*, **4**, 38–41.
21. Dallinger, D., Lehmann, H.r., Moseley, J.D., Stadler, A., and Kappe, C.O. (2011) *Org. Process Res. Dev.*, **15**, 841–854.
22. Shore, G., Yoo, W.-J., Li, C.-J., and Organ, M.G. (2010) *Chem. Eur. J.*, **16**, 126–133.
23. Shore, G., Morin, S., and Organ, M.G. (2006) *Angew. Chem. Int. Ed.*, **45**, 2761–2766.
24. Shore, G., Tsimerman, M., and Organ, M.G. (2009) *Beilstein J. Org. Chem.*, **5**, 35.
25. Cravotto, G., Beggiato, M., Penoni, A., Palmisano, G., Tollari, S., Lévêque, J.-M., and Bonrath, W. (2005) *Tetrahedron Lett.*, **46**, 2267–2271.
26. Bergamelli, F., Iannelli, M., Marafie, J.A., and Moseley, J.D.A. (2010) *Org. Process Res. Dev.*, **14**, 926–930.
27. Moseley, J.D. and Woodman, E.K. (2008) *Org. Process Res. Dev.*, **12**, 967–981.
28. Patil, N.G., Benaskar, F., Rebrov, E.V., Meuldijk, J., Hulshof, L.A., Hessel, V., and Schouten, J.C. (2014) *Org. Process Res. Dev.*, **18**, 1400–1407
29. Wirth, T. (2008) *Microreactors in Organic Synthesis and Catalysis*, Wiley-VCH Verlag GmbH.
30. Stankiewicz, A. (2007) *Ind. Eng. Chem. Res.*, **46**, 4232–4235.
31. Yoshida, J.-i., Kim, H., Nagaki, A., Illg, T., Hessel, V., Löb, P., Schouten, J.C., Irfan, M., Glasnov, T.N., and Kappe, C.O. (2011) *ChemSusChem*, **4**, 285.
32. Hessel, V., Cortese, B., and de Croon, M.H.J.M. (2011) *Chem. Eng. Sci.*, **66**, 1426–1448.
33. Zuidhof, K.T. (2010) PhD thesis. The Beckmann rearrangement of cyclohexanone oxime to e-caprolactam in micromixers and microchannels. Eindhoven University of Technology.
34. Engels, V., Benaskar, F., Patil, N.G., Rebrov, E.V., Hessel, V., Hulshof, L.A., Jefferson, D.A., Vekemans, J.A.J.M., Karwal, S., Schouten, J.C., and Wheatley, A.E.H. (2010) *Org. Process Res. Dev.*, **14**, 644–649.
35. Benaskar, F., Engels, V., Rebrov, E.V., Patil, N.G., Meuldijk, J., Thüne, P.C., Magusin, P.C.M.M., Mezari, B., Hessel, V., Hulshof, L.A., Hensen, E.J.M., Wheatley, A.E.H., and Schouten, J.C. (2012) *Chem. Eur. J.*, **18**, 1800–1810.
36. Geyer, K., Codée, J.D.C., and Seeberger, P.H. (2006) *Chem. Eur. J.*, **12**, 8434–8442.
37. Jähnisch, K., Hessel, V., Löwe, H., and Baerns, M. (2004) *Angew. Chem. Int. Ed.*, **43**, 406–446.
38. Hessel, V., Lob, P., and Lowe, H. (2005) *Curr. Org. Chem.*, **9**, 765–787.
39. Hessel, V., Löwe, H., Müller, A., and Kolb, G. (2005) *Chemical Micro Process Engineering*, Wiley-VCH Verlag GmbH & Co. KGaA, pp. 505–638.
40. Pennemann, H., Watts, P., Haswell, S.J., Hessel, V., and Löwe, H. (2004) *Org. Process Res. Dev.*, **8**, 422–439.
41. Zhang, Y., Jamison, T.F., Patel, S., and Mainolfi, N. (2010) *Org. Lett.*, **13**, 280–283.

42. Takayama, S., Link, G., Miksch, S., Sato, M., Ichikawa, J., and Thumm, M. (2006) *Powder Metall.*, **49**, 274–280.
43. Yoshikawa, N. (2010) *J. Microwave Power Electromagn. Energy*, **44**, 4–13.
44. Bond, G., Moyes, R.B., and Whan, D.A. (1993) *Catal. Today*, **17**, 427–437.
45. Anklekar, R.M., Bauer, K., Agrawal, D.K., and Roy, R. (2005) *Powder Metall.*, **48**, 39–46.
46. Roy, R., Agrawal, D., Cheng, J., and Gedevanishvili, S. (1999) *Nature*, **399**, 668–670.
47. Walton, D. (2004) *J. Appl. Phys.*, **95**, 5247–5248.
48. Perry, W.L., Cooke, D.W., Katz, J.D., and Datye, A.K. (1997) *Catal. Lett.*, **47**, 1–4.
49. Tse, M.Y., Depew, M.C., and Wan, J.S. (1990) *Res. Chem. Intermed.*, **13**, 221–236.
50. Whittaker, A.G. and Mingos, D.M.P. (1995) *J. Chem. Soc., Dalton Trans.*, 2073–2079.
51. Whittaker, A.G. and Mingos, D.M.P. (2000) *J. Chem. Soc., Dalton Trans.*, 1521–1526.
52. Benaskar, F., Patil, N.G., Engels, V., Rebrov, E.V., Meuldijk, J., Hulshof, L.A., Hessel, V., Wheatley, A.E.H., and Schouten, J.C. (2012) *Chem. Eng. J.*, **207–208**, 426–439.
53. Thomas, J.R. (1997) *Catal. Lett.*, **49**, 137–141.
54. Walton, D., Boehnel, H., and Dunlop, D.J. (2004) *Appl. Phys. Lett.*, **85**, 5367–5369.
55. Ma, J., Diehl, J.F., Johnson, E.J., Martin, K.R., Miskovsky, N.M., Smith, C.T., Weisel, G.J., Weiss, B.L., and Zimmerman, D.T. (2007) *J. Appl. Phys.*, **101**, 074906.
56. Rybakov, K.I., Semenov, V.E., Egorov, S.V., Eremeev, A.G., Plotnikov, I.V., and Bykov, Y.V. (2006) *J. Appl. Phys.*, **99**, 023506.
57. Mishra, P., Upadhyaya, A., and Sethi, G. (2006) *Metall. Mater. Trans. B*, **37**, 839–845.
58. Mondal, A., Agrawal, D., and Upadhyaya, A. (2009) *J. Microwave Power Electromagn. Energy*, **43**, 5–10.
59. Lee, C.-L. and Jou, C.-J.G. (2012) *Environ. Pollut.*, **1**, 159–168.
60. Engels, V., Benaskar, F., Jefferson, D.A., Johnson, B.F.G., and Wheatley, A.E.H. (2010) *J. Chem. Soc., Dalton Trans.*, **39**, 6496–6502.
61. Engels, V., Jefferson, D.A., Benaskar, F., Thüne, P.C., Berenguer-Murcia, A., Johnson, B.F.G., and Wheatley, A.E.H. (2011) *Nanotechnology*, **22**, 205701.
62. D'Angelo, N.D., Peterson, J.J., Booker, S.K., Fellows, I., Dominguez, C., Hungate, R., Reider, P.J., and Kim, T.-S. (2006) *Tetrahedron Lett.*, **47**, 5045–5048.
63. Zhao, F., Zhang, L., Liu, H., Zhou, S., and Liu, H. (2013) *Beilstein J. Org. Chem.*, **9**, 2463–2469.
64. Benaskar, F., Engels, V., Patil, N., Rebrov, E.V., Meuldijk, J., Hessel, V., Hulshof, L.A., Jefferson, D.A., Schouten, J.C., and Wheatley, A.E.H. (2010) *Tetrahedron Lett.*, **51**, 248–251.
65. Liu, Z.-J., Vors, J.-P., Gesing, E.R.F., and Bolm, C. (2011) *Green Chem.*, **13**, 42–45.
66. Correa, A. and Bolm, C. (2007) *Adv. Synth. Catal.*, **349**, 2673–2676.
67. Larsson, P.-F., Bolm, C., and Norrby, P.-O. (2010) *Chem. Eur. J.*, **16**, 13613–13616.
68. Ke, F., Chen, X., Li, Z., Xiang, H., and Zhou, X. (2013) *RSC Adv.*, **3**, 22837–22840.
69. He, H. and Wu, Y.-J. (2003) *Tetrahedron Lett.*, **44**, 3445–3446.
70. Li, F., Wang, Q., Ding, Z., and Tao, F. (2003) *Org. Lett.*, **5**, 2169–2171.
71. Narendar, N. and Velmathi, S. (2009) *Tetrahedron Lett.*, **50**, 5159–5161.
72. Wu, Y.-J., He, H., and L'Heureux, A. (2003) *Tetrahedron Lett.*, **44**, 4217–4218.
73. Kale, S., Kahandal, S., Disale, S., and Jayaram, R. (2012) *Curr. Chem. Lett.*, **1**, 69–80.
74. Baqi, Y. and Müller, C.E. (2007) *Org. Lett.*, **9**, 1271–1274.
75. Lin, Y.-G., Hsu, Y.-K., Chen, S.-Y., Chen, L.-C., and Chen, K.-H. (2011) *J. Mater. Chem.*, **21**, 324–326.
76. Cross, H.E., Parkes, G., and Brown, D.R. (2012) *Appl. Catal. A: Gen.*, **429–430**, 24–30.
77. Mallikarjuna, N.N. and Varma, R.S. (2007) *Cryst. Growth Des.*, **7**, 686–690.

78. Benaskar, F., Ben-Abdelmoumen, A., Patil, N.G., Rebrov, E.V., Meuldijk, J., Hulshof, L.A., Hessel, V., Krtschil, U., and Schouten, J.C. (2011) *J. Flow Chem.*, **1**, 74–89.
79. Singh, B.K., Kaval, N., Tomar, S., Eycken, E.V.d., and Parmar, V.S. (2008) *Org. Process Res. Dev.*, **12**, 468–474.
80. Jachuck, R.J.J., Selvaraj, D.K., and Varma, R.S. (2006) *Green Chem.*, **8**, 29–33.
81. Savin, K., Robertson, M., Gernert, D., Green, S., Hembre, E., and Bishop, J. (2003) *Mol. Divers.*, **7**, 171–174.
82. Patil, N.G., Benaskar, F., Rebrov, E.V., Meuldijk, J., Hulshof, L.A., Hessel, V., and Schouten, J.C. (2014) *Chem. Eng. Technol.*, **37**, 1645–1653.
83. Hoogenboom, R., Wilms, T.F.A., Erdmenger, T., and Schubert, U.S. (2009) *Aust. J. Chem.*, **62**, 236–243.
84. Esveld, E., Chemat, F., and Haveren, J.v. (2000) *Chem. Eng. Technol.*, **23**, 429–435.
85. Esveld, E., Chemat, F., and Haveren, J.v. (2000) *Chem. Eng. Technol.*, **23**, 279–283.
86. Patil, N.G., Rebrov, E.V., Eränen, K., Benaskar, F., Meuldijk, J., Mikkola, J.-P., Hessel, V., Hulshof, L.A., Murzin, D.Y., and Schouten, J.C. (2012) *J. Microwave Power Electromagn. Energy*, **46**, 83–92.
87. Chemat, F., Poux, M., de Martino, J.-L., and Berlan, J. (1996) *Chem. Eng. Technol.*, **19**, 420–424.
88. Plazl, I., Pipus, G., and Koloini, T. (1997) *AlChE J.*, **43**, 754–760.
89. Pipus, G., Plazl, I., and Koloini, T. (2000) *Chem. Eng. J.*, **76**, 239–245.
90. Wilson, N.S., Sarko, C.R., and Roth, G.P. (2004) *Org. Process Res. Dev.*, **8**, 535–538.
91. Roberts, B.A. and Strauss, C.R. (2005) *Acc. Chem. Res.*, **38**, 653–661.
92. Cecilia, R., Kunz, U., and Turek, T. (2007) *Chem. Eng. Process. Process Intensif.*, **46**, 870–881.
93. Herrero, M.A., Kremsner, J.M., and Kappe, C.O. (2007) *J. Org. Chem.*, **73**, 36–47.
94. He, P., Haswell, S.J., and Fletcher, P.D.I. (2005) *Sens. Actuators, B*, **105**, 516–520.
95. Datta, A.K., Geedipalli, S.S.R., and Almeida, M.F. (2005) *Food Technol.*, **59**, 36–46.
96. McMinn, W.A.M., McLoughlin, C.M., and Magee, T.R.A. (2005) *Drying Technol.*, **23**, 513–532.
97. Bonnet, C., Estel, L., Ledoux, A., Mazari, B., and Louis, A. (2004) *Chem. Eng. Process. Process Intensif.*, **43**, 1435–1440.
98. Saillard, R., Poux, M., Berlan, J., and Audhuy-Peaudecerf, M. (1995) *Tetrahedron*, **51**, 4033–4042.
99. Robinson, J., Kingman, S., Irvine, D., Licence, P., Smith, A., Dimitrakis, G., Obermayer, D., and Kappe, C.O. (2010) *Phys. Chem. Chem. Phys.*, **12**, 4750–4758.
100. Sturm, G.S.J., Stefanidis, G.D., Verweij, M.D., Van Gerven, T.D.T., and Stankiewicz, A.I. (2010) *Chem. Eng. Process. Process Intensif.*, **49**, 912–922.
101. Rakesh, V., Seo, Y., Datta, A.K., McCarthy, K.L., and McCarthy, M.J. (2010) *AlChE J.*, **56**, 2468–2478.
102. Rakesh, V., Datta, A.K., Walton, J.H., McCarthy, K.L., and McCarthy, M.J. (2012) *AlChE J.*, **58**, 1262–1278.
103. Patil, N.G., Benaskar, F., Meuldijk, J., Hulshof, L.A., Hessel, V., Schouten, J.C., Esveld, E.D.C., and Rebrov, E.V. (2014) *AlChE J.*, **60**, 3824–3832.
104. Shore, G. and Organ, M.G. (2008) *Chem. Commun.*, 838–840.
105. Amore, K.M. and Leadbeater, N.E. (2007) *Macromol. Rapid Commun.*, **28**, 473–477.
106. Kremsner, J.M., Stadler, A., and Kappe, C.O. (2006) *Top. Curr. Chem.*, **266**, 233–278.
107. Lehmann, H. and LaVecchia, L. (2010) *Org. Process Res. Dev.*, **14**, 650–656.
108. Nakamura, T., Nagahata, R., Kunii, K., Soga, H., Sugimoto, S., and Takeuchi, K. (2010) *Org. Process Res. Dev.*, **14**, 781–786.
109. Bowman, M.D., Schmink, J.R., McGowan, C.M., Kormos, C.M., and Leadbeater, N.E. (2008) *Org. Process Res. Dev.*, **12**, 1078–1088.
110. Marafie, J.A. and Moseley, J.D. (2010) *Org. Biomol. Chem.*, **8**, 2219–2227.
111. Bowman, M.D., Holcomb, J.L., Kormos, C.M., Leadbeater, N.E., and

Williams, V.A. (2007) *Org. Process Res. Dev.*, **12**, 41–57.

112. Moseley, J.D. and Lawton, S.J. (2007) *Chim. Oggi*, **25**, 16–19.

113. Schmink, J.R., Kormos, C.M., Devine, W.G., and Leadbeater, N.E. (2010) *Org. Process Res. Dev.*, **14**, 205–214.

114. Bagley, M.C., Fusillo, V., Jenkins, R.L., Lubinu, M.C., and Mason, C. (2010) *Org. Biomol. Chem.*, **8**, 2245–2251.

115. Bierbaum, R., Nüchter, M., and Ondruschka, B. (2005) *Chem. Eng. Technol.*, **28**, 427–431.

7
Pilot Plant for Continuous Flow Microwave-Assisted Chemical Reactions

Mitsuhiro Matsuzawa and Shigenori Togashi

7.1
Introduction

Chemical reaction processes using microwaves have attracted attention. Numerous effects caused by microwaves in organic synthesis have been reported [1–5], for example, reductions in chemical reaction times from hours to minutes, reductions in side reactions, and improvements in selectivity. Chemical–synthetic procedures using microwaves are expected to conserve energy, decrease the burden imposed on the environment, and simplify reaction processes.

Most chemical reactors that use microwaves are batch-type reactors. However, one of the problems with these reactors is the difficulty in scaling them up from the laboratory to the production scale, because the depth that microwaves penetrate into absorbing materials is limited; for example, the penetration depth generally used at 2.45 GHz is in the order of a few centimeters, depending on the dielectric properties of the absorbing materials. Consequently, when we try to scale these batch types up, microwaves are only absorbed at the surface of the batch reactors, and it is difficult to transmit microwaves inside them. In addition, the mixing efficiency of reactant solutions decreases when the batch-type reactors are scaled up.

A continuous flow process is expected to be used to solve the penetration depth problem [6–8]. Namely, irradiating microwaves to a reactant solution flowing through a reactor tube without increasing the size of the tube is expected to solve the problem of the limited penetration depth of microwaves.

One example of the research on continuous flow processing using microwaves is the study conducted by He *et al.* [9, 10] on Suzuki–Miyaura coupling reactions using a microcapillary tube through which reactant solution flowed and was irradiated by microwaves. They reported that they obtained a 70% yield of the product in less than 60 s. Organ *et al.* [11, 12] conducted microscale organic synthesis using microwaves and reported that excellent conversion was observed in a variety of cross-coupling and ring-closing metathesis reactions. However, the microwave-assisted chemical reactors used in these studies were commercially available ones that were not designed especially for continuous flow processing,

Microwaves in Catalysis: Methodology and Applications, First Edition.
Edited by Satoshi Horikoshi and Nick Serpone.

so the energy absorption efficiency was not high, and the throughput was limited to laboratory scale, for example, the throughputs were about several dozen $mm^3 min^{-1}$.

Accordingly, we developed a continuous flow microwave-assisted chemical reactor in which a reactant solution can absorb microwave energy efficiently, and which can also be used in chemical processes using microreactors [13]. A microreactor is a device that enables chemical reactions to be performed on a microscale and has a feature to mix fluids rapidly [14–17].

Additionally, we developed a pilot plant for continuous flow microwave-assisted chemical reactions using the concept of numbering-up, in which microwaves generated from a single microwave generator were divided into multiple reaction fields, and a reactant solution can be irradiated simultaneously in each reaction field by microwaves for increasing the throughput [18].

7.2
Continuous Flow Microwave-Assisted Chemical Reactor

7.2.1
Basic Structure

We developed a continuous flow microwave-assisted chemical reactor in which a reactant solution flowing through the reactor tube is irradiated by microwaves and can absorb microwaves efficiently. The reactor is expected to be able to solve the problem of the limited penetration depth of microwaves since the reactor tube is not large. Figure 7.1 outlines the basic structure of the apparatus. It consists of a microwave generator that generates microwaves at 2.45 GHz (HPP121A-INV-02, Hitachi Power Solutions Co., Ltd), a power monitor that measures the power of incident and reflected waves, a three-stub tuner that matches impedance in the apparatus, an applicator that the reactor tube penetrates vertically. It is possible to control the impedance in the apparatus by controlling the insertion length of the stub tuners and the position of the short-circuit plane. Therefore, it is possible to optimize the efficiency of microwave energy absorbed by the reactant solution in the reactor tube. Since the amount of microwave energy absorbed by a heated material is proportional to the square of electric intensity, the reactor tube needs to be placed in the area where the electric intensity is strong in order for the reactant solution to absorb microwaves efficiently. In this study, electromagnetic simulation was used to design the shape and location of the reactor tube. MicroStripes™ (CST) was the electromagnetic simulator used. In this apparatus, incident waves from the microwave generator and reflected waves that are reflected at the short-circuit plane encounter interference and standing waves are formed. Accordingly, it was necessary to set the reactor tube in the area where the electric intensity was strong for the standing waves. Table 7.1 lists the dielectric properties used in the simulation.

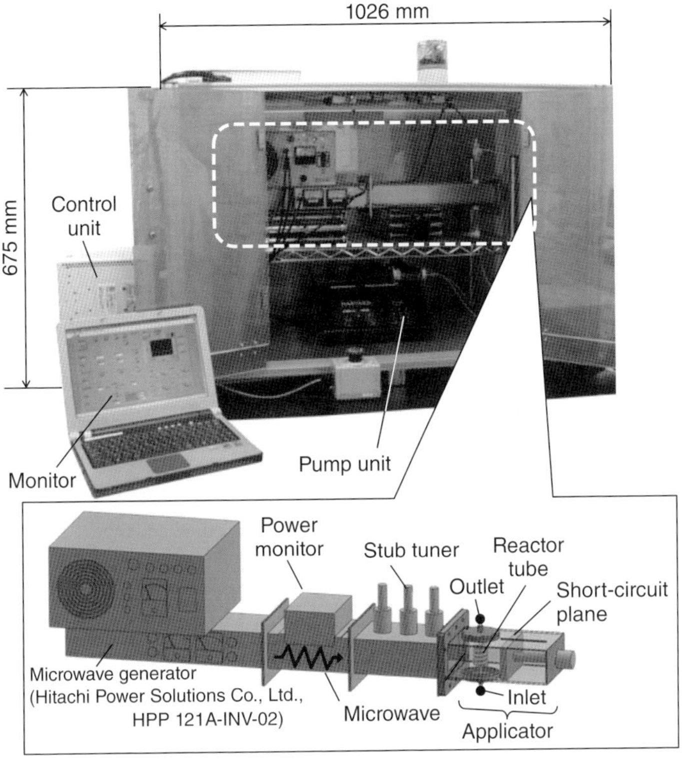

Figure 7.1 Basic structure of the continuous flow microwave-assisted chemical reactor.

Table 7.1 Dielectric properties used in simulation.

	Material	**ε_r' (relative permittivity)**	**ε_r'' (relative dielectric loss factor)**
Reactant solution	Water	77	11
Reactor tube	Quartz	2.9	5.2×10^{-3}
Applicator, tuner	Aluminum	—	—

Figure 7.2 shows the electric field intensity distribution in the apparatus when water flowed through a reactor tube that had an inner diameter of 3.0 mm, an outer diameter of 5.0 mm, and a coil diameter of 20.0 mm. The reactor tube was fabricated from quartz. Figure 7.2 shows (a) the electric field intensity distribution when the stub tuners were adjusted and (b) the distribution without the stub tuners. It is clear from the figures that the reactor tube was placed in an area where the electric field intensity was strong. Here, the energy absorption efficiency is defined as the ratio of the absorbed energy by the water to the output energy from the microwave generator. In this case, the energy absorption efficiency by

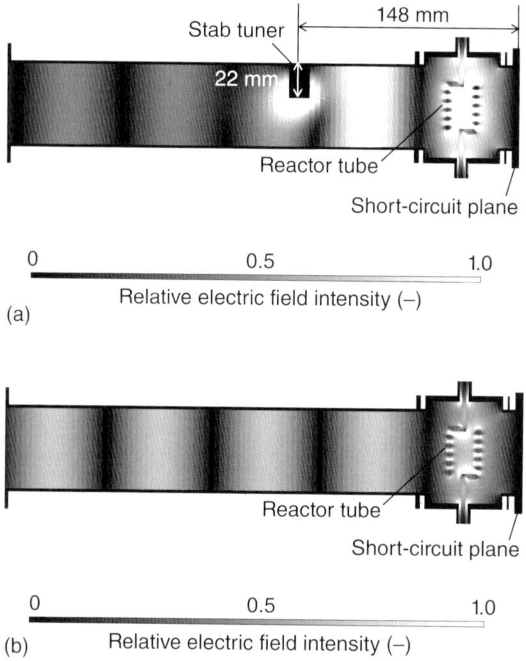

Figure 7.2 Electric field intensity distribution in the apparatus (a) with adjusting by stub tuner and (b) without stub tuner.

the water was about 95% when the stub tuners were adjusted in the simulation, whereas it was about 63% when the stub tuners were not used.

The energy absorption efficiency in the developed apparatus was examined by conducting water heating tests. The energy absorption efficiency was calculated from the difference in temperature of the water, ΔT, between the inlet and outlet

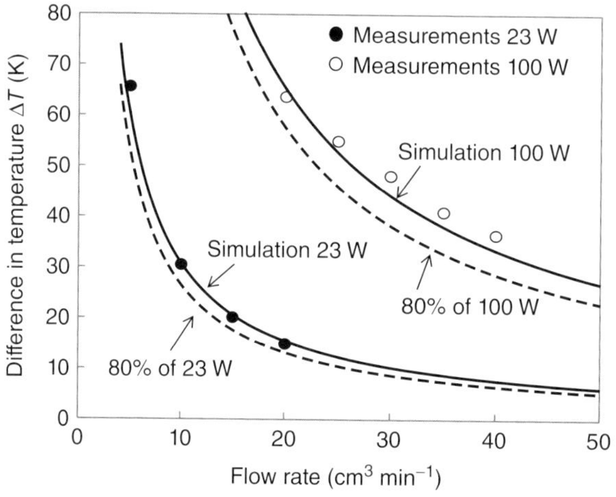

Figure 7.3 Result of water heating test when microwaves at 23 and 100 W were irradiated.

as shown in Figure 7.1. The difference in temperature ΔT was measured when water flowed at 5–40 $cm^3\ min^{-1}$ and was microwave-irradiated at 23 and 100 W. Figure 7.3 plots the results. The solid lines indicate the results of simulation, and the symbols indicate the measurements. The broken lines indicate the line corresponding to 80% of 23 and 100 W, respectively. The simulation corresponded to the measurement within 9%, and the energy absorption efficiency was above 80% at the least.

7.3 Pilot Plant

7.3.1 Design of Waveguide

A pilot plant for continuous flow microwave-assisted chemical reactions combined with microreactors using the concept of numbering-up was developed, in which microwaves generated from a single microwave generator were divided into multiple reaction fields, and a reactant solution was simultaneously irradiated in the respective reaction fields by microwaves. The configuration of the waveguide needed to be designed so that it would be able to heat the reactant solutions efficiently in the multiple reaction fields using a single microwave generator. Accordingly, a branch waveguide was designed by using electromagnetic simulation in this study.

A configuration of the branch waveguide was investigated in which incident waves were divided into two directions uniformly and without losing energy. As a result of our investigation, we found that a waveguide with a 45° inclined plane, as shown in Figure 7.4, could transmit microwaves efficiently. In this waveguide,

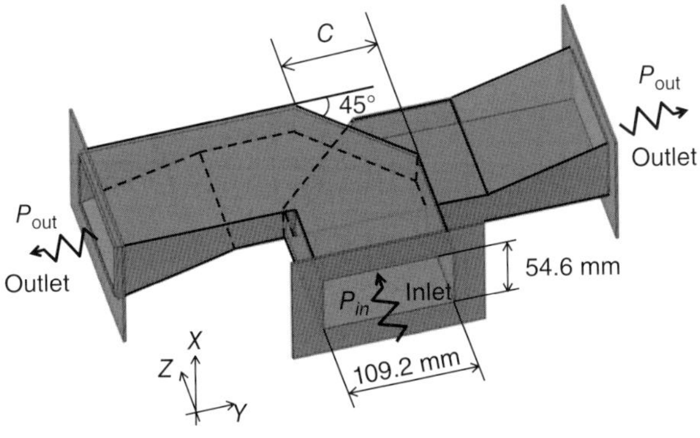

Figure 7.4 Configuration of the two-branch waveguide used in this study.

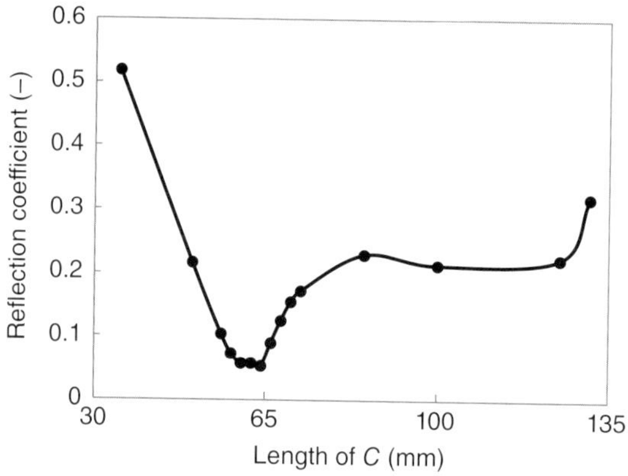

Figure 7.5 Calculation results of reflection coefficient when the length of *C* was changed.

the incident waves are halved along the *X*-axis and the divided waves transmit to each outlet. We optimized the length of *C* as shown in Figure 7.4 by using electromagnetic simulation. Figure 7.5 plots the calculation results of the reflection coefficient when the length of *C* was changed. The reflection coefficient describes the intensity of a reflected wave relative to an incident wave, and the transmission efficiency increases when the reflection coefficient decreases. The results show that the reflection coefficient was smallest when the length of *C* was 64 mm. The transmission efficiency was above 99% due to the simulation when the length of *C* was 64 mm.

Also investigated was a four-branch waveguide configuration that was based on the above two-branch waveguide design. This configuration was designed to increase throughput by heating the reactant solution in four reaction fields

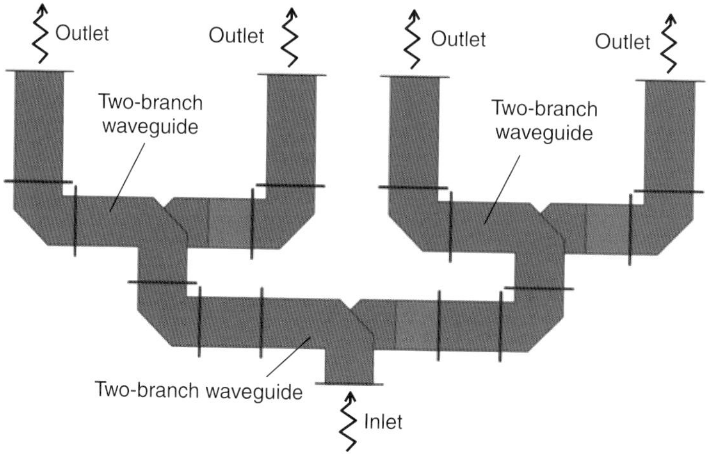

Figure 7.6 Schematic of the four-branch waveguide.

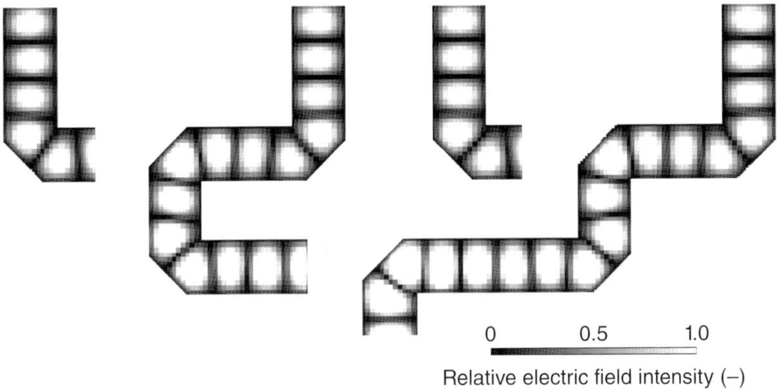

Figure 7.7 Electric field intensity distribution in the four-branch waveguide.

simultaneously. Figure 7.6 outlines the schematic of the four-branch waveguide. It consists of three two-branch waveguides. Two of the two-branch waveguides are connected to an outlet of each branch of the first waveguide. Figure 7.7 shows the relative electric field intensity distribution in the four-branch waveguide using the two-branch waveguides designed earlier. This figure shows that microwaves transmitted to the outlets of the waveguides efficiently when the designed two-branch waveguides were used. The results of calculating the transmission efficiency indicated that the transmission efficiency was 99% when the designed two-branch waveguides were used.

Based on the four-branch waveguide, we developed a microwave apparatus by which the reactant solutions could be heated simultaneously by microwaves in four reaction fields using a single microwave generator. Figure 7.8 shows the configuration of the microwave apparatus. It consists of a single microwave generator that generates microwaves at 2.45 GHz, the designed two-branch waveguides, isolators that absorb only reflected waves, stub tuners that control the impedance in the waveguides, and applicators. The reflected waves generated from one reaction field may influence reactions occurring in other reaction fields if the reflected waves diffract to other reaction fields. The isolators set in the respective reaction fields can absorb the reflected waves, so this apparatus can suppress diffraction of the reflected waves generated from one reaction filed to other reaction fields.

7.3.2 Configuration of Pilot Plant

Based on the microwave apparatus described in Section 7.3.1, we developed a pilot plant for continuous flow microwave-assisted chemical reactions combined with microreactors. Figure 7.9 outlines an overview of the plant. It consists of a flow control unit, the microwave apparatus, four microreactors, a control system, and

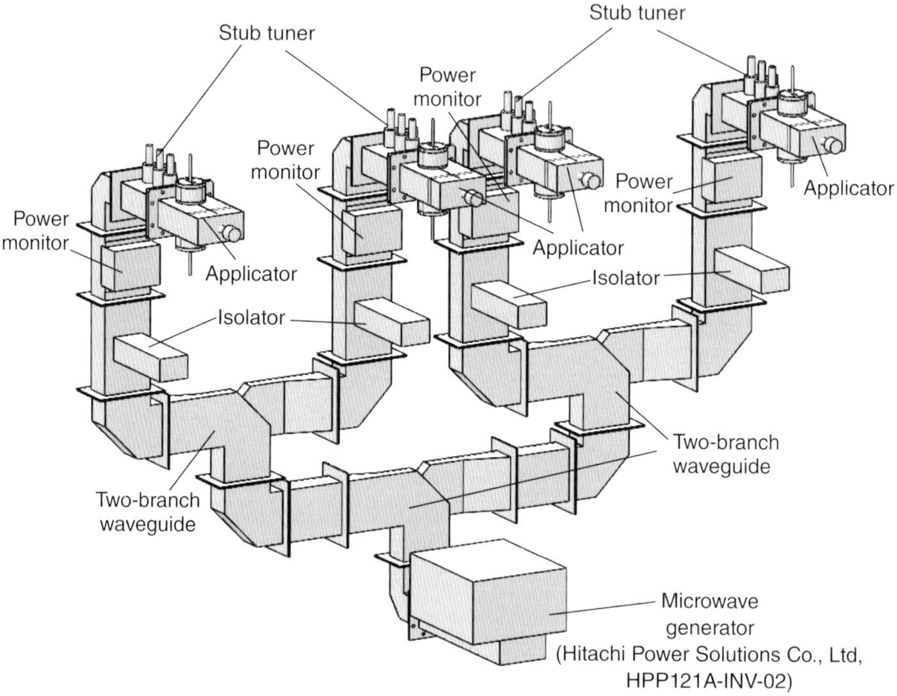

Figure 7.8 Configuration of the microwave apparatus.

a monitoring system. Figure 7.10 illustrates a block diagram of the pilot plant. Two reagents flow into the microreactors and are mixed there; then they flow into the reactor tube in the applicator of the microwave apparatus, and the reactant solutions are microwave-irradiated and heated. The flow control unit consists of two nonpulsatile pumps. Pressure sensors were set in the upstream of the microreactors, and fiber optic probes for measuring temperatures were set at the inlets and outlets of the reactor tubes in the applicator. Flow sensors were positioned at the downstream of the reactor tubes. These components make it possible to monitor the temperatures, pressures, and flow rates during reactions.

7.3.3 Water Heating Test

Water heating tests were conducted to evaluate the temperature control ability of the pilot plant. Water flowed at 10 $cm^3\ min^{-1}$ in the respective reaction fields; microwaves at 168 W were irradiated (i.e., microwaves at 42 W were irradiated in each of the respective reaction fields); and the reactant solution temperatures of the inlets and outlets of the respective reactor tubes were measured by the fiber optic probes.

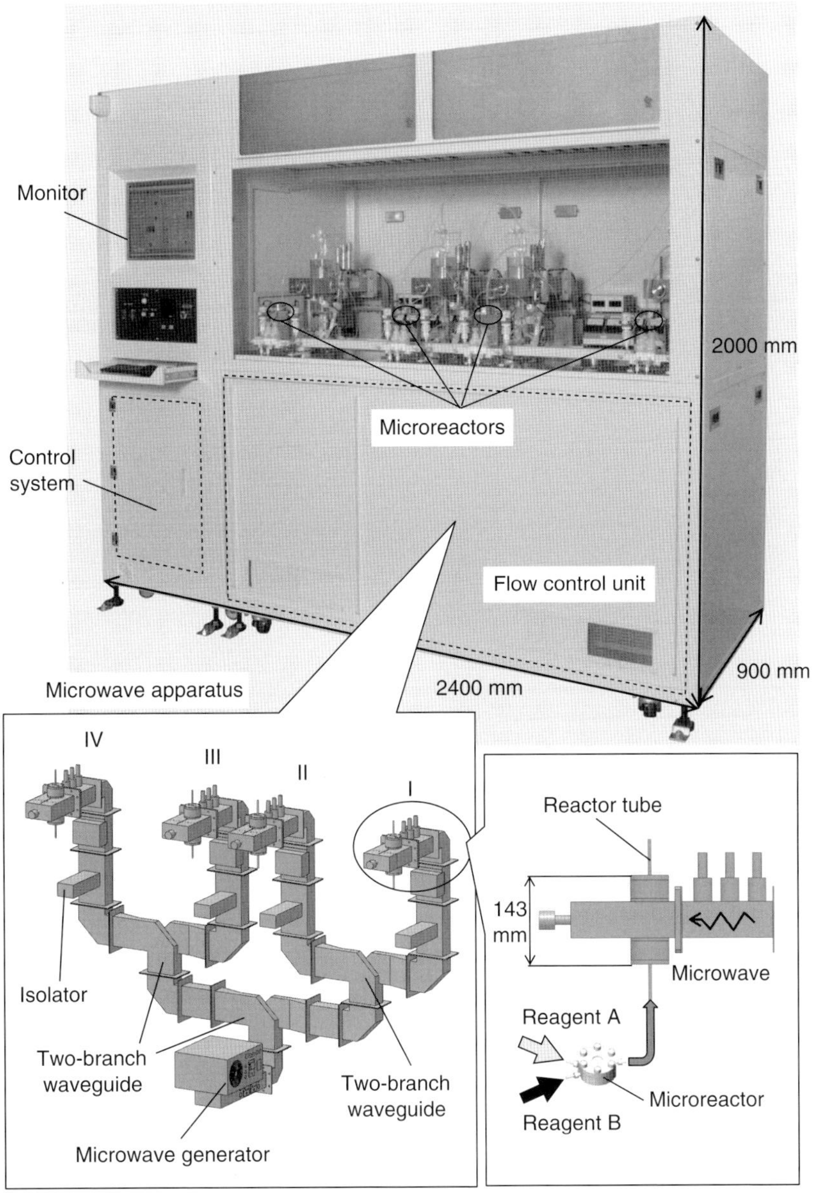

Figure 7.9 Overview of the pilot plant for continuous flow microwave-assisted chemical reactions combined with microreactors.

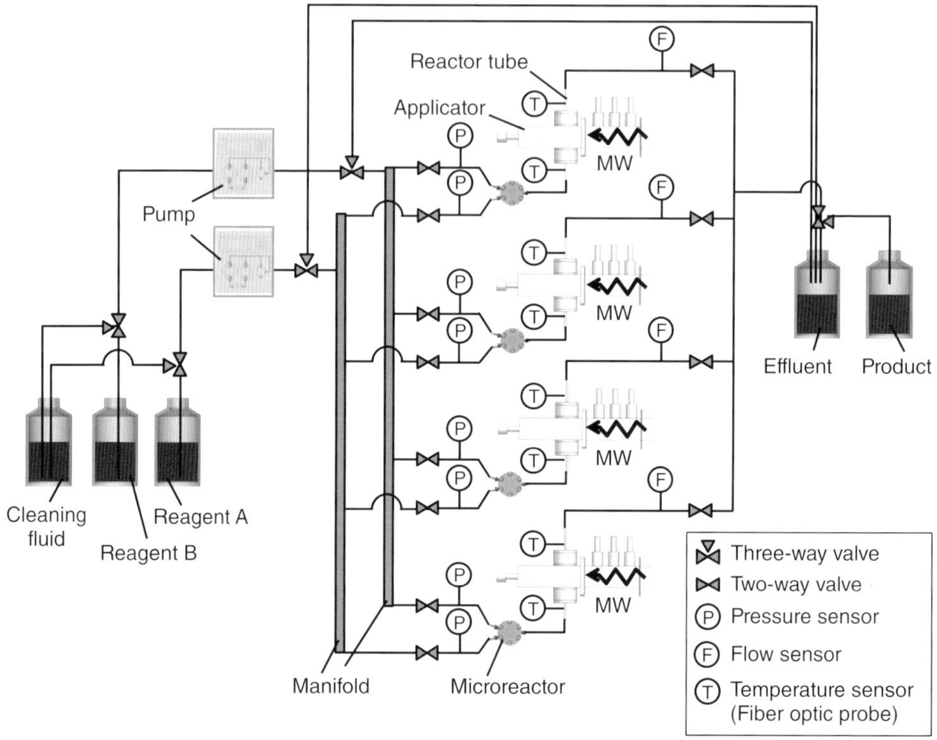

Figure 7.10 Block diagram of the pilot plant.

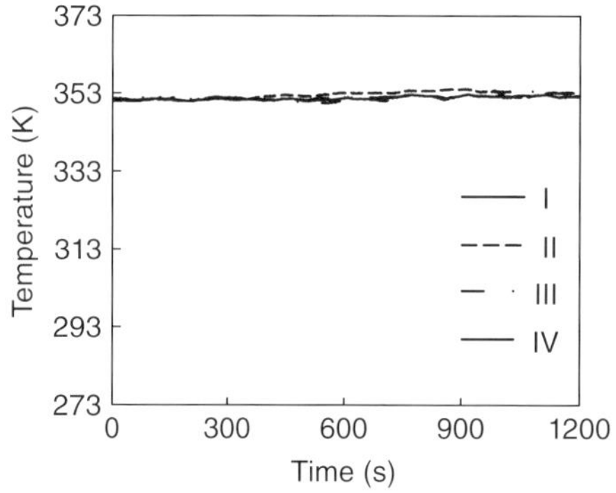

Figure 7.11 Result of water heating tests.

The results are plotted in Figure 7.11. The temperatures of the outlets were controlled within ±1.1 K at 353.2 K. Additionally, Figure 7.12 plots the results of the energy absorption efficiency, which were about 90% in the respective reaction fields, whereas the energy absorption efficiency was about 40% when 100 cm^3 of water was heated by a commercially available multimode microwave chemical reactor. The evaluation test demonstrated that the reactant solution was able to

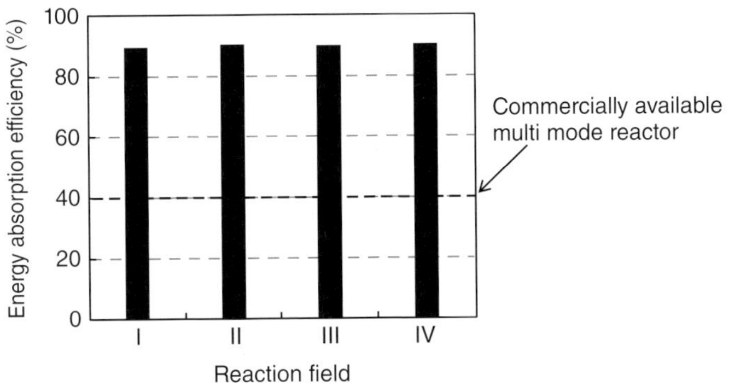

Figure 7.12 Energy absorption efficiency by water in the four reaction fields.

absorb microwaves efficiently, and the pilot plant was able to control the reactant solution temperatures uniformly as designed by the simulation.

7.3.4 Sonogashira Coupling Reaction

The Sonogashira coupling reaction [19, 20] was also conducted to evaluate the developed pilot plant. Reaction (7.1) shows the reaction formula. Two reactant solutions were prepared. One consisted of 0.20 mol l^{-1} of 1-iodo-4-nitrobenzene and 0.80 mol l^{-1} of triethylamine and *N*,*N*-dimethylacetamide (DMA) as a solvent. The other consisted of 1.0 mol% of $Pd(PPh_3)Cl_2$ as a catalyst and 0.40 mol l^{-1} of phenylacetylene and DMA as a solvent.

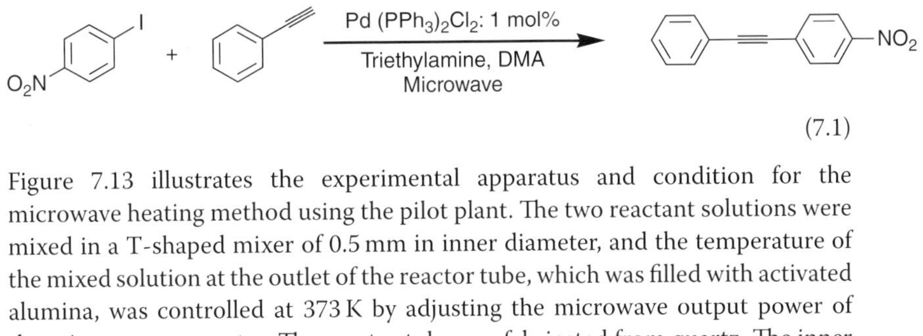

(7.1)

Figure 7.13 illustrates the experimental apparatus and condition for the microwave heating method using the pilot plant. The two reactant solutions were mixed in a T-shaped mixer of 0.5 mm in inner diameter, and the temperature of the mixed solution at the outlet of the reactor tube, which was filled with activated alumina, was controlled at 373 K by adjusting the microwave output power of the microwave generator. The reactor tube was fabricated from quartz. The inner diameter was 10 mm, and the outer diameter was 12 mm. For comparison, we also conducted experiments by using the oil-bath heating method as shown in Figure 7.14, which is a conventional heating method. We conducted the experiments by immersing a stainless tube of 1.0 mm in inner diameter in an oil-bath at 373 K.

Figure 7.15 shows the experimental results. The reaction conducted by microwave heating was accelerated, and the reaction time, which was equal to the residence time in this study, was reduced by approximately half of that by

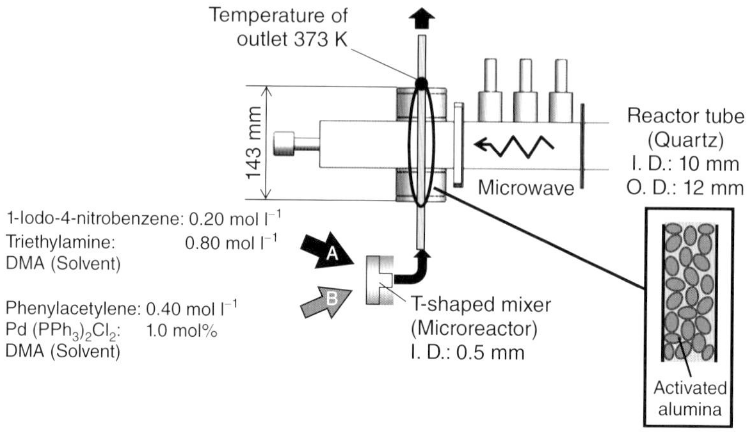

Figure 7.13 Experimental apparatus and condition for the Sonogashira coupling reaction by microwave heating method using the pilot plant.

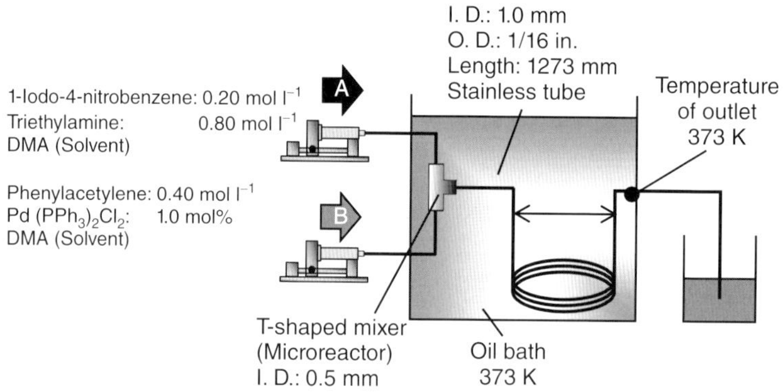

Figure 7.14 Experimental apparatus and condition for the Sonogashira coupling reaction by oil-bath heating method.

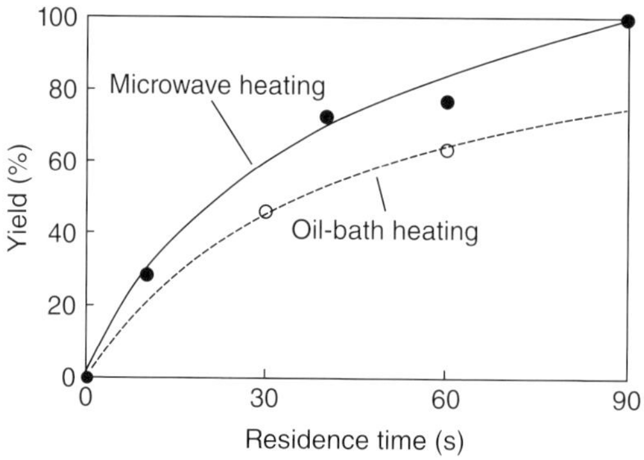

Figure 7.15 Relation between yield and residence time in the Sonogashira coupling reaction.

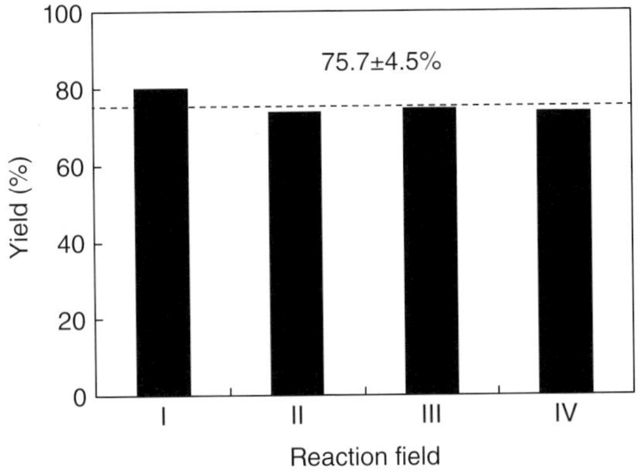

Figure 7.16 Yields obtained by microwave heating method in the four reaction fields.

oil-bath heating. It is believed that the reaction was accelerated due to the high temperature of the catalyst close to the activated alumina, which was heated by microwaves locally and selectively. The yield arrived at 100% when the reaction time was 90 s in microwave heating.

Figure 7.16 shows each yield obtained in the four reaction fields when the reaction time was 60 s. The yields in the four fields were within ±4.5% at 75.5% when the reaction time was 60 s, and we confirmed that the reaction progressed almost uniformly in the four fields. Additionally, each yield in the four fields was 100% when the reaction time was 90 s. The flow rate in each field was equal to 4.8 $cm^3\ min^{-1}$ when the reaction time was 90 s, and it corresponded to 5.8 $t\ year^{-1}$ of throughput when the operating time per year of the pilot plant is assumed to be 5000 h. It was demonstrated that throughput was able to be increased without deteriorating the quality of the products by heating the reactant solution simultaneously in multiple reaction fields with this method.

7.4
Conclusions

A pilot plant for continuous flow microwave-assisted chemical reactions combined with microreactors was developed, and water heating tests were conducted to evaluate the developed plant. We first designed a reactor by using an electromagnetic simulation and found that the energy absorption efficiency of the water was more than 80% at the least by conducting a water heating test. Additionally, we developed a microwave apparatus having a single microwave generator that can heat reactant solutions in four reaction fields simultaneously in order to increase throughput. We also designed a four-branch waveguide using an electromagnetic simulation and found that the transmission efficiency was 99%. Finally, we developed the pilot plant by using the microwave apparatus and conducted water heating tests. We found that the parallel flows were uniformly controlled within ±5%, and the temperatures were controlled within ±1.1 K at 353.2 K. Moreover, we found that the energy absorption efficiency was about 90% in the respective reaction fields, whereas the energy absorption efficiency was about 40% when $100\,cm^3$ of water was heated by a commercially available multimode microwave chemical reactor. We also conducted the Sonogashira coupling reaction by using the developed pilot plant, and the reaction progressed almost uniformly in the four reaction fields. The yield was 100% when the reaction time was 90 s. The flow rate in each reaction field was equal to $4.8\,cm^3\,min^{-1}$ when the reaction time was 90 s, and it corresponded to $5.8\,t\,year^{-1}$ of throughput when the yearly operating time of the pilot plant was assumed to be 5000 h. It was demonstrated that throughput was able to be increased without deteriorating the quality of the products by heating the reactant solution simultaneously in multiple reaction fields in this study.

Acknowledgment

This research was financially supported by the Project of "Development of Microspace and Nanospace Reaction Environment Technology for Functional Materials" of New Energy and Industrial Technology Development Organization (NEDO), Japan.

References

1. Lidström, P., Tierney, J., Wathey, B., and Westman, J. (2001) *Tetrahedron*, **57**, 9225–9283.
2. Kappe, C.O. (2004) *Angew. Chem. Int. Ed.*, **43**, 6250–6284.
3. Hoz, A.D.L., Ortiz, A.D., and Moreno, A. (2007) *J. Microwave Power Electromagn. Energy*, **41**, 45–66.
4. Hoz, A., Ortiz, A.D., and Moreno, A. (2005) *Chem. Soc. Rev.*, **34**, 164–178.
5. Gawande, M.B., Shelke, S.N., Zboril, R., and Varma, R.S. (2014) *Acc. Chem. Res.*, **47**, 1338–1348.
6. Glasnov, T.N. and Kappe, C.O. (2007) *Macromol. Rapid Commun.*, **28**, 395–410.
7. Razzaq, T. and Kappe, C.O. (2010) *Chem. Asian J.*, **5**, 1274–1289.
8. Bowman, M.D., Holcomb, J.L., Kormos, C.M., Leadbeater, N.E., and Williams, V.A. (2008) *Org. Process Res. Dev.*, **12**, 41–57.
9. He, P., Haswell, S.J., and Fletcher, P.D.I. (2004) *Appl. Catal., A*, **274**, 111–114.
10. He, P., Haswell, S.J., and Fletcher, P.D.I. (2004) *Lab Chip*, **4**, 38–41.
11. Comer, E. and Organ, M.G. (2005) *Chem. Eur. J.*, **11**, 7223–7227.
12. Comer, E. and Organ, M.G. (2005) *J. Am. Chem. Soc.*, **127**, 8160–8167.
13. Matsuzawa, M., Togashi, S., and Hasebe, S. (2009) *Trans. JSME*, **75**, 316–322 (in Japanese).
14. Togashi, S., Miyamoto, T., Asano, Y., and Endo, Y. (2009) *J. Chem. Eng. Jpn.*, **42**, 512–519.
15. Ehrfeld, W., Golbig, K., Hessel, V., Löwe, H., and Richter, T. (1999) *Ind. Eng. Chem. Res.*, **38**, 1075–1082.
16. Mae, K. (2007) *Chem. Eng. Sci.*, **62**, 4842–4851.
17. Hasebe, S. (2004) *Comput. Chem. Eng.*, **29**, 57–64.
18. Matsuzawa, M., Togashi, S., and Hasebe, S. (2011) *J. Therm. Sci. Technol.*, **6**, 69–79.
19. Sonogashira, K., Tohda, Y., and Hagihara, N. (1975) *Tetrahedron Lett.*, **16**, 4467–4470.
20. Fukuyama, T., Shinmen, M., Nishitani, S., Sato, M., and Ryu, I. (2002) *Org. Lett.*, **4**, 1691–1694.

8
Efficient Catalysis by Combining Microwaves with Other Enabling Technologies

Giancarlo Cravotto, Laura Rinaldi, and Diego Carnaroglio

8.1
Introduction

Scientific and technological progress now occurs at the interface between two and more scientific domains, and chemistry overlaps almost all fields. In spite of chemistry's multidisciplinary interactions, chemists are often extremely conservative when designing experimental protocols. Meanwhile, chemical laboratories and industries are heavily involved in the development of mild, simple, environmental friendly, and inexpensive catalytic processes that respect the principles of green chemistry of process intensification and competitive production requirements [1]. The result is a huge gap between classic production processes and new green protocols, which are based on integrated technologies, higher efficiency, and sustainability [2].

Microwaves (MWs), ultrasound (US), and other enabling technologies play a pivotal role in catalysis [3], both in catalyst preparation and in efficiently promoting almost all organic reactions [4]. Numerous examples of catalytic conversion and environmentally friendly applications, which make use of hyphenated and tandem techniques, are described in the current literature [5].

Complimentary and synergistic effects have been described when MW are combined with other enabling technologies such as US, hydrodynamic cavitation (HC), high-shear mixers, UV rays, plasma, microfluidic reactors, and mechanochemistry. Although sequential treatments can be more easily applied, hybrid reactors for simultaneous irradiation and flowing multiunit systems have become more prominent as a means to optimize mass/heat transfer in heterogeneous and homogeneous catalyses [6].

Power US is one of the most common techniques in green and applied chemistry. The formation of bubbles is promoted by the transfer of acoustic waves in the liquid media while their consequent collapse is accompanied by shock waves and shear forces, which contribute to mixing and particle fragmentation when solids are present. Several mechanical actions account for enhanced mass and energy transfer and result in significant effects on mixing, rapid emulsification, polymer scission, and even supramolecular organization [7], inducing faster and

Microwaves in Catalysis: Methodology and Applications, First Edition.
Edited by Satoshi Horikoshi and Nick Serpone.

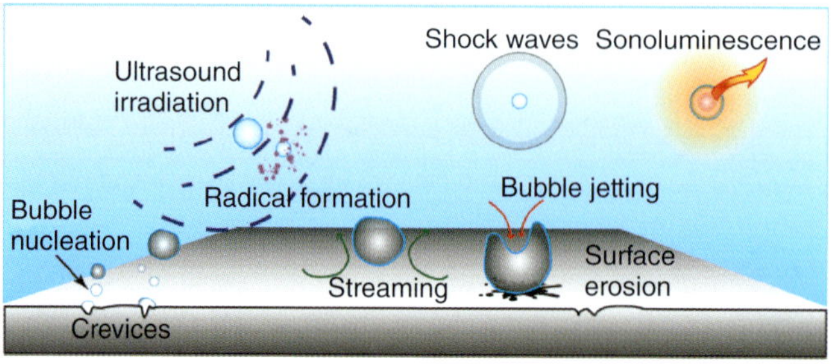

Figure 8.1 Phenomena related to acoustic cavitation. Reproduced with permission from Ref. [9]. © 2012 David Fernández Rivas, Enschede, The Netherlands.

more selective chemical transformations [8]. Figure 8.1 summarizes the multiple phenomena that occur under acoustic cavitation [10].

Of particular importance are the incorporation of combined MW and US technologies into flow systems and the emerging field of microfluidics, that is, microreactors for chemical reactions, which are considerably enhanced under the effects of these energy sources, notably in terms of mass transfer, rapid heating, and the prevention of clogging. Though little US horns equipped with thin tips have been coupled with microreactors, the most suitable approach is the sonication via transducers equipped with metallic plates that act on a wide surface (Figure 8.2).

Photocatalysis with visible light and ultraviolet (UV) irradiation can boast of a long history in chemical synthesis and in the degradation of light-sensitive pollutants. The energy involved in the UV part of the electromagnetic spectrum is able to break or reorganize most covalent bonds. The resulting excited states are rich in energy and antibonding orbitals are occupied, suitable conditions for reactions that are highly endothermic in the ground state. Photochemical reactions can include singlet and triplet states while thermal reactions usually only show singlet states. For this reason, some intermediates are not accessible via thermal treatment in photochemical reactions.

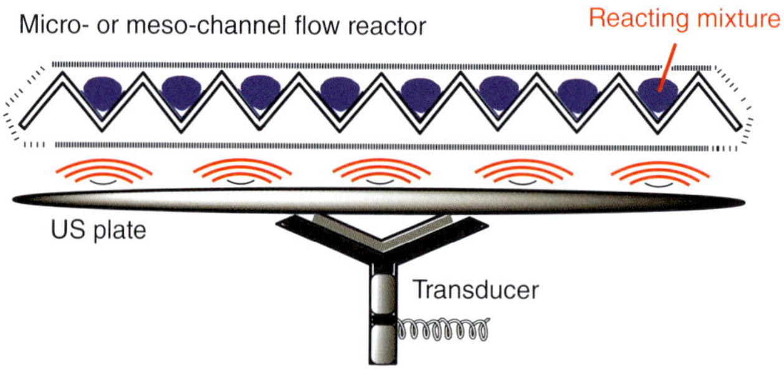

Figure 8.2 Hybrid metallic microreactor tightly fixed with an ultrasonic plate.

In recent years, chemists have paid increased attention to mechanochemistry mainly as a means to carry out solvent-free reactions. The name mechanochemistry refers to reactions performed using mechanical energy. The narrower field of reactions generated by friction during the milling of solid reagents is known as *tribochemistry* [9]. The occurrence of these reactions has been ascribed to the heat generated in the milling process and the large area of contact between the solids. At the end of nineteenth century, however, it was noticed that mechanochemical processes were different from thermal processes [11]. Mechanical defects act as high-energy structures and the role they play in chemical transformations was recognized later. The grinding of two solid substances leads to a complex series of physical and chemical transformations. Once the edges reach the point of collision, the solids deform and even melt, generating hot points where the molecules can reach very high vibrational excitation states leading to bond breakage. The use of a mortar and pestle is the simplest way to carry out mechanochemical reactions in the laboratory. When higher energy is required, ball mills and vibrators are employed. Prolonged high-energy milling as in mechanical alloying or hard crystalline solid amorphization can benefit from the use of very high-energy vibrators such as high-speed attritors or stainless steel ball mills of high impact [12].

8.2 Catalysis with Hyphenated and Tandem Techniques

Nonconventional enabling technologies such as MW, US, and HC aid process intensification, entail safer protocols, and result in cost reductions and energy savings [13]. The combination of MW and other enabling techniques gives remarkable improvements in terms of mass transfer in heterogeneous catalysis and the preparation of the supported catalyst itself [14].

Sonochemistry has a longer history than other techniques and has found a growing number of relevant applications in the last two decades. While popular wisdom simply associates MW with superior heating and US with efficient agitation, these techniques are capable of doing so much more and this potential provides additional impulse to their expansion in catalysis [15]. Cavitational reactors (US and HC) and MW heating can be coupled in numerous ways to best exploit the hot spots generated at a molecular level by both energy sources.

Sequential and simultaneous irradiation with these sources entailed technical and safety considerations in pioneering works over the last decade; their coupling has been performed efficiently at the laboratory and pilot-scale levels [16]. MW and US can be used in simultaneous and sequential modes (Figure 8.3), which either involve the use of loop reactors, with two separate compartments, or two-step treatments (pre-sonication and MW-assisted reaction) [17].

All this commercially available equipment has been used in organic synthesis with a focus on metal-assisted catalytic reactions [18].

Mechanochemical protocols, including mechanical milling and alloying, have been developed for the preparation of a wide selection of nanomaterials. Ball mills

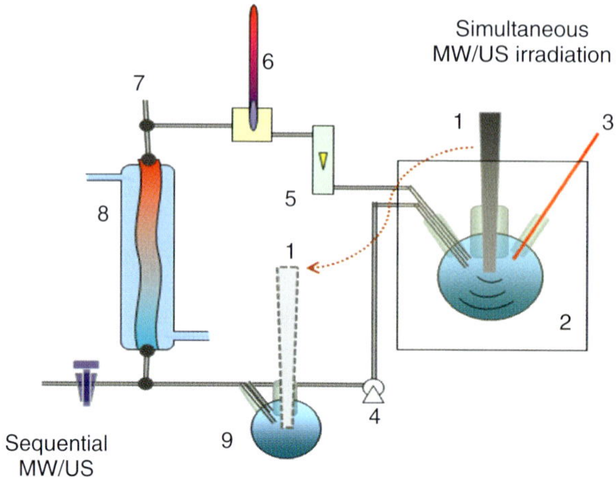

Figure 8.3 Loop MW/US reactor for simultaneous and sequential irradiation. 1, US non-metallic horn; 2, MW oven; 3, optical fiber thermometer; 4, pump; 5, flow meter; 6, thermometer; 7, inlet and sampler; 8, heat exchanger; and 9, external flask.

have been found to efficiently favor the synthesis of nanostructures, thanks to the processes of welding, particle deformation, and fracture generated by repeated ball–powder collisions. The combination of multiple collisions and high temperatures is perfect for creating fine nanoparticles/nanopowders of various materials, including a varied range of metallic substrates. Ball-milled materials have shown enhanced effects in particular when used in sequence with MW. In fact, recent years have seen the coupling of ball mill pretreatment and MW synthesis ever more used as a means to discover sustainable and alternative pathways for the preparation of efficient catalysts. Moreover, the ball mill metal-loaded catalysts usually require less metal for high activities. This characteristic is probably due to the increase in surface area and more intimate mixing generated by ball mill grinding.

MW and UV–vis irradiation are part of the electromagnetic spectrum and involve different energies and frequencies. The energy of MW is only able to promote molecular rotation and is not able to disrupt the bonds of common organic molecules, while UV holds higher energy than MW irradiation. Essentially, photoionization is responsible for chemical change, and MW radiation subsequently affects the course of the subsequent reaction. Direct and indirect effects are present in catalytic MW/UV processes. MW can interact with the catalyst, thus activating it, and UV can have a direct effect on the formation of new bonds in the course chemical reactions, dramatically enhancing the reaction rate and selectivity. MW-assisted photochemistry is frequently, but not necessarily, connected with the electrodeless discharge lamp (EDL), which generates UV radiation when placed in the MW field. EDL [19] consists of a glass tube filled with an inert gas and an excitable substance and is sealed under a lower pressure noble gas. MW can trigger gas discharge and cause the emission

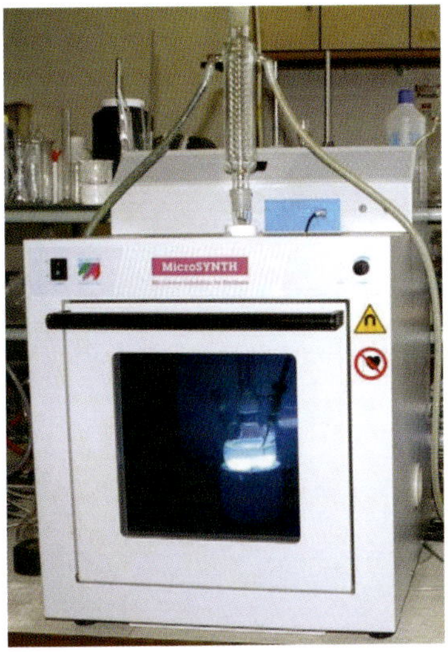

Figure 8.4 MW/UV combined technology with EDL lamps. Reproduced with permission from Ref. [22].

of electromagnetic radiation [20]. The use of MW as the energy source for electrodeless lamps was remarkably discovered by Tesla [21] in the 1890s. Only in recent years have microwave discharge electrodeless lamps (MDELs) started to find practical applications, thanks to their miniaturization as well as the lower costs of magnetrons and semiconductors [22].

The EDL lamp can be introduced into the MW cavity or directly into the reaction mixture and give simultaneous sample MW/UV irradiation. High temperatures and pressure photochemistry applications can be performed with good photochemical efficiency (as the EDL is "inside" the sample). The simplicity of the experimental arrangement and low cost of the EDL lamp make the MW/UV combined technology an efficient solution for stubborn reactions. Moreover, different EDL materials may modify its spectral output (Figure 8.4).

The use of hyphenated and tandem techniques has become popular in catalysis as it satisfies the ultimate goal of ever more selective and efficient processes, in terms of product yields and energy savings, while still meeting other green criteria, such as safety and null toxicity.

8.3 Microwave and Mechanochemical Activation

Although the simultaneous use of MW and ball mills is hindered by technical problems, caused by incompatible reactor materials, sequences of grinding plus dielectric heating are well documented, mainly in the preparation of catalysts. Lorenson *et al.* [23] showed that a decrease in particle size results in greater polarizability, higher dielectric constant, and thus higher MW penetration depth. In addition, ball mill treated samples usually display better disposition for further treatment. Li *et al.* [24] have studied the synthesis of perovskite

Figure 8.5 SEM images of Ta_2O_5 powder, ball milled for (a) 0 h, (b) 6 h, (c) 8 h, and (d) 12 h. Reproduced with permission from Ref. [25].

$NaTaO_3$ nanocrystals with Ta_2O_5 and NaOH, as starting materials, using a microwave-assisted hydrothermal (MHT) technique. They showed that, when ball-milled Ta_2O_5 was used as a starting material, the reaction occurred along a different reaction pathway with a lower thermodynamic energy barrier. Ball mill particles displayed smaller size (see Figure 8.5) and high surface area, making them easier to dissolve in alkaline solutions at certain hydrothermal temperatures and changing the common reaction pathway. $NaTaO_3$ photocatalysts obtained via the sequential use of a ball mill and MW show higher overall activity in water splitting under UV irradiation.

In general, ball milling catalyst preparation is advantageous because of the higher activity it provides, in particular in MW-assisted catalysis. In 2011, Luque *et al.* [25] proposed the first one-step mechanochemical synthesis of highly active low-loaded iron nanoparticles supported on mesoporous aluminosilicates. The catalyst was tested in the oxidation of benzyl alcohol and alkylation of toluene with benzyl alcohol under MW irradiation.

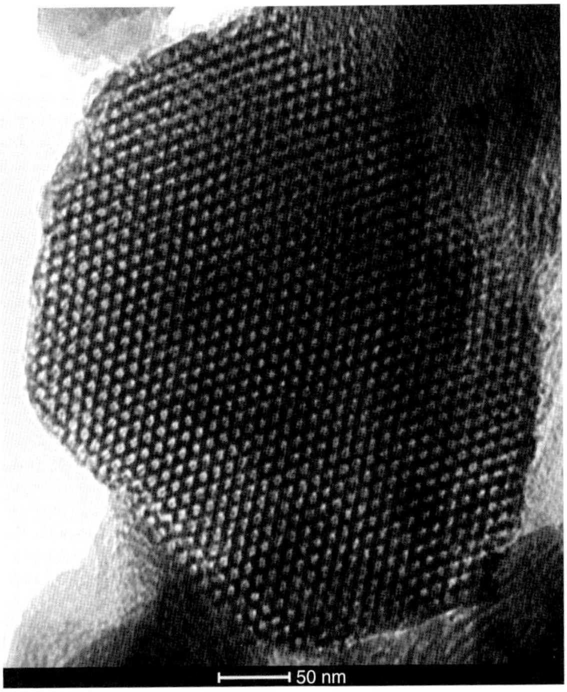

Figure 8.6 TEM micrographs of iron oxide nanoparticles on mesoporous aluminosilicates synthesized via a dry-milling protocol. Reproduced with permission from Ref. [25].

Despite the extremely low iron loading, the activity of the mechanochemically prepared catalyst was comparable to the MW prepared catalyst, but gave superior turnover frequency (TOF) values. The mechanochemical treatment allowed for the formation of highly dispersed and accessible iron oxide NPs on the edges of the support (Figure 8.6), granting surprising good catalyzing properties [26].

The same authors developed a wide range of supported metal nanoparticles on mesoporous aluminosilicates synthesized via ball mill protocols [27]. These catalysts have been proven to be highly active and selective in MW-assisted reactions of phenylacetylene and different phenols under mild reaction conditions. The outstanding value of the catalysts is increased as it is possible to reuse them at least three times while maintaining almost 90% of their initial activity (Scheme 8.1).

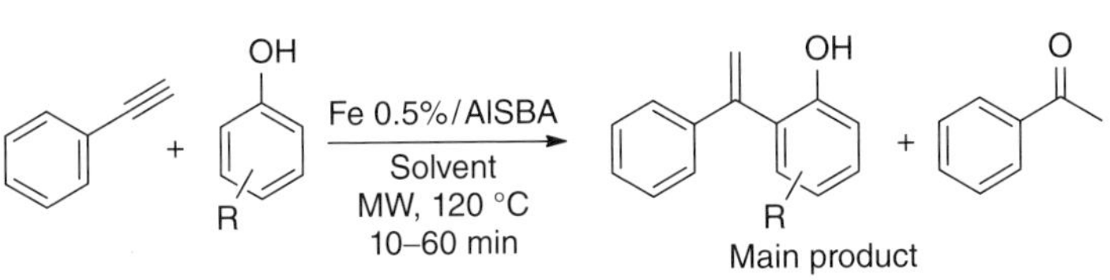

Scheme 8.1 MW-assisted preparation of 1,1-diarylsubstituted alkenes.

Nowadays, environmentally harmless energy carriers such as hydrogen play a crucial role for a renewable energy-based future [28]. The MW-assisted dehydrogenation of some metal hydrides (LiH, NaH, MgH_2, CaH_2, and TiH_2) and alkali borohydrides ($LiBH_4$, $NaBH_4$, and KBH_4) have been studied for hydrogen storage applications [29]. Ball mills have been introduced in order to add an appropriate catalyst into the system and improve kinetics and decomposition temperature. Dobbins *et al.* [30] have studied the effect of MW irradiation on the reversible desorption reaction in $NaAlH_4$ doped with 2 mol% $TiCl_2$ as catalyst and preactivated by high-energy ball milling. These promising results have recently prompted Awad *et al.* [31] to use the ball mill technique to mix carbon materials with magnesium and prevent the magnesium particles from sticking. Carbon materials acted as catalysts and improved hydrogen absorption/desorption and also worked as MW receptors. Since ball milling produced cracks in the fibers gave an intimate mixture of C/MgH_2 and introduced defects, the synthesized materials show enhanced surface properties and improved sorption kinetics [32]. Consequently, these materials presented some interesting features that may see them used as promising hydrogen storage systems.

8.4
Microwave and UV Irradiation

Over the last decade, synergistic MW and UV light have been used for the photocatalytic degradation of organic pollutants. Many studies have set up MW/UV-assisted protocols as convenient alternatives for treating/remediating environmental pollutants and provide improvements in terms of high degradation levels, short reaction times, and costs [33].

Titanium dioxide (TiO_2) is the most frequently studied stable semiconductor to be used as a photocatalyst and has already been used in photo-assisted treatments for commercial air pollution applications [34]. The method was found not to be suitable for large-scale wastewater treatment because of the poor degradation observed. Surprisingly, MW/UV coupling by MDELs gave better results. Many papers showed the use of MW-activated EDL treatments for the decomposition of different kinds of wastewater pollutants. Table 8.1 summarizes the performance of the MW/UV method in the presence of TiO_2 in the degradation treatments.

Horikoshi and coworkers [59] have published one of the most meaningful studies in this field: the efficiency of MW-TiO_2 photo-assisted reactions was improved by the selective MW absorption. Only when the aqueous TiO_2 dispersion was exposed to simultaneous MW and UV light irradiation, the degradation of the dye was fast and significant. The same group have set up an integrated MW/photoreactor system for wastewaters in order to better understand the degradation pathway and the nonthermal MW effect [60] (see Chapter 19).

These authors demonstrated that the MW's influence on the degradation process was due to the MW's peculiar ability to modify the hydrophilic/hydrophobic

Table 8.1 List of pollutants subjected to degradation treatments using microwave/UV irradiation (MW/UV) in the presence of TiO_2.

Pollutants	MDEL	MW power (W)[a]	Degradation extent (%)	Treatment time (min)	Enhancement factor	References
Acetaldehyde[b] (1000 mg m^{-3})	On	700	77	21	—[c]	[35]
Acid orange (750 ml, 100 mg l^{-1})	On	700	100	150	—[c]	[36]
Alizarin green (1000 ml, 50 mg l^{-1})	On	700	100	90	—[c]	[37]
Atrazine (50 ml, 20 mg l^{-1})	On	—	100	5	—[c]	[38]
Atrazine (50 ml, 12 mg l^{-1})	On	900	97	14	—[c]	[39]
Benzoic acid (30 ml, 0.1 mmol l^{-1})	Off	220	40[d]	120	—[c]	[40]
Bisphenol A (30 ml, 0.1 mmol l^{-1})	Off	300	90	120	1.4	[41]
Chloroacetic acid (150 ml, 0.1 mol l^{-1})	On	1000	100	490	—[c]	[42]
Chloroacetic acid (150 ml, 25 mmol l^{-1})	On	1000	100	140	—[c]	[43]
4-Chlorophenol (30 ml, 0.1 mol l^{-1})	Off	220	90	120	1.4	[44]
4-Chlorophenol (500 ml, 30 mg l^{-1})	On	750	85	120	2.2	[45]
p-Cresol (0.35 l, 0.1 mmol l^{-1})	Off	300	90	120	3	[46]
2,4-Dichlorophenoxyacetic acid (30 ml, 0.05 mmol l^{-1})	On	80	88	10	4	[47]
Malachite green (50 ml, 20 mg l^{-1})	On	900	99	3	—[c]	[48]
Malachite green (30 ml, 20 mg l^{-1})	On	750	98	10	2.25[e]	[49]
Methylene blue (50 ml, 100 mg l^{-1})	On	900	100	15	—[c]	[50]
Pentachlorophenol (5 ml, 40 mg $l^{-1)}$	On	900	94	20	—[c]	[51]
Phenol (0.35 l, 0.1 mmol l^{-1})	On	300	78	120	2.2	[52]
Reactive brilliant red (750 ml, 400 mg l^{-1})	On	700	100	180	1.8	[53, 54]
Rhodamine B (30 ml, 0.05 mmol l^{-1})	Off	300	100	180	3	[55]
Rhodamine B (150 ml, 0.05 mmol l^{-1})	On	180	100	30	—[c]	[56]
Rhodamine B (30 ml, 0.05 mmol l^{-1})	Off	250	100	100	1.4	[57]
Rhodamine B (600 ml, 50 mg l^{-1})	On	15	95	360	—[c]	[58]

a) MWs generated at 2.45 GHz.
b) Volume not reported.
c) No direct comparison was made regarding the UV/TiO_2 process alone.
d) Measured at 198 nm.
e) Compared with naked $BiWO_4$.

characteristics of the TiO_2 particle surface, which changes the manner in which the target molecules were adsorbed onto the surface. Evidence of the efficient rhodamine B degradation was clear, thanks to solution color intensity after different treatments. This hybrid technique may play an important physical catalyst role in advanced oxidation technologies (AOTs).

8.5
Microwave and Ultrasound

The combined use of MW and US as efficient hybrid technologies was first employed in the mid-1990s [61]. The main fields of application are biodiesel production, biowaste valorization, preparation of nanomaterials, and organic synthesis.

Biodiesel can be produced from a variety of feedstock that contains triglycerides via a well-known process that usually involves alkaline-catalyzed transesterification with methanol [62]. In 2010, Lin *et al.* [63] used sequential US and MW and achieved very high conversion rates, namely 97.7%, after 1 min sonication and 2 min MW irradiation in a closed vessel (NaOH 1.0 wt%, 60 °C, 6 : 1 methanol/oil molar ratio) [64]. Biodiesel synthesis from nonedible oils and waste cooking oils has been carried out using sequential MW–US treatments [65]. The synergistic interaction between the physical effects of US microemulsification, acoustic streaming, and rapid MW heating afforded a significant degree of process intensification. Gnaneswar Gude *et al.* have recently proposed a novel application for simultaneous MW/US irradiation in the transesterification of waste vegetable oil. Two different catalysts were tested under the same conditions: 6 : 1 ratio MeOH/oil, 60 °C, MW (100 W), and US (100 W) for only 2 min reaction time. A 98% yield was obtained in homogeneous catalysis conditions using NaOH 0.75 wt% [66], whereas the yield was slightly lower (96%) in the heterogeneous phase when barium oxide 0.75 wt% was used [67]. Moreover, this technique usually provides reduced biodiesel synthesis operating costs and overall favorable economics.

Biomass is a valuable feedstock material for industry. An interesting study has presented the catalytic conversion of starch-based industrial waste into sugars using a simultaneous MW/US-assisted procedure [68]. The three raw materials (potato starch and wet/dry potato sludge) were irradiated for 2 h at 60 °C in H_2SO_4 3 or 13 mol l^{-1}. Depolymerization was attributed to a thermally controlled process since it gave the highest total reduced sugars value, whereas US irradiation displayed poor results. A fascinating feature of combined MW/US irradiation, by means of a Pyrex horn inserted into a multimode MW oven from the top, was the process ability to selectively furnish glucose. A synergistic effect on depolymerization rate was detected and let to a higher yield and a shorter reaction time.

Great interest has recently been shown in ecocompatible micro- or nanomaterial synthesis. The use of nontoxic chemicals, environmental friendly solvents, and renewable resources are meaningful points that merit important consideration in green synthetic strategy [69]. The last 10 years has seen several papers tackle these issues using a US/MW system [70]. The two main features of the sequential and simultaneous MW/US techniques are the possibility to achieve geometry control and increase the photocatalytic activity of the synthesized nanoparticle. In some cases, US/MW-assisted synthesis has replaced conventional heating methods and has afforded energy savings and shorter reaction times. For example, $SrWO_4:Ln^{3+}$ (Eu^{3+}, Tb^{3+}, Sm^{3+}, and Dy^{3+}) phosphors have been successfully synthesized via US/MW treatments at very low temperatures (70 °C) in only 45 min, avoiding the high temperatures and long reaction times typically used in conventional heating methods, such as sintering (above 500 °C for more than 4 h) and hydrothermal protocols (above 100 °C overnight) [71]. In addition, the photocatalytic performance and emission properties of MW/US synthesized particles were improved as compared to their traditionally prepared analogs. These enhanced characteristics have been shown in many studies for different catalyst nanoparticle types such as $V_2O_5/MgF_2:Eu^{3+}$ [72], monoclinic $BiVO_4$ [73], ZnO [74], and silver vanadates [75]. In 2014, Su *et al.* [76] reported an interesting example of controlled individual CuO nanosheet and nanosheet-built dendrite synthesis. MW hydrothermal treatment contributed to the disassembly of sheet-built CuO dendrites for the formation of discrete CuO nanosheets. Such a disassembly process was, however, inhibited by US heat pretreatment (Figure 8.7).

MW/US had already been proposed for the preparation of porous square-brick-like CuO nanostructures via a CuC_2O_4 intermediate by Zhu *et al.* [77]. The authors underlined how significant US processing's role was on the formation of the square-brick-like morphology.

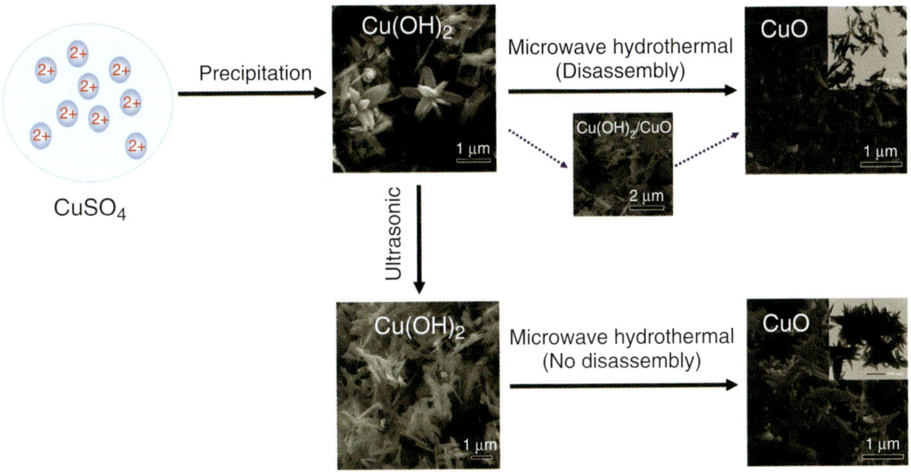

Figure 8.7 Schematic illustration of CuO nanosheets and dendrites prepared using MW hydrothermal synthesis. Reproduced with permission from Ref. [76].

The combined use of MW and US has also been used in several C–C coupling (Suzuki, Heck reaction) and click chemistry (CuAAC, Cu-catalyzed azide-alkyne cycloaddition) reactions. A representative example of sequential irradiation has been proposed in which biaryls in aqueous media were obtained from a flow reactor that was thermostated at 45 °C, under combined US (20.5 kHz, 40 W cm^{-2}) and MW (700 W) irradiation and using 10% Pd/C as catalyst. The Pd-catalyzed homo- and cross-couplings of boronic acids and aryl halides provide the desired product after only 60 min irradiation [78]. The use of simultaneous MW/US in a Heck reaction afforded the product in high yields and very low ligandless catalyst loads [3] (Scheme 8.2).

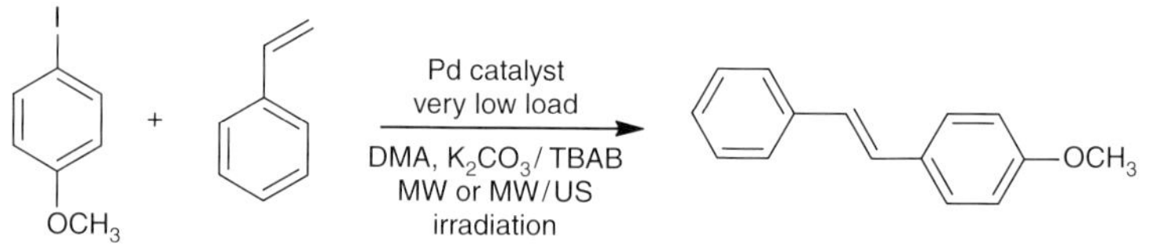

Scheme 8.2 Heck reactions under MW or MW/US irradiation.

The combined MW/US system was once again successfully employed in CuAAC, which made possible the use of metallic copper powder as the catalyst instead of copper salts. US irradiation has also been applied to activating copper metal surfaces [79]. This synthetic protocol was also used to functionalize more complex molecules, such as cyclodextrins, without losing efficiency, giving high yields (>80%) in much shorter times than under conventional heating [80].

8.6 Conclusions

Complimentary and synergistic effects obtained from the combination of MW and other enabling technologies such as US, UV rays, and mechanochemistry have been described in this chapter. The combination of these technologies has helped scientists to find answers to new synthetic challenges and favored the development of readily available efficient catalysts. The design of new catalytic and stoichiometric strategies in synthesis will also benefit from green and mild technologies which are capable of increasing yields and catalytic turnovers, as well as potentially modifying the reactivity of substrates and the role of solvents.

References

1. Wu, Z., Borretto, E., Cravotto, G., Medlock, J., and Bonrath, W. (2014) *ChemCatChem*, **6** (10), 2762–2783.
2. Cravotto, G., Bonrath, W., Tagliapietra, S., Speranza, C., Calcio Gaudino, E., and Barge, A. (2010) *Chem. Eng. Process. Process Intensif.*, **49**, 930–935.
3. Palmisano, G., Bonrath, W., Boffa, L., Garella, D., Barge, A., and Cravotto, G.

(2007) *Adv. Synth. Catal.*, **349**, 2338–2344.
4. Barge, A., Tagliapietra, S., Tei, L., Cintas, P., and Cravotto, G. (2008) *Curr. Org. Chem.*, **12**, 1588–1612.
5. Cravotto, G. and Cintas, P. (2012) in *Microwaves in Organic Synthesis*, Chapter 3, 3rd edn, vol. **1** (eds A. De La Hoz and A. Loupy), Springer Science, New York, pp. 541–562.
6. Cravotto, G. and Cintas, P. (2007) *Chem. Eur. J.*, **13** (7), 1902–1909.
7. Cravotto, G. and Cintas, P. (2009) *Chem. Soc. Rev.*, **38**, 2684–2697.
8. Cravotto, G. and Cintas, P. (2006) *Chem. Soc. Rev.*, **35** (2), 180–196.
9. (a) Rodriguez, B., Bruckmann, A., Rantanen, T., and Bolm, C. (2007) *Adv. Synth. Catal.*, **349**, 2213–2233; (b) Kauppa, G. (2008) *J. Phys. Org. Chem.*, **21**, 630–643; (c) Stolle, A., Szuppa, T., Leonhardt, S.E.S., and Ondruschka, B. (2011) *Chem. Soc. Rev.*, **40**, 2317–2329.
10. Fernández Rivas, D. (2012) Taming acoustic cavitation. PhD thesis. University of Twente (NL), Chapter 2, p. 10, doi: 10.3990/1.9789036534192.
11. Heinicke, G. (1984) *Tribochemistry*, Akademic-Verlag, Berlin.
12. Schneider, F., Szuppa, T., Stolle, A., Ondruschka, B., and Hopf, H. (2009) *Green Chem.*, **11**, 1894–1899.
13. Cravotto, G., Tagliapietra, S., Caporaso, M., Garella, D., Borretto, E., and Stilo, A. (2013) *Chem. Heterocycl. Compd.*, **49**, 811–826.
14. Wu, Z., Cherkasov, N., Cravotto, G., Borretto, E., Ibhadon, A.O., Medlock, J., and Bonrath, W. (2015) *ChemCatChem*, 7, 952–959.
15. Leonelli, C. and Mason, T.J. (2010) *Chem. Eng. Process.*, **49**, 885–900.
16. Cintas, P., Cravotto, G., and Canals, A. (2012) in *Handbook on Applications of Ultrasound: Sonochemistry for Sustainability*, Chapter 25 (eds D. Chen, S.K. Sharma, and A. Mudhoo), CRC Press, Boca Raton, FL, pp. 659–673.
17. Cravotto, G., Garella, D., Calcio Gaudino, E., and Lévêque, J.-M. (2008) *Chem. Today*, **26**, 39–41.
18. Cintas, P., Carnaroglio, D., Rinaldi, L., and Cravotto, G. (2012) *Chem. Today*, **30**, 33–35.
19. Fehsenfeld, F.C., Evenson, K.M., and Broida, H.P. (1965) *Rev. Sci. Instrum.*, **36**, 294–298.
20. Brown, S.C. (1996) *Introduction to Electrical Discharges in Gases*, John Wiley & Sons, Inc.
21. Tesla, N. (1891) *Lecture delivered before the American Institute of Electrical Engineers*, Columbia College, New York.
22. Horikoshi, S., Abe, M., and Serpone, N. (2009) *Photochem. Photobiol. Sci.*, **8**, 1087–1104.
23. Claude, P., Lorenson, C.L., Patterson, M., Risto, G., and Kimber, R. (1992) *Mater. Res. Soc. Symp. Proc.*, **269**, 129–135.
24. Shi, J., Liu, G., Wangab, N., and Li, C. (2012) *J. Mater. Chem.*, **22**, 18808–18813.
25. Pineda, A., Balu, A.M., Campelo, J., Romero, A.A., Carmona, D., Balas, F., Santamaria, J., and Luque, R. (2011) *ChemSusChem*, **4**, 1561–1565.
26. Pineda, A., Balu, A.M., Campelo, J.M., Luque, R., Romero, A.A., and Serrano-Ruiz, J.C. (2012) *Catal. Today*, **187**, 65–69.
27. Hosseinpour, R., Pineda, A., Garcia, A., Romero, A.A., and Luque, R. (2014) *Catal. Commun.*, **48**, 73–77.
28. Lee, D.-H. and Hung, C.-P. (2012) *Int. J. Hydrogen Energy*, **37**, 15753–15765.
29. Nakamori, Y., Orimo, S., and Tsutaoka, T. (2006) *Appl. Phys. Lett.*, **88**, 1–3.
30. Krishnan, R., Agrawal, D., and Dobbins, T. (2009) *J. Alloys Compd.*, **470**, 250–255.
31. Awad, A.S., Tayeh, T., Nakhl, M., Zakhour, M., Ourane, B., Le Troëdec, M., and Bobet, J.-L. (2014) *J. Alloys Compd.*, **607**, 223–229.
32. Ouyang, L.Z., Cao, Z.J., Wang, H., Liu, J.W., Sun, D.L., Zhang, Q.A., and Zhu, M. (2014) *J. Alloys Compd.*, **586**, 113–117.
33. Nascimento, U.M. and Azevedo, E.B. (2013) *J. Environ. Sci. Health., Part A*, **48**, 1056–1072.
34. Fujishima, A., Hashimoto, K., and Watanabe, T. (1999) *TiO_2 Photocatalysis Fundamentals and Applications*, BKC Publications, Tokyo.
35. Horikoshi, S., Kajitani, M., Horikoshi, N., Dillert, R., and Bahnemann, D.W.

(2008) *J. Photochem. Photobiol., A*, **193**, 284–287.
36. Zhang, X.W., Sun, D.D., Li, G.T., and Wang, Y.Z. (2008) *J. Photochem. Photobiol., A*, **199**, 311–315.
37. Xiong, Z., Xu, A., Li, H., Ruan, X., Xia, D., and Zeng, Q. (2013) *Ind. Eng. Chem. Res.*, **52**, 362–369.
38. Gao, Z., Yang, S., Na, T., and Sun, C. (2007) *J. Hazard. Mater.*, **145**, 424–430.
39. Qin, C., Yang, S., Sun, C., Zhan, M., Wang, R., Cai, H., and Zhou, J. (2010) *Front. Environ. Sci. Eng. China*, **4**, 321–328.
40. Horikoshi, S., Hojo, F., Hidaka, H., and Serpone, N. (2004) *Environ. Sci. Technol.*, **38**, 2198–2208.
41. Horikoshi, S., Tokunaga, A., Hidaka, H., and Serpone, N. (2004) *J. Photochem. Photobiol., A*, **162**, 33–40.
42. Cırkva, V., Zabova, H., and Hıajek, M. (2008) *J. Photochem. Photobiol., A*, **198**, 13–17.
43. Kmentova, H. and Cırkva, V. (2013) *J. Chem. Technol. Biotechnol.*, **88**, 1109–1113.
44. Dowey, W.F. Jr., (1995) Water decontamination using a peroxide photolysis ionizer. US Patent 5, 439, 595.
45. Ai, Z.H., Yang, P., and Lu, X.H. (2005) *Chemosphere*, **60**, 824–927.
46. Karthikeyan, S. and Gopalakrishnan, A.N. (2011) *Indian J. Chem. Technol.*, **18**, 458–462.
47. Horikoshi, S., Kajitani, M., Sato, S., and Serpone, N. (2007) *J. Photochem. Photobiol., A*, **189**, 355–363.
48. Ju, Y.M., Yang, S.G., Ding, Y.C., Sun, C., Zhang, A., and Wang, L. (2008) *J. Phys. Chem. A*, **112**, 11172–11177.
49. Ju, Y., Hong, J., Zhang, X., Xu, Z., Wei, D., Sang, Y., Fang, X., Fang, J., and Wang, Z. (2012) *J. Environ. Sci.*, **24** (12), 2180–2190.
50. Hong, J., Sun, C., Yang, S., and Liu, Y. (2006) *J. Hazard. Mater.*, **133**, 162–166.
51. Gao, Z., Yang, S., Sun, C., and Hong, J. (2007) *Sep. Purif. Technol.*, **58**, 24–31.
52. Karthikeyan, S. and Gopalakrishnan, A.N. (2011) *J. Sci. Ind. Res.*, **70**, 71–76.
53. Zhang, X.W., Wang, Y.Z., and Li, G.T. (2005) *J. Mol. Catal. A: Chem.*, **237**, 199–205.
54. Zhang, X.W., Li, G.T., and Wang, Y.Z. (2007) *Dyes Pigm.*, **74**, 536–544.
55. Horikoshi, S., Hidaka, H., and Serpone, N. (2002) *Environ. Sci. Technol.*, **36**, 1357–1366.
56. Horikoshi, S., Hidaka, H., and Serpone, N. (2002) *J. Photochem. Photobiol., A*, **153**, 185–189.
57. Horikoshi, S., Hidaka, H., and Serpone, N. (2003) *J. Photochem. Photobiol., A*, **159**, 289–300.
58. Li, L., Wang, L., Hu, T., Zhang, W., Zhang, X., and Chen, X. (2014) *J. Solid State Chem.*, **218**, 81–89.
59. Horikoshi, S., Kajitani, M., Hidaka, H., and Serpone, N. (2008) *J. Photochem. Photobiol., A*, **196**, 159–164.
60. Horikoshi, S. and Serpone, N. (2014) *Catal. Today*, **224**, 225–235.
61. Maeda, M. and Amemiya, H. (1995) *New J. Chem.*, **19**, 1023–1028.
62. Sharma, Y.C., Singh, B., and Upadhyay, S.N. (2008) *Fuel*, **87**, 2355–2373.
63. Hsiao, M.-C., Lin, C.-C., Chang, Y.-H., and Chen, L.-C. (2010) *Fuel*, **89**, 3618–3622.
64. Hsiao, M.-C. and Lin, C.-C. (2013) *Int. J. Energy Power Eng.*, **2** (2), 54–59.
65. Vitthal, L., Parag, G., and Gogate, R. (2013) *Fuel Process. Technol.*, **106**, 62–69.
66. Martinez-Guerra, E. and Gnaneswar Gude, V. (2014) *Fuel*, **137**, 100–108.
67. Martinez-Guerra, E. and Gnaneswar Gude, V. (2014) *Energy Convers. Manage.*, **88**, 633–640.
68. Hernoux-Villiere, A., Lassi, U., Hu, T., Paquet, A., Rinaldi, L., Cravotto, G., Molina-Boisseau, S., Marais, M.-F., and Leveque, J.-M. (2013) *ACS Sustainable Chem. Eng.*, **1**, 995–1002.
69. Raveendran, P., Fu, J., and Wallen, S.L. (2003) *J. Am. Chem. Soc.*, **125**, 13940–13941.
70. Boffa, L., Tagliapietra, S., and Cravotto, G. (2013) in *Microwaves in Nanoparticles Synthesis*, Chapter 11 (eds S. Horikoshi and N. Serpone), Wiley-VCH Verlag GmbH, Weinheim, p. 55.
71. Zheng, Y., Lin, J., and Wang, Q. (2012) *Photochem. Photobiol. Sci.*, **11**, 1567–1574.

72. Lin, J., Pang, S., and Wang, Q. (2013) *Mater. Lett.*, **98**, 12–14.
73. Zhang, Y., Li, G., Yang, X., Yang, H., Lu, Z., and Chen, R. (2013) *J. Alloys Compd.*, **551**, 544–550.
74. Li, Q., Li, H., Wang, R., Li, G., Yang, H., and Chen, R. (2013) *J. Alloys Compd.*, **567**, 1–9.
75. Vu, T.A., Dao, C.D., Hoang, T.T.T., Dang, P.T., Tran, H.T.K., Nguyen, K.T., Le, G.H., Nguyen, T.V., and Lee, G.D. (2014) *Mater. Lett.*, **123**, 176–180.
76. Yang, C., Wang, J., Xiao, F., and Su, X. (2014) *Powder Technol.*, **264**, 36–42.
77. Qiu, M., Zhu, L., Zhang, T., Li, H., Sun, Y., and Liu, K. (2012) *Mater. Res. Bull.*, **47**, 2437–2441.
78. Cravotto, G., Beggiato, M., Penoni, A., Palmisano, G., Tollari, S., Leveque, J.-M., and Bonrath, W. (2005) *Tetrahedron Lett.*, **46**, 2267–2271.
79. Cintas, P., Barge, A., Tagliapietra, S., Boffa, L., and Cravotto, G. (2010) *Nat. Protoc.*, **5**, 607–616.
80. Barge, A., Tagliapietra, S., Binello, A., and Cravotto, G. (2011) *Curr. Org. Chem.*, **15**, 189–203.

Part IV
Applications – Organic Reactions

Microwaves in Catalysis: Methodology and Applications, First Edition.
Edited by Satoshi Horikoshi and Nick Serpone.

9
Applications of Microwave Chemistry in Various Catalyzed Organic Reactions

Rick Arneil Desabille Arancon, Antonio Angel Romero, and Rafael Luque

9.1
Introduction

Microwave chemistry is an emerging field of science mainly dealing with the diverse applications of microwave energy to chemical processes. Recent publications in the field have revealed that microwave energy can significantly increase the rate of product formation of a chemical reaction under specific conditions. The improved heating rate of microwave-assisted reactions results in higher product yield, greater catalyst turnover, milder reaction conditions, and faster reaction times. The behavior of chemical species under microwave conditions cannot be explained solely by rapid heating of the system, it is believed that the "microwave effect" is a synergistic effect of both thermal and nonthermal effects [1].

Microwave reactors, as well as commercial microwaves ovens, operate within ~2.4 GHz. At this frequency, some microwave energy absorbed by the species present in a sample can be dissipated as heat (dielectric loss, Figure 9.1) which, in turn, causes the energy transmission among molecules [2]. As opposed to the convection phenomenon observed in conventional heating, this energy transmission is therefore dependent on factors such as polarizability and electrical conductivity. Because of these, microwave-assisted reactions are mostly performed in the presence of ionic/polar solvents to facilitate the energy transfer [3]. And since the effectivity of microwave energy to cause rapid heating is dependent on specific compound properties, reactions can therefore be made selective and performed under high pressures without damaging the microwave-transparent reaction vessel. Table 9.1 shows the differences between microwave and conventional heating.

Many researchers argue that the improved yield and turnover obtained with microwave-assisted reactions is a combination of both thermal and nonthermal effects. Recent studies, however, show that there could be no known nonthermal microwave effect that produces the desired results [4]. In elegant studies conducted by Kappe *et al.* [5], it was found that the significant rate enhancements in microwave-assisted reactions are only attributed to the selective heating of

Microwaves in Catalysis: Methodology and Applications, First Edition.
Edited by Satoshi Horikoshi and Nick Serpone.

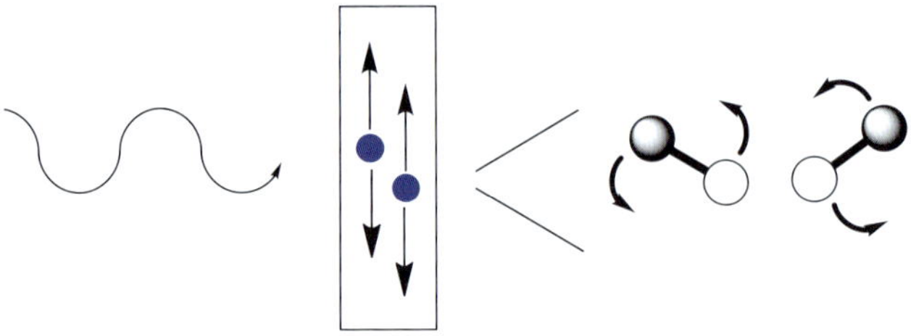

Figure 9.1 The introduction of microwaves to a sample causes increased vibrations among the molecules, which in turn leads to improved heat transfer. Image reproduced from Ref. [2] with permission.

Table 9.1 The differences between the properties of microwave heating and conventional heating.

Microwave heating	Conventional heating
Energetic coupling	Conduction/convection
Coupling at the molecular level	Superficial heating
Rapid	Slow
Volumetric	Superficial
Selective	Nonselective
Dependent on the properties of the material	Less dependent

Reproduced from Ref. [1] with permission.

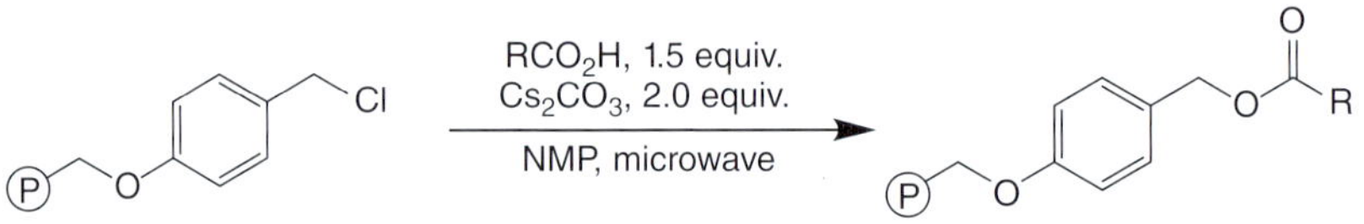

Figure 9.2 Reaction scheme used to compare microwave heating from conventional heating (Biginelli reaction).

the solvent, and not on specific nonthermal effects (Figure 9.2). To arrive at this conclusion, carboxylation of several *chloromethylated* polymeric resins were studied under microwave and under conventional heating using the same temperature (80 °C) and other reaction parameters. Obtaining a similar kinetic profile in both with almost no differences, it is apparent that the enhancement of rate for the microwave-assisted reaction is caused solely by a more rapid heating of the system. The main critical factor affecting this is the dielectric loss of the solvent – its ability to convert microwave energy into heat.

In spite of the current debate about the existence of microwave nonthermal effects in assisted reactions, the technique has become a widely applied method in the syntheses and transformations of different organic compounds. In this chapter, we aim to provide a survey of microwave-assisted organic reactions performed in both organic and aqueous solvents.

Table 9.2 List of solvents commonly used for microwave and their corresponding dielectric constants [7, 8].

Solvent	tan δ
Ethylene glycol	1.350
Ethanol	0.941
DMSO	0.825
Methanol	0.659
1,2-Dichlorobenzene	0.280
Acetic acid	0.174
DMF	0.161
1,2-Dichloroethane	0.127
Water	0.123
Chloroform	0.091
Acetonitrile	0.062
Acetone	0.054
Tetrahydrofuran	0.047
Dichloromethane	0.042
Toluene	0.040
Hexane	0.020

9.1.1 Homogeneous Catalysis

Traditionally, most organic reactions are performed using a homogeneous/heterogeneous catalyst for a few hours. The convenience of microwave irradiation has allowed these types of syntheses to be performed in only a few minutes. As previously discussed, because microwave heating is greatly dependent on polarity, organic reactions can also be controlled by carefully selecting the dielectric properties of solvents and other reaction contents. Apart from this, the ability of microwaves to increase the solubility of organic reactants to aqueous media [6] has provided chemists with the opportunity to produce methods solely based on water as a solvent. Water itself is an excellent solvent for organic reactions because it can easily be separated from the organic products, and it can easily be purified and recycled. Although water has a not so high tangent loss (Table 9.2), it is essential to use a more benign solvent such as water to avoid more dangerous organic ones.

9.2 Microwave-Assisted Reactions in Organic Solvents

One of the most practical reasons for employing microwave irradiation in an organic reaction is to vastly accelerate the process from hours to only within minutes. This ability of microwave-assisted processes has given chemists more efficiency in studying less kinetically favored reactions. An example is the gold(I)-catalyzed intramolecular hydroamination reaction reported by Liu *et al.* [9]. In this work, the reaction was run under normal heating and microwave. For both setups, the conditions involved the use of the (PR3)AuOTf complex and toluene solvent (Figure 9.3). Comparing microwave-assisted and direct heating, improvement of the reaction is very noticeable with respect to both time and

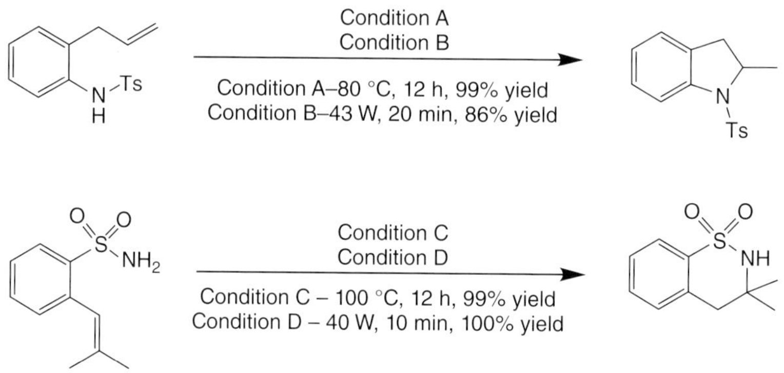

Figure 9.3 The gold-catalyzed cyclization of various aromatic compounds comparing direct heating and microwave irradiation.

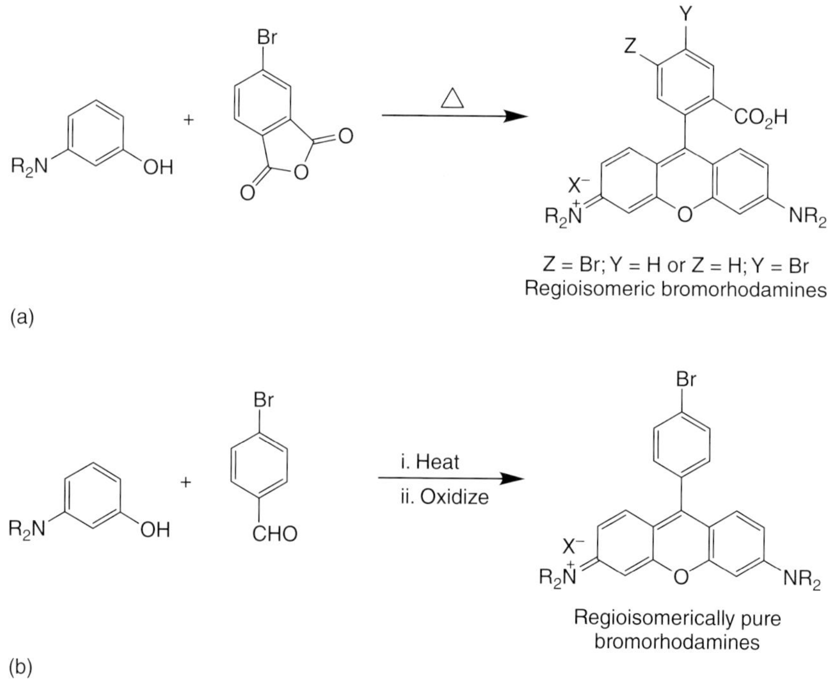

Figure 9.4 The use of microwave in the synthesis of bromorhodamines was made regiospecific by employing microwave irradiation.

yield. When direct heating was used, almost quantitative yield was obtained after 12 h of reaction time (at 100 °C); but this was cut to only 10 min (at 140 °C) using microwave (Conditions C and D).

As previously mentioned, employing microwave to an organic reaction could also improve the selectivity of a reaction to a specific product. Such ability of microwave radiation is excellently illustrated by the work of Burgess *et al.* [10] in which they were able to synthesize specific bromorhodamine regioisomers using the technique.

As seen in Figure 9.4, by changing the starting bromobenzene substrate, a specific regioisomer was obtained; and by employing microwave, the reaction time and yield of the product in Figure 9.4b was drastically improved (Table 9.3).

Table 9.3 Isolated yields of the dyes under microwave and direct heating (dyes drawn below).

Dye	Microwave		Thermal	
	Temperature (°C), time (min)	Yield (%)	Temperature (°C), time (min)	Yield (%)
1	150, 20	27	160, 24	8
2	90, 30	41	160, 22	12
3	90, 30	38	90, 18	35
	150, 10	73		
4	150, 10	53	160, 24	5

Thermal conditions performed with 60% H_2SO_4. After microwave, 2 equiv. of chloranil was added to the mixture to ensure complete oxidation and then the products were isolated via flash chromatography.

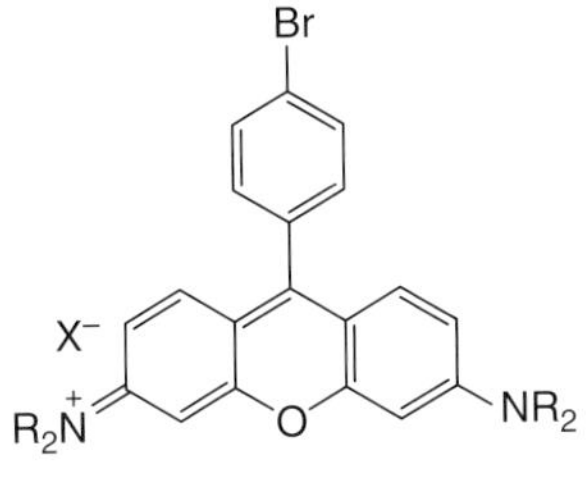

1R = H; 2R = Me

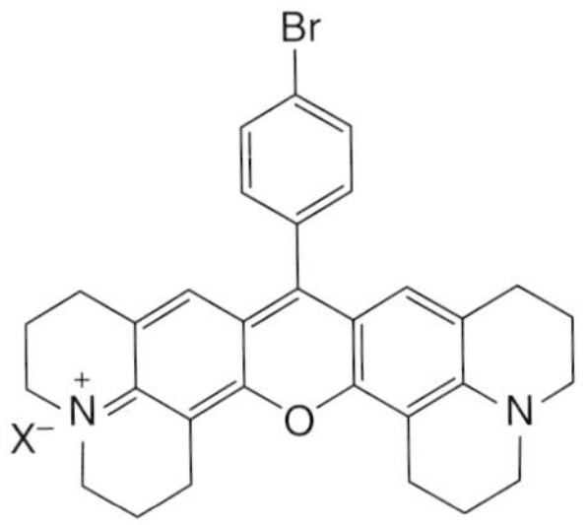

3

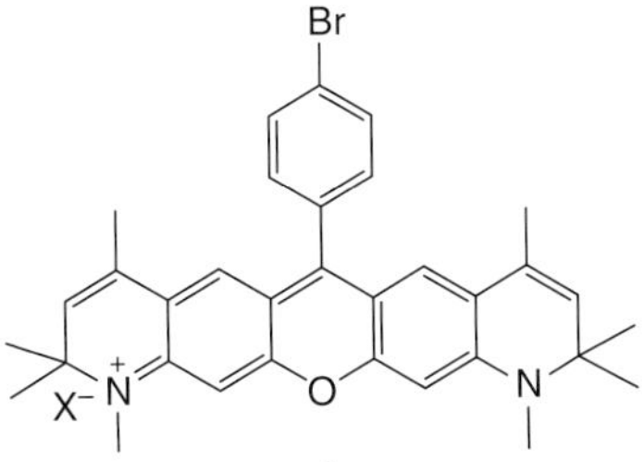

4

The high-throughput turnover of microwave-assisted organic synthesis (MAOS) had also advanced the area of drug discovery in many ways [11]. For one, the very short reaction times of MAOS have allowed the synthesis of many potential drugs and treatments with very small half-life such as isotope-labeled compounds. Also, the high rate and selectivity of MAOS opened the door for the rapid synthesis of drug candidates cutting the synthesis time to more than 50%. Case in point is the work of Besson *et al.* [12] about the synthesis of thiazolo[5,4-*f*]quinazolines. As seen in Figure 9.5, the reactions were performed using the same steps, but the reaction times to produce a good yield were reduced from hours to only within minutes [12].

Similarly, drug precursors such as flavones and chromones have also been synthesized using microwaves (Figure 9.6). In the work of Kabalka and Mereddy [13], these polyketide natural products are in very high yields for only 5 min as compared to 2–6 h of synthesis reported with direct heating [14].

Further highlighting the effect of microwave on natural product synthesis and drug discovery, the technique has also been used for the one-pot synthesis of more complex heterocycles [15]. One example is the three-component synthesis of dihydroazolopyrimidines, which are known cardiovascular vasodilators, and calcium and potassium-channel inhibitors. In the work of Kappe *et al.* [15],

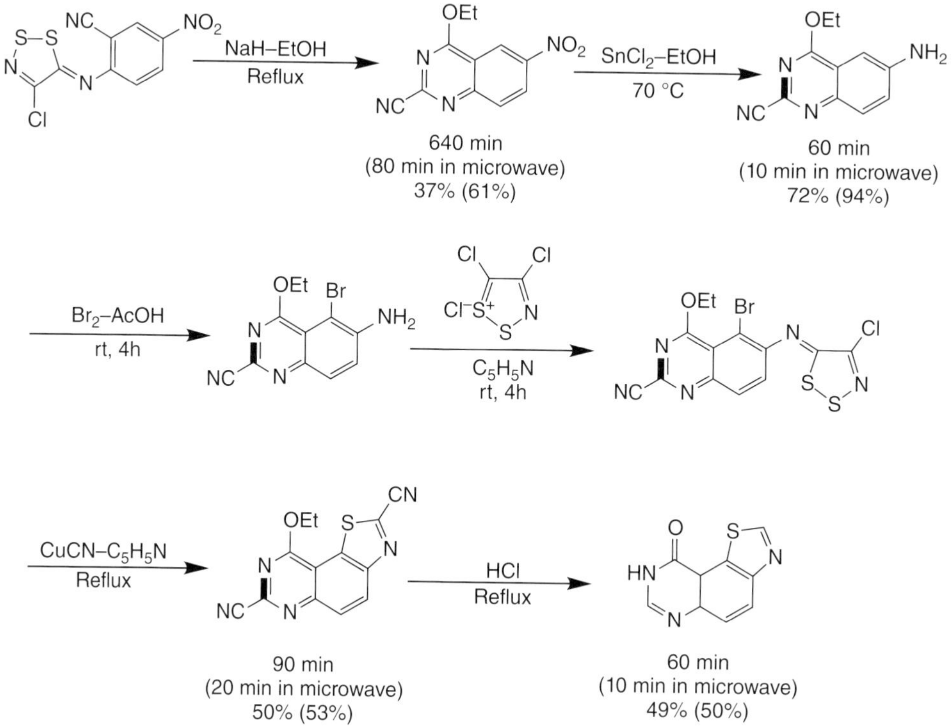

Figure 9.5 Synthesis of thiazol[5,4-*f*]quinazoline (numbers in parentheses indicate the time and yield for microwave-assisted reactions).

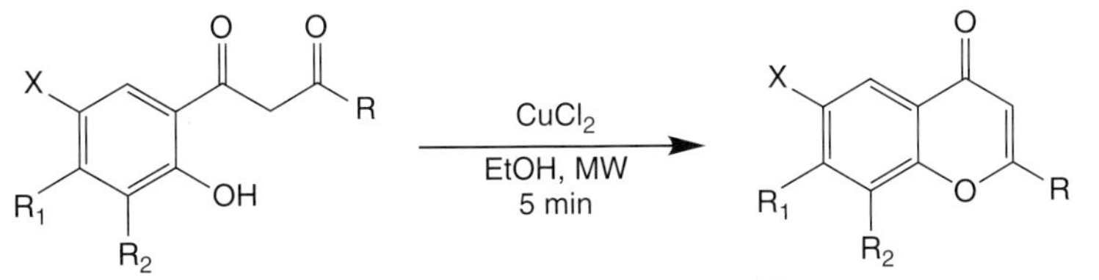

Figure 9.6 Synthesis of flavones and chromones.

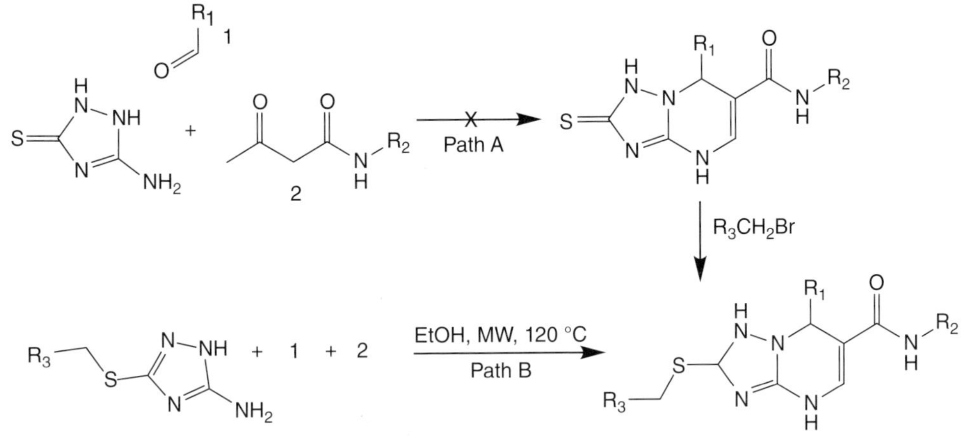

Figure 9.7 Synthesis of 7-aryl-2-alkylthio-4,7-dihydro-1,2,4-triazolo[1,5-*a*]pyrimidine-6-carboxamide using microwave irradiation.

7-aryl-2-alkylthio-4,7-dihydro-1,2,4-triazolo[1,5-*a*]pyrimidine-6-carboxamide systems were synthesized by the condensation of 3-amino-1,2,4-triazol-5-thione derivatives with aromatic aldehydes and acetoacetamides under microwave conditions. These compounds are very closely related to the mentioned drugs with biological activities and they have functional groups that can be readily modified for diversity (Figure 9.7). In this work, the selectivity of microwave irradiation is also exemplified when the reactions with microwave were compared to directly heated reactions. The time needed for microwave irradiation was set at 5 min, while the reflux under conventional heating only lasted for 10 min. There may not be large differences in yield, but the purity of the obtained crystals were higher for microwave assisted than conventionally heated.

9.3
Microwave-Assisted Reactions in Water-Coupling Reactions

So far, the reactions mentioned in the discussion have been MAOS in organic media. However, it is widely known within the scientific community that microwave technology has also opened the avenue for aqueous reactions. Among the many studied reactions, the most notable results obtained so far have been for coupling reactions. The application of microwave to coupling has been so diverse that not only this group moved from organic solvents to aqueous, it is also heading toward being metal free. As will be discussed in this section, the development of aqueous homogeneous organic reactions (with respect to coupling) is due to the ability of microwave energy to increase the solubility of organic compounds in water [16, 17]. Also, the very rapid heating under microwave could very well account for some reports on transition-metal- free coupling.

9.3.1
The Heck Reactions

The Heck reaction is one of the most widely studied coupling system in organic chemistry today. This very versatile organic transformation was first reported in the 1970s by scientists Mizoroki and Heck [18, 19]. The reaction involves the direct coupling of 2sp^2 carbons involving aryl and/or vinyl halides (Figure 9.8) [20]. Traditionally, this coupling is carried out under Pd(0), usually generated from a palladium complex, and base, and usually runs for 4–24 h depending on the starting substrate.

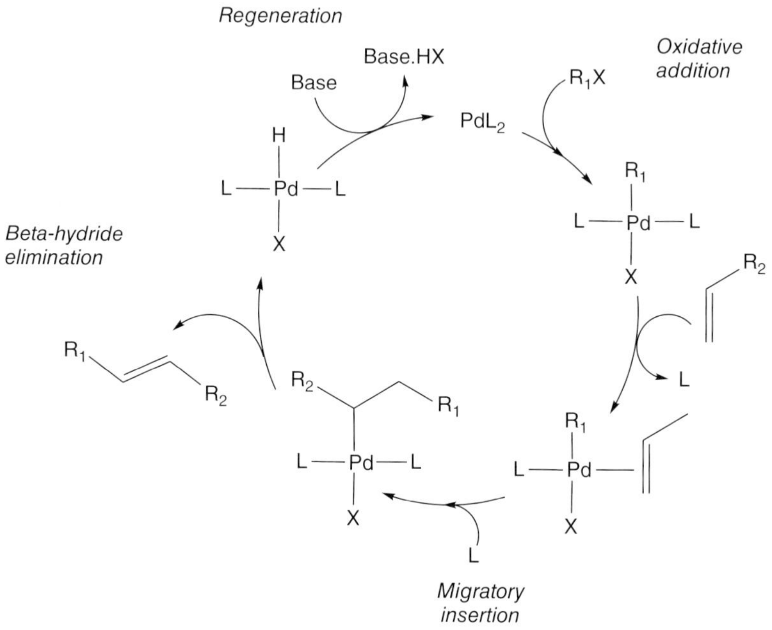

Figure 9.8 Heck reaction scheme and catalytic cycle.

The scope of the reaction has also been widely studied. Generally, the reaction favors the use of more activated aryl halides because of the reactants' greater ability to oxidatively add to the metal. Also, the rate of the first step of the catalytic could also be dependent on the leaving ability of the halide. The reactivity of the aryl halide for oxidative addition to the Pd complex follows the order: Ar–Cl < Ar–Br ≪ Ar–I. Because of this, many reported classical Heck coupling reactions are centered on the use of iodobenzene. Although this is more favorable kinetically, Ar–Cl and Ar–Br are more abundant and cheaper than Ar–I, making the method more valuable. Therefore, one of the challenges of the field today is the development of novel protocols for Heck reactions that do not just depend on iodine [21].

The development of aqueous Heck reactions started as early as 1991 in the seminal reports of Kiji *et al.* [22]. In this work, they probed for possible conditions that could lead to the development of an aqueous coupling reaction. For this purpose, they explored three possible conditions: (1) using water-soluble haloarene and catalyst, (2) using a water-soluble haloarene and an insoluble catalyst, and (3) using a water-soluble catalyst and insoluble haloarene. For all three conditions, bromobenzoic acid and gaseous ethylene were tested. For the first conditions, catalyst deactivation by coordination and subsequent precipitation were observed. Also, the very low solubility of ethylene in an aqueous media did not favor the reaction, while increasing the pressure of the system allowed for the formation of ethylene–Pd(II) complex that forms a black precipitate. The same low solubility was observed for condition 2. For the third condition however, by substituting the bromobenzoic acid with iodophenol and using a triethylamine–water system and $NaHCO_3$, 50.1% of the coupling product was obtained. This early work highlighted the possibility of replacing organic solvent with water for coupling. In 1994, the yield of aqueous coupling was reported to be up to 98% when water–acetonitrile was used as a system with an added organic salt [23]. The following year, Heck coupling was reported to have been carried out using superheated water [24, 25] (Figure 9.9).

To highlight the necessity of developing aqueous-based microwave protocols for organic synthesis, Nájera *et al.* [26] reported the synthesis of methylated resveratrol analogs in both organic solvents and water under microwave. Methylated resveratrol analogs are important molecules in the pharmaceutical industry because they are usually precursors of common chemotherapeutic agents. In this work, the biologically active molecules were synthesized from disubstituted haloarenes and aromatic styrenes in the presence of either (dicyclohexyl)amine, triethylamine, or potassium carbonate as base in water. As seen in Figure 9.10, 99% yield was obtained with route B using microwave power, while as much as 96% was obtained when the Pd was decreased (route C). Interestingly, only 2% yield was reported for the compound when normal heating was used in the reaction. Comparatively, excellent yields were also reported when Pd homogeneous catalysts were immobilized in a solid support such as a glass vessel [27]. By attaching a newly developed benzothiole-oxime Pd precatalyst to a Raschip

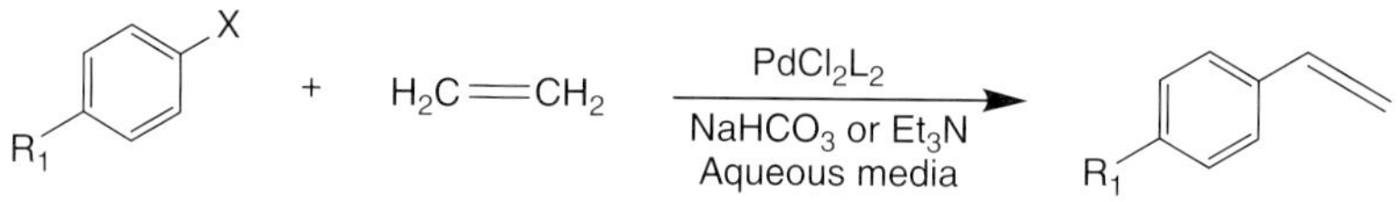

Haloarene	Catalyst	Phase	Base	Yield
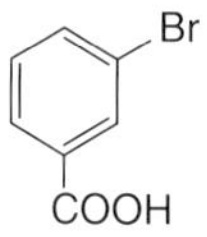	$Pd(OAc)_2$	H_2O	Et_3N	7.6%
	$Pd(OAc)_2$	H_2O/THF	Et_3N	34%
	$PdCl_2$	H_2O/THF	Et_3N	82.0%

Figure 9.9 Reaction scheme for early reports of coupling reactions performed in water.

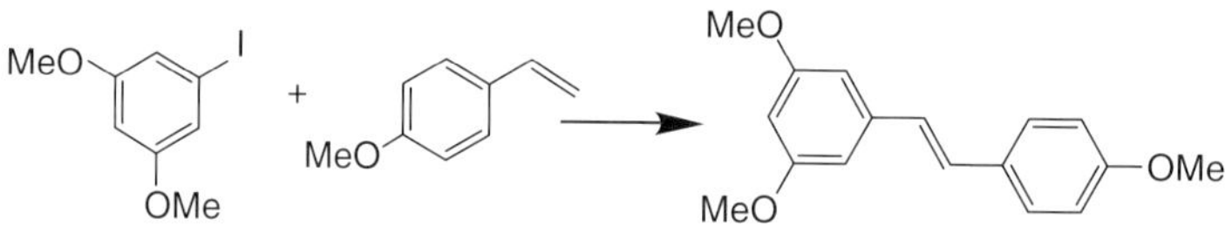

Route A (2% yield)
Route B (99% yield)
Route C (96% yield)

Route A – Pd $(OAc)_2$ (0.5 mol% Pd); 14 h reaction time
4:1 DMA/H_2O, Cy_2NMe, TBAB, 120 °C bath temp.

Route B – Palladacycle (0.5 mol% Pd); 10 min reaction time
4:1 DMA/H_2O, Cy_2NMe, TBAB, 120 °C microwave temp.

Route C – Palladacycle (0.011 mol% Pd); 13 h reaction time
4:1 DMA/H_2O, Cy_2NMe, TBAB, 120 °C bath temp.

Figure 9.10 Microwave-mediated synthesis of resveratrol using Heck coupling reaction.

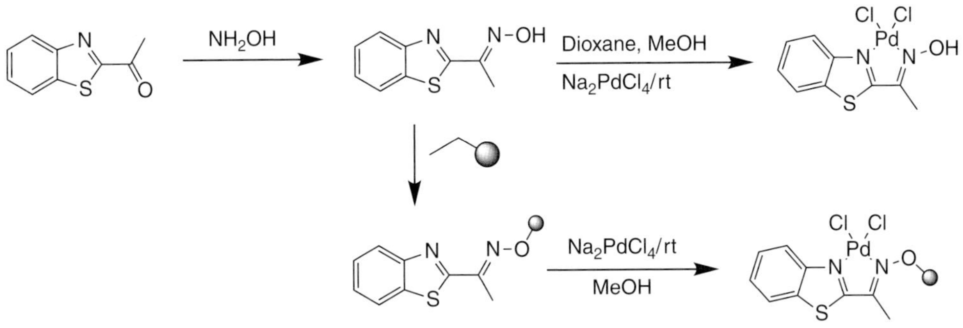

Figure 9.11 Illustration for the immobilization of benzothiazolate-oxime Pd catalyst for the microwave-mediated Heck coupling.

ring-shaped glass/polymer composite, up to 98% of Heck coupling yield was obtained (Figure 9.11).

The move to use water for coupling has progressed significantly through the years, parallel to the development of microwave chemistry. Currently, Heck reactions have moved toward the development of metal-free protocols. Leadbeater *et al.* [28] in 2004 reported Heck coupling reaction between 4-bromoanisole and styrene in the presence of very low Pd-concentrations (500 ppb), water, transfer agent TBAB, and K_2CO_3 (Figure 9.10). Interestingly for this protocol, good yields were observed when the reaction mixtures were not mixed during microwave exposure. It is believed that in the absence of mechanical stirring, the water dissolves the base and the coupling reaction happens within the organic–aqueous interface. The same group of Leadbeater also reported the possibility of using an automated process for microwave-assisted reactions [29]. Using an automated stop-flow voyager microwave apparatus, Heck reactions were successfully scaled from 1 to 10 mmol (with styrene and *p*-bromoanisole as substrates), therefore providing proof of concept of an industrial scaleup (Figure 9.12).

The development of Heck coupling has also been extended in the use of aryl boronic acids instead of haloarenes [30]. Arylboronic acids are commonly used for Suzuki-type couplings, but Heck reactions can also be performed using this substrate in an oxidative manner (Figures 9.13 and 9.14). Performed under air and ambient temperature, the coupling reaction was smooth and efficient with hours of reaction times reduced to only within minutes using the microwave.

Aside from the development of the Heck coupling in water, other possible green solvents are also explored to expand the scope of the synthesis. Ionic liquids are ionic organic compounds that have very low vapor pressure at ambient temperatures. Because of their property as high-boiling point organic compounds, they have become attractive solvents in many synthetic protocols [31–33]. Although relatively expensive than water, ILs provide ease of separation and more increased efficiency in catalysis, with some scientists believing that the organic portion of these molecules also act as catalytic ligands [34]. In the work of Larhed *et al.* [35], the ionic liquid 1-butyl-3-methylimidazolium hexafluorophosphate ($bmimPF_6$)

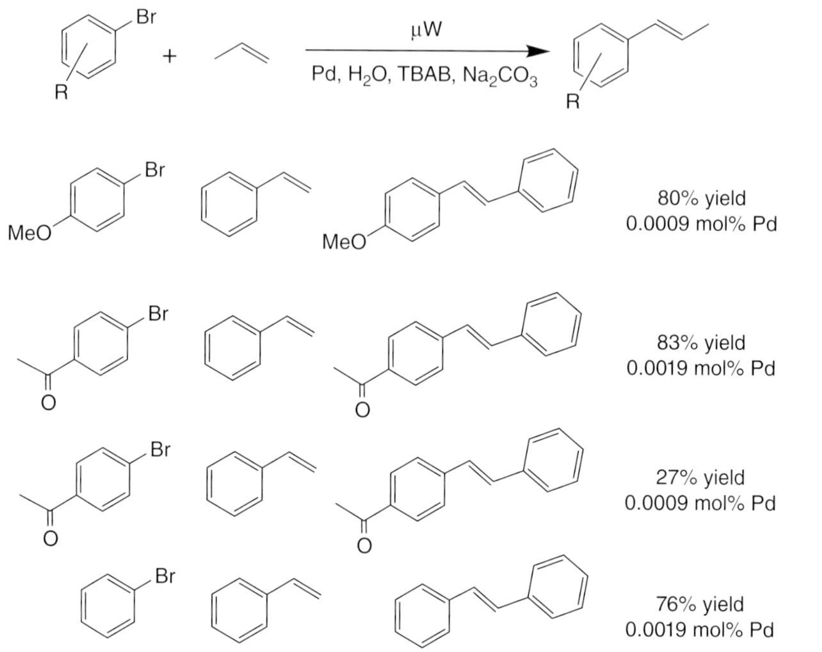

Figure 9.12 Heck coupling reaction with ultralow Pd as reported by Leadbeater *et al.* [28].

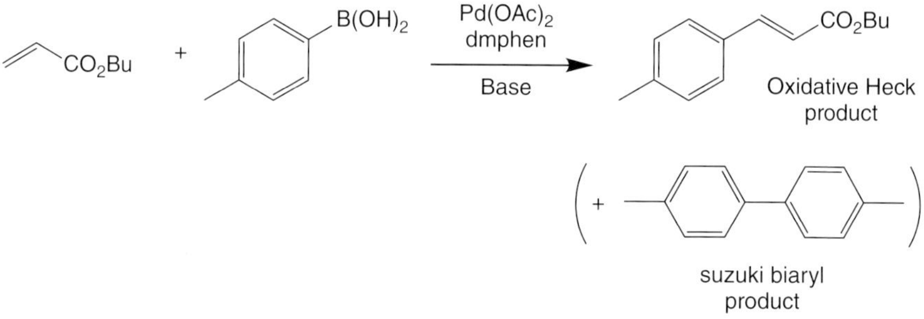

Figure 9.13 Heck coupling of arylboronic acid and an electron-poor olefin.

were used to carry out the Heck coupling between *p*-bromoanisole and olefin butyl acetate in the presence of Pd and Et_3N as base. Distillation was used to purify the ionic liquids after the reactions, which led up to five times recyclability (20 min reactions) at 180 °C (Figure 9.15).

Similar reports on IL were presented by Pan *et al.* [36] using the ionic liquid 1-octanyl-3-methylimidazolium tetrafluoroborate ([OMIm]BF4) also in the absence of phosphine ligands, obtaining yields up to 86% for only 2 min (Figure 9.16).

The developments mentioned herein for Heck reactions highlight the real industrial value of this organic transformation in industries. Because of its application in the facile synthesis of many drug precursors, it is of importance to

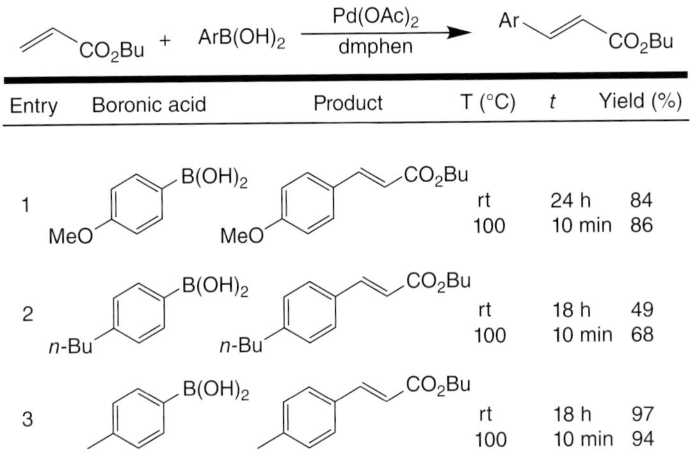

Figure 9.14 Suzuki coupling reactions comparing those put only in room temperature from those subjected under microwave.

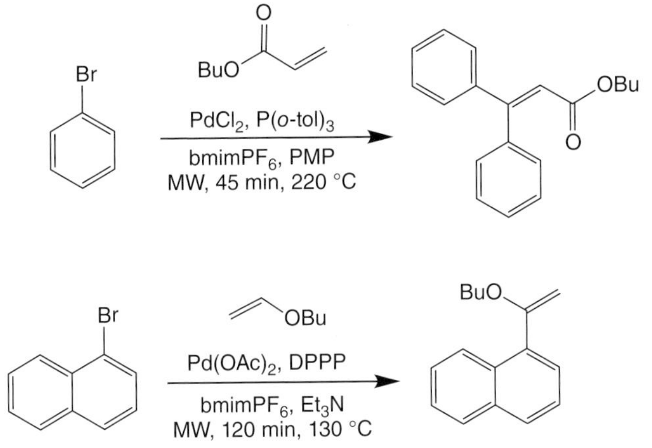

Figure 9.15 IL and microwave assisted of functionalized aryl compounds in the presence of Pd and base Et_3N.

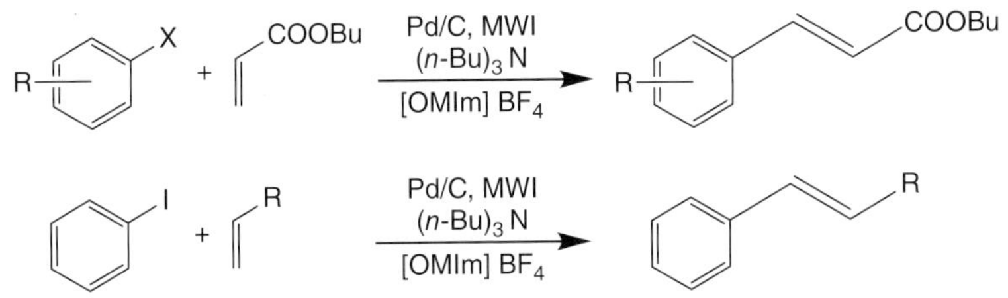

Figure 9.16 IL and microwave-assisted coupling reactions between a halobenzene and an alkene.

develop cleaner methodologies with the Heck coupling. On a similar note, other coupling reactions, such as the widely popular Suzuki coupling, have been tested for aqueous conditions under microwave.

9.3.2
The Suzuki Reaction

The microwave-assisted conditions for Suzuki reactions are almost always similar to that of Heck, except for the substrates. In a typical Suzuki coupling, haloarenes and arylboronic acids are used in the presence of homogeneous Pd. The Suzuki coupling is considered to be one of the most important C–C forming reactions in organic chemistry today, most especially in the synthesis of biaryl compounds (Figure 9.17).

Within the late 1990s, after reports of microwave-mediated coupling reactions in aqueous medium, many other protocols started appearing. One of the early researches on this field is that of Schotten *et al.* [37] where they performed ligandless Suzuki coupling reactions of polyethylene glycol (PEG) ester of bromo-, iodo-, and triflate-para-substituted benzoates and arylboronic acids in both conventional heating and microwave. The conventionally heated samples were without an organic cosolvent up to 70 °C for 2 h, while the reactions with microwave irradiation were ran at 75 W and 2–4 min. Interestingly, as illustrated in this study, polymer structures can be preserved with microwave as opposed to direct heating. A polymer cleavage of up to 45% was observed for the directly heated samples (Figure 9.18).

The microwave-assisted Suzuki coupling reaction had also been shown to be effective in the synthesis of thiophene oligomers [38]. Thiophenes are attractive

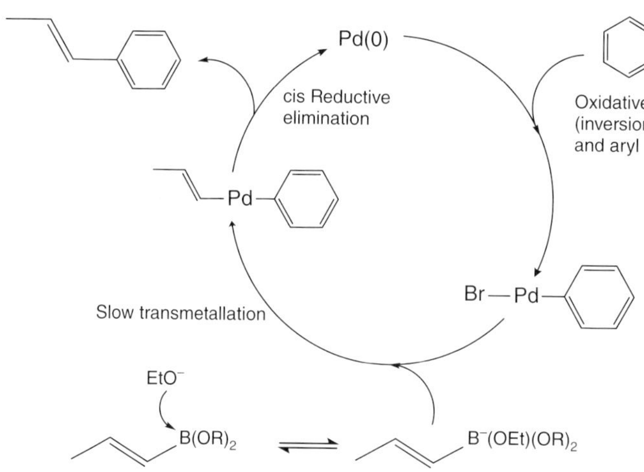

Figure 9.17 Proposed catalytic cycle for the Suzuki coupling between a bromobenzene and an alkeneboronic acid.

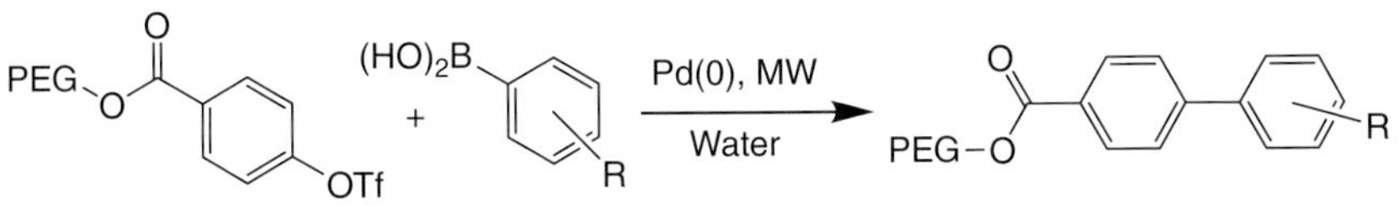

Figure 9.18 Direct coupling between a boronic acid and a PEG-functionalized benzene triflate in the presence of Pd catalyst.

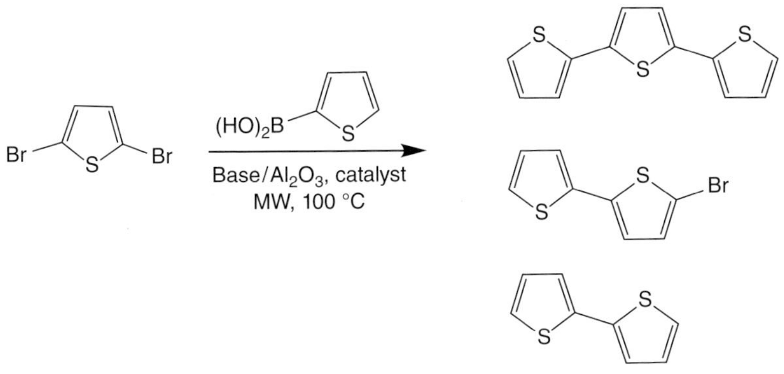

Figure 9.19 Solvent-free microwave-assisted synthesis of 2,5′ : 2′,5″-terthiophene 3.

molecules because they have multifunctional properties fit for organic electronics and semiconducting applications. In optimizing the properties of a thiophene, the polymerization (commonly uses Stille or Suzuki coupling) should be regioselective. Every thiophene monomer may also be functionalized in some other parts to produce a more specific oligomer (Figure 9.19).

It was found that in order to favor oligomer 3 (Figure 9.19), the complex $PdCl_2$(dppf) with KF base should be used. In this case, apart from oligomer 3, oligomers 4 and 5 were also obtained but in smaller ratios (3 : 4 : 5 is 73 : 1 : 22). Extending this to the synthesis of longer thiophene oligomers such as Terthiophene (3), Quaterthiophene (7), and Quinquethiophene (12), it was found that purification time of the products becomes much easier if the synthesis of these long chains were performed in a stepwise manner as opposed to building big molecules at once. Oligomer 7 was obtained in this study after 4 min of microwave exposure at 80 °C (with halide 6 and boronic derivative 2a at 5 mol% of catalyst and 5–10 equiv. of KF and 0.2 ml of KOH) (Figure 9.20).

Early reports of metal-free Suzuki coupling were reported by Leadbeater *et al.* [39] and claimed that they were metal free. After reassessment, it was found that these microwave-assisted coupling was catalyzed by very low concentrations of Pd, up to 50 ppb, present in trace amounts in the sodium carbonate used [39]. The coupling of 4-bromoacetophenone with phenylacetylene was studied under different conditions in the presence of a transfer agent, base, and microwave. This study has also shown that Heck-type coupling reactions of aryl halides and alkynes are possible (Figure 9.21) [40].

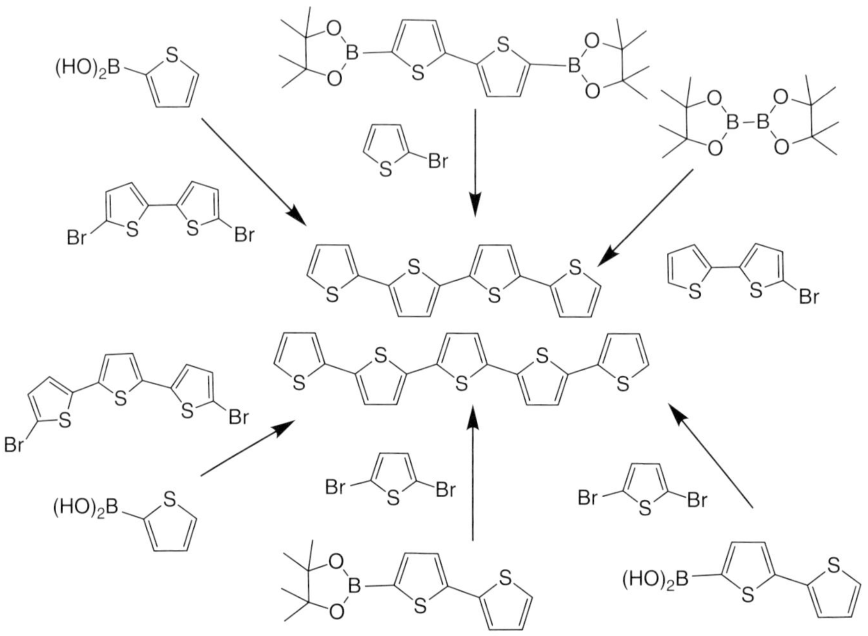

Figure 9.20 Using microwave for the synthesis of polythiophenes could lead to a myriad of possible products with different electronic properties.

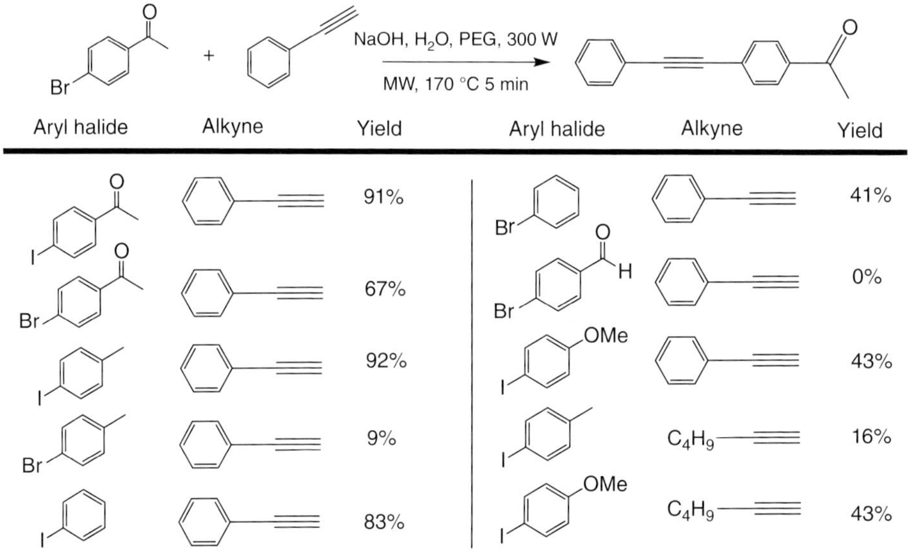

Figure 9.21 Synthetic scheme for the coupling between a bromobenzene and an aromatic alkyne using microwave and water without any metal.

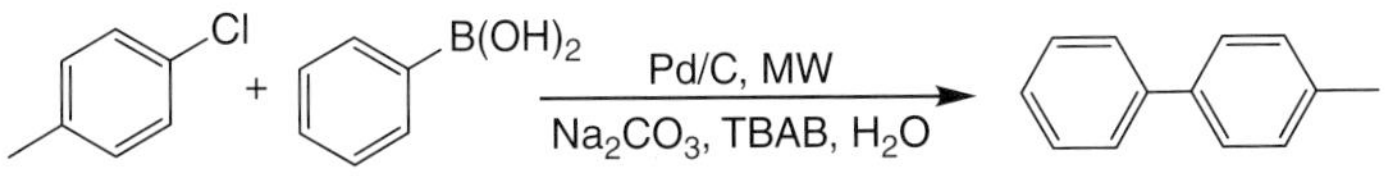

Entry	Temp (°C)	Simultaneous cooling	Product (mmol)	Material recovered (mmol)	
				4-Chlorotoluene	Total
1	120	No	0.40	0.23	0.63
2	100	No	0.19	0.19	0.38
3	135	No	0.56	0.08	0.64
4	150	No	0.61	0.10	0.71
5	120	Yes	0.75	0.07	0.82
6	100	Yes	0.40	0.12	0.52
7	135	Yes	0.71	0.10	0.81
8	150	Yes	0.56	0.19	0.75

Figure 9.22 Simultaneously cooling the products of a coupling reaction directly from the microwave reactor could lead to improved product ratios.

As with the case for Heck reactions, organic bases have also been tested for the microwave-promoted Suzuki coupling of various substrates [41]. It was found that using 1,8-diazabicycloundec-7-ene (DBU) or 1,1,3,3-tetramethylguanidine could increase the coupling product yield up to 99%. Although there is always an issue with solubility in aqueous medium, the yield is possibly increased mainly due to the selective heating of microwave. Also, when the use of amine bases for coupling was extended to couplings in larger scale, similar yields were obtained. One way to also increase the yield of an aqueous microwave assisted is to perform simultaneous cooling [42]. When cooling was performed simultaneously, as seen in Figure 9.22, higher product to reactant ratio was recovered that is mainly attributed to the increased lifetime of aryl halides in the reaction mixture.

One interesting way to address the need for downstream separation of coupling products is on the use of perfluorooctylsulfonates instead of aryl halides [43]. In the work of Nagashima *et al.* [43], perfluorooctylsulfonates were synthesized from phenol via base hydrolysis and attached to the perfluorooctane surface of fluorous solid-phase extraction (F-SPE). In this way, when the surface was exposed to the arylboronic acid and microwave for coupling (and with an appropriate elutant), the product simply washes out of the column. Another possibility is to attach the homogeneous catalyst to a solid support. In the work of Lee *et al.* [44], chitosan-supported Pd(0) was prepared by the adsorption of Pd(II) and then by the subsequent reduction process. In the presence of TBAB as the transfer agent, the catalyst was used for the Suzuki coupling of aryl halides and boronic acids (Figure 9.23).

The development of microwave-assisted coupling has progressed so much in the past years that many reports of transition-metal-free coupling have appeared in scientific literature [45–47]. Although the experiments involved here are elegant,

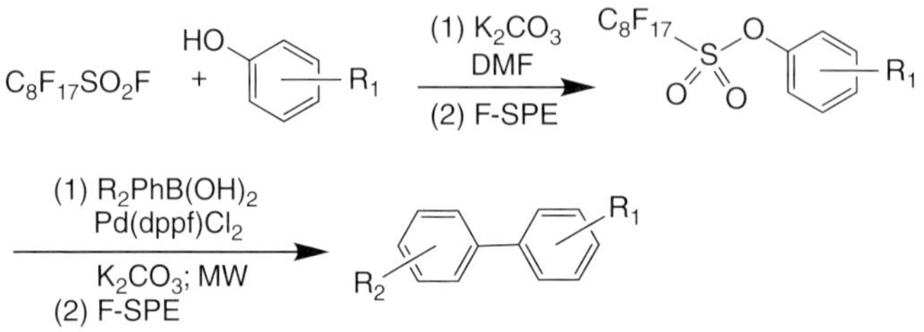

Figure 9.23 Fluorous synthesis of biaryls using microwave.

one cannot completely be sure that a microwave-assisted coupling is completely metal free because, as shown previously, even very small concentrations of the metal can catalyze the entire process.

9.4 Conclusions and Prospects

This contribution has been aimed to discuss a number of different examples of homogeneous-catalyzed reactions under microwave. The effect of microwave to reaction systems can be attributed to its ability to induce more molecular vibration, and thus rapid heating, being able to reduce reaction times from hours to only within minutes. Because of this, microwave-assisted reactions are also selective and can heat only specific components of a mixture, and leaving others untouched. Most of the homogeneous-catalyzed systems reported are that of coupling processes. We have also discussed here two of the most popular coupling reactions and several microwave-assisted protocols, including those performed in water. In all of these reports, the primary motivations of using both water and microwave are to increase the turnover of the reaction and make separation much easier (which is also considered green). The increased solubility of organic reactants in water under microwave is very advantageous, since they can easily then be separated under normal conditions. Other ways of increasing separation efficiency have been reported. As we have previously pointed out, in many ways, MAOS had provided a different dimension to organic synthesis by providing avenues for faster reaction and better yields. For drug discovery, this technology is very important because it will also lead to increased efficiency of synthesis and biological testing, thus possibly greatly reducing the time necessary for drug development.

Acknowledgments

Rafael Luque gratefully acknowledges Spanish MICINN for financial support via the concession of a RyC contract (ref: RYC-2009-04199) and funding under project CTQ2011-28954-C02-02 (MEC). Consejeria de Ciencia e Innovacion, Junta de Andalucia is also gratefully acknowledged for funding project P10-FQM-6711.

References

1. de la Hoz, A., Diaz-Ortiz, A., and Moreno, A. (2005) *Chem. Soc. Rev.*, **34**, 164.
2. Lidström, P., Tierney, J., Wathey, B., and Westman, J. (2001) *Tetrahedron Lett.*, **57**, 9225.
3. Larhed, M., Moberg, C., and Hallberg, A. (2002) *Acc. Chem. Res.*, **35**, 717.
4. Kuhnert, N. (2002) *Angew. Chem. Int. Ed.*, **41**, 1863.
5. Stadler, A. and Kappe, C.O. (2001) *Eur. J. Org. Chem.*, **2001**, 919.
6. Mehta, V.P. and Van der Eycken, E.V. (2011) *Chem. Soc. Rev.*, **40**, 4925.
7. Dallinger, D. and Kappe, C.O. (2007) *Chem. Rev.*, **107**, 2563.
8. Hayes, B.L. (2002) *Microwave Synthesis: Chemistry at the Speed of Light*, CEM Publishing, Matthews, NC.
9. Liu, X.Y., Li, C.H., and Che, C.M. (2006) *Org. Lett.*, **8**, 2707.
10. Jiao, G.S., Castro, J.C., Thoresen, L.H., and Burgess, K. (2003) *Org. Lett.*, **5**, 3675.
11. Wathey, B., Tierney, J., Lidström, P., and Westman, J. (2002) *Drug Discov. Today*, **7**, 373.
12. Besson, T., Guillard, J., and Rees, C.W. (2000) *Tetrahedron Lett.*, **41**, 1027.
13. Kabalka, G.W. and Mereddy, A.R. (2005) *Tetrahedron Lett.*, **46**, 6315.
14. Valenti, P., Bisi, A., Rampa, A., Belluti, F., Gobbi, S., Zampiron, A., and Carrara, M. (2000) *Bioorg. Med. Chem.*, **8**, 239–246.
15. Chebanov, V.A., Muravyova, E.A., Desenko, S.M., Musatov, V.I., Knyazeva, I.V., Shishkina, S.V., Shishkin, O.V., and Kappe, C.O. (2006) *J. Comb. Chem.*, **8**, 427.
16. Polshettiwar, V. and Varma, R.S. (2008) *Acc. Chem. Res.*, **41**, 629–639.
17. Arancon, R.A., Lin, C.S.K., Vargas, C., and Luque, R. (2014) *Org. Biomol. Chem.*, **12**, 10.
18. Mizoroki, T., Mori, K., and Ozaki, A. (1971) *Bull. Chem. Soc. Jpn.*, **44**, 581.
19. Heck, R.F. and Nolley, J.P. (1972) *J. Org. Chem.*, **37**, 2320.
20. Whitcombe, N.J., Hii, K.K.M., and Gibson, S.E. (2001) *Tetrahedron Lett.*, **57**, 7449.
21. Biffis, A., Zecca, M., and Basato, M. (2001) *J. Mol. Catal. A-Chem.*, **173**, 249–274.
22. (a) Kiji, J. and Okano, T. (1991) *Trends Org. Chem.*, **2**, 41; (b) Kiji, J. and Okano, T. (1994) *J. Synth. Org. Chem. Jpn.*, **52**, 276.
23. Jeffery, T. (1994) *Tetrahedron Lett.*, **35**, 3051.
24. Diminnie, J., Metts, S., and Parsons, E.J. (1995) *Organometallics*, **14**, 4023.
25. Reardon, P., Metts, S., Crittendon, C., Daugherity, P., and Parsons, E.J. (1995) *Organometallics*, **14**, 3810.
26. Botella, L. and Nájera, C. (2004) *Tetrahedron Lett.*, **60**, 5563.
27. Dawood, K.M. (2007) *Tetrahedron Lett.*, **63**, 9642.
28. Arvela, R.K. and Leadbeater, N.E. (2005) *J. Org. Chem.*, **70**, 1786.
29. Kappe, C.O. and Dallinger, D. (2005) *Nat. Rev. Drug Discov.*, **5**, 51.
30. Lindh, J., Enquist, P.A., Pilotti, Å., Nilsson, P., and Mats, L. (2007) *J. Org. Chem.*, **72**, 7957.
31. Welton, T. (1999) *Chem. Rev.*, **99**, 2071.
32. Sheldon, R. (2001) *Chem. Commun.*, **23**, 2399.
33. Wasserscheid, P. and Keim, W. (2000) *Angew. Chem. Int. Ed.*, **39**, 3772.
34. Fukuyama, M., Shinmen, M., Nishitani, S., Sato, M., and Ryu, I. (2002) *Org. Lett.*, **4**, 1691.
35. Vallin, K.S.A., Emilsson, P., Larhed, M., and Hallberg, A. (2002) *J. Org. Chem.*, **67**, 6243.
36. Xie, X., Lu, J., Chen, B., Han, J., She, X., and Pan, X. (2004) *Tetrahedron Lett.*, **45**, 809.
37. Blettner, C.G., König, W.A., Stenzel, W., and Schotten, T. (1999) *J. Org. Chem.*, **64**, 3885.
38. Melucci, M., Barbarella, G., and Sotgiu, G. (2002) *J. Org. Chem.*, **67**, 8877.
39. Leadbeater, N.E. (2010) *Nat. Chem.*, **2**, 1007.
40. Leadbeater, N.E., Marco, M., and Tominack, B.J. (2003) *Org. Lett.*, **21**, 3919.
41. Chanthavong, F. and Leadbeater, N.E. (2006) *Tetrahedron Lett.*, **47**, 1909.

42. Arvela, R.K. and Leadbeater, N.E. (2005) *Org. Lett.*, 7, 2101.
43. Zhang, W., Chen, C.H.T., Lu, Y., and Nagashima, T. (2004) *Org. Lett.*, **6**, 1473.
44. Yi, S.S., Lee, D.H., Sin, E., and Lee, Y.S. (2007) *Tetrahedron Lett.*, **48**, 6771.
45. Leadbeater, N.E. and Marco, M. (2003) *Angew. Chem. Int. Ed.*, **115**, 1445.
46. Leadbeater, N.E. (2005) *Chem. Commun.*, **23**, 2881.
47. Singh, B.K., Kaval, N., Tomar, S., Van der Eycken, E., and Parmar, V.S. (2008) *Org. Process Res. Dev.*, **12**, 468–474.

10
Microwave-Assisted Solid Acid Catalysis

Hyejin Cho, Christian Schäfer, and Béla Török

10.1
Introduction

As a result of the ever strengthening environmental regulations, solid acids are becoming the acid catalysts of choice to replace the corrosive and harmful traditional mineral acids for organic synthesis. While a broad range of materials fulfill the role of a solid acid, these catalysts possess either Brønsted or Lewis acidic sites and often possess both. Heterogeneous Brønsted acids include acidic ion-exchange resins (e.g., Nafion-H, Amberlyst, Dowex, Deloxane), heteropoly acids (HPAs) and their derivatives (e.g., $H_{0.5}Cs_{2.5}[PW_{12}O_{40}]$). Solid Lewis acids include metal chlorides (e.g., $ZnCl_2$, $AlCl_3$), triflates, metal oxides, and their sulfated derivatives (e.g., sulfated zirconia). Catalysts with mixed Brønsted and Lewis acid centers, including clays and zeolites, form the most common group. Due to the focus of the present work, catalytic aspects such as catalyst preparation and characterization or surface science tools will not be discussed. The application of solid acids has been frequently reviewed, including several accounts with a focus on microwave-assisted solid acid catalysis [1]. There are several excellent reviews on related topics where further details can be obtained [2]. With the development of reliable and accurate laboratory microwave reactors, the application of microwave-assisted organic synthesis (MAOS) in solid acid-catalyzed reactions became widespread. As these solid acids possess significant polar character, they commonly are strong microwave absorbers [3]. Thus, when such solids are applied as catalyst, the catalytic material also heats up very quickly in addition to the reactants providing energy for the surface reactions. Based on the goal of this work, the emphasis will be on more recent applications, for more information on older processes the reader should consult the large set of available accounts [4, 5].

10.2
Microwave-Assisted Clay Catalysis

Clays are aluminosilicates of layered structures that provide opportunity to broad applications. While natural clays are usually weak acids ($-H_0 \approx 3$) and

Microwaves in Catalysis: Methodology and Applications, First Edition.
Edited by Satoshi Horikoshi and Nick Serpone.

not particularly active in acid catalysis, several semisynthetic clay products are excellent solid acids ($-H_0 \approx 8$). The overwhelming majority of the applications describes the use of such materials [6]. Microwave-assisted clay catalysis was first reported by Varma [7] and became a useful approach for the synthesis of a broad range of products.

The most common use of such catalysts is in the synthesis of heterocyclic compounds. Several five-membered N-heterocycles, including the line of pyrrole-to-indole-to-carbazole, were synthesized by K-10 montmorillonite-catalyzed ring closure and subsequent electrophilic annelations. The method provided good to excellent yields and remarkably works with a wide variety of NH_2-containing units including amines or sulfonamides (Scheme 10.1) [8].

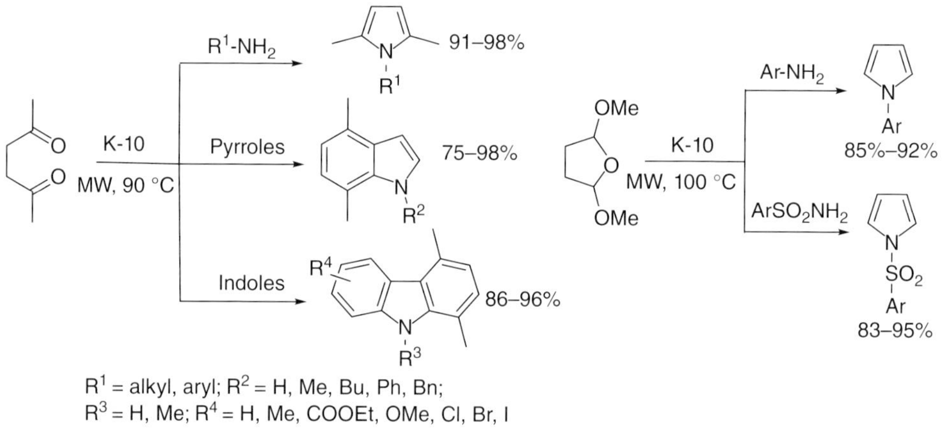

Scheme 10.1

Similar cyclizations were applied for the synthesis of pyrimidine-indoles and carbazoles [9]. While the conventionally heated reaction required over 24 h for completion, the microwave-assisted reactions resulted in ~75% yield for the products after only 1 h irradiation (Scheme 10.2).

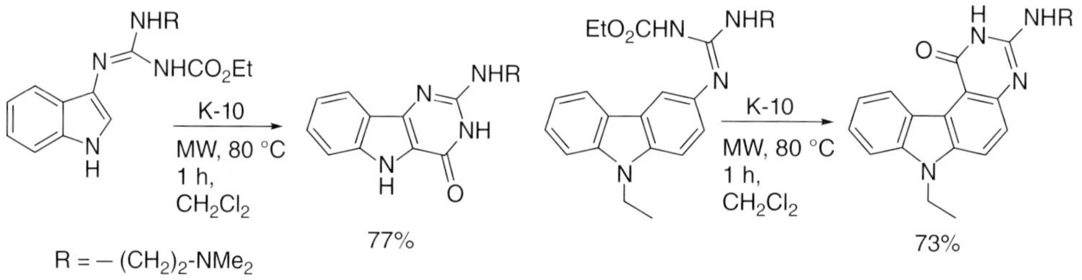

Scheme 10.2

De Paolis *et al.* reported the synthesis of quinolines applying a modified version of the classical Doebner–Miller synthesis (Scheme 10.3). Cinnamaldehydes and

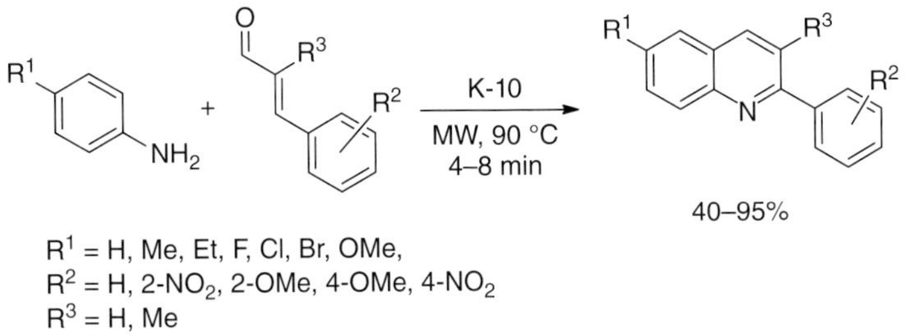

Scheme 10.3

anilines readily undergo a condensation–cyclization–aromatization domino pathway in a K-10 catalyzed one-pot process [10]. The K-10 catalyst has a dual role in the sequence, acting as an acid catalyst in the condensation and cyclization steps and as an oxidation catalyst in the final aromatization producing the product quinolines in 40–95% yield during rapid reactions (4–8 min).

5-Exo-dig cyclocondensation of alk-3-yn-1-ones with diversely substituted hydrazines readily yields 1,3,5-triaryl pyrazoles when K-10 montmorillonite is applied as a catalyst (Scheme 10.4). The reaction occurs with high selectivity and nearly 100% atom economy making the process environmentally benign. The products form in short (10 min) reactions with good to excellent yields [11]. Using a similar design, phthalazinones can be obtained from phthalaldehydic acid and hydrazines [12].

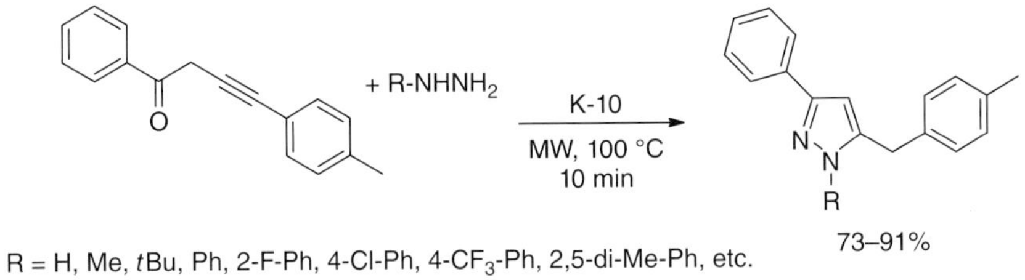

Scheme 10.4

The synthesis of indolyl-benzimidazoles is achieved by a solvent-free methodology via the condensation and cyclization of cyanoacetyl-indoles and 2-phenylenediamines (Scheme 10.5). The reaction occurs readily in 10 min reaction times providing 70–90% yields, representing approximately one order of magnitude decrease in reaction time and a 50% increase in yields [13]. Similar condensation reactions with phenylenediamines also provide benzodiazepines, benzimidazoles, and quinoxalinones [14].

The cyclization reactions appear extendable to O-containing heterocycles. The K-10 catalyzed ring closure of several flavonoid derivatives was applied during the total synthesis of baicalein derivatives that were proposed as antitumor agents (Scheme 10.6). While the microwave-assisted reactions usually provided higher yields, often double than that of the conventional reactions (up to 77%), in several

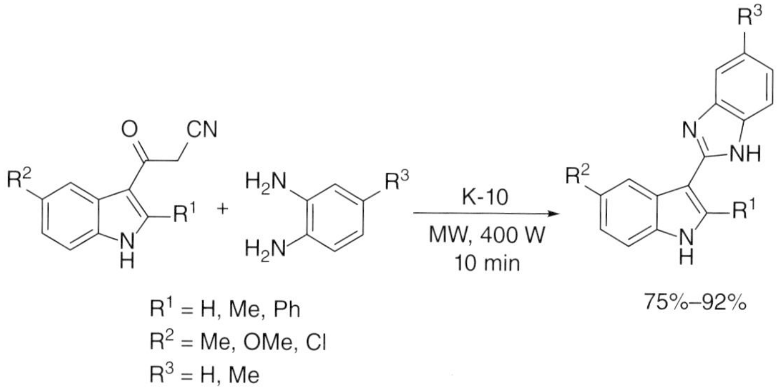

Scheme 10.5

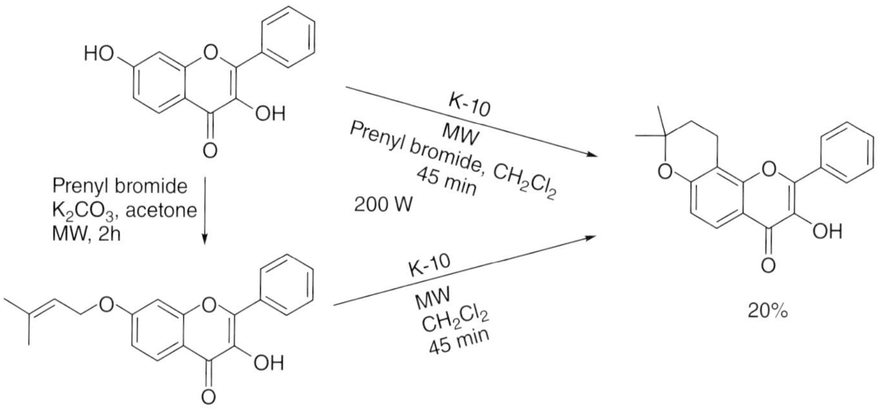

Scheme 10.6

cases the yield was only 6% (1% under conventional heating), or the product did not form despite the fact that the conventional reaction resulted in product formation [15].

The synthesis of combined N and O-containing heterocycles was achieved by Suresh *et al.* The application of the reusable Al^{3+}-doped K-10 montmorillonite allows the formation of aryl-substituted 1,2,4- and 1,3,4-oxadiazoles in a one-pot multicomponent reaction, using simple and readily available starting materials. The reaction is versatile, with over 40 examples reported, and tolerates a wide range of starting materials [16] (Scheme 10.7).

2,4-Diarylquinolines are synthesized by a K-10 montmorillonite catalyzed multicomponent domino reaction (Scheme 10.8) [17]. A diverse group of 2,4-disubstituted quinolines was prepared using anilines, benzaldehydes, and terminal phenylacetylenes. The process appeared to have no limit in its potential extension to combinatorial synthesis. The reaction did not alter the catalyst that was reused five times without a noticeable loss in its activity.

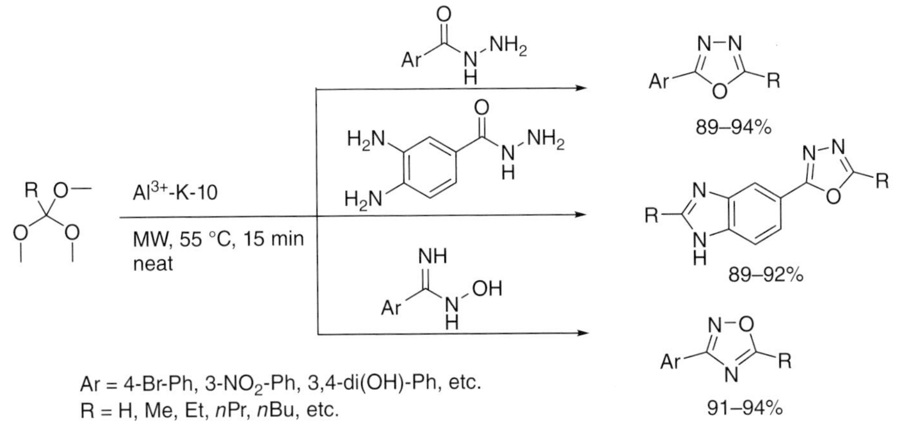

Scheme 10.7

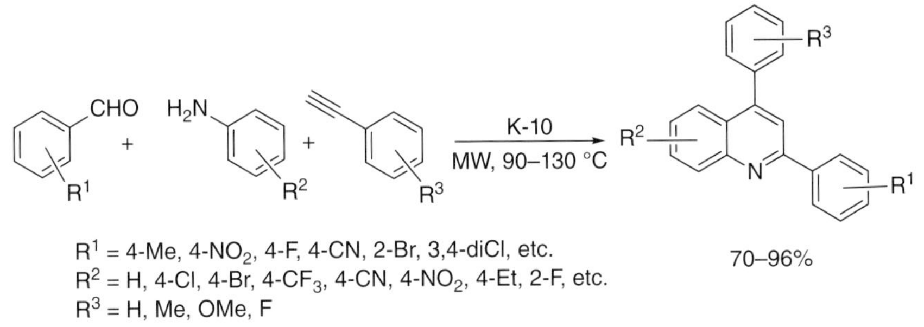

Scheme 10.8

Microwave-assisted clay catalysis is also widely applicable in aromatic electrophilic substitutions. The alkylation of indoles with ninhydrin readily yields the corresponding bis-indolyl-1,3-indanediones in a solvent-free K-10 catalyzed process [18]. The reactions require only 5–6 min reaction times and provide the products in nearly quantitative yields (90–96%) (Scheme 10.9). Other

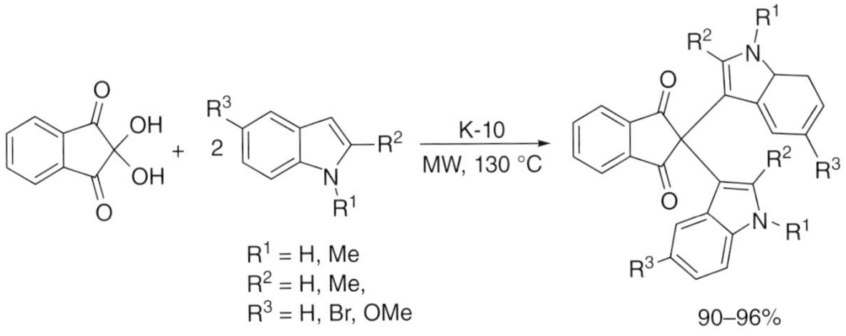

Scheme 10.9

aromatic alkylations include alkylations with alkenes, carbonyl compounds, or alcohols [19].

The Friedel–Crafts hydroxyalkylation of indoles with ethyl trifluoropyruvate and ethyl trifluoroacetoacetate (Scheme 10.10) with solvent-free K-10 catalysis provided the corresponding *N*-heteroaryl(trifluoromethyl)hydroxyalkanoic acid esters in excellent yields (up to 98%) [20].

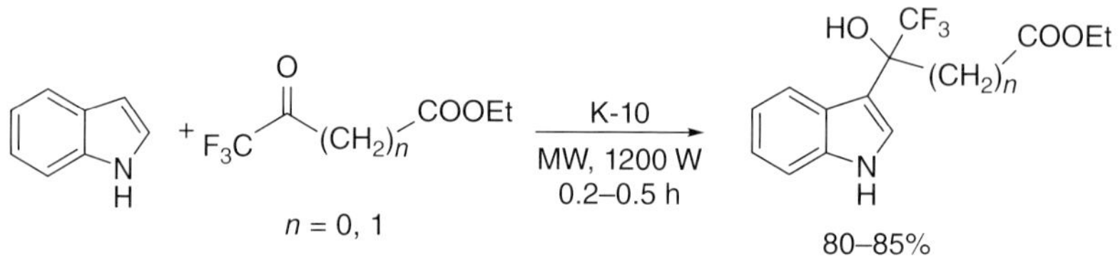

Scheme 10.10

Microwave-mediated hydroxyl protection/deprotection has been successfully achieved on K-10 montmorillonite. Unlike in the case of several mineral acids that debenzylated the protected sugars, the solid acid catalysts resulted in the deacetylation of the substrate in 78–85% yield [21] (Scheme 10.11).

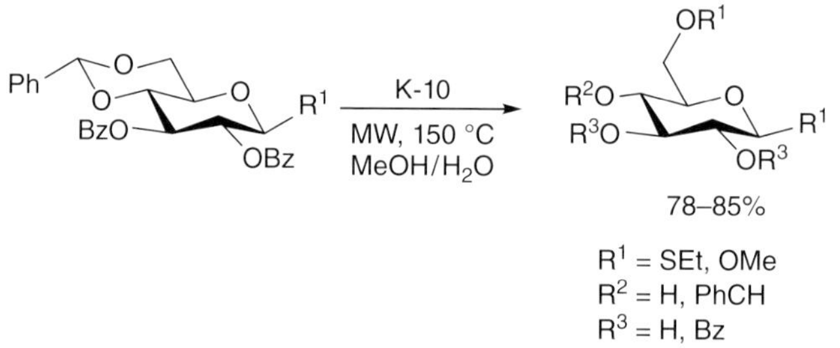

Scheme 10.11

A KSF montmorillonite catalyzed cross-aldol condensation of acetophenones and benzaldehydes readily yielded the corresponding chalcone derivatives in short reaction times and moderate to high yields (Scheme 10.12). The reaction tolerates a broad range of substituted aryl groups in both starting materials, and the scope

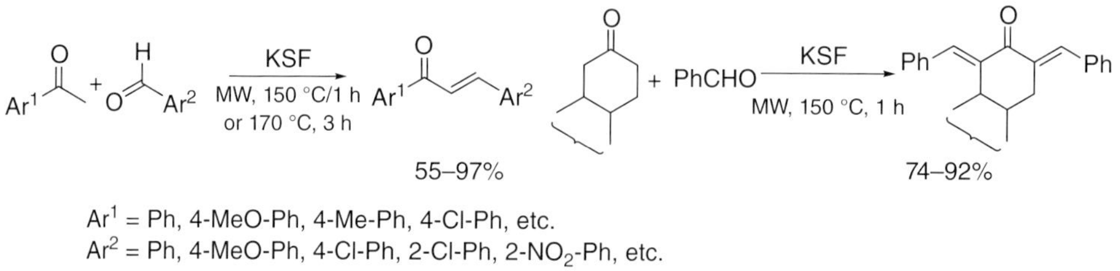

Scheme 10.12

is also extended to the synthesis of symmetrical, diarylmethylene-α,β-unsaturated ketones from aromatic aldehydes and ketones [22]. The proposed explanation of the mechanism involves a selective anchoring and activation of the aldehyde by the Lewis acid centers of the clay.

As a reverse version of the above reaction, KSF montmorillonite $BiNO_3 \cdot 5H_2O$ supported on KSF montmorillonite catalyzed to provide vanillin in a simple, solvent-free, and rapid reaction in 77% yield (Scheme 10.13) [23].

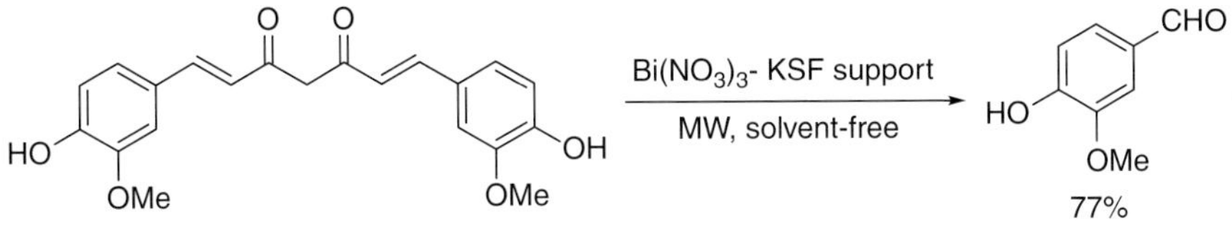

Scheme 10.13

Alkenes can be successfully transformed to iodohydrins using a copper(II) acetate doped K-10 montmorillonite catalyst with I_2 and alcohols as reagents. A broad variety of alkenes can be applied in the reaction providing the products in short reaction times and, with a few exceptions, in excellent yields [24] (Scheme 10.14).

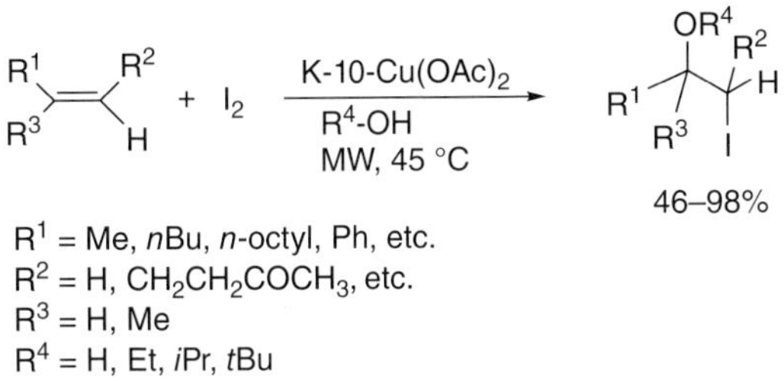

Scheme 10.14

Clay catalysts can also contribute to cooperative catalysis with added noble metals. Due to the small particle size of the applied metal, the undesirable arching phenomenon does not occur under these conditions. Several such reactions include an oxidative Hantzsch reaction, and the synthesis of pyrazoles and β-carbolines have been performed through multicomponent domino reactions [25].

10.3 Zeolites in Microwave Catalysis

Zeolites are aluminosilicates of the generic formula $M^{n+}_{x/n}[(AlO_2)^-_x(SiO_2)_y]\cdot z\,H_2O$. Zeolites occur naturally, but can also be prepared synthetically. Due to the well-defined, manipulable microporous framework and the possibility of exchanging the intercalated cation, they have become attractive materials for various applications. Some of their major industrial applications include their use as sorbents (e.g., H_2O removal) or as catalysts (e.g., cracking of petrochemicals). Zeolites are well-known solid acids, with a highly temperature-dependent acidity. Their H_0 value at room temperature is ~3–4, with several studies describing superacidic behavior at elevated temperatures [26]. Due to their highly polar structure, they absorb microwave energy readily and are thus excellent catalytic materials for MAOS.

Their acidic properties make zeolites attractive catalysts for the formation of ester [27] and amide bonds [27a, 28] (Scheme 10.15). Typically, a carboxylic acid is mixed with the catalyst and the alcohol or amine, and heated under microwave conditions to yield the corresponding ester or amide. The reactions proceed without the use of a solvent and more rapidly while providing higher yields as compared to conventional heating.

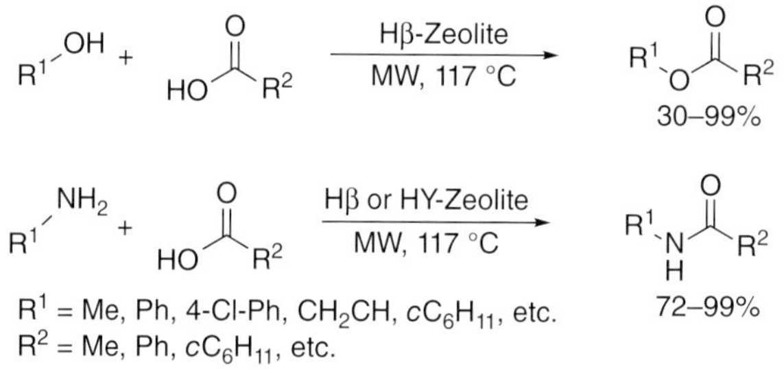

Scheme 10.15

It has also been shown that amide formation can occur intramolecularly [29]. Similar to the above intermolecular reactions, both the yield and reaction time are greatly improved under microwave irradiation.

The synthesis of underivatized amides can also be achieved through the hydrolysis of nitriles [30] (Scheme 10.16). The hydrolysis proceeds with improved yields in shorter reaction times than under conventional heating.

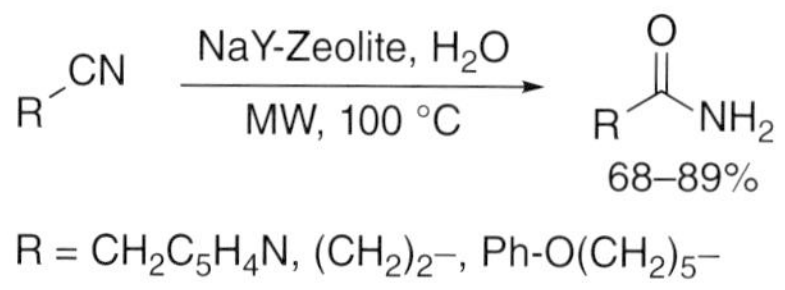

Scheme 10.16

Ketones can also be transformed to amides under microwave conditions and zeolite catalysis [28b] (Scheme 10.17). Under the conditions used, first the oxime is formed which then undergoes a Beckmann rearrangement to produce the amide.

Camargo-Ordonez *et al.* have shown that the sulfonamide formation is improved under microwave conditions [31]. Aromatic sulfonyl chlorides are

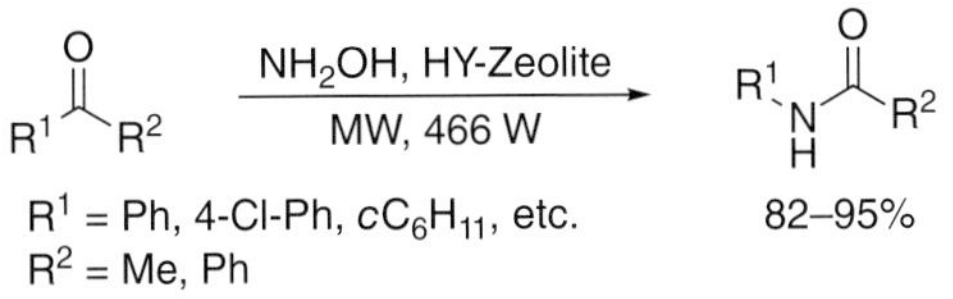

Scheme 10.17

coupled with aromatic amines using 4Å MS as a catalyst in good to excellent yields.

The use of microwave irradiation has been demonstrated to positively affect zeolite-catalyzed Friedel–Crafts acylation and alkylations with respect to yield, reaction time, and product selectivity. Acylations can be performed with carboxylic acids [32] (Scheme 10.18), anhydrides [33], or acid chlorides [33a] as acylating agents. All reactions resulted in improved yields and shorter reaction time when compared to conventional heating.

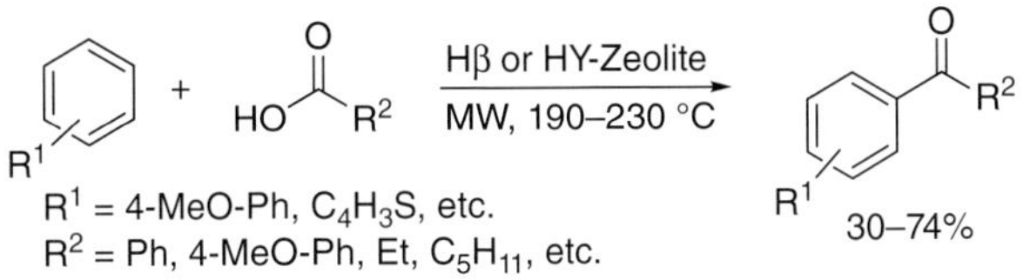

Scheme 10.18

Hiroki *et al.* demonstrated that acylation is not limited to intermolecular reactions; intramolecular versions are also possible, as shown in the synthesis of 1-Tetralone derivatives from 4-arylbutyric acids [34].

Friedel–Crafts alkylation has been investigated using methyl-*tert*-butyl ether (MTBE) as an alkylating agent [35]. In the reactions of hydroquinone, monoalkylation occurred exclusively; most likely due to limited space in the pores of the catalyst. It was found that the reaction can be enhanced if SO_3H functionalized zeolites are used as a catalyst.

Investigations of microwave-assisted catalytic oxidations and alkylation reactions point out that the morphology of zeolites with identical molecular formulae plays a role in catalytic activity and reusability of the material. It is reasoned that this different behavior is due to the different accessibility of the active sites of the catalyst [36].

It has been demonstrated that zeolites can efficiently catalyze the formation of various types of heterocycles. Cyanopyridines [37] (Scheme 10.19) and pyrimidobenzimidazoles [38] can be obtained in higher yields and shorter reaction times without a solvent under microwave heating than under conventional heating. It has been shown for the synthesis of pyrimidobenzimidazoles that the regiochemistry of the reaction can be determined through the appropriate selection of the zeolite catalyst, most likely due to the different cage sizes allowing for the formation of one product only.

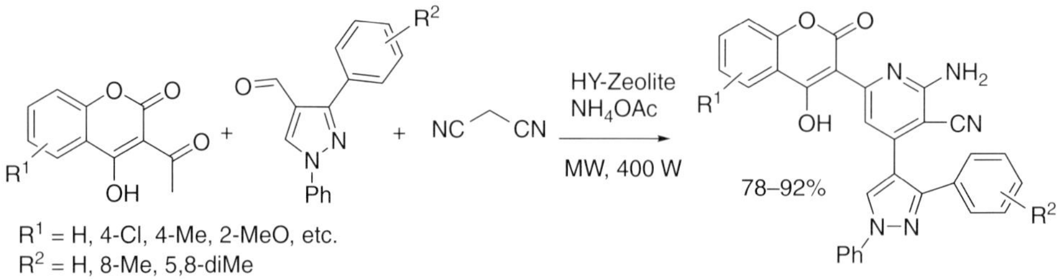

Scheme 10.19

It was also shown that zeolites catalyze the formation of imidazoles and benzimidazoles from various precursors. Balalaie *et al.* have shown that substituted imidazoles can be obtained in a one-pot multicomponent reaction (Scheme 10.20) [39].

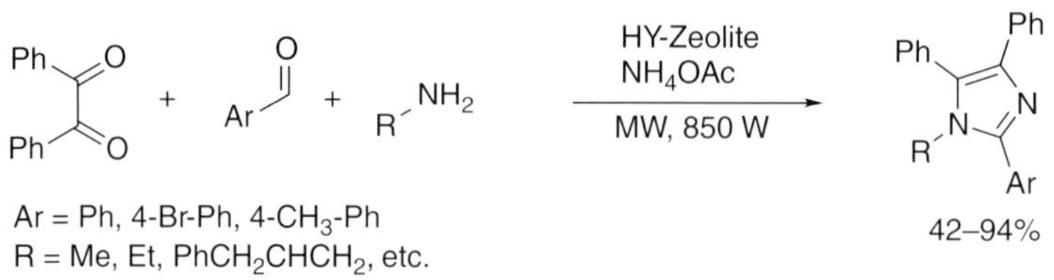

Scheme 10.20

Fused aminoimidazoles have been synthesized by Thompson *et al.* using HY-zeolite as a catalyst under microwave irradiation (Scheme 10.21) [40]. The reaction proceeds in short reaction times (minutes) in acceptable yields and with good regioselectivity.

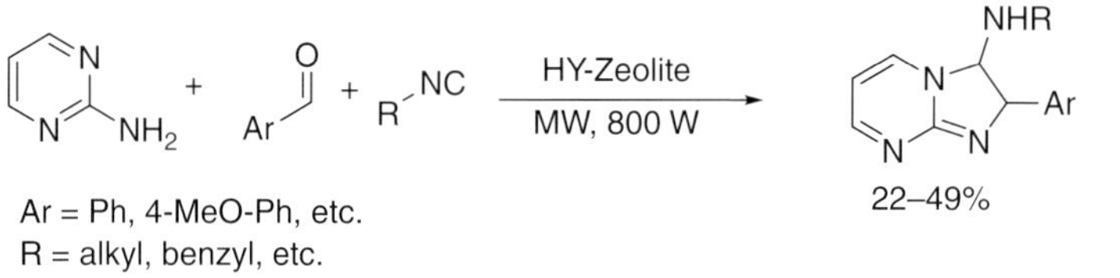

Scheme 10.21

In a similar reaction, benzimidazoles can be synthesized from *o*-phenylenediamines and a suitable carbonyl compound, which could be an orthoester [41], an anhydride [41], formaldehyde [41], or a carboxylic acid [42] (Scheme 10.22).

The same methodology can be extended to the synthesis of 2,3-disubstituted-4(3*H*)-quinazolinones [43] (Scheme 10.23). The reactions do not require the use of a solvent and perform better regarding the reaction time and yield under microwave irradiation than using conventional heating.

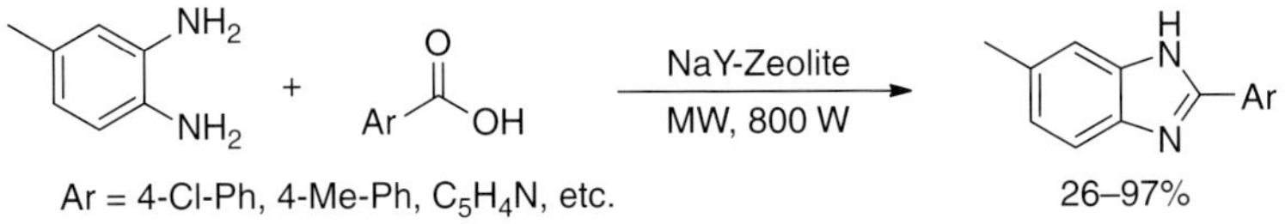

Scheme 10.22

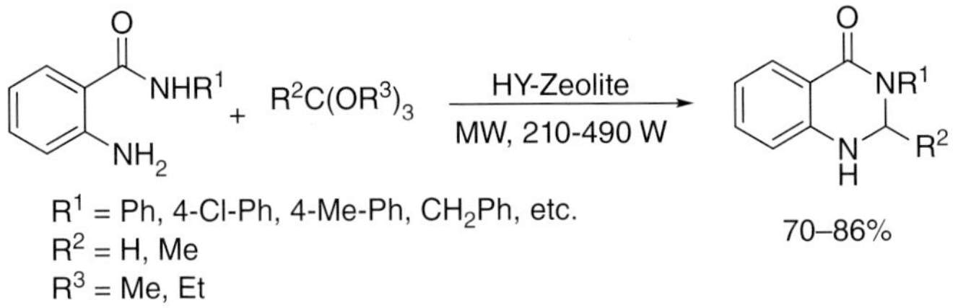

Scheme 10.23

Polyhydroquinolines have been synthesized using a ZnO-modified β-zeolite under microwave heating [44] (Scheme 10.24). The reaction of a β-ketoester, a cyclic diketone, and an aldehyde proceeds under solvent-free conditions and tolerates a variety of aromatic aldehydes. It can also be performed with unmodified H-β-zeolite albeit with lower yield and longer reaction times.

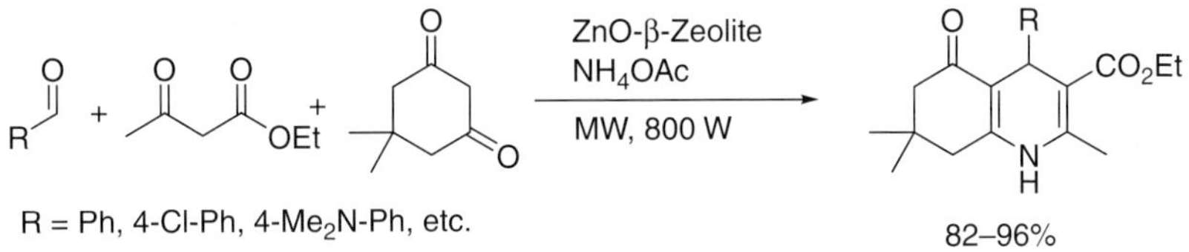

Scheme 10.24

Sucheta *et al.* described the synthesis of 1,5-benzodiazepine derivatives using a zeolite-like catalyst under microwave conditions (Scheme 10.25) [45]. The reaction proceeds without a solvent and gives the products in better yields under microwave versus conventional heating.

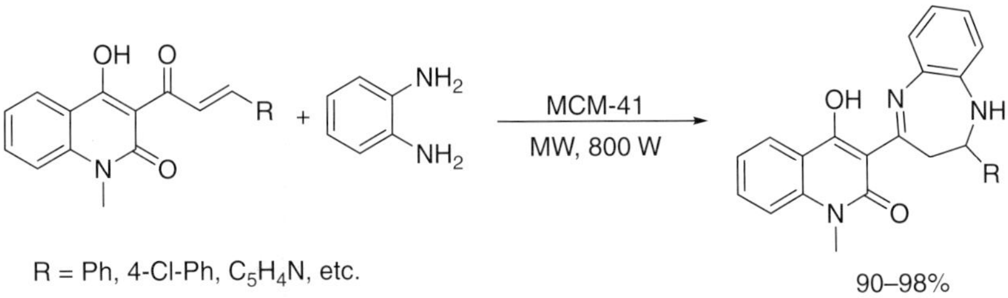

Scheme 10.25

Biginelli reactions yielding 4-aryl-3,4-dihydropyrimidine-2-(1*H*)-ones have also been performed under microwave conditions using zeolites as catalysts [46] (Scheme 10.26). While earlier investigations found the $FeCl_3$-modified catalyst to be best [46b], recent investigations showed that this is not necessary if another type of zeolite is used [46a].

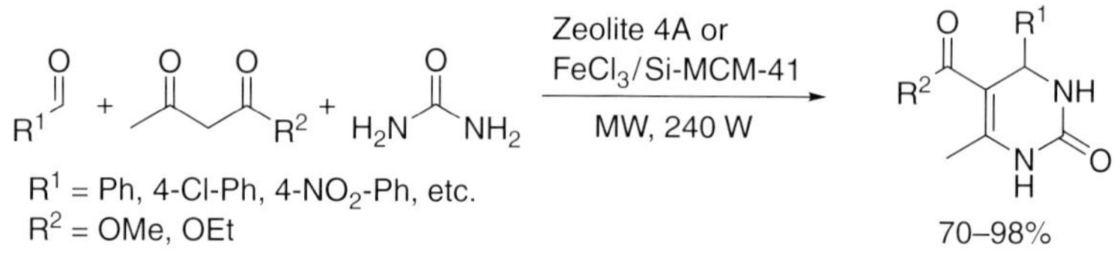

Scheme 10.26

N-Arylisatins have been successfully prepared from *in situ* prepared arynes and methyl-2-oxo-2-(arylamino)acetates [47] (Scheme 10.27). The reaction proceeds with different types of zeolites as catalysts and with a variety of substrates. It is believed that the catalyst assists in the aryne formation as well as in activating the amine for the attack on the aryne.

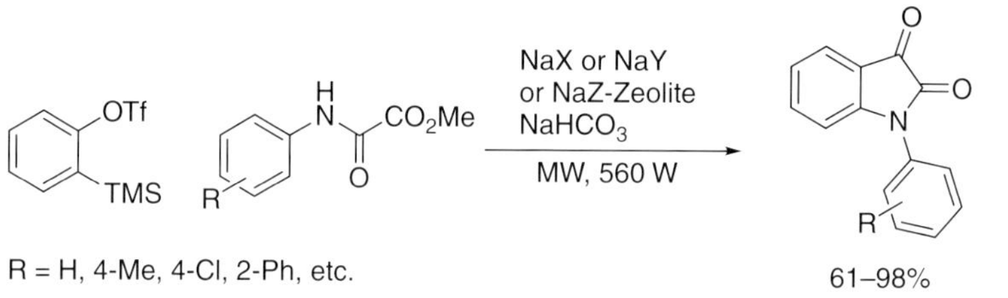

Scheme 10.27

Sulfur-containing heterocycles can also be prepared under microwave heating as demonstrated by the synthesis of thiazolidin-4-ones [48] (Scheme 10.28). The thiazolidinones are formed through the reaction of an imine derivative and thioglycolic acid.

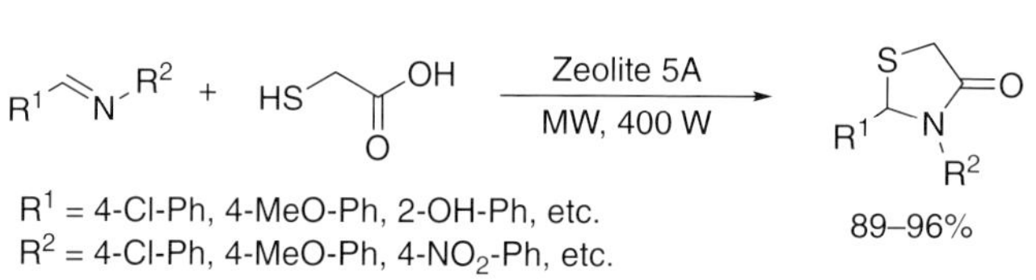

Scheme 10.28

10.4 Microwave Application of Other Solid Acid Catalysts

The above-discussed catalysts were all natural or semisynthetic aluminosilicates of different structures. While the examples clearly dominate microwave-assisted solid acid catalysis, in this chapter, several less structurally homogeneous examples will be summarized. The major classes include HPAs, acidic ion-exchange resins, and metal oxides.

10.4.1 Heteropoly Acids

HPAs are well-known acid catalysts, and due to their stability under aqueous conditions, they are good candidates for the replacement of the environmentally hazardous and corrosive mineral acids [49]. While a broad variety of HPAs are known, the most common and commercially available type is Keggin-type structures such as $H_4[SiMo_{12}O_{40}]$, $H_3[PMo_{12}O_{40}]$, $H_4[SiW_{12}O_{40}]$, and $H_3[PW_{12}O_{40}]$ and their derivatives [50].

The synthesis of esters from alcohols and carboxylic acids was accomplished by novel and rapid methods using HPAs as catalysts under microwave irradiation in the absence of solvent (Scheme 10.29) [51]. The proposed method allows shortening of the reaction time from hours to minutes while maintaining excellent yields.

$$R^1COOH + R^2OH \xrightarrow[\text{MW, 5 min}]{0.05\text{ mol \% } H_3PW_{12}O_{40}} R^1COOR^2 + H_2O$$

R^1 = Me, Ph — 98–99%

R^2 = Bu, C_7H_{15}, $PhCH_2$, C_7H_{15}, C_8H_{17}

Scheme 10.29

A rapid and easy route for the protection of alcohols via microwave-assisted acetylation is a further example for ester formation (Scheme 10.30) [52]. Using a catalytic amount of HPAs (especially V-containing examples), microwave irradiation furnished the corresponding acetates in high yields with reduced reaction times under solvent-free conditions. Effects of P, Si, and transition-metal substitution on the catalytic activity of HPAs of Keggin and Preyssler structures were

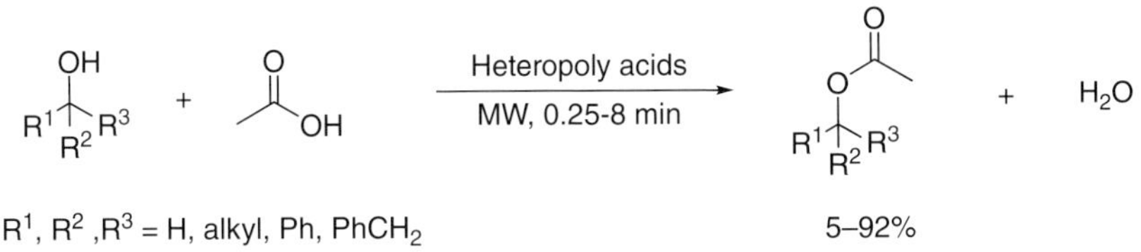

Scheme 10.30

investigated. The P-substituted $H_5PM_{10}V_2O_{40}$ (M = W, Mo) catalysts show excellent activity in the acetylation and appear to be more active than the Si-substituted $H_7SiW_9V_3O_{40}$, $H_4SiW_9Mo_3O_{40}$, and $H_5SiW_9Mo_2VO_{40}$ catalysts.

HPA catalysts such as $H_3PW_{12}O_{40}$, $H_3PMo_{12}O_{40}$, and $H_3PW_{12}O_{40}/SiO_2$ have been used for the synthesis of 4-aminopyrazolo[3,4-*d*]pyrimidine derivatives from 1-substituted-5-amino-4-cyano-pyrazoles and formamide under classical heating and microwave irradiation (Scheme 10.31) [53]. The microwave–solid acid combination leads to a convenient catalytic synthesis of 4-aminopyrazolo[3,4-*d*]pyrimidines with remarkably reduced reaction times from 5–8 h to 8–12 min as compared to traditional conditions. Among the above catalysts, $H_3PW_{12}O_{40}$ showed the best catalytic activity in terms of yields and reaction times.

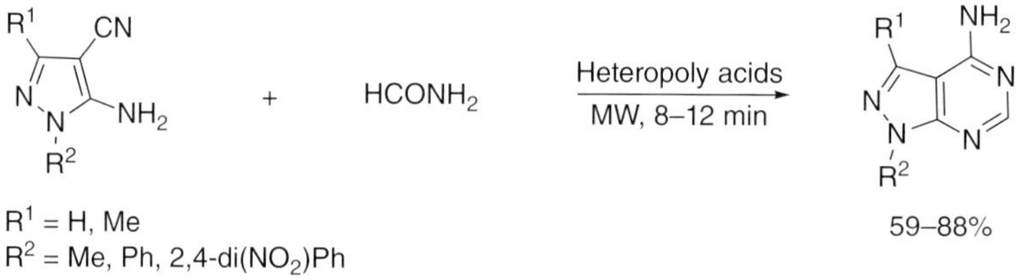

Scheme 10.31

A highly diastereoselective microwave-assisted three-component synthesis of azabicyclo[2.2.2]octan-5-ones by a silicotungstic acid-catalyzed aza-Diels–Alder cyclization is also achieved (Scheme 10.32) [54]. The one-pot process involves the formation of the *in situ* generated Schiff base and its immediate cyclization with cyclohex-2-enone. The short reaction times, good yields, and excellent diastereoselectivity make this annulation a practical and environmentally attractive method for the synthesis of the target compounds.

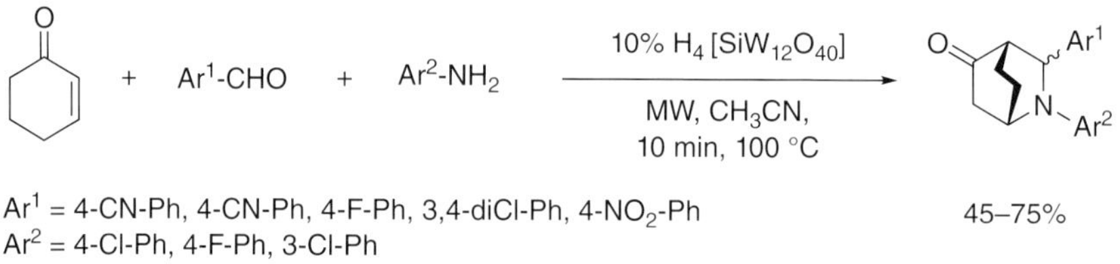

Scheme 10.32

10.4.2 Acidic Ion-Exchange Resins (Nafion-H, Amberlyst, Dowex)

10.4.2.1 Nafion-H

Perfluorinated resin-supported sulfonic acid (Nafion-H) has been extensively used as an ecofriendly catalyst in organic synthesis. Nafion-H is an insoluble resin, inert to corrosive environments, stable up to 210 °C, and easy to recover and recycle [55].

Nafion-H was used as a catalyst for a multicomponent condensation reaction to obtain 4-(3*H*)-quinazolinones in a single pot under microwave irradiation conditions (Scheme 10.33) [56]. The reaction occurs within a few minutes under solvent-free conditions to afford the products in good yields. The catalyst was recovered and could be recycled up to six times without any decline in activity.

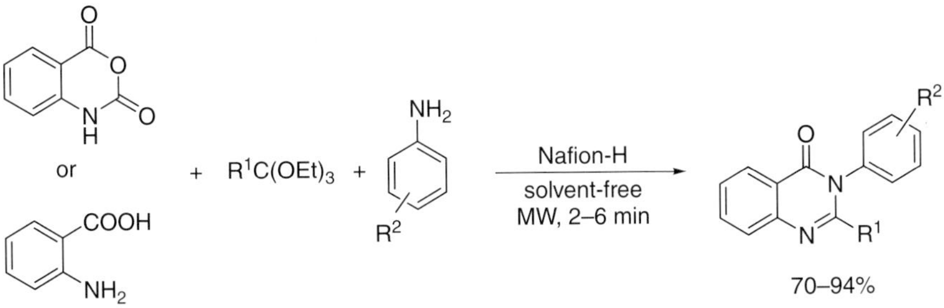

R^1 = H, Me, Et, Ph
R^2 = 3-CF_3, 4-CF_3, 2-Cl-5-CF_3, 2-NO_2-4-CF_3, 2,6-$(NO_2)_2$-4-CF_3, 4-Me

Scheme 10.33

A new Nafion-H catalyzed microwave-assisted synthesis of trifluoromethylimines was described [57]. 1,1,1-Trifluoroacetophenone reacts readily with benzylamine to produce the corresponding imines (Scheme 10.34). The application of microwave irradiation coupled with solvent-free Nafion-H catalysis was employed. The products were isolated in good to excellent yields and high selectivities. Additionally, this process significantly cuts down on the reaction time compared to the traditional heating in toluene, which required 48 h.

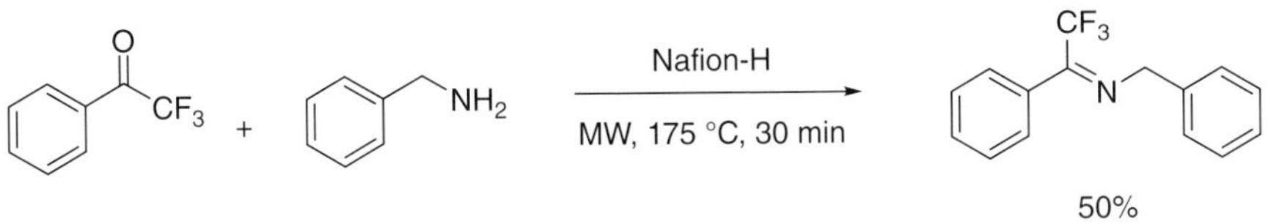

Scheme 10.34

10.4.2.2 Amberlyst

Amberlyst-type resins are prepared in the form of cation- or anion-exchange resins. The H-form of the cation-exchange resins shows acid catalytic activities. The thermal stability of Amberlyst, however, is limited up to 120 °C. One of the most popular and commercially available Amberlysts, Amberlyst-15, is an acidic variation, which functions as a macroreticular-type cation-exchange resin.

Studies indicate that Amberlyst-15 can also be used for nonaqueous catalysis. It is known that the hydrogen form of Amberlyst-15 displays sulfonic acid functionalities [58].

Isatoic anhydride, treated with various kinds of anilines and benzaldehydes, gives dihydroquinazolin-4(1*H*)-ones in the presence of Amberyst-15 catalyst under microwave irradiation (Scheme 10.35) [59]. The catalyst is reusable even after four consecutive runs.

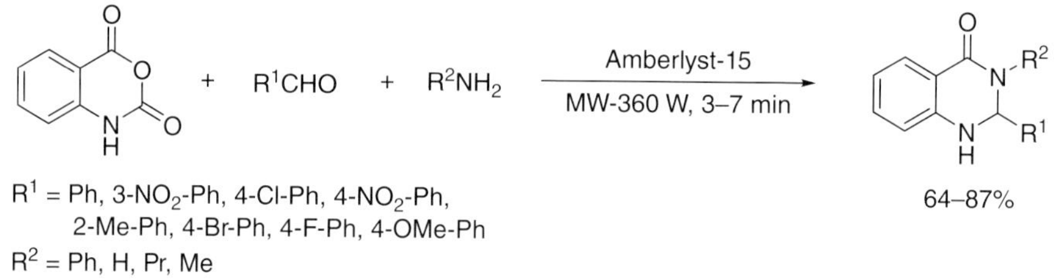

Scheme 10.35

10.4.2.3 Dowex

Similar to Amberlyst, Dowex is a cation- or anion-exchange resin with a polystyrene backbone and sulfonic acid groups.

The microwave-assisted catalytic dehydration of fructose into 5-hydroxymethylfurfural (5-HMF) with Dowex 50WX8 as a catalyst in a mixed organic–aqueous solvent system was reported (Scheme 10.36) [60]. Since the resin is heat sensitive, the applicability of the resin at high temperatures ranging between 100 and 180 °C was investigated under microwave irradiation. The resin works well at temperatures up to 150 °C, without decrease in catalytic activity even after five times of reuse. Compared to conventional sand bath heating, microwave irradiation had a remarkable accelerating effect not only on the fructose conversion but also on selectivity, thus the 5-HMF yield.

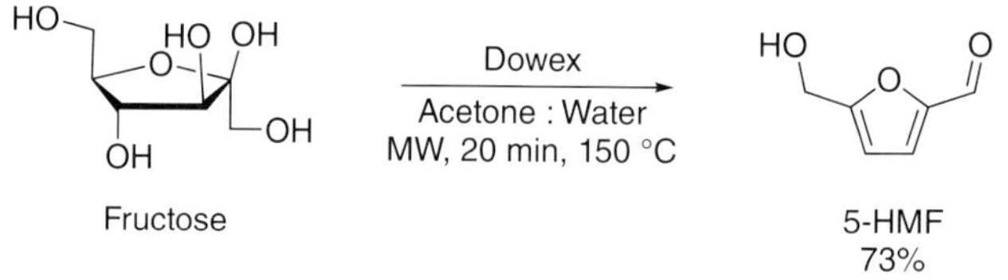

Scheme 10.36

Structurally different epoxides were efficiently converted to the corresponding thiiranes using thiourea on Dowex 50WX8 under solvent-free conditions (Scheme 10.37) [61]. The reactions were carried out either in an oil-bath or under microwave irradiation to give the thiiranes in 75–98% yields within 0.5–120 min. It is shown that the irradiation of a mixture of styrene oxide resulted in a complete reaction within 90 s with 96% yield under microwave irradiation. The scope of the

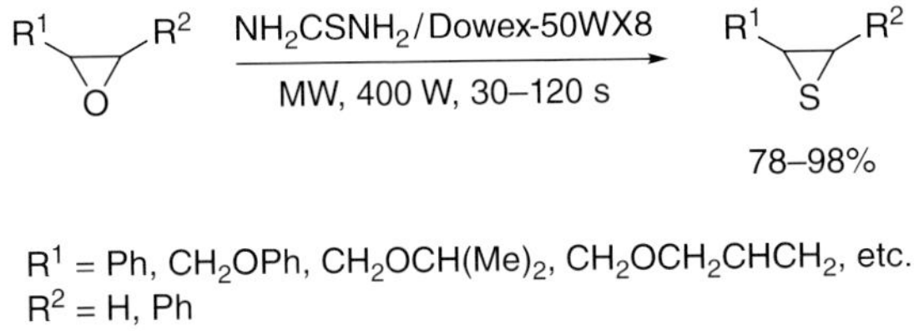

Scheme 10.37

reaction is extended to various epoxides under optimized conditions. A distinct rate enhancement under microwave conditions is observed in addition to better yields (78–98%); the reactions are completed within 0.5–2 min as compared to 15–120 min under conventional heating.

10.5 Conclusions and Outlook

The past two decades have seen an explosion-like development in the application of solid acids; microwave-assisted use is no exception. The widespread use of these catalysts in MAOS is largely aided by environmental concerns and the fact that these materials are usually strong microwave absorbers; hence they could serve as a reaction as well as a heating medium. These facts and the current trends in making the industrial production of chemicals greener initiate significant efforts from microwave instrument manufacturers to produce reactors that are able to significantly scale up existing processes. It is expected that these efforts will result in the extension of MAOS to industrial level in the future.

References

1. (a) Olah, G.A., Iyer, P.S., and Prakash, G.K.S. (1986) *Synthesis*, 513–531; (b) Török, B. and Molnár, Á. (1998) *C. R. Acad. Sci. Paris, Ser. II c*, 381–396; (c) Balogh, M. and Laszlo, P. (1993) *Organic Chemistry Using Clays*, Springer-Verlag, Berlin, Heidelberg; (d) Kozhevnikov, I.V. (2002) *Catalysis by Polyoxometalates*, John Wiley & Sons, Ltd, Chichester; (e) Gates, B.C. (2003) in *Encyclopedia of Catalysis*, vol. **2** (ed I.T. Horvath), John Wiley & Sons, Inc., New York, pp. 104–142.
2. (a) Abid, M., Török, B., and Huang, X. (2009) *Aust. J. Chem.*, **62**, 208–222; (b) Kulkarni, A. and Török, B. (2011) *Curr. Org. Synth.*, **8**, 187–207; (c) Bag, S., Dasgupta, S., and Török, B. (2011) *Curr. Org. Synth.*, **8**, 237–261.
3. (a) Tierney, J.P. and Lidstrom, P. (2009) *Microwave-Assisted Organic Synthesis*, 2nd edn, Wiley-Blackwell, Oxford; (b) de la Hoz, A. and Loupy, A. (eds) (2012) *Microwaves in Organic Synthesis*, 3rd edn, Wiley-VCH Verlag GmbH, Weinheim; (c) Leadbeater, N.E. (2011) *Microwave Heating as a Tool for Sustainable Chemistry*, CRC Press, Boca Raton, FL; (d) Kappe, C.O., Stadler, A., and Dollinger, D. (2012) *Microwaves in Organic and Medicinal Chemistry*, 2nd edn, Wiley-VCH Verlag GmbH, Weinheim.
4. (a) Polshettiwar, V. and Varma, R.S. (2008) *Pure Appl. Chem.*, **80**, 777–790; (b) Mehta, V.P., Appukkuttan, P., and Van der Eycken, E. (2011) *Curr. Org. Chem.*, **15**, 265–283; (c) Kruithof, A.,

Ruijter, E., and Orru, R.V.A. (2011) *Curr. Org. Chem.*, **15**, 204–236; (d) Patil, S.A., Patil, R., and Miller, D.D. (2011) *Curr. Med. Chem.*, **18**, 615–637.
5. (a) Dasgupta, S. and Török, B. (2008) *Org. Prep. Proced. Int.*, **40**, 1–65; (b) Dasgupta, S. and Török, B. (2008) *Curr. Org. Synth.*, **5**, 321–342; (c) Martins, M.A.P., Frizzo, C.P., Moreira, D.N., Dayse, N., Buriol, L., and Machado, P. (2009) *Chem. Rev.*, **109**, 4140–4182.
6. (a) Theng, B.K.G. (1974) *The Chemistry of Clay-Organic Reactions*, Halsted Press-a Wiley Division, New York; (b) Benesi, H.A. and Winquest, B.H.C. (1978) *Adv. Catal.*, **27**, 97–182; (c) Vaccari, A. (1998) *Appl. Clay Sci.*, **14**, 161–198.
7. (a) Varma, R.S. (2002) *Tetrahedron*, **58**, 1235–1255; (b) Polshettiwar, V. and Varma, R.S. (2008) *Acc. Chem. Res.*, **41**, 629–639; (c) Polshettiwar, V., Nadagouda, M.N., and Varma, R.S. (2009) *Aust. J. Chem.*, **62**, 16–26.
8. (a) Abid, M., Spaeth, A., and Török, B. (2006) *Adv. Synth. Catal.*, **348**, 2191–2196; (b) Abid, M., Landge, S.M., and Török, B. (2006) *Org. Prep. Proced. Int.*, **35**, 495–500.
9. Debray, J., Zeghida, W., Baldeyrou, B., Mahieu, C., Lansiaux, A., and Demeunynck, M. (2010) *Bioorg. Med. Chem. Lett.*, **20**, 4244–4247.
10. De Paolis, O., Teixeira, L., and Török, B. (2009) *Tetrahedron Lett.*, **50**, 2939–2942.
11. Borkin, D.A., Puscau, M., Carlson, A., Solan, A., Wheeler, K.A., Török, B., and Dembinski, R. (2012) *Org. Biomol. Chem.*, **10**, 4505–4508.
12. Outerbridge, V.M., Landge, S.M., Tamaki, H., and Török, B. (2009) *Synthesis*, 1801–1806.
13. Biradar, J.S. and Sharanbasappa, B. (2011) *Synth. Commun.*, **41**, 885–890.
14. (a) Landge, S.M. and Török, B. (2008) *Catal. Lett.*, **122**, 338–343; (b) Solan, A., Nişanci, B., Belcher, M., Young, J., Schäfer, C., Wheeler, K.A., Török, B., and Dembinski, R. (2014) *Green Chem.*, **16**, 1120–1124.
15. Perro Neves, M., Cidade, H., Pinto, M., Silva, A.M.S., Gales, L., Damas, A.M., Lima, R.T., Vasconcelos, M.H., and de São José Nascimento, M. (2011) *Eur. J. Med. Chem.*, **46**, 2562–2574.
16. Suresh, D., Kanagaraj, K., and Pitchumani, K. (2014) *Tetrahedron Lett.*, **55**, 3678–3682.
17. Kulkarni, A. and Török, B. (2010) *Green Chem.*, **12**, 875–878.
18. Naskar, S., Paira, P., Paira, R., Mondal, S., Maity, A., Hazra, A., Sahu, K.B., Saha, P., Banerjee, S., Luger, P., Webe, M., and Mondal, N.B. (2010) *Tetrahedron*, **66**, 5196–5203.
19. (a) Baran, P.S. and Richter, J.M. (2005) *J. Am. Chem. Soc.*, **127**, 15394–15396; (b) Kulkarni, A., Quang, P., and Török, B. (2009) *Synthesis*, 4010–4014.
20. (a) Abid, M. and Török, B. (2005) *Adv. Synth. Catal.*, **347**, 1797–1803; (b) Török, B., Sood, A., Bag, S., Kulkarni, A., Borkin, D.A., Lawler, E., Dasgupta, S., Landge, S., Abid, M., Zhou, W., Foster, M., LeVine, H. III,, and Török, M. (2012) *ChemMedChem*, **7**, 910–919.
21. Richela, A., Laurenta, P., Watheleta, B., Watheletb, J.P., and Paquot, M. (2011) *Catal. Today*, **167**, 141–147.
22. Rocchi, D., González, J.F., and Menéndez, J.C. (2014) *Molecules*, **19**, 7317–7326.
23. Bandyopadhyay, D. and Banik, B.K. (2012) *Org. Med. Chem. Lett.*, **2**, 15.
24. Sharma, S.M.L. and Singh, J. (2012) *Synth. Commun.*, **42**, 1306–1324.
25. (a) De Paolis, O., Baffoe, J., Landge, S.M., and Török, B. (2008) *Synthesis*, 3423–3428; (b) Landge, S.M., Schmidt, A., Outerbridge, V.M., and Török, B. (2007) *Synlett*, 1600–1604.
26. (a) Mirodatos, C. and Barthomeuf, D. (1981) *J. Chem. Soc., Chem. Commun.*, 39–40; (b) Koltunov, K.Y., Walspurger, S., and Sommer, J. (2004) *Catal. Lett.*, **98**, 89–94.
27. (a) Krishna Mohan, K.V.V., Narender, N., and Kulkarni, S.J. (2006) *Green Chem.*, **8**, 368–372; (b) Chandra Shekara, B.M., Ravindra Reddy, C., Madhuranthakam, C.R., Jai Prakash, B.S., and Bhat, Y.S. (2011) *Ind. Eng. Chem. Res.*, **50**, 3829–3835.
28. (a) Gadhwal, S., Dutta, M.P., Boruah, A., Prajapati, D., and Sandhu, J.S. (1998) *Indian J. Chem., Sect. B: Org. Chem. Incl. Med. Chem.*, **37B**, 725–727; (b) Srinivas,

K.V.N.S., Reddy, E.B., and Das, B. (2002) *Synlett*, 625–627.

29. Mogilaiah, K. and Vasudeva Reddy, N. (2003) *Synth. Commun.*, **33**, 1067–1072.
30. (a) Singh, R. and Kumar, R. (2011) *Orient. J. Chem.*, **27**, 317–320; (b) Singh, R. and Kumar, R. (2012) *Asian J. Chem.*, **24**, 1403–1404.
31. Camargo-Ordonez, A., Moreno-Reyes, C., Olazaran-Santibanez, F., Martinez-Hernandez, S., Bocanegra-Garcia, V., and Rivera, G. (2011) *Quim. Nova*, **34**, 787–791.
32. (a) Bond, G., Gardner, J., McCabe, R., and Shorrock, D. (2007) *J. Mol. Catal. A: Chem.*, **278**, 1–5; (b) Yamashita, H., Mitsukura, Y., and Kobashi, H. (2010) *J. Mol. Catal. A: Chem.*, **327**, 80–86.
33. (a) Winé, G., Vanhaecke, E., Ivanova, S., Ziessel, R., and Pham-Huu, C. (2009) *Catal. Commun.*, **10**, 477–480; (b) Luque, R., Clark, J.H., and MacQuarrie, D.J. (2009) in *Experiments in Green and Sustainable Chemistry* (eds H.W. Roesky and D.K. Kennepohl), Wiley-VCH Verlag GmbH, Weinheim, pp. 3–6; (c) Bai, G., Li, T., Yang, Y., Zhang, H., Lan, X., Li, F., Han, J., Ma, Z., Chen, Q., and Chen, G. (2012) *Catal. Commun.*, **29**, 114–117.
34. Hiroki, K., Hatori, M., Yamashita, H., and Sugiyama, J.-I. (2008) *Chem. Lett.*, **37**, 320–321.
35. (a) Yamashita, H., Mitsukura, Y., Kobashi, H., Hiroki, K., Sugiyama, J.-I., Onishi, K., and Sakamoto, T. (2010) *Appl. Catal., A*, **381**, 145–149; (b) Ng, E.-P., Mohd Subari, S.N., Marie, O., Mukti, R.R., and Juan, J.-C. (2013) *Appl. Catal., A*, **450**, 34–41.
36. Carrillo, A.I., Serrano, E., Luque, R., and Garcia-Martinez, J. (2013) *Appl. Catal., A*, **453**, 383–390.
37. Thakrar, S., Bavishi, A., Radadiya, A., Vala, H., Parekh, S., Bhavsar, D., Chaniyara, R., and Shah, A. (2014) *J. Heterocycl. Chem.*, **51**, 555–561.
38. Arya, K. and Tomar, R. (2013) *Res. Chem. Intermed.*, 1–12.
39. (a) Balalaie, S. and Arabanian, A. (2000) *Green Chem.*, **2**, 274–276; (b) Balalaie, S., Arabanian, A., and Hashtroudi, M.S. (2000) *Monatsh. Chem.*, **131**, 945–948.
40. Thompson, M.J., Hurst, J.M., and Chen, B. (2008) *Synlett*, 3183–3187.
41. Hashtroudi, M.S., Nia, S.S., Asadollahi, H., and Balalaie, S. (2000) *Indian J. Heterocycl. Chem.*, **9**, 307–308.
42. (a) Saberi, A. and Rangappa, K.S. (2009) *Synth. React. Inorg., Met.-Org., Nano-Met. Chem.*, **39**, 425–427; (b) Mobinikhaledi, A., Zendehdel, M., and Jamshidi, F.H. (2007) *Synth. React. Inorg., Met.-Org., Nano-Met. Chem.*, **37**, 175–177.
43. (a) Montazeri, N., Pourshamsian, K., Fomani, A., and Kalantarian, S.J. (2012) *Asian J. Chem.*, **24**, 2805–2807; (b) Bakavoli, M., Sabzevari, O., and Rahimizadeh, M. (2007) *Chin. Chem. Lett.*, **18**, 533–535.
44. Katkar, S., Arbad, B., and Lande, M. (2011) *Arabian J. Sci. Eng.*, **36**, 39–46.
45. Sucheta, K. and Rao, B.V. (2005) *Indian J. Chem., Sect. B: Org. Chem. Incl. Med. Chem.*, **44B**, 2152–2154.
46. a) Meenakshi, C. and Christopher, E. (2010) *Asian J. Chem.*, **22**, 5784–5786; (b) Choudhary, V., Tillu, V., Narkhede, V., Borate, H., and Wakharkar, R. (2003) *Catal. Commun.*, **4**, 449–453.
47. Singh, R. and Kumar, R. (2013) *Asian J. Chem.*, **25**, 4935–4938.
48. (a) Vandana, T., Jyostna, M., and Parvez, A. (2010) *Pharma Chem.*, **2**, 187–195; (b) Meshram, J., Ali, P., and Tiwari, V. (2010) *Green Chem. Lett. Rev.*, **3**, 195–200; (c) Rawal, M.K., Shrimali, K., Sahu, A., Sitha, D., Sharma, V.K., and Punjabi, P.B. (2010) *Int. J. Chem. Sci.*, **8**, 943–950.
49. Olah, G.A., Prakash, G.K.S., Williams, R.E., Wade, K., and Molnar, A. (2011) *Hydrocarbon Chemistry*, 2nd edn, John Wiley & Sons, Inc., Hoboken, NJ.
50. Misono, M. (2003) in *Encyclopedia of Catalysis*, vol. **3** (ed I.T. Horvath), John Wiley & Sons, Inc., Hoboken, NJ, pp. 433–447.
51. Ozturk, G., Gumgum, B., and Akba, O. (2002) *Catal. Lett.*, **82**, 233–235.
52. Rezaei-Seresht, E., Zonoz, F., Estiri, M., and Tayebee, R. (2001) *Ind. Eng. Chem. Res.*, **50**, 1837–1846.
53. Heravi, M., Rajabzadeh, G., Bamoharra, F., and Seifi, N. (2006) *J. Mol. Catal. A*, **256**, 238–241.
54. Borkin, D., Morzhina, E., Datta, S., Rudnitskaya, A., Sood, A., Török, M.,

and Török, B. (2011) *Org. Biomol. Chem.*, **9**, 1394–1401.

55. Olah, G.A., Prakash, G.K.S., Sommer, J., and Molnár, Á. (2009) *Superacid Chemistry*, 2nd edn, John Wiley & Sons, Inc., Hoboken, NJ.
56. Lingaiah, B., Ezikiel, G., Yakaiah, T., Reddy, G., and Rao, P. (2006) *Synlett*, **15**, 2507–2509.
57. Abid, M., Savolainen, M., Landge, S.M., Hu, J., Prakash, G.K.S., Olah, G.A., and Török, B. (2007) *J. Fluorine Chem.*, **128**, 587–594.
58. (a) Ballini, R. and Petrini, M. (1988) *J. Chem. Soc., Perkin Trans. 1*, 2563; (b) Ramesh, C., Banerjee, J., and Pal, B. (2002) *Adv. Synth. Catal.*, **345**, 557; (c) Yu, W., Hidajat, K., and Ray, A. (2004) *Appl. Catal. A: Gen.*, **260**, 191–205; (d) Feng, X., Guan, C., and Zhao, C. (2004) *Synth. Commum.*, **34**, 487.
59. Surpur, M.P., Singh, P.R., Patil, S.B., and Samant, S.D. (2007) *Synth. Commum.*, **37**, 1965–1970.
60. Qi, X., Watanabe, M., Aida, T.M., and Smith, R.L. (2008) *Green Chem.*, **10**, 799–805.
61. Zeynizadeh, B. and Yeghaneh, S. (2009) *Phosphorus, Sulfur Silicon Relat. Elem.*, **184**, 362–368.

11
Microwave-Assisted Enzymatic Reactions

Takeo Yoshimura, Shigeru Mineki, and Shokichi Ohuchi

11.1
Introduction

An enzyme is a protein molecule that possesses a catalytic function. Specifically, enzymes catalyze a number of reactions in living organisms and play an important role in supporting life. An important feature of an enzyme is that it can only work in very limited reaction conditions, such as with particular substrates and at particular temperatures. A number of studies have been conducted on the use of enzymes as catalysts to efficiently obtain chiral molecules, which are employed as pharmaceutical raw materials. In particular, in 1986, Klibanov [1] revealed that enzymes can function as catalysts in organic solvents, since then the notions of the conditions required by enzymes changed drastically and their uses expanded dramatically. However, in enzyme reactions in nonaqueous systems such as organic solvents, reaction rates were shown to be reduced considerably, and their yields and recovery rates were low even when the specifics of the substrate were ideal; this prevented practical realizations. Therefore, supercritical reaction fields and ionic liquids (ILs) producing dielectric environments equivalent to aqueous systems have been tested as devices for enhancing reaction rates. In addition to this, microwave irradiation was expected to have the effect of promoting reactions and also attracted early attention. Initial characteristic research examples include a series of studies on lipase under microwave irradiation in 1996 [2]. Microwave heating research has advanced and microwave devices for chemical reactions have been developed in recent years, which have enabled high-accuracy temperature measurement and control. In particular, studies of microwave heating have been conducted on immobilized lipase in microwave enzyme reactions over the course of many years. Enzymatic reactions using microwave heating have been studied in a wide range of fields such as pharmaceutical synthesis, extraction, and decomposition, as well as biodiesel fuels and proteome [3].

The reaction temperature is an important factor for enzyme reactions. They are proteins comprising 20 types of amino acids, catalyzing various reactions in living organisms. Single-chain polypeptides, where amino acids

Microwaves in Catalysis: Methodology and Applications, First Edition.
Edited by Satoshi Horikoshi and Nick Serpone.

are polymerized, form characteristic structures such as spiral α helices and β sheets, called *secondary structures.* Furthermore, peptide chains containing these secondary structures form a specific three-dimensional tertiary structure as a whole, which is intensely involved in catalytic functions. The three-dimensional structure of this enzyme selects suitable forms of molecules at active sites, resulting in molecular recognition (substrate specificity and selectivity) – the property of enzymes that makes them decisively different from general acid–base catalysts. Reactions are catalyzed by synergetic actions of functional groups of plural amino acids in substrate molecules incorporated into active sites.

Additionally, it is necessary to use enzymes under limited reaction conditions, such as optimal temperatures and optimal pH, to demonstrate their catalytic functions. The optimal conditions for exhibiting this activity vary by enzyme, and they cannot exert their catalytic functions sufficiently if not under their optimal conditions. Temperature conditions must be regulated when comparing microwave and conventional heating for the accurate comparison of results. It should further be noted that enzymes cannot form any three-dimensional structures under high-temperature conditions and, therefore, lose their catalytic functions. This deactivation of enzymes is irreversible, and they can no longer re-form their original structures. Therefore, it is considered very important to properly set the appropriate reaction conditions, including the temperature, in microwave-assisted enzymatic reactions, as compared with organic chemistry or inorganic chemistry.

The significance of accurately observing the temperature, which is the most fundamental condition in chemical reactions, is indicated in particular as part of the recent evolution of research into microwave chemistry. Kappe *et al.* [4] measured the temperature in a container at three locations (top, middle, and bottom) with a fiber-optic (FO) probe using a microwave device from CEM Corporation and managing the temperature with an infrared (IR) sensor. As a result, it was found that a temperature gradient appeared even within the same reaction vessel. Moreover, by measurement of the container temperature through microwave heating, it was shown that the temperature measurement with the IR sensor did not provide an accurate temperature by comparing the result of an FO probe and a noncontact IR thermometer. Furthermore, they revealed the importance of stirring to reduce the temperature gradient within a single-mode cavity. They concluded that the difference between the results of the observed microwave heating and those of the conventional heating could be attributed to the inability to measure an accurate temperature because the IR sensor was used.

Mochizuki and Wada [5] reviewed accurate temperature measurements in microwave-assisted reactions. The temperature generated by microwave heating can be measured with a thermocouple, an IR thermometer, magnetic resonance imaging (MRI), Raman spectroscopy, and an FO thermometer. However, the thermocouple itself would be heated by the microwave energy. While MRI allows simultaneous measurements at several points of interest, it is expensive and difficult to use. Raman spectroscopy, which works by measuring the Raman peaks of solid surfaces, is hard to apply to the chemical reactions of a liquid. Therefore, measurements using FO and IR are frequently conducted; however, IR is only used to measure the surface temperature, as indicated by Kappe *et al.*, which is a different temperature from that in the reaction solution. On the other hand, the use of the FO thermometer ensures a more accurate measurement of the temperature, but it is not the ideal method for measuring the temperature gradient in the interior of the reaction. For these reasons, Mochizuki and Wada [5] put forward the need to reconsider the results of the previous microwave (MW) chemical reactions.

Kappe *et al.* [6] compared temperature measurement experiments and simulations – focusing on the types of solvents, the impacts of containers, the material properties of containers, and the effects of stirring – using the CEM Discover LabMate System, the network analyzer (Agilent 8753), and the COMSOL Multiphysics. They measured the loss tangent ($\tan\delta = \varepsilon''/\varepsilon'$) using the coaxial probe method for solvents and the cavity resonator method for Pyrex tubes. To investigate errors between experimental data and simulations, they measured the temperature for four types of solvents (bmimPF6, EtOH, MeCN, hexane) and used hexane, which had the smallest error, for the subsequent simulations. Kappe *et al.* demonstrated from the results of several simulations that electric field and power density gradients exist at the edges of stirrers, in internal temperature probes, and around other projections, generating large temperature gradients. They also indicated that the material properties of reaction vessels and the amount of solvents have significant effects on electric field distribution.

Sturm *et al.* [7] also demonstrated the existence of uneven heating in single-mode microwave heating from experiments for examining heating changes using thermal Fax paper and simulations using COMSOL Multiphysics. Using the CEM Discover system, with a hole for observing the interior portions from the front and left sides, they filled three test tubes of differing volumes with water from the bottom to a height of 40 mm, attaching a Fax paper (thermal paper) to each tube, and conducted microwave heating until the water came to a boil. They also measured the temperature at four locations in the interior of the test tubes with an FO sensor. As a result, they demonstrated that the heating patterns of the Fax papers were different depending on the container, and that the heating was uneven. Moreover, it was shown from the simulation by COSMOL that the heating in the test tube interiors was uneven.

Additionally, Kappe *et al.* [8] reported the importance of temperature measurement and stirring under a microwave environment, using the microwave device of Anton Paar, Inc. – Microwave 300 – enabling simultaneous temperature measurement with the FO probe and IR sensor. Using the FO probe and IR sensor, authors examined the difference in measured temperature due to the stirring speed of *N*-methyl-2-pyrrolidone (NMP), which has high microwave absorbency, high viscosity, and high boiling point. No differences were observed in the temperature profiles between IR and FO when comparing the temperature measurement of water at 200 °C (3 ml, 100 rpm). On the other hand, as a result of measuring the temperature with IR when the temperature was turned to 200 °C by changing the stirring speed or the capacity of NMP, it was shown that the temperature did not reach 200 °C when the stirring speed was low (200 rpm or less), no stirring was performed, or the capacity was large (5 ml). In other words, it was demonstrated that attention should also be given to stirring conditions for achieving a constant temperature in reaction solutions because temperature measurements by stirring can vary depending on the content of microwave heating. Furthermore, it was shown that the measured temperature varied between FO and IR when products turned into polymers by chemical reactions and became viscous.

Additionally, Kappe [9] compiled likely errors during temperature measurements and monitoring in reactions under microwave irradiation during their tutorial review using three microwave devices: CEM Discover, the Anton Paar Monowave 300, and Biotage Initiator. One of the notable results was to show that the measured temperature was different for each microwave device when conducting microwave heating, while maintaining 3 ml of bmibBr at 100 °C, as the measurement temperature of IR from the outputs of the three microwave devices and the results of the temperature profiles of IR/FO. This result suggests that it is not possible to reproduce the same experimental conditions through IR control when reactions are conducted with different devices. Moreover, the IR temperature sensor for measuring the container surface would not reflect the reaction temperature accurately, whereas the FO thermometer is of favorable sensitivity and high accuracy but is used only for the temperature measured near the measurement section. By taking measurements inside and outside a sample using both IR and FO, it is possible to read the reaction temperature accurately. It was concluded that temperature data with higher reliability can derive more reproducible microwave chemistry.

From the above studies, temperature measurement methods in microwave chemistry, existence of uneven heating, importance of stirring, and temperature issues among microwave devices were indicated. Microwave heating entails more issues that must be considered compared with conventional heating, including the mode (multi or single), the frequency, the output, the shape and material property of containers, the reaction volumes, the stirring rates, the viscosity of products, the thermometers (thermocouple, IR, FO probe), the presence or absence of cooling devices, and the pressure.

Therefore, we consider that microwave heating can be made into a useful tool to help enzyme reactions by combining the studies for each device used to generate microwave heating. It is possible to limit output sources and the frequency of the microwave reaction as well as vessels and reaction scales by bringing together papers for each device, enabling a more effective comparison of respective findings. Furthermore, to demonstrate the microwave effect for enzymatic reactions, we introduced that enzyme reaction in this paper was limited to lipases, glucosidase, aldolase, and β-galactosidase.

11.2 Synthewave (ProLabo)

11.2.1 Lipase

The group of Legoy reported that microwave radiation can increase the rate of enzyme-catalyzed reactions (Scheme 11.1) in organic media [2a]. The reaction mixture in a quartz vessel (capacity, 20 ml) was placed in a microwave oven (Synthewave S402). A glass paddle stirrer was used to stir the reaction at a stirring rate similar to that used for the classically heated system. The temperature of the reaction mixture was typically set to 50 °C and measured using a noncontact IR continuous feedback temperature system. Microwave irradiation (2.45 GHz, 50 °C) of the hydrated lipase enzyme suspended in an organic medium enhanced the reaction rate by two to three times compared to classical heating. However, they showed that when the water activity (a_w) was increased to 0.97, the reverse effect was observed.

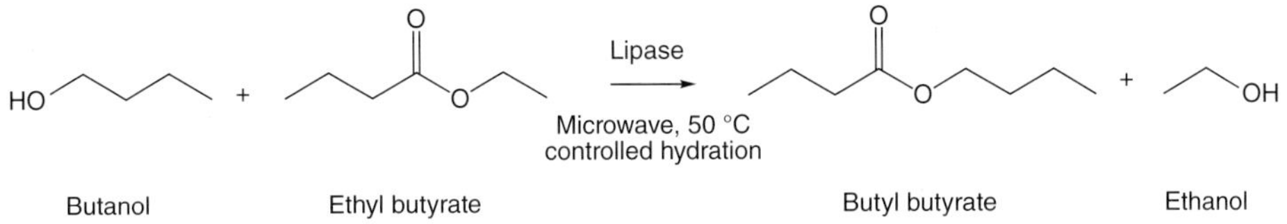

Scheme 11.1 Synthesis of butyl butyrate between butanol and ethyl butyrate.

Loupy and coworkers [2b] demonstrated the lipase-catalyzed esterification of some α-D-glucopyranosides in dry media under focused microwave irradiation. Immobilized *Candida antarctica* lipase (Novozym 435) was employed to catalyze the esterifications of methyl α-D-glucopyranoside, α-D-glucose, and α,α-trehalose with dodecanoic acid in dry media under focused microwave irradiation and classical heating conditions at 95–110 °C (Scheme 11.2). The reactions were performed in a Synthewave S402 with a focused electromagnetic field. The system was fitted with a stirring system and an IR temperature detector to indicate the surface temperature. Consequently, the advantages of performing the reactions in the microwave reactor were evident in all the cases in terms of yield and/or product purity.

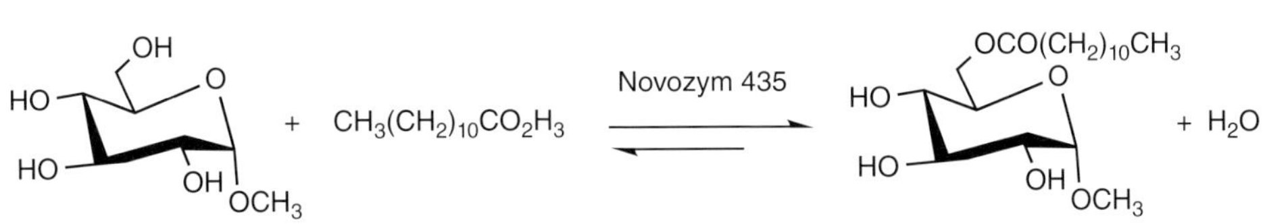

Scheme 11.2 Regioselective esterification of methyl-α-D-glucopyranoside.

Petit *et al.* [2c] showed that microwave irradiation increased the enzymatic affinity and selectivity of supported lipases in esterification and transesterification reactions performed in dry media at temperatures near 100 °C (Scheme 11.3). The reactions were performed in a Synthewave S402 fitted with a stirring system and an IR temperature detector to measure surface temperature. The lipases selected for our study were the *Pseudomonas cepacia* lipase (LP) and *C. antarctica* lipase (SP 435). At the optimal temperature (90 °C for water or 85 °C for ethanol elimination), the initial rates and enantiomeric ratios *E* were significantly enhanced under microwave irradiation. The specificity of microwave effects as compared to classical heating is evident.

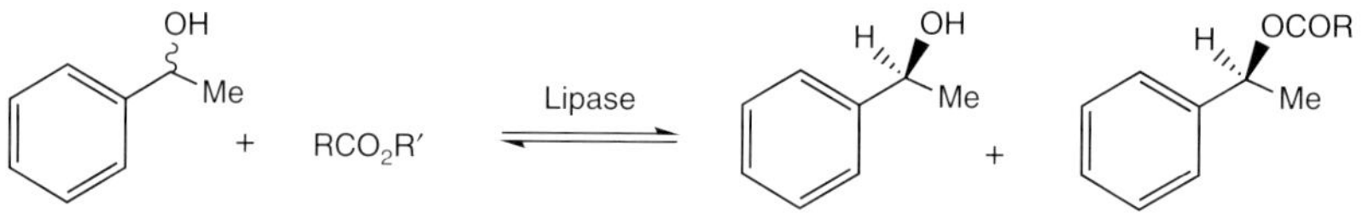

1 *R, S* **1** *S* **2** *R*

Scheme 11.3 Resolution of racemic 1-phenylethanol.

Legoy *et al.* [10] conducted the selective enzymatic esterification of free fatty acids obtained from blackcurrant oil by *n*-butanol using four immobilized lipases (immobilized lipases from *Candida cylindracea*, *Mucor miehei*, and *P. cepacia* and Lipozyme) under microwave irradiation and classical heating. The Synthewave S402 was employed to perform enzymatic reactions under microwave irradiation. The quartz reactor was equipped with a variable speed rotation system, visual control system, and IR temperature detector. Microwave irradiation had positive effects on the chemical yields of the enzymatic reaction products as well as the specificity of the lipases compared to control heating in an incubator equipped with shaking applied during the identical enzyme-mediated processes. The maximum quantity of γ-linolenic acid (30%) was obtained with Lipozyme under microwave heating.

Roy's group reported the first example of a lipase-catalyzed chemoselective reduction of various azidoarenes to areneamines in organic media under microwave irradiation [11]. PhN_3 (0.11 g) in MeOH (10 ml) was placed in the reaction vessel of a Synthewave S402 reactor. Lipase (0.5 g) was then added and the mixture was stirred. The program for the reaction was adjusted to limit the temperature to almost 50 °C for 5 min. After the usual workup and chromatographic purification, the corresponding amine was obtained in 80% yield. Similar treatment of other azides gave the corresponding amines in 70–80% yields.

Réjasse and coworkers studied both "immobilized" and "free" lipase B from *C. antarctica* under microwave irradiation. Both the reactions were performed in a Synthewave S402 microwave oven (300 W monomode system) equipped with a variable-speed rotation system and an IR temperature detector. The influence of microwave heating on the stability of "immobilized" *C. antarctica* lipase B was studied at 100 °C in an organic medium [12]. *C. antarctica* lipase

B was immobilized on macroporous polyacrylic resin beads (Novozym 435). The model reaction chosen to test the enzymatic activity after incubation at 100 °C was the alcoholysis between ethyl butyrate and butanol conducted in equimolar conditions without adding any solvent (Scheme 11.4). The microwave irradiation was carried out before the enzymatic reaction or during the enzymatic catalysis. In both the cases, enzymatic stability was higher under microwave heating than that under conventional heating. Furthermore, the enhancement of enzymatic stability under microwave heating appears to be higher in a more polar solvent.

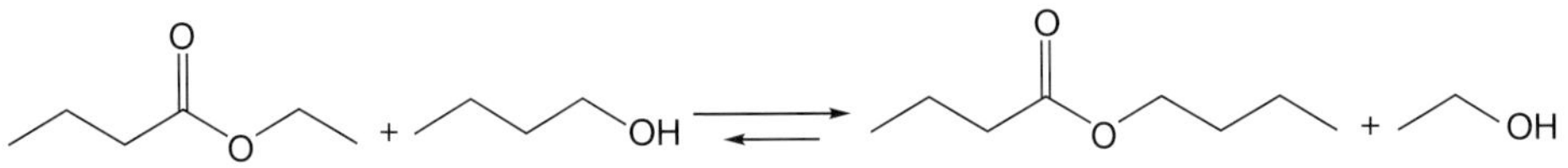

Scheme 11.4 Synthesis of butyl butyrate.

Réjasse *et al.* [13] also studied the influence of the activity and stability of "free" *C. antarctica* lipase B under microwave irradiation. In a previous study, they investigated the effect of microwave heating on "immobilized" *C. antarctic*a lipase B activity at 100 °C in a nonaqueous medium. Free forms of lipase are more thermosensitive than the immobilized forms, enabling us to vary the temperature in this study from 40–110 °C. Lipase activity was tested with a transesterification model reaction between ethyl butyrate and butanol in a solvent-free medium (Scheme 11.4). With respect to the activity, identical initial rates and conversion yields were obtained under microwave radiation and classical thermal heating for the alcoholysis between ethyl butyrate and butanol in a solvent-free system. On the other hand, the kinetics of the free lipase inactivation in butanol appears to be influenced by the heating mode. The Arrhenius plot obtained under classical heating was linear for all studied temperatures, whereas a biphasic Arrhenius plot was obtained under microwave irradiation.

Réjasse and coworkers [14] further reviewed the influence of microwave irradiation on enzymatic properties in aqueous and nonaqueous media. This review is important because the use of microwave irradiation in enzymatic synthesis was still limited at that time. They focused on the influence of microwave irradiation on enzymatic properties such as enzymatic activity, selectivity, and stability. The first part of the review described the early studies on enzymatic reactions under microwave irradiation. The second part described the use of microwave energy in biocatalysis to discuss the future potential of microwave-assisted enzyme-catalyzed synthesis. The authors concluded that the properties of irradiated enzymes are identical to those heated traditionally in aqueous media [15], and that the activity, selectivity, and stability of enzymes can be improved by microwave heating in nonaqueous media. The authors suggested that the specific effect of microwaves may have a thermal origin. The temperature measured during the studies is a macroscopic average parameter of the system. Microwave-assisted reactions could be the consequence of different thermal profiles at the microscopic level. This review showed that microwave heating appears to be a promising technology for improving the enzymatic catalysis sector.

11.2.2
Glucosidase

Maugard *et al.* [16] synthesized galacto-oligosaccharides (GOSs) using immobilized and free β-galactosidase from *Kluyveromyces lactis* (Lactozym 3000 L HP-G) under focused microwave irradiation and conventional heating (Scheme 11.5). Focused microwave irradiations were performed at atmospheric pressure with a Synthewave S402 microwave reactor. The temperature of the reaction was measured using a noncontact IR continuous feedback temperature system. The addition of solvent and performing the reaction under microwave irradiation increased the production of GOS. In combination with the use of immobilized β-galactosidase and cosolvents, such as hexanol, microwave irradiation results in a 217-fold increase in the selectivity for GOS synthesis.

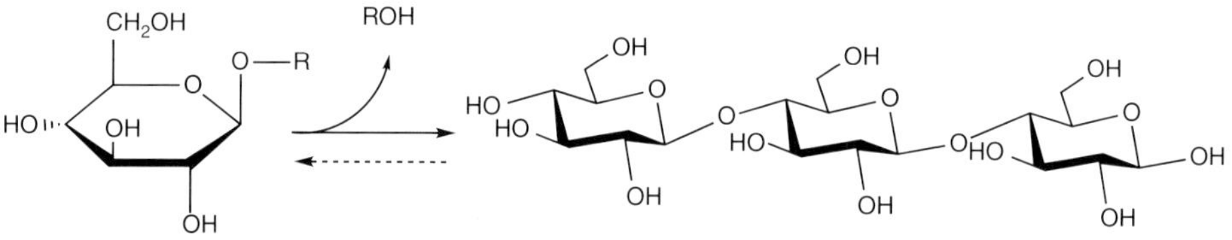

Scheme 11.5 Synthesis of galacto-oligosaccharides (GOS) from lactose using β-galactosidase from *Kluyveromyces lactis* (free and immobilized).

11.3
Discover Series (CEM)

The Discover series instruments of the CEM Corporation are currently the most widely used analytical tools for microwave enzyme reactions. The future of these instruments is based on a single-mode microwave applicator using self-turning circular waveguide technology. The Discover instruments span a range between 0 and 300 W, which is sufficient for rapid heating of most reaction mixtures. An IR sensor performs temperature measurements at the bottom cavity located below the vessel, or alternatively, the assembly can be equipped with an optical-fiber sensor for internal temperature measurement.

11.3.1
Lipase (Synthesis, Esterification)

Yadav and Lathi [17] investigated the transesterification of methyl acetoacetate, a β-keto ester, with various alcohols in the presence of immobilized lipases and microwave irradiation (Scheme 11.6). Novozym 435, Lipozyme TLIM,

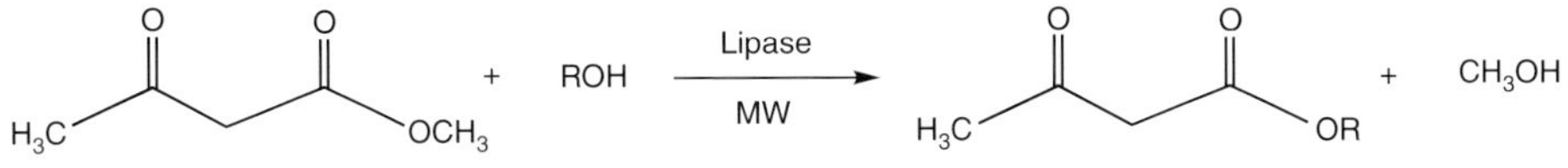

Scheme 11.6 Transesterification of methyl acetoacetate with an alcohol ROH.

and Lipozyme RMIM were screened in the experiment and Novozym 435 was found to be the most active under conventional heating. Primary and secondary alcohols were used to study the effect of chain length on the reaction. The results showed that as the chain length of the alcohol increases, the conversion increases in the presence of microwaves as compared to conventional heating. It was also found that there was a 2.2–4.6 times enhancement of initial activity for microwave-irradiated reactions over conventional reactions. A ping-pong bi–bi mechanism was found to fit the data well for enzyme catalysis alone as well as for microwave-irradiated enzyme catalysis.

Kerep and Ritter [18] investigated the influence of microwave irradiation on lipase-catalyzed ring-opening polymerization of ε-caprolactone in boiling solvents (Scheme 11.7). Five grams of caprolactone was mixed in a flask with 20 g of solvent, and the resulting monomer solution was divided into two parts. Novozym 435 (125 mg) was applied to each of the solutions and placed in the Discover instrument and oil-bath, respectively. Temperature control was achieved using a CEM fiber-optic temperature sensor. For the case of boiling toluene or benzene, the MW-assisted reaction proceeded significantly slower as compared to oil-bath heating. In contrast, boiling with diethyl ester as the solvent showed an increase in the polymerization rate due to MW irradiation.

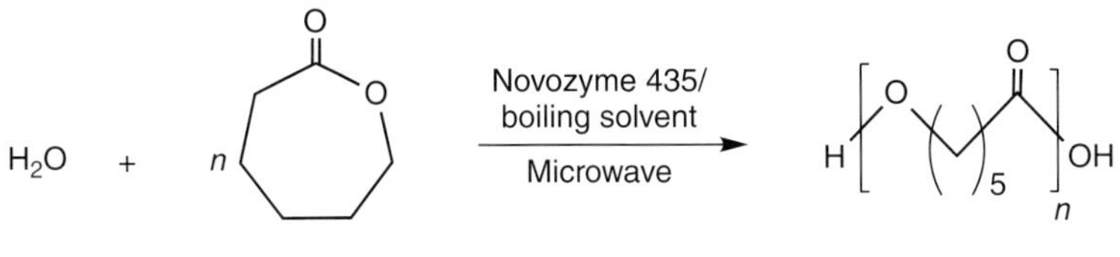

Scheme 11.7 Ring-opening polymerization of ε-caprolactone.

Yadav and Lathi also carried out the microwave-assisted enzymatic synthesis of propylene glycol monolaurate from 1,2-propanediol and lauric acid under microwave irradiation (Scheme 11.8) and investigated the kinetic modeling of microwave-assisted chemoenzymatic epoxidation of styrene. Both experiments were performed using the CEM-SP 1245 with mechanical stirring in a cylindrical

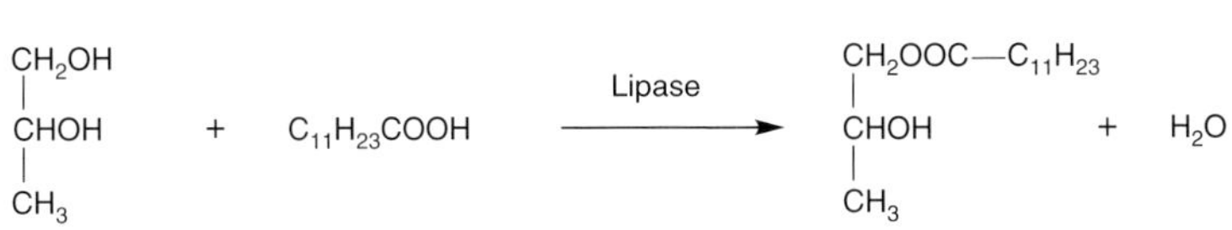

Scheme 11.8 Synthesis of propylene glycol monolaurate from 1,2-propanediol and lauric acid.

glass vessel (how temperature was monitored in the experiments was not mentioned). The synthesis of propylene glycol monolaurate was performed using Novozym 435 (*C. antarctica* lipase B immobilized on polyarylic resin) under conventional heating and microwave irradiation [19]. Reaction conditions were optimized with respect to speed of agitation, catalyst loading, reactant mole ratios, and effect of temperature under the microwave irradiation as well as conventional heating. A synergistic effect was observed on reaction rate and specificity of lipases with microwave irradiation as compared to conventional heating. It was determined that there was no change in the reaction mechanism but an enhancement of 2.84 times in the pre-exponential factor under microwave irradiation at 60 °C with Novozym 435.

The synthesis of styrene oxide by lipase-catalyzed formation of perlauric acid was carried out under conventional and microwave heating (Scheme 11.9) [20]. Microwave-assisted enzymatic perhydrolysis occurred, and both the overall conversion and rate of reaction were higher with microwave irradiation than under conventional heating. Various kinetic parameters affecting the conversion and initial rates of styrene to styrene-oxide were studied including mass transfer, mechanism, kinetic modeling, and deactivation. Under microwave irradiation, Yadav concluded that there was an increase in the frequency factor resulting from enhanced molecular collisions, which can in turn be attributed to the increasing entropy of the system under microwave irradiation.

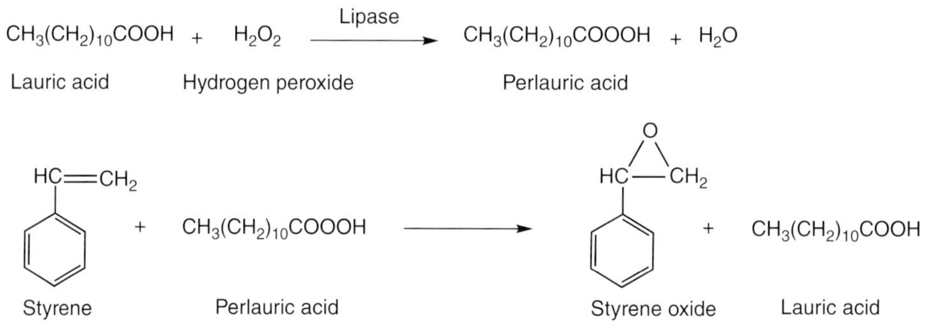

Scheme 11.9 Synthesis of styrene oxide by lipase-catalyzed formation of perlauric acid.

Yadav and Lathi [21] also reported that the synthesis of *n*-butyl diphenyl methyl mercaptyl acetate was performed using lipases (Novozym 435, Lipozyme RMIM, Lipozyme TLIM, and Amano lipase PS-C 1 (PS-C)) under microwave irradiation (Scheme 11.10). The studies were performed in a CEM-SP 1245 reactor with a mechanical stirring and IR temperature sensor. The reactor was a cylindrical glass vessel. Novozym 435 was found to be the best lipase for the reaction, with a conversion rate of 34% in 24 h at 60 °C. The authors reported a ping-pong bi–bi mechanism with substrate *n*-butanol inhibition at the higher concentration suitable for the reaction kinetics.

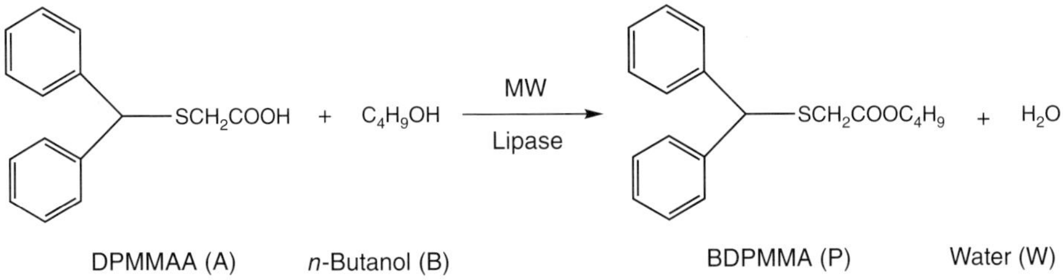

Scheme 11.10 Synthesis of *n*-butyl diphenyl methyl mercapto acetate.

The synthesis of isoniazid from ethyl isonicotinate and hydrazine hydrate in nonaqueous media via lipase-catalyzed hydrazinolysis under both conventional heating and microwave irradiation using different supported lipases was studied (Scheme 11.11). The study was performed in a CEM-SP 1245 with mechanical stirring at appropriate temperature and pressure control. Comparison of the three lipases, Novozym 435, Lipozyme RMIM (*Rhizomucor miehei* lipase), and Lipozyme TMIM (*Thermomyces lanuginosus* lipase), showed that Novozym 435 was the most effective, with a conversion of 54% for equimolar concentrations at 50 °C in 4 h. The rate of reaction and that of final conversion increased synergistically under microwave irradiation in comparison with conventional heating, which showed 36.4% conversion, even after 24 h, in a control experiment.

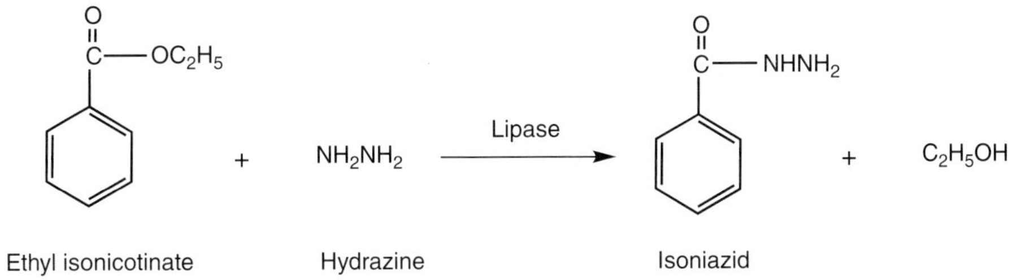

Scheme 11.11 Synthesis of isoniazide using lipase as a catalyst.

The effect of microwave irradiation on lipases has been reported using the Discover series and other microwave applicators since 1996. On the other hand, Wood *et al.* [22] showed noteworthy results of microwave-assisted lipase reactions (Scheme 11.12). The lipase-catalyzed transesterification of methyl acetoacetate in toluene as a solvent was studied using carefully controlled conditions. Lipase acrylic resin from *C. antarctica* was used and microwave reactions were conducted using the CEM Discover equipped with magnetic

Scheme 11.12 Lipase-catalyzed transesterification reaction of methyl acetoacetate.

Table 11.1 Effect of microwaves on lipase-catalyzed transesterification reactions.

Entry	Reaction conditions	Product conversion (%)
1	Conventional heating	31
2	Microwave heating, no cooling	32
3	Microwave heating, air cooling	28
4	Microwave heating, cryogenic cooling	30

stirring. The temperature of the vessel was monitored using a calibrated "fiber-optic probe" inserted into the reaction vessel by means of a sapphire immersion well. The reaction vessels were either round-bottomed flasks or galls tubes. For low-temperature reactions, a specially designed jacketed reaction vessel was used through which a microwave transparent fluid was passed continuously by means of a pump (CEM CoolMate). For these enzyme-catalyzed transformations, no differences were detected between conventional and microwave heating (Table 11.1).

Yadav and Borkar [23] employed Novozym 435 (*C. antarctica* lipase B immobilized on polyacrylic resin) to synthesize citronellol acetate directly from citronellol and vinyl acetate under microwave irradiation and conventional heating (Scheme 11.13). The reactions were performed in a CEM-1245. The reactor was a 120-ml capacity, fully baffled cylindrical glass vessel with provision for mechanical stirring and 4.5-cm internal diameter. The authors concluded that lipase activity increased under microwave heating, owing to enhanced collision of molecules and reduction of the inhibition constant, and that microwave irradiation resulted in an increase in the affinity of the substrate to the active site of lipase. They provided suggestions for increasing the entropy of the system.

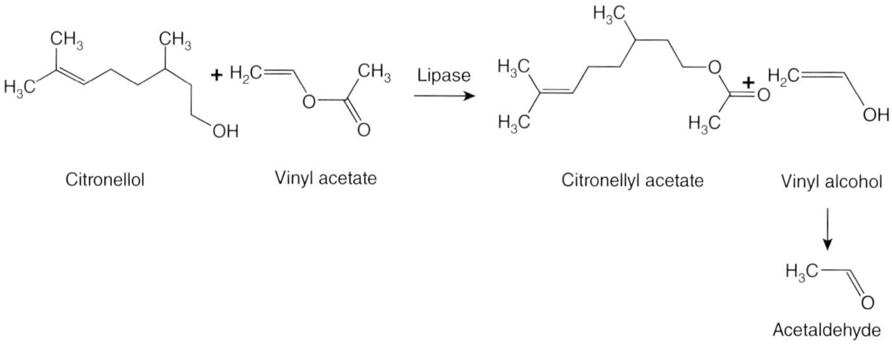

Scheme 11.13 Synthesis of citronellyl acetate from citronellol and vinyl acetate.

Major *et al.* [24] synthesized the ethyl lactate by using an immobilized candida antarctica lipase B (CALB) (Novozym 435) in IL under microwave (CEM BenchMate microwave apparatus equipped with a magnetic stirrer and fiber-optic sensor). The enzymatic synthesis of ethyl lactate was achieved in IL (Cyphos 104 and Cyphos 202). The microwave irradiation resulted in a shorter reaction time (7 h) compared with the conventional conditions (24 h).

To assess whether non-themal microwave effects play a role in the biodiesel synthesis, de Souza and coworkers demonstrated that microwave-assisted biodiesel synthesis is catalyzed by Novozym 435 (Scheme 11.14) [25]. The transesterifications were performed at 40 °C, in both CEM Discover and Monowave 300 (Anton Paar) systems, both of which allow accurate internal reaction temperature measurements using fiber-optic probes. The results did not reveal any differences in the reaction times or yields.

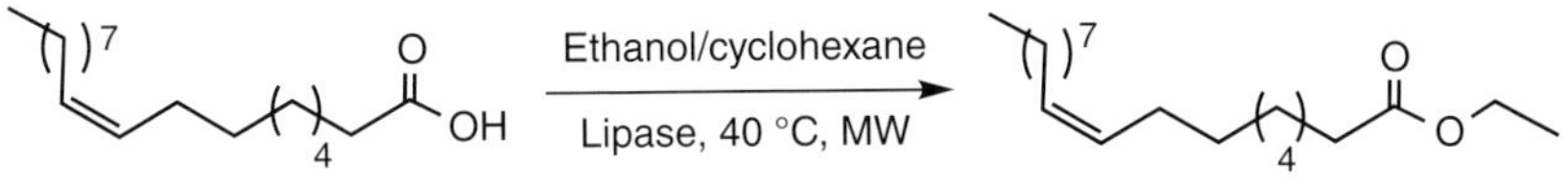

Scheme 11.14 Esterification reaction of oleic acid by Novozym 435.

Yadav and Thorat [26] reported that microwave-assisted lipase (Novozym 435) catalyzed the synthesis of isoamyl myristate in a solvent-free system. A monomode microwave instrument (CEM-SP 1245) was employed to maintain the desired temperature at the set temperature within an accuracy of ±1 °C (the temperature measurement system was not described). The frequency of the microwave generated in the magnetron was 2455 MHz. Isoamyl myristate was successfully synthesized from isoamyl alcohol and myristic acid (Scheme 11.15) via a solvent-free esterification reaction under microwave irradiation. The authors claimed that this solvent-free microwave-assisted enzymatic reaction followed a ping-pong bi–bi mechanism.

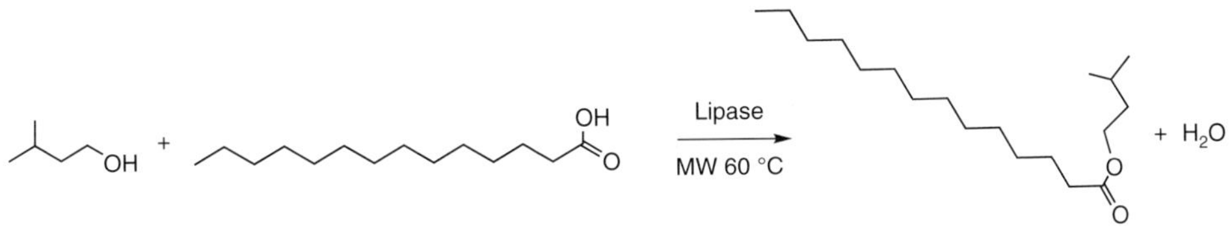

Scheme 11.15 Synthesis of isoamyl myristate by Novozym 435 in solvent-free system.

Ziaullah and Rupasinghe [27] reported an efficient microwave-assisted enzyme-catalyzed regioselective synthesis of long-chain acylated derivatives of flavonoid glycosides (Scheme 11.16). Reactions were performed by three methods: a 10–24-h long conventional method, a solution-phase method, and a solvent-free method. In the microwave-assisted solution-phase method, the reaction mixtures were irradiated for 120–160 s during 12–16 intervals (each interval of 10–20 s) in a CEM microwave reactor with an operating frequency of 2.5 GHz and a noncontact IR temperature sensor. The temperature of the reaction was maintained at 45–60 °C. In the microwave-assisted solvent-free method, flame-dried molecular

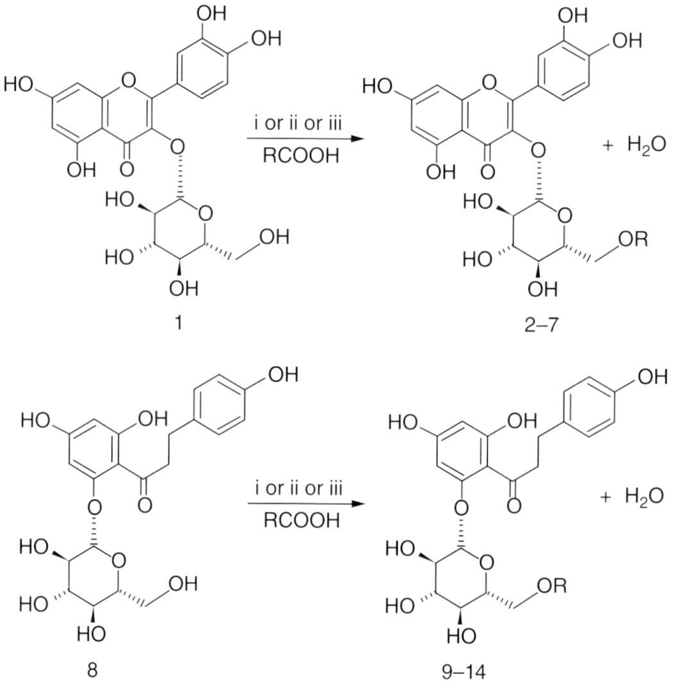

Scheme 11.16 Reagents and conditions: (i) acetone, 4 Å molecular sieves, Novozym 435, 45–60 °C, stirring, 18–24 h. (ii) Acetone, 4 Å molecular sieves, Novozym 435, microwave irradiation, 45–60 °C, stirring, 120–160 s. (iii) Novozym 435, 4 Å molecular sieves, 45–60 °C, microwave irradiation, 75–105 s. R = oleic, stearic, linoleic, α-linolenic, eicosapentaenoic (EPA), docosahexaenoic acids (DHAs), or their esters.

sieves along with four sufficient intervals of time (approximately 75–105 s in 5–7 intervals (15 s each)) were used and the temperature was maintained at 45–60 °C. Microwave irradiation under solvent-free conditions was found to be more efficient than conventional conditions.

Shinde and Yadav [28] recently synthesized alkyl benzoate esters using Novozym 435 as a biocatalyst under the synergistic effect of microwave irradiation. The CEM-1245 model was employed to perform enzymatic reactions under microwave irradiation. The experimental system is shown in Figure 11.1. The synthesis of *n*-hexyl benzoate employed a model reaction (Scheme 11.17). The optimized kinetic parameters were 300 rpm speed of agitation, *n*-hexane as the solvent, 0.02-g cm^{-3} Novozym 435 loading, and 60 °C temperature. Under microwave synergism, an increase in initial rates up to 6.5-fold was observed. The authors found that the reaction followed a ternary complex ordered bi–bi mechanism with inhibition by *n*-hexanol.

Sontakke and Yadav [29] performed the synthesis of ethyl 2-(4-aminophenyl) acetate using Novozym 435 in toluene as a solvent under microwave irradiation

Figure 11.1 Experimental setup for lipase-catalyzed methyl benzoate transesterification under microwave heating. (1) CEM microwave reactor, (2) glass slurry reactor, (3) glass stirrer, (4) Remi's lab stirrer, (5) computer control unit, and (6) Remi's speed regulator. (Reproduced from Ref. [28]. Copyright 2014 by American Chemical Society.)

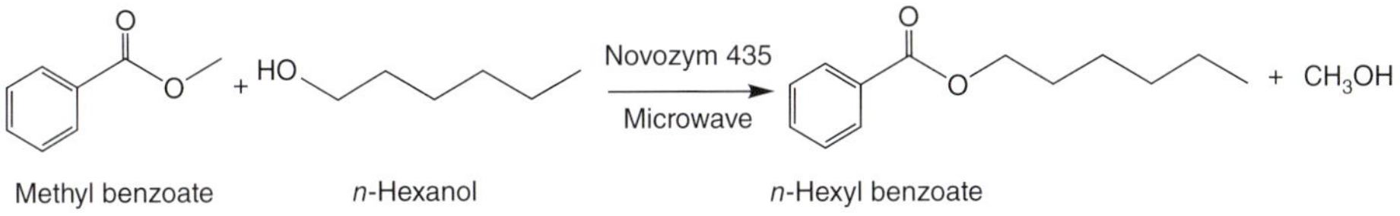

Scheme 11.17 Novozym 435 catalyzed synthesis of *n*-hexyl benzoate.

(CEM-SP 1245) and conventional heating (Scheme 11.18). Reaction parameters such as speed of agitation, temperature (40, 50, 60, and 70 °C), catalyst (Lipozyme TLIM, Lipozyme RMIM, and Novozym 435), enzyme loading, and mole ratio of the reactants were optimized. As a result, a synergistic effect of microwave irradiation and lipases on reaction rate was observed in comparison with conventional heating. The authors concluded that the reaction followed a ping-pong bi–bi mechanism.

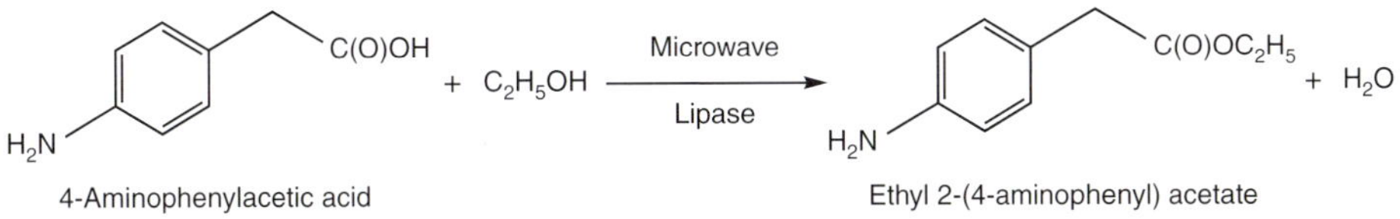

Scheme 11.18 Lipase-catalyzed synthesis of ethyl 2-(4-aminophenyl)acetate.

11.3.2
Enzymatic Resolution

Yu *et al.* investigated the enzymatic resolution of (*R*,*S*)-2-octanol with vinyl acetate as the acyl donor through transesterification under microwave irradiation. A CEM-SP 1245 model was used as the microwave reactor (Scheme 11.19) [30]. The reaction temperature was controlled to ±1 °C with the use of an IR temperature system. It was found that the activity and enantioselectivity of Novozym 435 was enhanced dramatically under microwave irradiation (see Table 11.2). Increased thermal stability and reusability of Novozym 435 under microwave irradiation was also observed. Under optimum conditions, (*S*)-2-octanol was obtained at 50.5% conversion with 99% enatiomeric excess in 2 h under microwave irradiation.

Yu *et al.* [31] used lipase from *Pseudomonas* sp. Lipase (PSL) for sequential kinetic resolution of (*R*,*S*)-2-octanol under microwave irradiation (Scheme 11.19). The supports for immobilization of the catalysts were screened and SBA-15 was selected for PSL immobilization. Microwave reactor CEM-SP 1245 was employed with a noncontact IR temperature system and contained a reaction chamber composed of cylindrical quartz vessels. The *E* value of the immobilized PSL was increased from 23 to 40 under conventional heating, with microwave irradiation for the resolution of (*R*,*S*)-2-octanol (Table 11.3). The immobilized catalysts were then used in sequential kinetic resolution of (*R*,*S*)-2-octanol under microwave irradiation after optimizing the reaction media (Scheme 11.19). Under optimal conditions (*R*,*S*)-2-octanolacetate was obtained at 99% enantiomeric excess with 84% yield in 2 h.

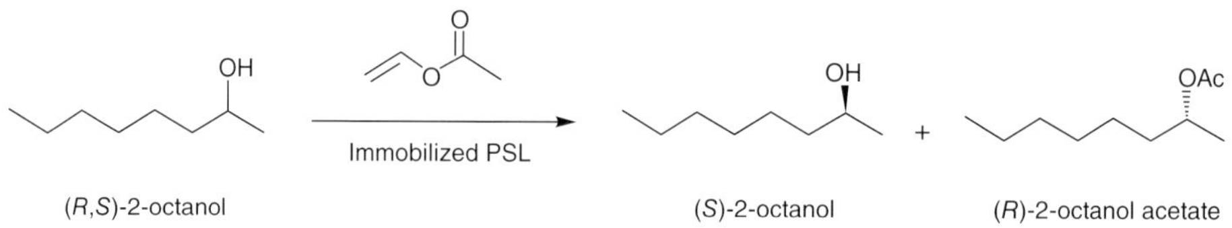

Scheme 11.19 Resolution of (*R*,*S*)-2-octanol by the immobilized PSL.

Table 11.2 Comparison of enzyme activity under microwave and conventional heating.

Substrate	Microwave irradiation		Conventional heating	
	Enzyme activity (μmol min^{-1} mg^{-1})	*E* value	Enzyme activity (μmol min^{-1} mg^{-1})	*E* value
(*R*)-2-Octanol	202.80	332	65.5	109
(*S*)-2-Octanol	0.64	—	0.47	—
(*R*,*S*)-2-Octanol	203.25	328	66.18	104

Table 11.3 Comparison of both activity and enantioselectivity of Novozym 435 in transesterification under microwave irradiation and conventional heating.

Conventional heating					Microwave irradiation				
Time (h)	Conversion (%)	ee_g	ee_p	*E* value	Time (h)	Conversion (%)	ee_g	ee_p	*E* value
16	33.0	43	88	—	0.5	22.4	27	94	—
32	45.8	70	85	23	1	36.9	54	92	40
48	54.9	92	75	—	2	55.8	99	78	—

Bachu *et al.* [32] studied the influence of microwave irradiation on the lipase-catalyzed kinetic resolution of racemic secondary alcohols (Scheme 11.20). Novozym 435 was used as a biocatalyst. *p*-Chlorophenyl acetate was used as the acyl donor and toluene as the solvent. The reaction mixture was stirred at 60 °C in a microwave reactor (CEM Discover) at 70 W at various times. The systems (±)-1-phenyl-1-propanol (**1**), (±)-1-(4-bromophenyl)-propan-1-ol (**3**), (±)-1-phenylbut-3-en-1-ol (**5**), and (±)-3-bromo-2-(2-hydroxypropyl)-1,4-dimethoxy-naphthalene (**7**) were successfully resolved into their respective (*S*)-alcohols and (*R*)-esters, with good enantiomeric excess. Resolution of (±)-ethyl-5-(4-methoxybenzyloxy)-3-hydroxypentanoate (**9**) into (*R*)-alcohol and (*S*)-ester was successful by this method. In addition, microwave-assisted lipase transesterification of meso-symmetric diol **11** affected desymmetrization to ester **12** with high enantiomeric excess. The conversion value for the

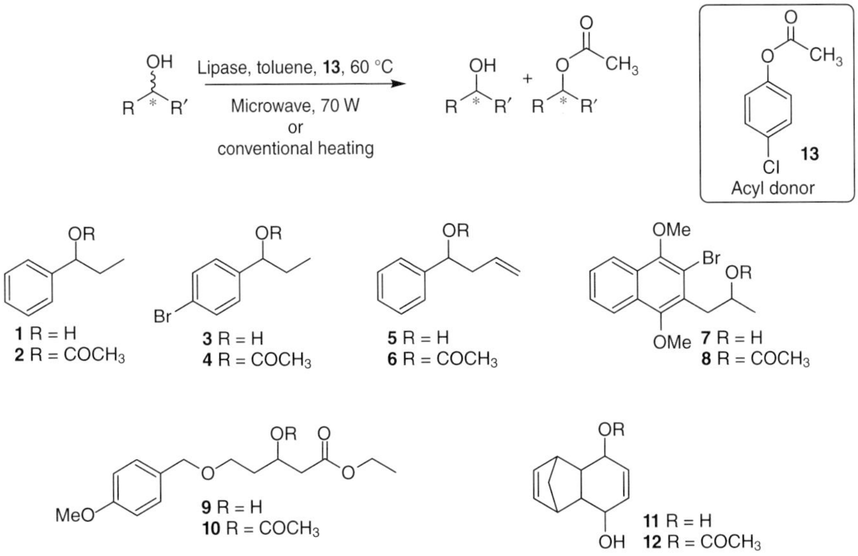

Scheme 11.20 Lipase-catalyzed kinetic resolution of racemic secondary alcohols.

microwave-assisted lipase kinetic resolution of secondary alcohols was higher than that obtained using conventional heating.

Parvulescu *et al.* [33] applied microwave irradiation to the racemization and dynamic kinetic resolution (DKR) of primary benzylic amines. Microwave reactions were performed in a CEM Discover reactor and magnetically stirred. Temperature control was ensured with a vertically focused IR temperature sensor. Racemization reactions catalyzed by 5% Pd/$BaSO_4$ and 5% Pd/$CaCO_3$ were faster and more selective when performed under microwave conditions. The use of microwave irradiation inhibited the formation of side products, such as secondary amines and ethylbenzene. The authors proposed that this result was associated with the selective heating of the metal sites under microwave heating. The use of microwave irradiation for kinetic resolution showed no influence on the activity and enantioselectivity of immobilized *C. antarctica* lipase B. The racemization catalysts were combined in one pot with the biocatalyst Novozym 435 to perform the DKR of benzylic amines under microwave irradiation (Scheme 11.21). Microwave irradiation proved to be an efficient tool for the racemization and DKR of chiral benzylic amines over Pd on alkaline earth metals and lipases.

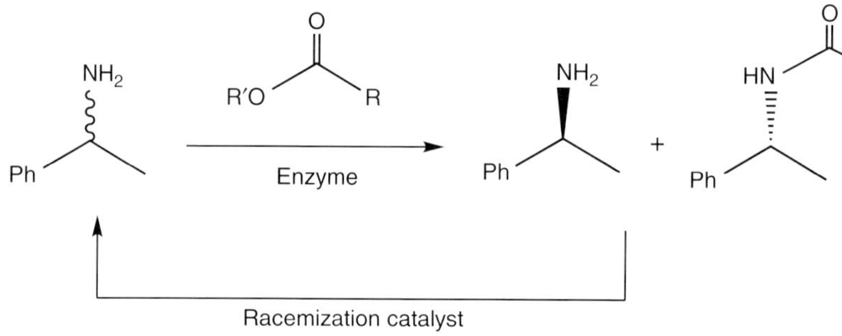

Scheme 11.21 Dynamic kinetic resolution of 1-phenylethylamine.

To probe the existence of nonthermal microwave effects, Kappe *et al.* [34] performed the lipase-catalyzed kinetic resolution of *rac*-1-phenylethanol with vinyl acetate as an acyl donor and cyclohexane as solvent under microwave dielectric heating and conventional heating (Scheme 11.22). Reactions were conducted at 40 °C in dedicated reactor setup that allowed accurate internal reaction temperature measurements with fiber-optic probes. Five immobilized lipases (Novozym

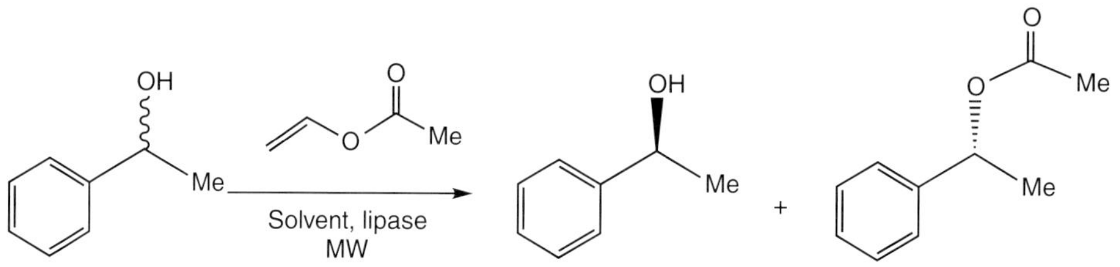

Scheme 11.22 Enzyme-catalyzed kinetic resolution of *rac*-1-phenylethanol.

435, Amano PS-C1, Amano AK, Amano, Novozym TL IM, Novozym RL IM) were examined. The observed reactivities and enantioselectivities in microwave and conventional heating were identical. Kappe *et al.* concluded that the change in both activity and selectivity of the immobilized lipases under irradiation was not observed.

Yadav and Devendran [35] studied the microwave-assisted lipase-catalyzed kinetic resolution of 1-(1-naphthyl) ethanol via transesterification with vinyl acetate in nonaqueous media (Scheme 11.23). A CEM-SP 1245 model microwave reactor, with a cylindrical glass vessel and mechanical stirring, was used. Three lipases (Novozym 435, Lipozyme RMIM, and Lipozyme TLIM) were employed. The reaction and conversion rates were enhanced under microwave irradiation compared with those of conventional heating. The maximum conversion of 47.74% with 90.05% enantioselectivity was obtained in 3 h using 30-mg enzyme loading with equimolar concentration of alcohol and ester at 60 °C under microwave irradiation.

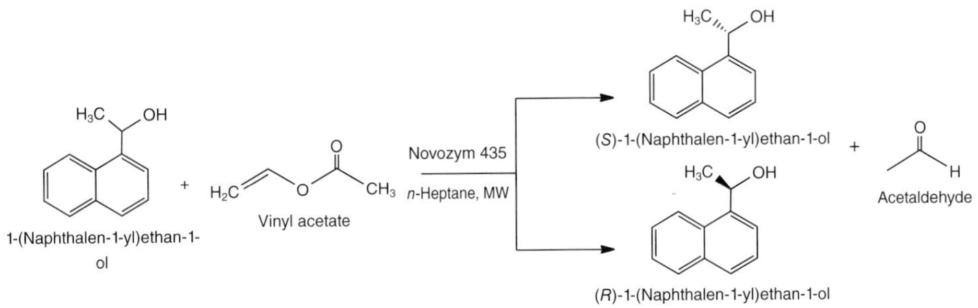

Scheme 11.23 Lipase-catalyzed kinetic resolution of (*R,S*)-1-(1-naphthyl)ethanol.

In addition, enantioselective resolution of DL-(±)-3-phenyllaceric acid using Novozym 435 under microwave irradiation was studied by Yadav and Pawar [36] (Scheme 11.24). Microwave applicator was CEM-SP 1245 model. To hold the constant temperature, microwave power was carried out 30–40 W (however, temperature measurement methods were not described). The effect of temperature on activity of Novozym 435 was attempted in the range of 40, 50, 60, 70 °C by conventional heating and microwave irradiation. The overall conversion was higher under microwave irradiation compared to conventional heating. On the other hand, the microwave effect for enantioselectivity was not shown.

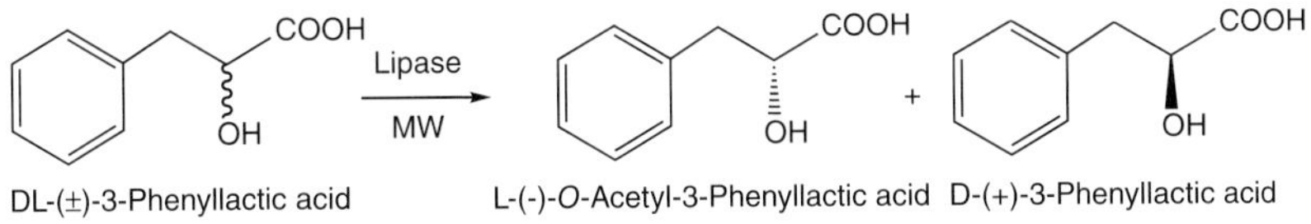

Scheme 11.24 Enzymatic resolution of dl-(±)-3-phenyllactic acid.

Devendran and Yadav [37] performed the microwave-assisted enzymatic kinetic resolution of (±)-1-phenyl-2-propyn-1-ol in nonaqueous media (Scheme 11.25). This microwave study used a CEM-SP 1245 and Novozym 435 as biocatalyst. The authors showed that microwave irradiation results in increased affinity of the substrate to the active site of lipase and enhances the collision frequency of reactant molecules, thus resulting in increased entropy of the system. The maximum conversion of 48.78% was obtained in 2 h using 10-mg enzyme loading with equimolar concentrations of alcohol and ester at 60 °C under microwave irradiation. From progress curve analysis, the authors concluded that the reaction followed the ping-pong bi–bi mechanism with dead-end inhibition of alcohol.

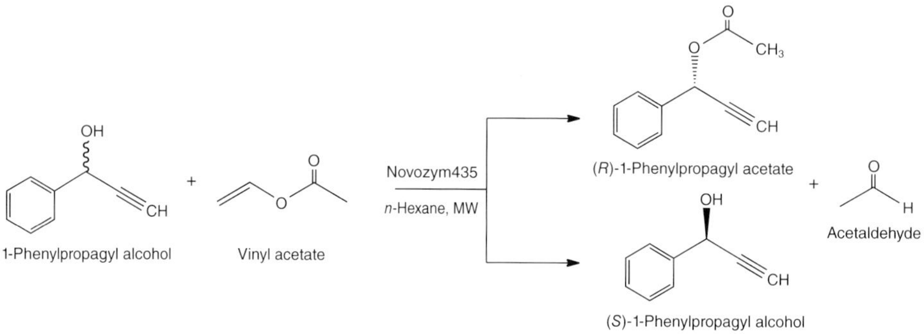

Scheme 11.25 Lipase-catalyzed transesterification of (*R*/*S*)-1-phenylpropagyl alcohol with vinyl acetate.

11.3.3 β-Glucosidase, β-Galactosidase

To clarify the microwave effects during the biocatalysis, Young and coworkers [38] attempted the microwave-assisted enzymatic catalysis (β-glucosidase, α-galactosidase, and carboxylesterase) in an aqueous environment. A β-glucosidase (CelB) from *Pyrococcus furiosus* catalyzed the digestion of exoglycosidic linkages. The reaction mixtures were applied until the reaction temperature reached 40 °C (CEM CoolMate; temperatures were measured using a fiber-optic sensor). As a result, an increase in the enzymatic activity of Pfu CelB of greater than four orders of magnitude was achieved using 300 W of microwave irradiation.

Elling and coworkers [39] reported that the synthesis of nucleotide activated oligosaccharides by transglycosylation with β-galactosidase from *Bacillus circulans* with the application of both microwave and conventional heating. The CEM Discover microwave synthesizer was employed with an FO temperature control system. To effectively cool the reaction mixture, a CEM CoolMate Accessory System was used. The enzymatic hydrolysis of the product was significantly decreased under microwave irradiation leading to stable product formation. The authors demonstrated that the obtained yields and absence of product hydrolysis under microwave irradiation could be explained by the time-dependent activation and inactivation of the β-galactosidase.

11.3.4 Aldolase

Xie *et al.* [40] studied the immobilization technique of 2-deoxy-D-ribose-5-phosphated aldolase (DERA) on mesocellular siliceous foams under consecutive microwave irradiation at low temperature. Aldolase catalyzes the aldol reaction between the two aldehydes. The consecutive irradiations were induced by controlling the temperature using a Discover CoolMate coupled with a cooling module (the method of temperature measurement was not described). At an output power of 30 W, 88.4% of the enzyme was immobilized after 3 min, and the specific activity of the immobilized DERA was 2.24 U mg^{-1}, which is, 149.2% higher than that of free aldolase and 157.0% higher than that of the nonmicrowave-assisted immobilized enzyme.

11.4 Mechanism of the Microwave-Assisted Enzymatic Reaction

Collins and Leadbeater [41] hypothesized that the microwave effects on enzymatic digestion might be because of the increased dipole moments of the α-helices of proteins. They illustrated the relationship between the dipole moment of the α-helix and the three-dimensional structure of the protein (Figure 11.2). The authors speculated that if the microwave energy interacts with the dipole of

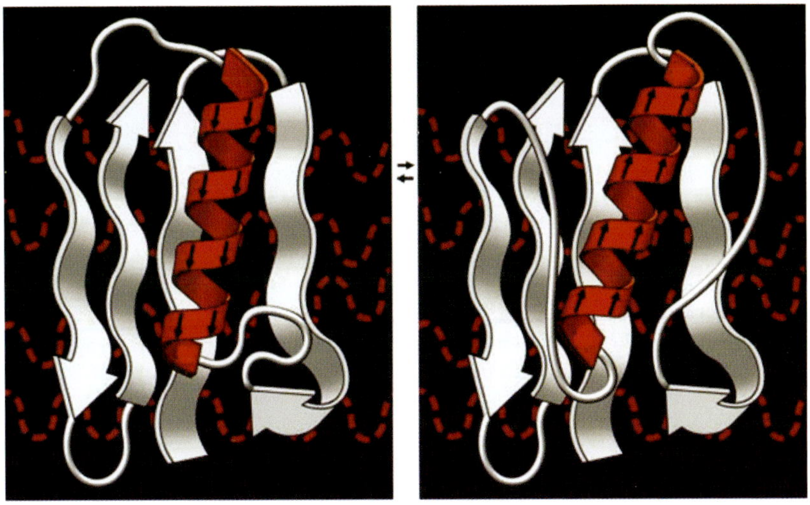

Figure 11.2 Dipole movement across α-helices and interaction with microwave radiation. (Reproduced from Ref. [41]. Copyright 2007 by Royal Society of Chemistry.)

α-helix, the perturbation of the protein structure could facilitate the digestion of the previously enclosed areas of the protein.

To understand enzyme stabilization and activity in ILs, Campbell *et al.* [42] examined the effects of IL properties on lipase stabilization under microwave irradiation. The transesterification of ethyl butyrate and 1-butanol was performed using Novozym 435, which was a thermal lipase B from *C. antarctica* (CALB) immobilized on acrylic resin in ILs under microwave heating (Figure 11.3). A Discover LabMate equipped with both an IR temperature probe and pressure-monitoring sensor was used as the microwave reactor. No superheating of the enzyme's surface or rate enhancement was observed under microwave irradiation. The authors suggested a model of water-induced superheating mechanism on the free or immobilized enzyme in dry media (Figure 11.3a) or in a small amount of dispersed water (Figure 11.3b). The authors concluded that the effect of microwave irradiation on the enzyme could be explained by the superheating of the water layer near the enzymes surface.

Microwave technology has been used for enzymatic protein digestions in the proteomics. In 2012, to reveal nonthermal microwave effects in proteomics, Kappe *et al.* [43] re-valuated the tryptic digests of three proteins (i.e., bovine serum albumin (BSA), cytochrome *c*, and β-casein) at 37 and 50 °C under both microwave and conventional heating. Furthermore, they simulated the tertiary structure of trypsin and BSA in an electromagnetic field. A Monowave 300 single-mode microwave reactor (Anton Paar) equipped with a fiber-optic thermometer was used. The effect of microwave irradiation on protein degradation was analyzed by SDS-PAGE and MALDI-TOF-MS. A comparison of BSA digestion under microwave irradiation with that under conventional heating is shown in Figure 11.4, which was created using PyMOL software. The results of their combined experimental and computational studies revealed that neither the enzyme stability itself nor the tryptic digest was directly affected by the electromagnetic field. They concluded that an incorrect temperature

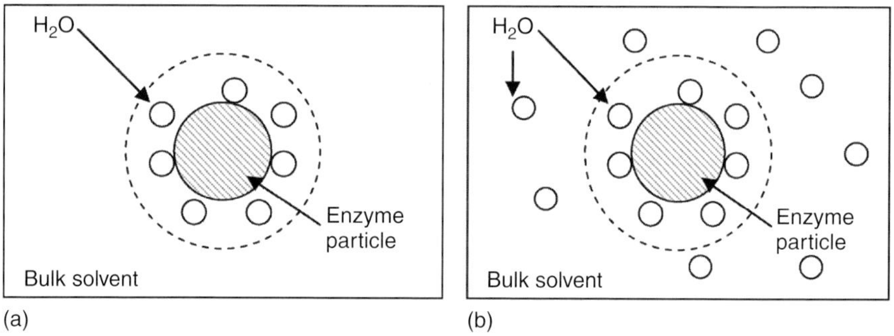

Figure 11.3 Illustrations of water-induced superheating mechanism: (a) the free or immobilized enzyme particle is surrounded by a layer of water molecules while the bulk hydrophobic solvent is dry; and (b) the enzyme particle is surrounded by a layer of water molecules while the bulk solvent contains a small amount of dispersed water. (Reproduced from Ref. [42]. Copyright 2009 by Elsevier.)

led to the erroneous conclusions about the non-thermal effects observed in microwave-assisted proteomics experiments.

Suzuki *et al.* [44] described a method for evaluating protein hydration in an aqueous solution using microwave dielectric measurements. The measurements were carried out using a precision microwave network analyzer and a thermostated glass cell at $20.0 \pm 0.01\,°C$ with an open-end flat-surface coaxial probe. The measured proteins were cytochrome *c*, myoglobin, ovalbumin, BSA, and hemoglobin. The dependence of the relative complex permittivity of a BSA on the microwave frequency was measured in the frequency range of 0.2 to 20 GHz. The Cole-Cole plot of the complex dielectric constant of BSA in water or buffer indicated that the Debye fittings were made in the frequency range 2.5-8 GHz based on the considerable observed effects of ionic conduction.

Shimizu and coworkers [45] reported the efficiency of microwave irradiation at 2.45 and 5.80 GHz in the hydrolysis reaction of thermostable β-glucosidase HT1. Microwave heating was performed using EYELA MWS-1000 (2.45 GHz) and Amil ATMW500B-(5.8 GHz) microwave reactors. 4-Methoxyphenyl glucopyranoside was employed as the substrate for β-glucosidase HT1. Three

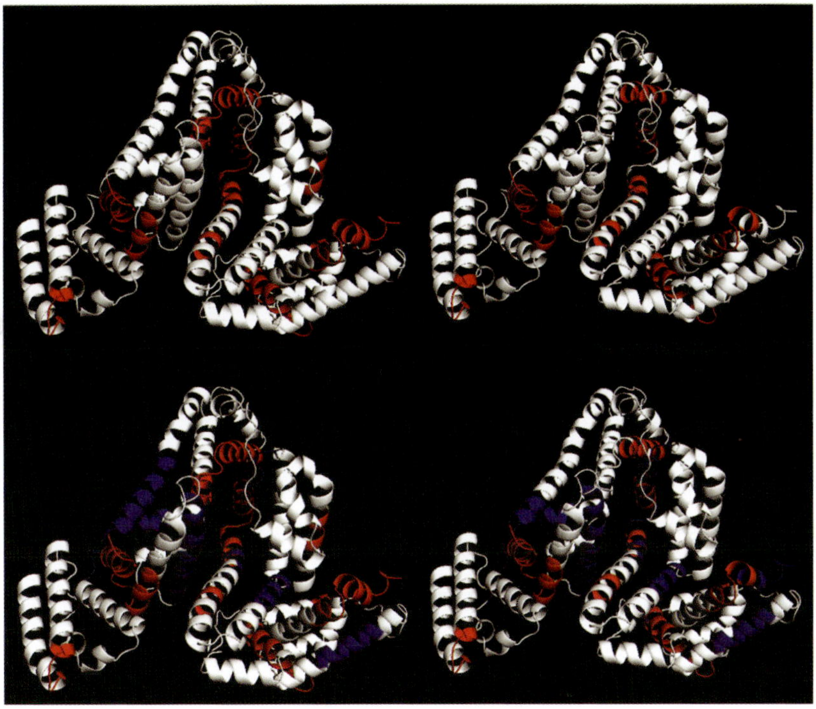

Figure 11.4 Comparison of time-dependent BSA digestion under conventional and microwave heating illustrated using the PyMOL software. The tryptic peptides obtained after a 5-min heating period at 37 °C are shown in red while additional fragment generated after 16 h at 37 °C are highlighted in blue. (Reproduced from Ref. [43]. Copyright 2012 by Elsevier.)

types of heating (i.e., conventional heating, 2.45 GHz, and 5.8 GHz, during = 60 min) were compared at four different temperatures. The results indicated that heating with the 2.45-GHz microwave was more effective than conventional heating, this method lowering the optimum temperature and accelerating the reaction. In contrast, the use of 5.80-GHz microwave irradiation did not improve this reaction. To determine the cause of this difference, relative complex permittivity was measured from 200 MHz to 14 GHz using the reflection probe method. The approximate curve (distilled water) was semicircular at all measured frequencies, which suggesting that pure water behaved only as a dielectric under microwave irradiation. On the other hand, the buffer solution results indicated the rectilinear locus as an electric conductor from 200 MHz to 2.45 GHz and the semicircular locus as dielectric from 2.45 to 14 GHz. Concerning the heating mechanism, the authors suggested that 5.8 GHz microwave irradiation affected only the water molecules in buffer solution, whereas the 2.45 GHz irradiation affected both water molecules and buffering ions.

References

1. (a) Klibanov, A.M. (1989) *Trends Biochem. Sci.*, **14**, 141; (b) Koskinen, A.M. and Klibanov, A.M. (1996) *Enzymatic Reactions in Organic Media*, Springer.
2. (a) Parker, M.-C., Besson, T., Lamare, S., and Legoy, M.-D. (1996) *Tetrahedron Lett.*, **37**, 8383; (b) Gelo-Pujic, M., Guibé-Jampel, E., Loupy, A., Galema, S.A., and Mathé, D. (1996) *J. Chem. Soc., Perkin Trans. 1*, 2777; (c) Carrillo-Munoz, J.-R., Bouvet, D., Guibé-Jampel, E., Loupy, A., and Petit, A. (1996) *J. Org. Chem.*, **61**, 7746.
3. (a) Chemat, F. and Cravotto, G. (2013) *Microwave-Assisted Extraction for Bioactive Compounds*, Springer; (b) Lill, J.R. (2009) *Microwave Assisted Proteomics*, Royal Society of Chemistry; (c) Polshettiwar, V. and Varma, R.S. (2010) *Aqueous Microwave Assisted Chemistry: Synthesis and Catalysis*, Royal Society of Chemistry; (d) Tsubaki, S. and Azuma, J.-I. (2011) *Advances in Induction and Microwave Heating of Mineral and Organic Materials*, Tech Education and Publishing, Vienna, p. 697; (e) Motasemi, F. and Ani, F.N. (2012) *Renewable Sustainable Energy Rev.*, **16**, 4719; (f) Gude, V.G., Patil, P., Martinez-Guerra, E., Deng, S., and Nirmalakhandan, N. (2013) *Sustainable Chem. Processes*, **1**, 5; (g) Zhong, H., Zhang, Y., Wen, Z., and Li, L. (2004) *Nat. Biotechnol.*, **22**, 1291; (h) Lill, J.R., Ingle, E.S., Liu, P.S., Pham, V., and Sandoval, W.N. (2007) *Mass Spectrom. Rev.*, **26**, 657.
4. Obermayer, D. and Kappe, C.O. (2010) *Org. Biomol. Chem.*, **8**, 114.
5. Mochizuki, D. and Wada, Y. (2011) *Mini-Rev. Org. Chem.*, **8**, 294.
6. Robinson, J., Kingman, S., Irvine, D., Licence, P., Smith, A., Dimitrakis, G., Obermayer, D., and Kappe, C.O. (2010) *Phys. Chem. Chem. Phys.*, **12**, 4750.
7. Sturm, G.S., Verweij, M.D., Van Gerven, T., Stankiewicz, A.I., and Stefanidis, G.D. (2012) *Int. J. Heat Mass Transfer*, **55**, 3800.
8. Hayden, S., Damm, M., and Kappe, C.O. (2013) *Macromol. Chem. Phys.*, **214**, 423.
9. Kappe, C.O. (2013) *Chem. Soc. Rev.*, **42**, 4977.
10. Vacek, M., Zarevúcka, M., Wimmer, Z., Stránský, K., Demnerová, K., and Legoy, M.-D. (2000) *Biotechnol. Lett.*, **22**, 1565.
11. Mazumder, S., Laskar, D.D., Prajapati, D., and Roy, M.K. (2004) *Chem. Biodivers.*, **1**, 925.
12. Réjasse, B., Lamare, S., Legoy, M.-D., and Besson, T. (2004) *Org. Biomol. Chem.*, **2**, 1086.

13. Réjasse, B., Besson, T., Legoy, M.D., and Lamare, S. (2006) *Org. Biomol. Chem.*, **4**, 3703.
14. Réjasse, B., Lamare, S., Legoy, M.-D., and Besson, T. (2007) *J. Enzyme Inhib. Med. Chem.*, **22**, 519.
15. (a) Porcelli, M., Cacciapuoti, G., Fusco, S., Massa, R., d'Ambrosio, G., Bertoldo, C., De Rosa, M., and Zappia, V. (1997) *FEBS Lett.*, **402**, 102; (b) La Cara, F., d'Auria, S., Scarfi, M., Zeni, O., Massa, R., d'Ambrosio, G., Franceschetti, G., De Rosa, M., and Rossi, M. (1999) *Protein Pept. Lett.*, **6**, 155; (c) La Cara, F., Scarffi, M.R., D'Auria, S., Massa, R., d'Ambrosio, G., Franceschetti, G., Rossi, M., and De Rosa, M. (1999) *Bioelectromagnetics*, **20**, 172.
16. Maugard, T., Gaunt, D., Legoy, M.D., and Besson, T. (2003) *Biotechnol. Lett.*, **25**, 623.
17. Yadav, G.D. and Lathi, P.S. (2004) *J. Mol. Catal. A: Chem.*, **223**, 51.
18. Kerep, P. and Ritter, H. (2006) *Macromol. Rapid Commun.*, **27**, 707.
19. Yadav, G.D. and Lathi, P.S. (2006) *Enzyme Microb. Technol.*, **38**, 814.
20. Yadav, G.D. and Borkar, I.V. (2006) *AIChE J.*, **52**, 1235.
21. (a) Yadav, G.D. and Lathi, P.S. (2007) *Clean Technol. Environ. Policy*, **9**, 281; (b) Yadav, G.D. and Sajgure, A.D. (2007) *J. Chem. Technol. Biotechnol.*, **82**, 964.
22. Leadbeater, N.E., Stencel, L.M., and Wood, E.C. (2007) *Org. Biomol. Chem.*, **5**, 1052.
23. Yadav, G.D. and Borkar, I.V. (2008) *Ind. Eng. Chem. Res.*, **48**, 7915.
24. Major, B., Kelemen-Horváth, I., Csanádi, Z., Bélafi-Bakó, K., and Gubicza, L. (2009) *Green Chem.*, **11**, 614.
25. Costa, I.C., Leite, S.G., Leal, I.C., Miranda, L.S., and de Souza, R.O. (2011) *J. Braz. Chem. Soc.*, **22**, 1993.
26. Yadav, G.D. and Thorat, P.A. (2012) *J. Mol. Catal. B: Enzym.*, **83**, 16.
27. Ziaullah and Rupasinghe, H.P.V. (2013) *Tetrahedron Lett.*, **54**, 1933.
28. Shinde, S.D. and Yadav, G.D. (2014) *Ind. Eng. Chem. Res.*, **53**, 8706.
29. Sontakke, B.J. and Yadav, D.G. (2014) *Curr. Catal.*, **3**, 27.
30. Yu, D., Wang, Z., Chen, P., Jin, L., Cheng, Y., Zhou, J., and Cao, S. (2007) *J. Mol. Catal. B: Enzym.*, **48**, 51.
31. Yu, D., Chen, P., Wang, L., Gu, Q., Li, Y., Wang, Z., and Cao, S. (2007) *Process Biochem.*, **42**, 1312.
32. Bachu, P., Gibson, J.S., Sperry, J., and Brimble, M.A. (2007) *Tetrahedron: Asymmetry*, **18**, 1618.
33. Parvulescu, A., Vandereycken, E., Jacobs, P., and Devos, D. (2008) *J. Catal.*, **255**, 206.
34. de Souza, R.O., Antunes, O.A., Kroutil, W., and Kappe, C.O. (2009) *J. Org. Chem.*, **74**, 6157.
35. Yadav, G.D. and Devendran, S. (2012) *J. Mol. Catal. B: Enzym.*, **81**, 58.
36. Yadav, G.D. and Pawar, S.V. (2012) *Appl. Microbiol. Biotechnol.*, **96**, 69.
37. Devendran, S. and Yadav, G.D. (2014) *Biomed. Res. Int.*, **2014**, 482678.
38. Young, D.D., Nichols, J., Kelly, R.M., and Deiters, A. (2008) *J. Am. Chem. Soc.*, **130**, 10048.
39. Kamerke, C., Pattky, M., Huhn, C., and Elling, L. (2012) *J. Mol. Catal. B: Enzym.*, **79**, 27.
40. Wang, A., Wang, M., Wang, Q., Chen, F., Zhang, F., Li, H., Zeng, Z., and Xie, T. (2011) *Bioresour. Technol.*, **102**, 469.
41. Collins, J.M. and Leadbeater, N.E. (2007) *Org. Biomol. Chem.*, **5**, 1141.
42. Zhao, H., Baker, G.A., Song, Z., Olubajo, O., Zanders, L., and Campbell, S.M. (2009) *J. Mol. Catal. B: Enzym.*, **57**, 149.
43. Damm, M., Nusshold, C., Cantillo, D., Rechberger, G.N., Gruber, K., Sattler, W., and Kappe, C.O. (2012) *J. Proteomics*, **75**, 5533.
44. Suzuki, M., Shigematsu, J., and Kodama, T. (1996) *J. Phys. Chem.*, **100**, 7279.
45. Nagashima, I., Sugiyama, J., Sakuta, T., Sasaki, M., and Shimizu, H. (2014) *Biosci. Biotechnol., Biochem.*, **78**, 758.

Part V
Applications – Hydrogenation and Fuel Formation

Microwaves in Catalysis: Methodology and Applications, First Edition.
Edited by Satoshi Horikoshi and Nick Serpone.

12
Effects of Microwave Activation in Hydrogenation–Dehydrogenation Reactions

Leonid M. Kustov

12.1
Introduction

The activation of heterogeneous catalysts, both at the stage of their preparation and during the application, is of the greatest importance for increasing the activity and selectivity of catalytic processes. Traditional methods for the activation of heterogeneous catalysts, such as thermal treatment and oxidation or reduction at high temperatures in a medium of various gases, often do not provide the necessary catalyst activity, selectivity, or stability, because such treatments generate a wide spectrum of active centers of various natures and strengths, including active centers that catalyze undesirable side reactions. Currently, the catalysts differing from systems used earlier by more complex composition and organization of the structure and active centers at the molecular level, as well as polyfunctionality of action, find ever-increasing application in industry. This makes the problem of the development of new nontraditional methods of their activation and regeneration still more important.

If microwave (MW) activation is used in catalytic processes, it should be taken into account that the objects of study are nonuniform multiphase systems, including catalysts and gases (that can dissociate or be polarized at the catalyst surface) in which volumetric structural and property changes occur under the action of electromagnetic radiation. The main idea of microwave catalysis is to exert volumetric controllable electromagnetic action on the catalyst–reagent system. This action should change the state of the system and increase the effectiveness of the work of catalysts, the selectivity of substrate conversion into valuable products, and catalyst stability. One of the main directions toward solving this problem is a decrease in the temperature and other reaction parameters by process execution under mild conditions in an electromagnetic microwave field with the retention of or increase in process efficiency.

Clearly, the use of microwaves in catalysis will be developed in two directions. The first direction is related to the preparation and preliminary activation of catalysts with the use of microwaves for traditional reactors, where heating is performed by traditional (thermal) methods. The second direction is the *in situ*

Microwaves in Catalysis: Methodology and Applications, First Edition.
Edited by Satoshi Horikoshi and Nick Serpone.

action of microwaves on catalysts and reaction media (if they absorb microwaves) during the catalyzed chemical reactions.

The use of microwave action on heterogeneous catalysts during their preparation sometimes allows one to design the catalysts with a more uniform distribution of active sites or metal nanoparticles, to accelerate the procedure of catalyst preparation, and to achieve a uniform catalyst volume heating. With catalysts consisting of several phases, the replacement of traditional heating by microwave heating can contribute to the preferable formation of separate phases. When microwave activation is used during a catalytic reaction, a decrease in the reaction temperature, an increase in the activity or selectivity, and a change in the distribution of reaction products are often observed. A comparison of traditional conditions with the microwave regime is, however, not always possible because of difficulties related to true process temperature measurements in the latter case.

This chapter covers reactions in which hydrogen participates as a starting reagent (hydrogenation) or as a product (dehydrogenation).

12.2 Specific Features of Catalytic Reactions Involving Hydrogen

Processes involving hydrogen are considered separately from other chapters, because of the very specific interaction of hydrogen with catalysts. First, hydrogen can reduce metal ions or oxides to the metallic state and it is known that the electrophysical properties of the metal salts or oxides are very much different from zero-valent metal, especially when the metal is present as nanoparticles. Second, when the metal exists as an M^o species, dissociation of hydrogen may occur on the nanoparticle surface, which may change dramatically (usually, enhance) the electrophysical properties of the metal nanoparticles. Finally, the effects of hydrogen spillover from the metal to the oxide or carbon carrier are possible, and this, in turn, is reflected in the ability to absorb the microwave energy. Conductivity and related Ohmic losses determine one of the microwave heating mechanisms that can operate in the catalytic systems used in hydrogenation processes. When a microwave field acts on a medium containing mobile charge carriers, conductivity currents appear in the system. For high-conductivity materials (catalysts), there are such conductivity values that conductivity loss is much larger than dipole relaxation effects. For the worst case from the point of view of application of MW activation, for example, aluminum oxide known to be a poor MW-absorbing material, the conductivity loss at room temperature becomes noticeable at low MW frequencies only. The loss in the microwave range occurs largely because of dipole relaxation. The conductivity of a material typically increases as the temperature increases, and the fraction of the conductivity loss in the microwave range increases and becomes comparable with the polarization effects. An increase in the conductivity of alumina as the temperature rises is related to the thermal activation of electrons, which are excited from the 2p oxygen valence band to the 3s3p conduction band. In addition, as a rule the electronic conductivity increases

Table 12.1 Temperature of the solvent (50 ml) achieved by MW heating at 560 W, 2.45 GHz, 1 min (the starting temperature was 27 °C).

Solvent	*T* (°C)	Boiling point (°C)
Water	81	100
Methanol	65	65
Ethanol	78	78
1-Propanol	97	97
1-Butanol	109	117
1-Pentanol	106	137
1-Hexanol	92	158
1-Chlorobutane	76	78
1-Bromobutane	95	101
Acetic acid	110	119
Ethyl acetate	73	77
Chloroform	49	61
Acetone	56	56
Dimethylformamide	131	153
Dimethyl ether	32	35
Hexane	25	68
Heptane	26	98
CCl_4	28	77

in the presence of defects, which create additional levels in the band-gap. The concentration of defects can also increase substantially at high temperatures.

In the case of hydrogenation reactions, the process can be performed in the gas phase or in a solvent, so that the choice of a solvent is no less important to gain the best performance under MW conditions than the choice of a catalyst. The electrophysical properties of organic substances that are often used as solvents in organic syntheses, including hydrogenation reactions, are well known. Table 12.1 presents the data on the heating efficiency of different solvents. The best solvents from the point of view of MW energy absorption are indicated in bold.

Table 12.2 summarizes the data on dielectric constants and dielectric loss tangent, which is related to the ability of a compound to absorb the MW energy. Again, the most suitable solvents for MW conditions are marked in bold. It follows from these data that polar solvents should be considered as solvents of choice, but the temperature range where this particular solvent exists as a liquid (below the boiling point) should be taken into account. Therefore, the use of methanol and ethanol is not recommended, whereas propanol and butanol are better solvents, unlike higher alcohols that become less polar.

Water is a benign and green solvent, but its use is also limited by the solubility of the substrate to be hydrogenated and that of the hydrogenated product as well as by the solubility of hydrogen; finally, the temperature range for aqueous systems is also limited to 100 °C. The best solvents seem to be dimethylformamide, dimethylsulfoxide, acetic or formic acids, and C_3–C_4 alcohols.

Table 12.2 Dielectric constants and dielectric loss tangent for some solvents.

Solvent	Dielectric constant (ε')	Dielectric loss tangent ($\tan\delta = \varepsilon''/\varepsilon'$)
Hexane	1.9	—
Benzene	2.3	—
Carbon tetrachloride	2.2	—
Chloroform	4.8	0.091
Acetic acid	6.1	0.174
Ethyl acetate	6.2	0.059
Tetrahydrofuran	7.6	0.047
Methylene chloride	9.1	0.042
Acetone	20.6	0.054
Ethanol	24.6	0.941
Methanol	32.7	0.659
Acetonitrile	36.0	0.062
Dimethylformamide	36.7	0.161
Dimethylsulfoxide	47.0	0.825
Formic acid	58.0	0.722
Water	80.4	0.123

The operating frequency is another important characteristic that should be tuned while properly designing the MW-activated catalytic process. Once the best solvents are selected, it is important to compare their electrophysical properties at different MW frequencies. Table 12.3 presents some data for organic solvents of choice in comparison with water. It is seen that the dielectric loss tangent (the ratio $\varepsilon''/\varepsilon'$) for water and methanol increases with increasing operating frequency, while for ethanol and propanol it passes through a maximum. For ethylene glycol, the increase of frequency results in an increase in the loss tangent, but after reaching 10 GHz, the efficiency of heating starts to diminish. Therefore, the most efficient operating frequency for the MW processes should be perfectly tuned taking into account the choice of solvent.

Table 12.3 Electrophysical properties of solvents.

Solvent	Frequency (Hz)					
	3×10^8		3×10^9		1×10^{10}	
	ε'	ε''	ε'	ε''	ε'	ε''
Water	77.5	1.2	76.6	12.0	55.0	29.7
Methanol	30.9	2.5	23.9	15.3	8.9	7.2
Ethanol	22.3	6.0	6.5	1.6	1.7	0.11
n-Propanol	16.0	6.7	3.7	2.5	2.3	0.20
Ethylene glycol	39	6.2	12	12	7	5.5

Table 12.4 Comparison of MW heating of water and the ionic liquid BMIM-BF_4.

Power (absorbed) (W)	Maximum temperature that can be reached (°C)	
	H_2O	BMIM-BF_4
3	39	71
5	87	128
7	100	152

Ionic liquids, which are near-ambient temperature molten salts consisting of an organic cation (typically, imidazolium, pyridinium, alkylammonium, etc.) and typically (but not necessarily) an inorganic anion (BF_4^-, PF_6^-, NTf_2^-, HSO_4^-, etc.), provide new insights into the MW-stimulated reactions occurring in solution, because they demonstrate much more efficient heating under MW irradiation compared to water and other organic solvents. Table 12.4 presents our data on heating one of the ionic liquids, BMIM-BF_4 (BMIM denotes 3-butyl-1-methylimidazolium) in comparison with water in an MW resonator (a 100% efficiency of MW absorption) at low power levels.

The choice of catalyst, or carrier for the catalyst that is intended to be used in the MW-assisted process, is most important. Even the carrier can be chosen in such a way that the performance under MW conditions and the efficiency of heating can be quite different. Efficiencies of some solid materials, with some being suitable as carriers or catalysts for the MW processes, are presented in Table 12.5 (with most well-absorbing materials presented in bold).

Table 12.5 Maximum temperature of heating solid materials in MW fields (25 g, initial temperature 20 °C, household MW oven, power 1 kW, frequency 2.45 GHz).

Solid	Temperature (°C)	Time required to reach this temperature (min)	Solid	Temperature (°C)	Time required to reach this temperature (min)
Al	577	6	CeO_2	99	30
C	1283	1	**CuO**	701	0.5
Co_3O_4	1290	3	Fe_2O_3	88	30
CuCl	619	13	**Fe_3O_4**	510	2
$FeCl_3$	41	4	La_2O_3	107	30
$MnCl_2$	53	1.75	**MnO_2**	321	30
CaO	83	30	**PbO_2**	182	7
Ni	384	1	Pb_3O_4	122	30
NiO	1305	6.25	SnO_2	102	30
$SbCl_3$	224	1.75	TiO_2	122	30
$SnCl_2$	476	2	**V_2O_5**	701	9
$ZnCl_2$	609	7	**WO_3**	532	0.5

Table 12.6 Power required to heat some representative catalysts to 100 °C.

Catalyst	Input power required to heat the sample to 100 °C (W)			
	Argon	Benzene	Hydrogen	Benzene + hydrogen
Pt/Al_2O_3	>2000	800	800	800
Pd/TiO_2	>500	20	20	20
Pt/C	8	5	5	7

Obviously, the carbon materials and mixed transition metal oxides with good conductivity, or at least semiconductors, should be preferred as carriers with other parameters being the same, provided of course that the carrier does not catalyze side reactions. Table 12.6 presents the data on the effective power required to heat the chosen catalysts with different electrophysical properties by 100 °C in different reaction media (our own data obtained with a 2.5 kW MW flow setup).

While designing the MW reactor, especially, for gas-phase processes, attention should be given to the distribution of the electromagnetic field in the catalyst bed, in order to reach the maximum enhancement of the MW irradiation efficiency (a decrease in reaction temperature, decrease in the reaction duration, and improvement of selectivity by suppressing side reactions).

The thermodynamics of the reaction (endothermic or exothermic) is a crucial factor in designing the reactor system. Hydrogenation reactions are known to be characterized by moderate to high exothermicity, whereas dehydrogenation reactions require energy consumption (are endothermic).

The theoretical analysis of the utilization of MW radiation with the aim to decrease energy consumption in endothermic processes has been the subject of a study [1]. The use of microwave was shown to significantly save energy (up to 60%) with a proper design of the reactor. The efficient use of a metal-coated surface of the reactor can help to enhance energy savings for reactors with $2L/\lambda_{eff} = 0.5n - 0.25$, where $n = 1, 2, 3, \ldots$, $2L$ is the dimension of the reactor, and λ_{eff} is the MW wavelength. The energy saving increases by factors of 2 and 1.5 at $2L/\lambda_{eff} = 0.25$ and 0.75, respectively.

12.3
Hydrogenation Processes under MW Conditions

Hydrogenation and dehydrogenation processes typically occur with the participation of nonpolar hydrocarbon molecules though several examples are known in the state of the art related to the hydrogenation of polar groups (such as nitro and nitrile groups) mediated with microwave irradiation. Unlike the conversion of polar organic molecules, the transformation of nonpolar hydrocarbon compounds via hydrogenation or dehydrogenation routes cannot gain the advantages of dipolar polarization, which is the most commonly used practice in organic syntheses occurring under MW conditions. Hydrocarbons are not polarizable and therefore do not absorb MW energy. The only way to use the benefits of the MW activation is the application of catalysts that are capable of absorbing the MW energy and thus can enhance the performance under MW conditions. At this stage, it is relevant to discuss some recent literature data on the hydrogenation of diverse substrates on catalysts of different nature. This is a sketch-wise overview of the most representative examples of the use of MW irradiation in catalytic processes, rather than a comprehensive collection of all available data. Some MW-assisted reactions involving hydrogen have been discussed in the review of Ref. [2]. In the last few years, several reviews on catalytic hydrogenations have given a thorough overview of modern green and efficient processes; a collection of studies of reactions promoted by MW have been published [3–5]. A few reports also compared MW-promoted hydrogenation with other nonconventional techniques, such as ultrasound activation [6].

Before discussing hydrogenation–dehydrogenation processes, it is worth paying some attention to the hydrodechlorination process, which is not strictly a hydrogenation process, but rather a substitution reaction where chlorine atoms are substituted for hydrogen atoms. In the study of Ref. [7], and in several other publications by the same authors, microwave activation of the Pd–Fe/Al_2O_3 catalyst of hydrodechlorination has been investigated. The size of supported metal nanoparticles increased under the action of microwaves, with metal particles forming alloys. An increase in the activity of reduced catalysts for the hydrodechlorination of chlorobenzene was discovered under MW conditions compared to the thermal process. Similar results were observed for the microwave-irradiated Pd/Al_2O_3 and Pd/SiO_2 catalysts of the hydrogenation of benzene. The same effects were found for the Pd/Nb_2O_5 catalyst of the hydrodechlorination reaction [8].

Microwave heating was also used to perform other reactions, namely (i) the hydrogenation of chlorinated aromatics on Pd/C [9] and (ii) the hydrogenation of olefins [10]. Examples of the successful use of microwave radiation in the hydrogenation of alkenes and hydrogenolysis of alkanes are given in Ref. [11], as well as in the reactions of hydrogenation and hydrocracking of other hydrocarbons [12, 13]. Positive effects of the *in situ* microwave activation were also reported by Wan *et al.* in the hydrogenation of alkenes [11].

Hydrogenation reactions under MW activation were the subject of the European project devoted to microwave-, acoustic-, and plasma-assisted syntheses (MAPSYN) that aimed at the selective hydrogenation of triple bonds and nitro compounds assisted by microwaves using a Lindlar-type catalyst (Eq. (12.1)) [14]. Hydrogenation of aromatics, as well as debenzylation, azide hydrogenation, and the hydrogenation of strychnine were studied in a high-pressure MW reactor at pressures up to 25 atm [15]:

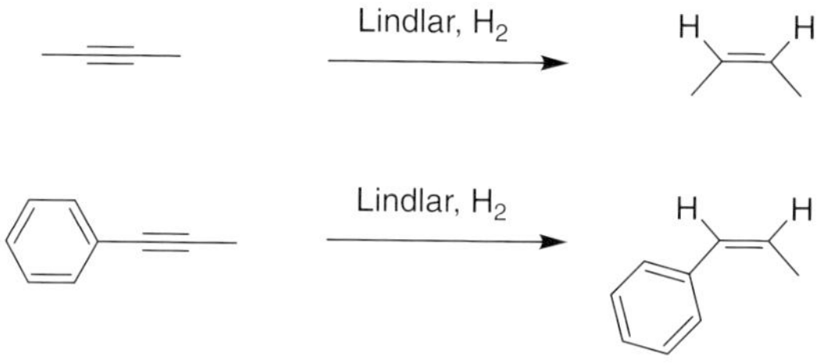

(12.1)

A series of electron-deficient alkenes was investigated in terms of efficient hydrogenation with a hydrogen donor supported on an ion-exchange resin and in the presence of Wilkinson's catalyst [10]. Fast and efficient hydrogenation over a Pd/C catalyst was also reported [16]. Deprotection of benzyl ethers and carboxybenzyl amides on a Pd/C catalyst turns out to be very efficient with 1,4-cyclohexadiene as the hydrogen transfer agent. The duration of the MW-assisted reactions was just 5 min at 100 °C. Pd/C also proved to be active in the hydrogenation of furfural in water at MW conditions [17]. Even low-loaded copper nanoparticles (0.5 wt%) on mesoporous silicates showed a very high efficiency in the hydrogenation of carbonyl compounds [18]. Substituted aromatic ketones and aldehydes were converted with a 100% yield of hydrogenated products within 5–10 min.

Another study compared the MW-assisted with the ultrasonically enhanced hydrogenation transfer of aromatic ketones using glycerol as the hydrogen transfer agent; also compared was the MW and ultrasound (US) performance with conventional heating, with both the MW and US approaches showing the intensification of the process (Eq. (12.2)) [19]:

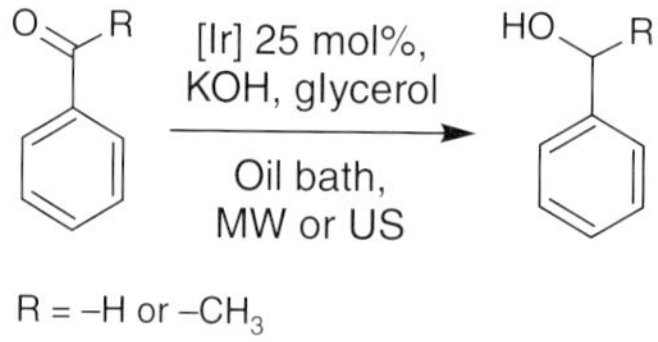

(12.2)

Glycerol can also be simultaneously used as both a solvent and a hydrogen donor in transfer hydrogenation of benzaldehyde catalyzed by carbonylchlorohydridotris(triphenyl-phosphine)ruthenium(II) [20]. Compared with traditional reflux heating, the reaction under MW irradiation was characterized by an improved turnover rate that increased from 280 to 6700 h^{-1} (Eq. (12.3)) [21]:

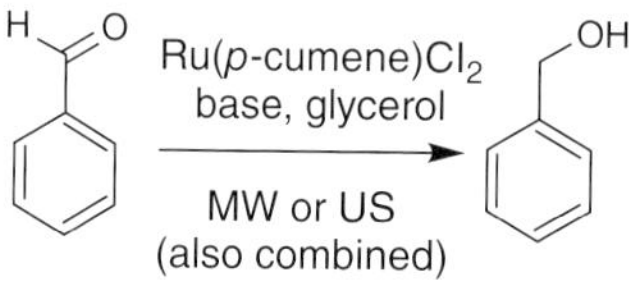

(12.3)

One frequently used substrate in hydrogenation reactions is ethyl pyruvate, because it is possible to estimate the enantiomeric excess (ee) [22]; Pt/Al_2O_3 was the catalyst (Eq. (12.4)):

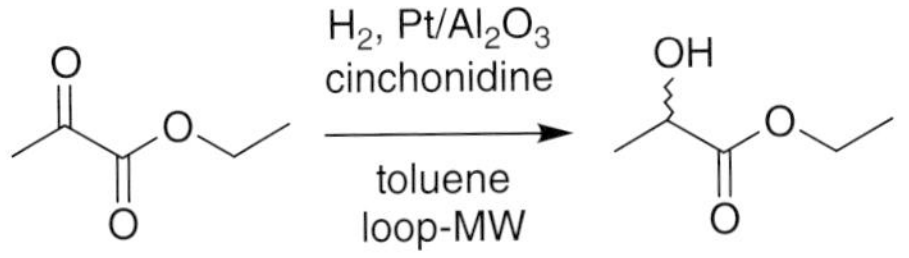

(12.4)

Kinetic studies have been carried out while varying the MW power and the nature of the solvent. The choice of solvent determined the reaction rate and the enantioselectivity, with the highest values (75%) being observed in toluene, even though this solvent does not absorb microwaves, which is consistent with the observed fact that the kinetics and enantioselectivity of the reaction in toluene are not affected by the mode of heating. The reaction rate in ethanol was the same for conventional and MW heating, but the enantioselectivity decreased significantly under microwave irradiation. According to the physicochemical studies, some sintering of the catalyst occurred under MW irradiation. Ethyl pyruvate hydrogenation was studied by the same team in a single-mode MW loop reactor over Pt/Al_2O_3 catalyst in the presence of cinchonidine as a chiral ligand. The highest ee was observed in experiments carried out in toluene (80%) [23]. A kinetic model was developed based on selective and unselective reaction routes. The temperature of the catalyst was estimated using a mathematical model for heat transfer and energy balances for solid and liquid phases. The mass and energy balances were decoupled because the heat effect of the hydrogenation reaction was much lower than the energy of the microwave irradiation.

Hydrogenation of more complicated organic molecules (e.g., sterols) has been reported [24]. Unsaturated sterols and bile alcohols were hydrogenated on Pd/C in a methylene chloride/propylene glycol mixture as the solvent in the presence of ammonium formate under microwave activation. Sterols, in particular, and cholesterol, campesterol, sitosterol, and bile alcohols with unsaturated side chains were converted into the hydrogenated products with high yields and selectivity.

Hydrogenation of nitro compounds is another important area of MW research [25]. Modeling of heat balances in a microwave reactor was performed using an electromagnetic simulation software HFSS®. In the presence of a catalyst, the catalyst heating is essential for homogeneous heating and industrial applications. A new continuous flow microwave reactor under pressure was developed with a fixed, fluidized, or circulating catalyst bed. Nitrobenzene was subjected to hydrogenation on a palladium catalyst.

Another example of the hydrogenation of nitro-aromatics into anilines is presented in [26]. Supported Rh metal nanoparticles were active in the chemo- and regio-selective reduction of nitroarenes functionalized with side groups containing multiple bonds, and dinitroarenes into amines with hydrazine hydrate as the reducing agent under MW conditions. The catalyst was recycled 13 times without deactivation.

Hydrogenation (via hydrogen transfer) of heteroaromatic nitro compounds was also quite efficient under MW operations [27]. Conventional Pd/C and Pt/C catalysts were shown to be very efficient with 1,4-cyclohexadiene as the hydrogen transfer agent. When the starting compound contained labile halogen substituents, Pt/C was more efficient with almost no dehalogenation; the reactions were completed within 5 min at 120 °C.

MW-stimulated asymmetric hydrogenation (via hydrogen transfer) of ketones was reported in Ref. [28]. TsCHDA chiral ligands supported onto a mesoporous silicate SBA-15 were used to prepare optically active secondary alcohols.

Cross-linked chitosan with Pd(II) ions reduced to Pd° with $NaBH_4$ [29] was used and recycled at least 10 times in the hydrogenation of α,β-unsaturated compounds. The selectivity was high, especially for the hydrogenation of cyclohex-2-enone in which only the C=C bond was hydrogenated. The hydrogenation of internal alkynes and imines was also successful under MW activation (Eq. (12.5)) [30]:

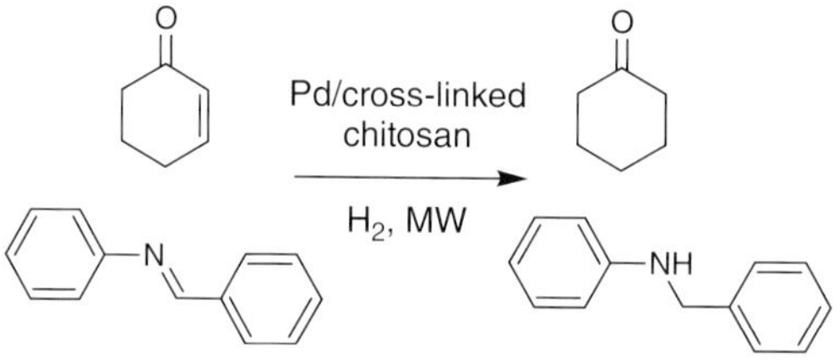

(12.5)

Microwave plasma can also be used in hydrogenation processes. One recent example is the hydrodechlorination of $SiCl_4$ in a low-temperature MW plasma [31]. A nonthermodynamic equilibrium plasma allowed the authors to achieve a high conversion and very high selectivity to $SiHCl_3$. The influence of $SiCl_4$/Ar and H_2/Ar ratios on the performance was studied and modeled. The maximum conversion was found at an $H_2/SiCl_4$ molar ratio of 1. The optimal pressure was in the range of 26.6 to 40 kPa, depending on the input power with the optimal feed molar energy input being 350 kJ mol^{-1}.

12.4 Dehydrogenation

Although oxidative dehydrogenation refers to the process that occurs formally without any participation (consumption or release) of hydrogen, it is worth saying a few words about this reaction, because once it was called simply dehydrogenation. Oxidative dehydrogenation will be discussed in more detail elsewhere (Section 14.3.8). A few publications reported on positive effects of microwave radiation in the oxidative dehydrogenation. For instance, oxidative dehydrogenation of ethylbenzene to styrene on iron oxide deposited on multilayer carbon nanotubes (CNTs) was enhanced under MW conditions [32, 33]. An increase in selectivity with respect to styrene was also observed under microwave activation conditions (380–450 °C); it was found that part of Fe_2O_3 is transformed into Fe^o nanocrystals encapsulated by polyhedral graphite shells, and this effect was observed only under microwave conditions. The optimal content of iron oxide supported on CNTs was shown to be 3 wt%.

Oxidative dehydrogenation of ethane into ethylene was thoroughly studied in Ref. [34]. A kinetic approach was successfully used to trace the MW effects in this reaction. The comparison of kinetic dependencies (the kinetic equations, selectivity or yield vs. conversion curves) with apparent parameters (activation energies) was made under thermal and MW regimes. For $VMoO_x$ and $VMoNbO_x$ catalysts, a distinct difference between ethane yields was observed. MW activation was revealed to modify the catalyst structure, including the phase distribution, which is different from that formed under conventional heating. These changes result in the variation of the catalytic behavior of $VMoO_x$ and $VMoNbO_x$ systems.

Nonoxidative dehydrogenation is studied in the literature more often than the oxidative dehydrogenation. Naphthenic molecules most frequently served the reaction substrates. A number of still unpublished data were obtained in our laboratory at N.D. Zelinsky Institute of Organic Chemistry, Moscow, Russia [35]. Some representative dependencies are shown in Figure 12.1 for the case of the dehydrogenation of cyclohexane to benzene. The electrophysical parameters of

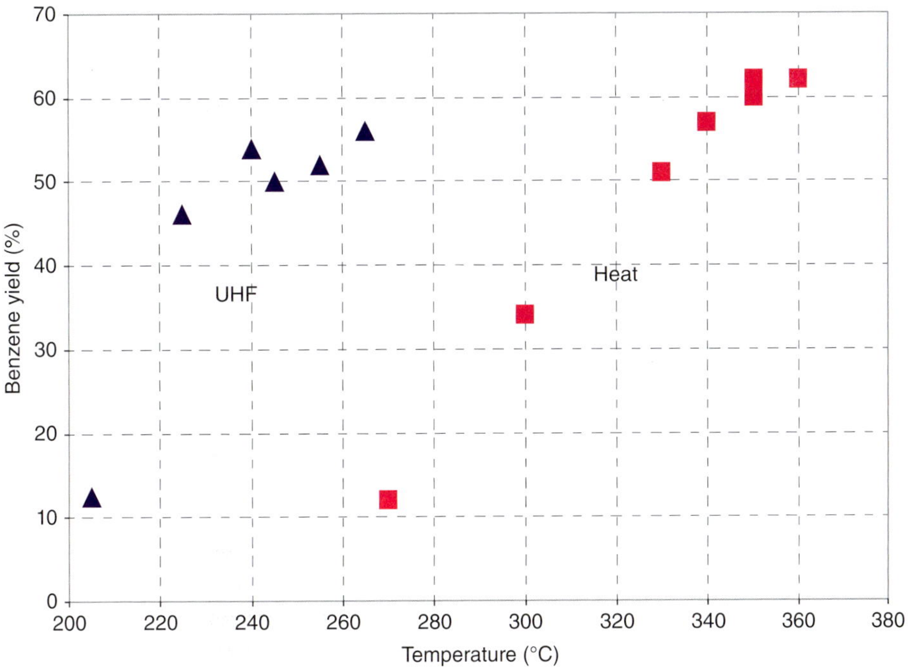

Figure 12.1 Cyclohexane dehydrogenation on a Pd/TiO_2 catalyst (flow conditions, 0.5 h^{-1}).

Table 12.7 Power density and strength of the electromagnetic field estimated for hydrogenation of benzene and dehydrogenation of cyclohexane on catalysts with different MW-absorbing ability.

Catalyst	Hydrogenation		Dehydrogenation	
	Power density ($W\ cm^{-3}$)	Strength of the electromagnetic field ($V\ cm^{-1}$)	Power density ($W\ cm^{-3}$)	Strength of the electromagnetic field ($V\ cm^{-1}$)
Pt/Al_2O_3	1–1.5 to 2	350–440		—
Pd/TiO_2	2–3	60–80	3–4	80–100
Pt/C	0.5 to 1	8–15	1–2	25–50

the dehydrogenation process logically are higher compared to the hydrogenation on the same catalysts (Table 12.7).

Decalin was used most often for dehydrogenation because of the higher boiling point and the possibility of extending and achieving a better control of the temperature range [36]. The study of different types of carbon materials, which included CNTs, carbon nanofibers (CNFs), carbon black (CB), activated carbon (AC), and graphite (G) as the supports for Pt was carried out for the dehydrogenation of decalin under microwave conditions. The dielectric properties and electroconductivities of the carbon supports and catalysts in a suspension in decalin were determined, and a correlation was found with the thermal behavior. CNTs turned out to be the best catalyst support under MW conditions. The high aspect ratio and low bulk density of this carrier resulted in a high electrical conductivity and enhanced dielectric loss. The catalytic performance with respect to H_2 evolution changed in the order: Pt/CNTs > Pt/CB > Pt/CNFs > Pt/AC ~ Pt/G. The two latter samples displayed zero activity in the dehydrogenation of decalin. The nanofibers, though being less active, demonstrated a higher selectivity (less side-product formation).

The MW heating characteristics of diverse carbon materials have been the subject of various studies [37–39]. Compared to spherical carbon materials, CNTs and nanofibers exhibited high aspect ratios, which resulted in the improvement of conductive loss according to the percolation theory [40].

12.5
Hydrogen Storage

The two reactions considered earlier, that is, hydrogenation of aromatic or heteroaromatic substrates and the reverse dehydrogenation of the obtained hydrogenated naphthenic molecules, were used in designing safe hydrogen storage materials because hydrogen is chemically bonded and the temperature of dehydrogenation may not be higher than 300 °C, whereas in the case of the hydrogenation of alkylcarbazoles the temperature can be decreased to 220–240 °C.

Horikoshi and coworkers have examined the MW effects on the gas-phase dehydrogenation of decalin and tetralin on Pt and Pt-supported overactivated carbon [41–43]. The enhanced conversion of tetralin under MW conditions was compared to the thermal heating; the effect was explained by a large-temperature gradient between the catalyst and the reaction media generated under the MW conditions. Enhancement of the heating ability of Pt/carbon particles and decrease in the temperature gradient (nonuniform heat distribution) in the tetralin solution under MW conditions were studied by three methods: (i) the use of a Dewar-like insulation reactor, (ii) the addition of an ionic liquid as a good MW-absorbing agent, and (iii) the optimization of the ratio of catalyst to reactant. In the case of the insulation reactor used to prevent the heat loss under microwave heating, the reaction could be heated to about 3 °C above the normal boiling point of tetralin. As a result, the tetralin conversion increased from 31% to 56% at an input MW power of 190 W. Addition of the ionic liquid as a heating agent significantly enhanced the heating rate of tetralin. However, the ionic liquid suppressed the activity of the catalyst. The third approach allowed the authors to improve the heat distribution; the highest tetralin conversion of 69% was achieved under the liquid film conditions. The stability of the catalyst also improved under MW conditions because MW heating generated the temperature gradient that favored faster molecule desorption or enhancement of species transport in the system. The strong adsorption of hydrocarbons can also be reduced, thereby resulting in a decrease in coke formation.

Tetralin dehydrogenation was also studied by the same authors in Ref. [44]. Carbon microcoils were applied as a new type of carbon support for Pt and Pd nanoparticles in comparison with activated carbon. In addition to tetralin dehydrogenation, the Suzuki–Miyaura coupling reaction between phenylboronic acid and 1-bromo-4-methylbenzene in toluene to yield 4-methyl-biphenyl was also investigated. The MW-absorbing ability of the new material turned out to be better than that of the activated carbon (in terms of dielectric constant, dielectric loss, and loss tangent). The possibility of microplasma generation (hot spots) on the catalyst surface was demonstrated. Conventional heating resulted in significantly lower yields of products for both reactions.

In several articles, the team of Kustov also used MW heating in the dehydrogenation of naphthenic molecules suitable for reversible and safe hydrogen storage [45]. The dehydrogenation of a few naphthene and polycyclic (including heterocyclic) compounds on supported metallic catalysts under microwave heating provides a hydrogen storage capacity of no less than 7–7.5 wt% based on the weight of the composite material, including the organic substrate and the catalyst. This is much higher than the capacity of intermetallic compounds traditionally used for hydrogen storage (3–4 wt%), metal organic frameworks (2–5%, high pressures, cryogenic conditions), or CNTs suggested earlier (1–2 wt%, also cryogenic). Two types of catalysts, namely traditional Pt/C and bifunctional Ni/silica/alumina, were used for the dehydrogenation of naphthenes under microwave irradiation.

The catalytic activity in the microwave regime increased more significantly than in the conventional heating mode under the same reaction conditions. Such an effect may result from the fact that the temperature of the metal particles (Pt, Ni) in the microwave mode is higher than the average temperature of the catalyst bed in the thermal mode. The best substrate from the viewpoint of a combination of properties (boiling points of the hydrogenated and dehydrogenated counterparts) was found to be meta-terphenyl (Eq. (12.6)):

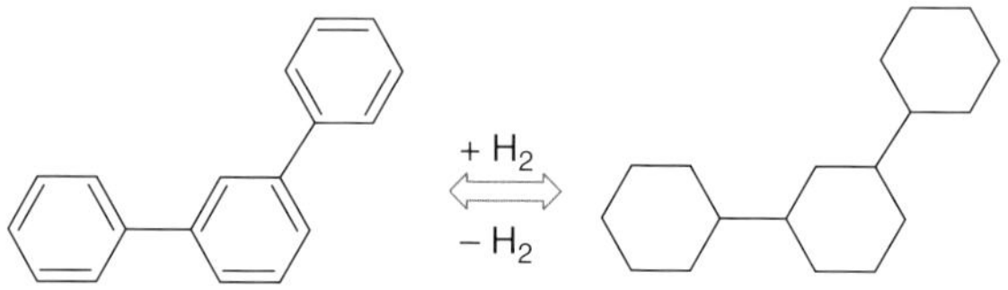

(12.6)

12.6 Hydrogenation of Coal

Hydrogenation of coal resulting in its liquefaction or gasification is a very important process and currently actively examined because of the development of new concepts of coal utilization. Also, this problem is closely related to the conversion of lignin and other raw materials. One of several approaches has been presented by Wang *et al.* [46]. MW-stimulated hydroconversion of demineralized coal liquefaction residues was performed using alcohols as solvents for the reaction and extraction. The Pd/C and Ni catalysts were used for hydroconversion.

Acknowledgment

The financial support of the Ministry of Education and Science of Russian Federation (project RFMEFI61614X0014) is acknowledged.

References

1. Bhattacharya, M. and Basak, T. (2013) A theoretical study on the use of microwaves in reducing energy consumption for an endothermic reaction: role of metal coated bounding surface. *Energy*, **55**, 278–294.
2. (a) Suib, S.L. (1998) Applications of microwaves in catalysis. *CATTECH*, **2**, 75; (b) Lidstrom, P., Tierney, J., Wathey, B., and Westman, J. (2001) Microwave assisted organic synthesis—a review. *Tetrahedron*, **57** (45), 9225–9283.
3. Banik, B.K., Barakat, K.J., Wagle, D.R., Manhas, M.S., and Bose, A.K. (1999) Microwave-assisted rapid simplified hydrogenation. *J. Org. Chem.*, **64**, 5746–5753.
4. Kustov, L.M. and Sinev, I.M. (2010) Microwave activation of catalysts and catalytic processes. *Russ. J. Phys. Chem.*, **84**, 1676–1694.
5. Nicks, F., Borguet, Y., Delfosse, S., Bicchielli, D., Delaude, L., Sauvage, X., and Demonceau, A. (2009) Microwave-assisted ruthenium-catalyzed reactions. *Aust. J. Chem.*, **62**, 184–207.
6. Barge, A., Tagliapietra, S., Tei, L., Cintas, P., and Cravotto, G. (2008) Pd-catalyzed reactions promoted by ultrasound

and/or microwave irradiation. *Curr. Org. Chem.*, **12**, 1588–1612.

7. Berry, F.J., Smart, L.E., Prasad, P.S.S., Lingaiah, N., and Rao, P.K. (2000) Microwave heating during catalyst preparation: influence on the dechlorination activity of alumina-supported palladium iron bimetallic catalysts. *Appl. Catal. A: Gen.*, **204**, 191–201.
8. Gopinath, R., Rao, K.N., Prasad, P.S.S., Madhavendra, S.S., Narayanan, S., and Vivekanandan, G. (2002) Hydrodechlorination of chlorobenzene on Nb_2O_5 supported Pd catalysts: influence of microwave irradiation during preparation on the stability of the catalyst. *J. Mol. Catal., A: Chem.*, **181**, 215–220.
9. Hidaka, H., Saitou, A., Honjou, H., Hosoda, K., Moriya, M., and Serpone, N. (2007) Microwave-assisted dechlorination of polychlorobenzenes by hypophosphite anions in aqueous alkaline media in the presence of Pd-loaded active carbon. *J. Hazard. Mater.*, **148**, 22–28.
10. Desai, B. and Danks, T.N. (2001) Thermal- and microwave-assisted hydrogenation of electron-deficient alkenes using a polymer-supported hydrogen donor. *Tetrahedron Lett.*, **42**, 5963–5965.
11. Wan, J.K.S., Wolf, K., and Heyding, R.D. (1984) Some chemical aspects of the microwave assisted catalytic hydrocracking processes. *Stud. Surf. Sci. Catal.*, **19**, 561–568.
12. Wan, J.K.S., Wolf, K., and Heyding, R.D. (1984) in *Catalysis on the Energy Scene* (eds S. Kaliaguine and A. Mahay), Elsevier, Amsterdam, pp. 439–455.
13. Wolf, K., Choi, H.K., and Wan, J.K.S. (1986) Microwave assisted catalytic conversion of cyclohexene. *Aust. J. Res.*, **3**, 53–59.
14. Hessel, V., Cravotto, G., Fitzpatrick, P., Patila, B.S., Lang, J., and Bonrath, W. (2013) Industrial applications of plasma, microwave and ultrasound techniques: Nitrogen-fixation and hydrogenation reactions. *Chem. Eng. Process.: Process Intensif.*, **71**, 19–30.
15. Heller, E., Lautenschläger, W., and Holzgrabe, U. (2005) Microwave-enhanced hydrogenations at medium pressure using a newly constructed reactor. *Tetrahedron Lett.*, **46**, 1247–1249.
16. Quinn, J.F., Razzano, D.A., Golden, K.C., and Gregg, B.T. (2008) 1,4-Cyclohexadiene with Pd/C as a rapid, safe transfer hydrogenation system with microwave heating. *Tetrahedron Lett.*, **49**, 6137–6140.
17. Garcia-Suarez, E.J., Balu, A.M., Tristany, M., García, A.B., Philippot, K., and Luque, R. (2012) Versatile dual hydrogenation–oxidation nanocatalysts for the aqueous transformation of biomass-derived platform molecules. *Green Chem.*, **14**, 1434–1439.
18. Yoshida, K., Gonzalez-Arellano, C., Luque, R., and Gai, P.L. (2010) Efficient hydrogenation of carbonyl compounds using low-loaded supported copper nanoparticles under microwave irradiation. *Appl. Catal., A: Gen.*, **379**, 38–44.
19. Azua, A., Mata, J.A., Peris, E., Lamaty, F., Martinez, J., and Colacino, E. (2012) Alternative energy input for transfer hydrogenation using iridium NHC based catalysts in glycerol as hydrogen donor and solvent. *Organometallics*, **31**, 3911–3919.
20. Cravotto, G., Orio, L., Gaudino, E.C., Martina, K., Tavor, D., and Wolfson, A. (2011) Efficient synthetic protocols in glycerol under heterogeneous catalysis. *ChemSusChem*, **4**, 1130–1134.
21. Gordon, E.M., Gaba, D.C., Jebber, K.A., and Zacharias, D.M. (1993) Catalytic transfer hydrogenation of benzaldehyde in a microwave oven. *Organometallics*, **12**, 5020–5022.
22. Toukoniitty, B., Roche, O., Mikkola, J.-P., Toukoniitty, E., Klingstedt, F., Eränen, K., Salmi, T., and Murzin, D.Y. (2007) Ethyl pyruvate hydrogenation under microwave irradiation. *Chem. Eng. J.*, **126**, 103–109.
23. Toukoniitty, B., Wärnå, J., Mikkola, J.-P., Helle, M., Saxén, H., Murzina, D.Y., and Salmi, T. (2009) Modelling of enantioselective and racemic hydrogenation of ethyl pyruvate on a Pt/Al_2O_3 catalyst in the presence of microwave irradiation. *Chem. Eng. Process.: Process Intensif.*, **48**, 837–845.

24. Dayal, B., Ertel, N.H., Rapole, K.R., Asgaonkar, A., and Salen, G. (1997) Rapid hydrogenation of unsaturated sterols and bile alcohols using microwaves. *Steroids*, **62**, 451–454.
25. Bonnet, C., Estel, L., Ledoux, A., Mazari, B., and Louis, A. (2004) Study of the thermal repartition in a microwave reactor: application to the nitrobenzene hydrogenation. *Chem. Eng. Process.: Process Intensif.*, **43**, 1435–1440.
26. Guha, N.R., Bhattacherjee, D., and Das, P. (2014) Solid supported rhodium(0) nanoparticles: an efficient catalyst for chemo- and regio-selective transfer hydrogenation of nitroarenes to anilines under microwave irradiation. *Tetrahedron Lett.*, **55**, 2912–2916.
27. Quinn, J.F., Bryant, C.E., Golden, K.C., and Gregg, B.T. (2010) Rapid reduction of heteroaromatic nitro groups using catalytic transfer hydrogenation with microwave heating. *Tetrahedron Lett.*, **51**, 786–789.
28. Jin, M.-J., Sarkar, M.S., and Jung, J.-Y. (2007) Highly efficient microwave-assisted asymmetric transfer hydrogenation with SBA-15-supported TsCHDA chiral ligands. *Stud. Surf. Sci. Catal.*, **165**, 455–458.
29. Martina, K., Leonhardt, S.E.S., Ondruschka, B., Curinic, M., Binelloa, A., and Cravottoa, G. (2011) *In situ* cross-linked chitosan Cu(I) or Pd(II) complexes as a versatile, eco-friendly recyclable solid catalyst. *J. Mol. Catal., A: Chem.*, **334**, 60–64.
30. Schüßler, S., Blaubach, N., Stolle, A., Cravotto, G., and Ondruschka, B. (2012) Application of a cross-linked Pd–chitosan catalyst in liquid-phase-hydrogenation using molecular hydrogen. *Appl. Catal., A: Gen.*, **445–446**, 231–238.
31. Lu, Z. and Zhang, W. (2014) Hydrogenation of silicon tetrachloride in microwave plasma. *Chin. J. Chem. Eng.*, **22** (2), 227–233.
32. Nigrovski, B., Zavyalova, U., Scholz, P., Pollok, K., Müller, M., and Ondruschka, B. (2008) Microwave-assisted catalytic oxidative dehydrogenation of ethylbenzene on iron oxide loaded carbon nanotubes. *Carbon*, **46**, 1678–1686.
33. Nigrovski, B., Scholz, P., Krech, T., Qui, N.V., Pollok, K., Keller, T., and Ondruschka, B. (2009) The influence of microwave heating on the texture and catalytic properties of oxidized multi-walled carbon, nanotubes. *Catal. Commun.*, **10**, 1473–1477.
34. Sinev, I., Kardash, T., Kramareva, N., Sinev, M., Tkachenko, O., Kucherov, A., and Kustov, L.M. (2009) Interaction of vanadium containing catalysts with microwaves and their activation in oxidative dehydrogenation of ethane. *Catal. Today*, **141**, 300–305.
35. Kustov, L.M., Tarasov, A.L., Bogdan, V.I., and Kustov, A.L. (2005) Method of hydrogen storage by performing catalytic reactions of hydrogenation-dehydrogenation of aromatic substrates in composite materials under the action of microwave radiation, RF Patent 2005130340 (*Chem. Abstr.*, 146, 382750).
36. Li, X., Tuo, Y., Li, P., Duan, X., Jiang, H., and Zhou, X. (2014) Effects of carbon support on microwave-assisted catalytic dehydrogenation of decalin. *Carbon*, **67**, 775–783.
37. Menéndez, J.A., Arenillas, A., Fidalgo, B., Fernández, Y., Zubizarreta, L., Calvo, E.G., and Bermúdez, J.M. (2010) Microwave heating processes involving carbon materials. *Fuel Process. Technol.*, **91**, 1–8.
38. Li, G., Xie, T., Yang, S., Jin, J., and Jiang, J. (2012) Microwave absorption enhancement of porous carbon fibers compared with carbon nanofibers. *J. Phys. Chem. C*, **116**, 9196–9201.
39. Wu, H.J., Wang, L.D., Guo, S.L., Wang, Y., and Shen, Z. (2012) Electromagnetic and microwave-absorbing properties of highly ordered mesoporous carbon supported by gold nanoparticles. *Mater. Chem. Phys.*, **133**, 965–970.
40. Paton, K.R. and Windle, A.H. (2008) Efficient microwave energy absorption by carbon nanotubes. *Carbon*, **46**, 1935–1941.
41. Suttisawat, Y., Horikoshi, S., Sakai, H., and Abe, M. (2010) Hydrogen production from tetralin over microwave-accelerated Pt supported activated carbon. *Int. J. Hydrogen Energy*, **35**, 6179–6183.

42. Suttisawat, Y., Sakai, H., Abe, M., and Horikoshi, S. (2012) Microwave effect in the dehydrogenation of tetralin and decalin with a fixed-bed reactor. *Int. J. Hydrogen Energy*, **37**, 3242–3250.
43. Suttisawat, Y., Horikoshi, S., Sakai, H., Rangsunvigit, P., and Abe, M. (2012) Enhanced conversion of tetralin dehydrogenation under microwave heating: effects of temperature variation. *Fuel Process. Technol.*, **95**, 27–32.
44. Horikoshi, S., Suttisawat, Y., Osawa, A., Takayama, C., Chen, X., Yang, S., Sakai, H., Abe, M., and Serpone, N. (2012) Organic syntheses by microwave selective heating of novel metal/CMC catalysts – The Suzuki–Miyaura coupling reaction in toluene and the dehydrogenation of tetralin in solvent-free media. *J. Catal.*, **289**, 266–271.
45. Sung, J.S., Choo, K.Y., Kim, T.H., Tarasov, A.L., Soldatov, A.V., and Kustov, L.M. (2008) Comparative study on dehydrogenation of bulky, branched and polycondensed naphthenes for hydrogen storage in microwave and thermal modes. *Int. J. Hydrogen Energy*, **33**, 4116–4121.
46. Wang, T.-X., Zong, Z.-M., Zhang, J.-W., Wei, Y.-B., Zhao, W., Li, B.-M., and Wei, X.-Y. (2008) Microwave-assisted hydroconversions of demineralized coal liquefaction residues from Shenfu and Shengli coals. *Fuel*, **87**, 498–507.

13
Hydrogen Evolution from Organic Hydrides through Microwave Selective Heating in Heterogeneous Catalytic Systems

Satoshi Horikoshi and Nick Serpone

13.1
Situation of Hydrogen Energy and Feature of Stage Methods

Building a sustainable energy society is mankind's common objective. Hydrogen is expected to provide a significant contribution as a clean energy carrier to the upcoming sustainable society [1]. Hydrogen is a regenerative and an environmental friendly energy vector, which reacts cleanly with oxygen in a highly exothermic reaction giving pure water as the unique exhaust by-product, with hydrogen being returned to water after being used in a fuel cell (Eq. (13.1)):

$$2H_2(g) + O_2(g) \rightarrow 2H_2O\,(l) + 572\ \text{kJ (i.e., 286 kJ mol}^{-1}) \qquad (13.1)$$

Japan has proposed as its major target to reduce by 50% the emission of the greenhouse gas CO_2 by the year 2050. The introduction of renewable energy sources is indispensable to achieve this goal. A hydrogen-based fuel cell technology should contribute considerably to achieve this goal. Accordingly, fuel cells have been the object of extensive studies in Japan. Certain public institutions (e.g., in the transportation sector such as buses) already make good use of fuel cells as a source of power. Moreover, private cars that drive using a fuel cell technology (fuel cell vehicle, FCV) are already commercially available (Toyota Motor Co., since December 2014). In addition, power supplied by fuel cells is widely used in Japanese houses. The hydrogen gas is first extracted from liquefied petroleum gas (LPG) and its energy is subsequently converted by the fuel cell system into electricity (Ene Farm System, Toshiba Co. and Panasonic Co.) [2]. The process occurring in a fuel cell is exothermic, the heat from which is used to boil water that is then used in bathrooms and in other household applications. Even now hydrogen energy permeates our daily activities.

Hydrogen is the most abundant element in the universe, yet on Earth it has to be produced first because it only occurs in the form of water and hydrocarbons. This implies that we have to expend energy to produce energy resulting in a difficult economic dilemma because ever since the industrial revolution we have become accustomed to consume energy at relatively low costs [3]. The second difficulty with hydrogen as an energy carrier is its low critical temperature of 33 K.

Microwaves in Catalysis: Methodology and Applications, First Edition.
Edited by Satoshi Horikoshi and Nick Serpone.

The differences between hydrogen-based fuels and other fuels in terms of caloric value are summarized in Table 13.1. The energy densities per weight of hydrogen are high compared with other fuels, whereas the energy densities per volume are rather low. For example, the volume energy density of hydrogen is about a quarter compared with liquefied natural gas, whereas when compared on a weight basis it is about threefold to fourfold greater. A weak point of hydrogen energy has to do with the storage and transportation of hydrogen gas, as well as energy savings because the energy densities per volume are low.

The notion of using hydrogen energy to build a "Hydrogen Economy" is rather old. A project of large-scale hydrogen transfer was conducted in the European Union (EU) and in Quebec, Canada, in the years 1986–1998. The electricity made with Canadian rich hydroelectric power generation (100 MW) was converted to hydrogen by the electrolysis of water, and then transported as liquid hydrogen by seaway to Europe (Hamburg, Germany). Hydrogen was used as a fuel for public transportation (buses) in the EU [4]. The project was not continued because the output density of the energy from the fuel cell was (at the time) seen as being insufficient to power a car. However, improvements in energy conversion efficiency and the decrease in device cost of fuel cells have been made in recent years [5], which have led to a reconsideration of the use of hydrogen-based fuel cells.

Hydrogen can be stored using seven methods: (i) high-pressure gas cylinders (up to 800 bar), (ii) liquid hydrogen in cryogenic tanks (at 21 K), (iii) adsorbed hydrogen on materials with a large surface area (at 100 K), (iv) absorbed on interstitial sites in a host metallic matrix (at ambient pressure and temperature) [6], (v) chemically bonded in covalent and ionic compounds (at ambient pressure), (vi) the organic hydride method as a means to store the hydrogen as organic compounds [7, 8], and (vii) through oxidation of reactive metals, for example, Li, Na, Mg, Al, Zn with water. When hydrogen is liquefied, it is possible to reduce the volume by nearly a factor of 800. However, cooling the hydrogen gas to the extraordinarily low temperature of 21°K and the high pressures (up to 800 bar) needed to liquefy hydrogen necessitate a considerable amount of electrical energy. When using alloys to absorb hydrogen, it is possible to decrease the volume of hydrogen by a factor of more than 1000. A total of about 300 kg of hydrogen-absorbing

Table 13.1 Difference between the hydrogen and other fuels for the calorific power.

Refuel	Higher calorific value	
	$MJ\,m^{-3}$	$MJ\,kg^{-1}$
Hydrogen gas	11.9	142
Liquid hydrogen	10.1	142
Liquefied natural gas	39.0	55
Gasoline	34.6	49

alloys is necessary to store about 4 kg of hydrogen gas so that such a system is not mobile as a hydrogen storage medium.

The organic hydride storage method (no need for low temperatures and high pressures) can make use of the infrastructure currently used to transport (lorries and ships) and to store (reservoir tanks) gasoline. After storing the hydrogen gas, the aromatic hydrocarbon is collected by known infrastructure equipment, or else the infrastructure equipment to store the hydrogen energy can share the equipment used to store gasoline. Aromatic hydrocarbons can also be used as fuels in times of natural disasters.

Metal hydrides have also been proposed by the U.S. Department of Energy as possible storage media for hydrogen; they are MgH_2, $NaAlH_4$, $LiAlH_4$, LiH, $LaNi_5H_6$, $TiFeH_2$, and palladium hydride [9]; some of these are liquids at ambient temperature and pressure while others are solids that can be made in the form of pellets. Metal hydridic materials display good volume energy density, though their weight energy density tends to be inferior to leading hydrocarbon fuels. Because many of the metal hydrides bind hydrogen very strongly, high temperatures in the range 120–200 °C are required to release the hydrogen, an energy cost that can be reduced using alloys consisting of strong hydride formers and weak hydride formers (e.g., $LiNH_2$, $LiBH_4$, and $NaBH_4$) [10]. However, if the binding is too weak, the pressure needed for re-hydriding is high, which would eliminate any energy savings. Currently, the only hydrides capable of achieving a 9 wt% gravimetric goal appear to be limited to Li-, B-, and Al-based compounds; at least one of the first-row elements or Al must also be added.

Metal hydrides that have been considered for use in a hydrogen economy include the simple magnesium hydrides [11] or transition metals and complex metal hydrides that typically contain Na, Li or Ca, and Al or B. Hydrides chosen for storage applications provide low reactivity (high safety) and high hydrogen storage densities, with LiH, $NaBH_4$, $LiAlH_4$, and $NH_3{\cdot}BH_3$ as leading candidates.

13.2 Selection of Organic Hydrides as the Hydrogen Carriers

Some of the features of aromatic hydrocarbons as possible carriers of hydrogen are listed in Table 13.2 [12]. The cyclohexane ↔ benzene cycle is a system very capable of storing and releasing hydrogen. However, benzene possesses strong carcinogenic properties, so that even the benzene included in gasoline is regulated to less than 1%. Therefore, the use of the cyclohexane ↔ benzene cycle is not suitable as an organic hydride carrier. The decalin ↔ naphthalene cycle indicates a high pressure for hydrogen evolution. Therefore, it is possible to extract most of the hydrogen from the decalin molecules. Fundamental research on the organic hydrides decalin ↔ naphthalene has been carried out continually until recently. The melting point of naphthalene is 80 °C, so that when it is cooled in the reactor or in the flow pipe after the reaction, the naphthalene will solidify, thereby clogging the reactor or the pipes. On the other hand, the presence of a methyl group in

Table 13.2 Features of organic hydrides as hydrogen carriers.

Organic hydrides	H_2 carriers	H_2 ($N\,m^3\,l^{-1}$)	Reactivity	Toxicity	Boiling point of organic hydrides (°C)	Melting point of H_2 carriers (°C)	Result	Comprehensive evaluation
Cyclohexane	Benzene	0.62	Good	Bad	81	5.5	Good	Bad
Methylcyclohexane	Toluene	0.53	Good	Good	101	−95	Good	Good
Dimethylcyclohexane	Xylene	0.46	Good	Good	83	−48	Good	Middle
Tetralin	Naphthalene	0.36	Middle	Good	206	80	Good	Bad
Decalin	Naphthalene	0.71	Middle	Good	193	80	Good	Bad
1-Methyldecalin	1-Methyl naphthalene	0.65	Middle	Good	210	−31	Bad	Bad
2-Methyldecalin	2-Methyl naphthalene	0.65	Middle	Good	205	34	Bad	Bad
2-Ethyldecalin	2-Ethyl naphthalene	0.59	Middle	Good	225	15	Bad	Bad

Reproduced from Ref. [12].

naphthalene lowers the melting point compared with naphthalene. Unfortunately, large quantities of these materials as major industrial materials are not easily available. Moreover, a higher temperature reaction field is needed because the boiling point of condensation ring compounds is high. Therefore, the methylcyclohexane ↔ toluene cycle appears as the most suitable carrier for the organic hydride method in industry.

13.3
Dehydrogenation of Hydrocarbons with Microwaves in Heterogeneous Catalytic Media

13.3.1
Selective Heating by the Microwave Method

Hydrogen production from tetralin as the source of hydrogen has been carried out under microwave heating (MWH); the overall process is summarized in Figure 13.1 [13]. The most common catalyst in dehydrogenation reactions is platinum supported on activated carbon (Pt/AC) particulates. However, when carried out by the conventional heating method, this type of reaction is usually limited by the mass transfer rate of molecule desorption, because after formation of the product – naphthalene in the case of tetralin – the aromatic product tended to deposit on the active sites of the catalyst, thereby causing the catalyst to become inactivated, not to mention the formation of coke deposits on the catalyst surface and thus the need for higher reaction temperatures. Accordingly, the advantage of microwaves in selective heating was applied to resolve this problem [13].

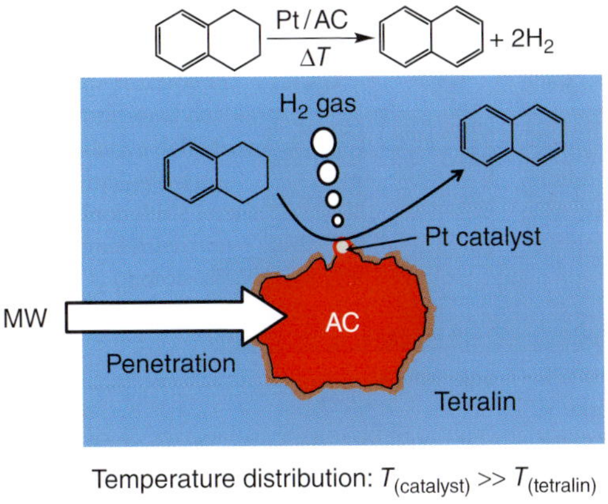

Figure 13.1 Cartoon illustrating the dehydrogenation of tetralin occurring on the Pt/AC (platinum supported on activated carbon) catalyst subjected to microwave heating. From Ref. [13].

Note that the most suitable organic hydride is the methylcyclohexane ↔ toluene system as alluded to in the previous section. In order to consider and examine some of the features of the microwave method, the tetralin ↔ naphthalene system has received most of the attention.

In the dehydrogenation of tetralin under MWH, the Pt/AC acted as both the catalyst and the microwave absorber. Being a nonpolar material, tetralin is a poor absorber of microwave radiation. Therefore, heat was generated selectively at the catalyst and subsequently transferred to the reactant. It is worth noting that when the heat flux is directed from the surface of the catalyst to the substrates, mass transfer will occur in the same direction owing to the coupling vector that facilitates the desorption of molecules from the active sites of the catalyst surface (Figure 13.1).

13.3.2 Dehydrogenation of Tetralin in a Pt/AC Heterogeneous Catalytic Dispersion Subjected to a Microwave Radiation Field

An air-equilibrated tetralin solution (1.5 ml) containing the Pt catalyst supported on activated carbon particles (loading, 300 mg) was introduced into the batch reactor, which was positioned in the microwave multimode applicator (continuous irradiation with 320-W 2.45-GHz microwaves). The reaction temperature was 207 °C. To consider the effect of the microwaves, an experiment was also carried out using a mantle heater for the dehydrogenation of tetralin in the Pt/AC heterogeneous catalyzed process. The temporal changes of temperature of a tetralin solution with and without the presence of the Pt/AC catalyst heated by microwave radiation have been investigated methodically [13]. The tetralin solution containing the catalyst was rapidly heated to the boiling point within 2 min of microwave irradiation; initial heating rate was 3.0 °C s^{-1}.

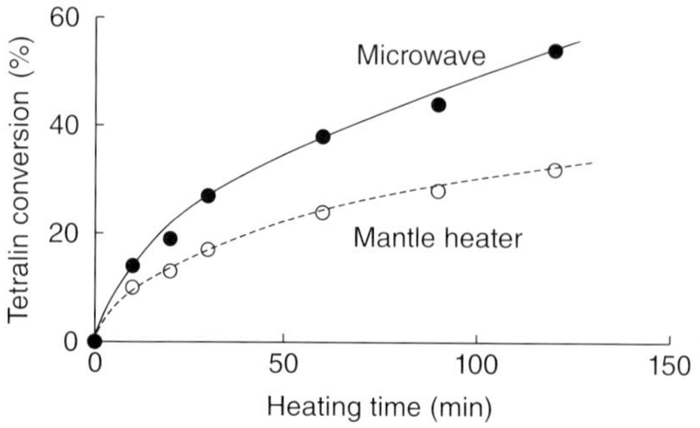

Figure 13.2 Conversion of tetralin to hydrogen gas and naphthalene during the dehydrogenation reaction under reflux conditions using microwave heating (MW power, 320 W) and conventional heating with a heat mantle at a temperature of 207 °C. Reproduced from Ref. [13]. Copyright 2010 by Elsevier B.V.

However, subjecting the tetralin solution to microwave radiation in the absence of the Pt/AC catalyst led to lower temperatures, thus demonstrating the excellent microwave absorbing ability of the activated carbon support.

The heating profile of pure distilled water and distilled water containing Pt/AC dispersion was also examined. The heating rate of pure distilled water was $4.43\,°C\,s^{-1}$, and distilled water containing Pt/AC was $7.83\,°C\,s^{-1}$. This showed that pure distilled water was heated by dipole polarization of H_2O, while the latter medium was heated by both dipole polarization (H_2O) and ionic conduction (Pt/AC) [14]. Even if the Pt/AC catalyst was dispersed in a polar solvent, the catalyst would nevertheless be heated by the microwave field.

A comparison of the dehydrogenation of tetralin between MWH and conventional heating in the suspension state is shown in Figure 13.2 [13]. The conversion yields of tetralin to naphthalene under MWH were greater than those observed under conventional heating (with a mantle heater) by nearly 19% under reflux conditions at 207 °C. Although the conversion by conventional heating seems to have reached a plateau after a reaction time of 2 h, the conversion of tetralin continued to increase in the microwave-assisted reaction. This result demonstrates the advantageous effect of MWH over conventional heating in the catalyzed dehydrogenation of tetralin in the presence of Pt on activated carbon support.

13.3.3
Effects of the Tetralin: Pt/AC Ratio on Tetralin Dehydrogenation

In the tetralin dehydrogenation reaction, the ratio of Pt/AC to tetralin was varied to investigate its effect on tetralin conversion. The catalyst was fixed at 0.2 g in each batch, while the amount of tetralin was varied between 0.55, 0.6, 0.7, 0.8, 1.0, and 1.2 ml (Figure 13.3) [15]. When the amount of tetralin in the system decreased (from 1.2 to 0.7 ml), both the average tetralin temperature under microwave irradiation for 60 min and tetralin conversion in the reaction increased linearly. Interestingly, in the batch with 0.6 ml tetralin, the conversion of tetralin showed

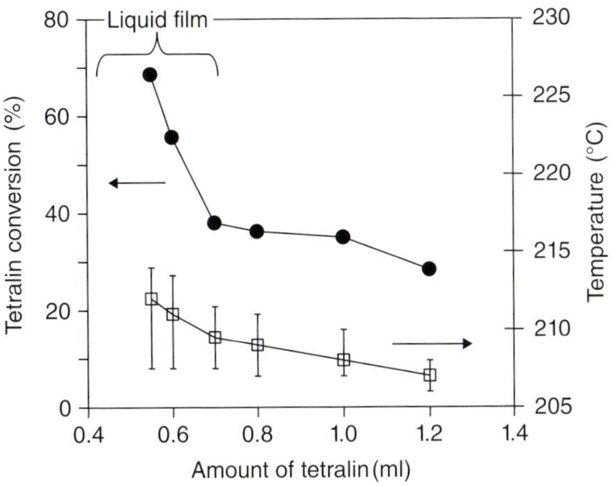

Figure 13.3 Correlation of (•) tetralin conversion and (□) average temperature from the tetralin dehydrogenation with 0.2 g of Pt/AC subjected to microwave irradiation for 60 min. Reproduced from Ref. [15]. Copyright 2012 by Elsevier B.V.

a remarkable increase from 39% to 56%. The conversion further increased by 69% with 0.2 g of Pt/AC and 0.55 ml of tetralin. Such results were explained by separating the reaction system into two stages: the *suspension state* and the *liquid-film state* [16]. For a batch reaction containing 0.7–1.2 ml tetralin and 0.2 g of Pt/AC, taken as the *suspension state*, any increase in tetralin conversion is likely due to an increase in tetralin accessibility to the active sites on the catalyst.

The images of the suspension state and the liquid-film state are displayed in Figure 13.4 [15]. The suspension state (loosely packed state) of the Pt/AC particles in tetralin showed a wider temperature difference between the catalyst

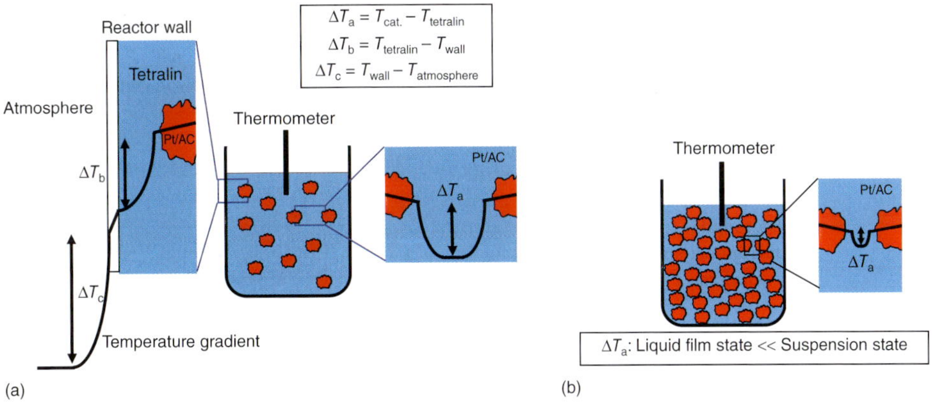

Figure 13.4 Models illustrating the temperature differences between tetralin and the Pt/AC catalytic system under microwave heating: (a) at a low ratio of the catalyst to tetralin and (b) at a high ratio of the catalyst to tetralin. Reproduced from Ref. [15]. Copyright 2012 by Elsevier B.V.

surface and the tetralin (Figure 13.4a). However, when the Pt/AC catalysts and tetralin were in the "liquid-film state" (Figure 13.4b), the temperature difference of the system was smaller than that of the "suspension state." This resulted in a much more uniform temperature of the tetralin mixture. Furthermore, the temperature of the catalyst particles as microwave absorbers could be maintained at a high temperature and be beneficial to the heating of tetralin molecules on the active sites by thermal conduction, usually referred to as "*localized superheating*" under microwave irradiation. This "localized superheating" on the catalyst active sites led to a microwave-induced enhancement of the reaction rate [17, 18]. This microwave effect on the enhancement of a chemical reaction can be attributed to the combination of thermal effects (overheating, hot spots) and nonthermal effects (polarization of molecules). As described by the Arrhenius reaction ($k = A\exp(-E_a/RT)$), the reaction rate depends on two factors: the preexponential factor (A) and the exponential factor $\{\exp(-E_a/RT)\}$.

In a liquid–solid phase reaction, the preexponential factor represents the collision frequency between the reactant molecules and the active sites, while the exponential factor represents the quantity of species that can overcome the activation energy. The magnitude of these parameters is temperature dependent [19]. The explanation of how microwaves can enhance tetralin dehydrogenation was described by the following mechanism. Tetralin (nonpolar liquid) was superheated by microwave-heated Pt/AC (temperature higher than its normal boiling point). This required a well-dissipated thermal energy; the molecules may have had large kinetic energy in order to create a high frequency of collisions of tetralin molecules on the catalyst active sites that led to an increase in the magnitude of the factors in the Arrhenius reaction rate equation [19, 20]. In addition, localized microplasma (sparks or hot spots) formed on the catalyst surface under MWH. It was obstructed by the liquid-film state condition. Details of hot spots are presented in Chapter 4.

13.3.4
Dehydrogenation of an Organic Carrier in a Continuous Flow System

The catalyzed dehydrogenation of tetralin over Pt/AC under MWH in a single-pass fixed-bed reactor was examined to affect improvements in reaction efficacy. Details of the experimental setup of the fixed-bed reactor in a single-mode microwave apparatus are schematically displayed in Figure 13.5 [21].

The microwave apparatus consisted of a single-mode TE_{103} cavity (transverse electric 103 mode), a short plunger, an iris, a three-stub tuner, a power monitor, an isolator, and a 2.45-GHz microwave generator [22]. The resonance of the microwaves was adjusted with the iris and the plunger at 1.5 cycles. The position of the sample was fixed at the maximal electric field (E-field) density within the waveguide. The Pt/AC catalyst was introduced in a 5-mm (inner diameter) quartz tube reactor [21]. The tetralin solution was subsequently introduced into the reactor with a syringe pump; a thermocouple measured the temperature of the catalyst inside the reactor. The yields of products were quantified by gas chromatography

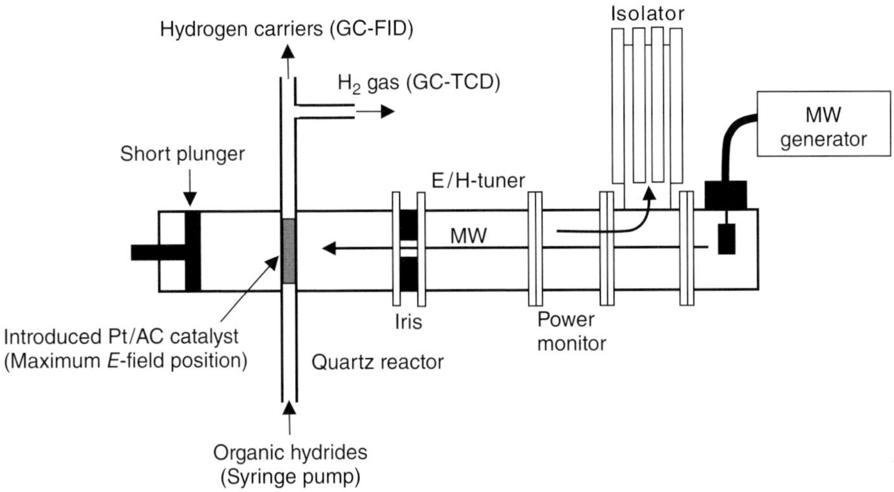

Figure 13.5 Experimental setup in the dehydrogenation of tetralin on a single-pass fixed-bed reactor involving a 2.45-GHz single-mode microwave apparatus. Reproduced from Ref. [21]. Copyright 2012 by Elsevier B.V.

(GC) using a thermal conductivity detector (TCD) for the hydrogen gas and a flame ionization detector (FID) for the analysis of other products.

The conversion yields of the dehydrogenation of tetralin at various temperatures under MWH and conventional heating (electric mantle heater) are reported in Figure 13.6a [21]. The prompt effect of the microwaves is clearly shown at low-temperature reaction conditions owing to the catalyst surface being at a higher temperature than the bulk under microwave irradiation. Conversion yields of the dehydrogenation of tetralin at various flow rates are summarized in Figure 13.6b; the reaction rates tended to decrease with an increase in flow

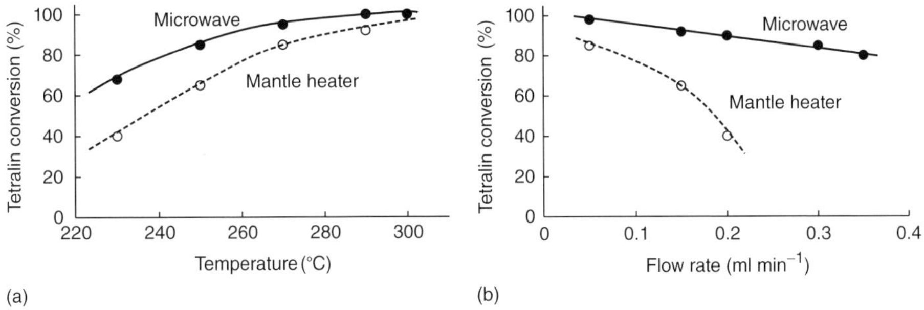

Figure 13.6 (a) Dependence of the conversion of tetralin to hydrogen gas and naphthalene over the Pt/AC catalyst surface on reaction temperature under microwave heating and conventional heating with a mantle heater. (b) Dependence of conversion of tetralin over Pt/AC catalyst on feed flow rate of tetralin under microwave heating and conventional heating at a reaction temperature of 207 °C and a microwave power output of 90–170 W. Reproduced from Ref. [21]. Copyright 2012 by Elsevier B.V.

rates. Reaction efficiency showed a remarkable decrease with an increase in flow rates at flow rates less than 0.2 ml min^{-1}, especially when heating was carried out with a mantle heater. In contrast with MWH, tetralin conversion occurred at conversion yields greater than 80%, even when the feed flow rate increased to about 0.4 ml min^{-1}. Such high conversion yields were attributed to the enhanced desorption (vs. adsorption) phenomenon taking place at the surface of the catalyst [21]. The bulk temperature was maintained at 207 °C throughout the process. From the observations of Figure 13.6, the significant beneficial effect of the microwaves on the catalyzed dehydrogenation of tetralin over Pt/AC was the driving force originating from the high-temperature gradients between the Pt/AC catalyst (strong microwave absorber) and tetralin (nonpolar reactant and a poor absorber of microwaves). Most significant, the microwaves enhanced desorption of product molecules from the catalyst surface and regenerated the catalytically active sites, thereby prolonging the activity of the catalyst and augmenting reaction efficacy.

The effect of a microwave field on the improvement in molecular diffusion or transport of species in a reaction system has been described [23, 24]. Antonio and Deam have illustrated the hypothesis that if the transport of an active species is a rate-limiting step in a reaction, the temperature gradient under MWH enhances the diffusion of that species, and the overall reaction rate is changed under MWH compared with conventional heating [24]. From the above results, when the feed flow rate was increased or the space time lowered, the reaction conversion percentage yield decreased dramatically. This may imply that the tetralin dehydrogenation reaction is limited by the rate of mass diffusion or species transport, in particular, and desorption of product species from the active sites of the catalyst. This assumption was supported by the behavior of naphthalene adsorption. Apart from hydrogen, naphthalene is a dominant product of the reaction. It is well known that naphthalene has strong adsorption energy because of the p-electrons in its aromatic rings. The strong adsorption of naphthalene on the catalyst active sites not only decelerated the reaction because of site blocking but also caused coke deposition. This led to the deactivation of the catalyst, as evidenced by a decrease in the dehydrogenation conversion time.

An advantageous effect of microwaves on an increase of species transport in the reaction correlated with the temperature gradient and the mass transfer whose direction was the reverse of that of conventional heating (Figure 13.7a,b) [21]. In the case of tetralin dehydrogenation under MWH, the heat transfer generated from the Pt/AC catalyst to the surrounding species also induced the driving force of mass transfer in the same direction because of the coupling vector. This phenomenon has two possible effects: (i) promote desorption of product species (naphthalene) from the catalyst active sites, thereby leading to facilitation of mass transfer in the system [23, 25] and (ii) prevention of a contiguous strong adsorption of hydrocarbon on the active sites, thereby causing a decrease in coke deposition. Hence, the heating phenomenon from the microwaves inverts that of conventional heating. Species transport in the system under MWH was better than that under conventional heating. Consequently, the overall reaction rate

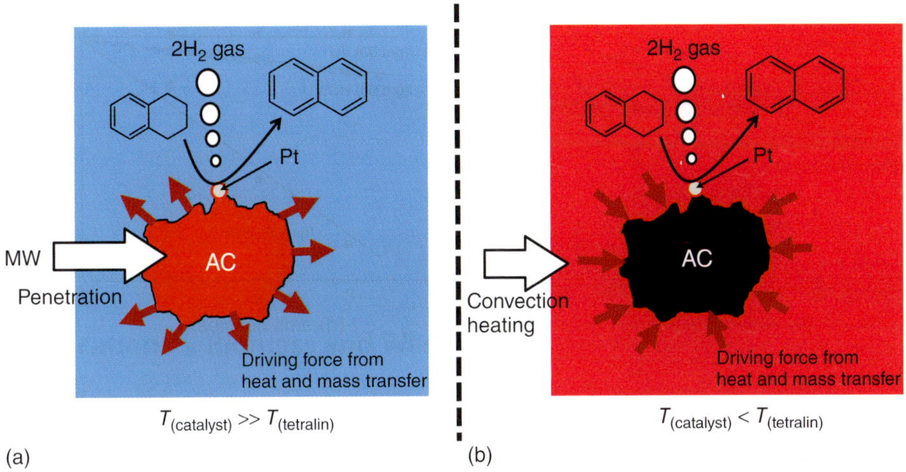

Figure 13.7 Model of heat transfer direction: (a) Under microwave heating (MWH): $T_{catalyst} > T_{surrounding}$, heat transfers from the catalyst to the surrounding matter; mass also transfers in the same direction because of the coupling vector. (b) Under classical convection heating (CH): $T_{catalyst} < T_{tetralin}$, heat transfers from the surrounding matter to the catalyst. Reproduced from Ref. [21]. Copyright 2012 by Elsevier B.V.

under MWH increased. This explained the MWH effect that enhanced tetralin dehydrogenation compared to conventional heating.

The boiling point of tetralin is 206 °C; the catalyst was heated to the high temperature and the continuous dehydrogenation reaction of tetralin was examined at about 750 °C. The evolution yield of hydrogen gas was at most 67% by heat conduction using a ceramics heater with an insulator [26]. However, its yield decreased remarkably with circulation time of the tetralin. Moreover, other kinds of by-products were formed by the CH method. On the other hand, the evolution yield of hydrogen was 78% by the MWH method. It is possible to explain these differences by the conditions of the introduced Pt/AC catalysts in the reactor at the completion of the reaction. The Pt/AC catalyst powder changed into a Pt/AC stick with sintering when using the ceramics heater (Figure 13.8) [26], with inactivation of the catalysts. On the other hand, the Pt/AC catalyst from the MWH method remained as a powder (Figure 13.8). The difference in these situations was considered by the temperature distribution in the reactor.

13.3.5 Dehydrogenation of Methylcyclohexane in a Microwave Fixed-Bed Reactor

Methylcyclohexane (MCH) is perhaps the most hopeful of the hydrogen carriers. The researcher's present attention is to enhance the efficiency of hydrogen evolution using conditions that require low applied energy for thermally activating the catalysts. The energy conservation for hydrogen evolution can be achieved by microwave selective heating of solid catalysts (e.g., Pd/AC). The

Figure 13.8 The state of the sintering of a catalyst after ceramics heater and microwave heating. From Horikoshi *et al.*, unpublished data [26].

features pertaining to the hydrogen evolution from MCH as the source with the microwave method is described in this section.

In general, the reaction temperature for the dehydrogenation of MCH begins at about 200 °C [27], following which the yield of hydrogen is saturated at about 340 °C. Palladium (5 wt%) deposited as particulates onto activated carbon as the support (Pd/AC) was used as the catalyst in the dehydrogenation reactions of MCH. The catalytic reactions were carried out under atmospheric pressure in a quartz fixed-bed tubular reactor (diameter: 17 mm; length: 55 mm) and loaded with 2.5 g of Pt/AC catalyst.

The methylcyclohexane was introduced in the reactor using a syringe pump at a flow rate of 0.1 ml min^{-1} (same apparatus as in Figure 13.5). Results from the dehydrogenation of MCH using MWH and conventional heating with a ceramics heater and insulator are reported in Figure 13.9 [28]. The added Pd/AC catalyst particulates were heated at 340 °C within 2 min by MWH, which resulted in over

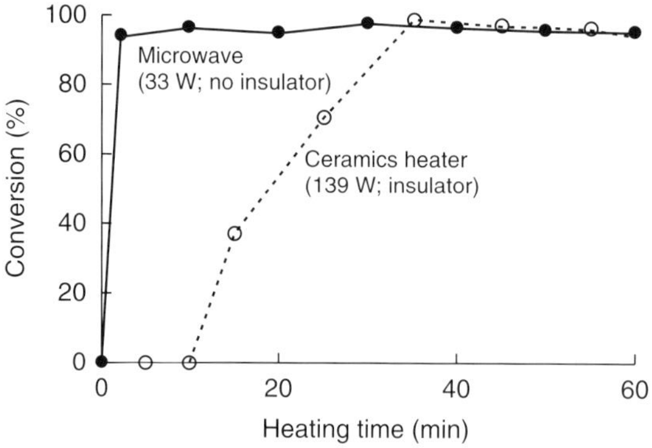

Figure 13.9 Dependence of the conversion of methylcyclohexane to hydrogen gas and toluene over the Pd/AC catalyst under microwave heating and conventional heating with a ceramics heater (temperature, 340 °C) [28].

94% dehydrogenation of MCH occurring in less than 2 min. On the other hand, when using a ceramics heater, dehydrogenation needed more than 30 min also at 340 °C even though an insulator was used. Clearly, MWH led to hydrogen evolution in a much shorter time. On suspension of microwave irradiation, the catalyst particulates promptly cool and hydrogen evolution stops, whereas suspension of conventional heating with the ceramics heater hydrogen gas continues to evolve for more than 35 min. The total power consumption of the microwave system was 33 W (microwave applied power, 12 W), while the ceramics heater expended 139 W of power. Evidently, MWH is also favored in terms of power consumption and thus an attractive profitable energy-saving system, in addition to the shorter times for hydrogen evolution. Recall that electric energy is being used to produce hydrogen energy.

13.3.6 Simulation Modeling for Microwave Heating of Pt/AC in the Methylcyclohexane Solution

The heat convection from Pd/AC in the methylcyclohexane solution was investigated and was simulated using the COMSOL Multiphysics software package (Version 4.3a). Selective heating of 180 particles of Pd/AC dispersed in methylcyclohexane was considered. The model geometry is displayed in Figure 13.10a [28]. The correlation between bulk temperature and the temperature of the Pd/AC particles is illustrated in Figure 13.10b. Results of the simulation demonstrated that to bring methylcyclohexane from room temperature (25 °C) to a temperature of 340 °C necessitates the temperature of the Pt/AC particles to be at 547 °C, resulting in a temperature gradient of 217 °C. Under microwave irradiation, the Pt/AC particles were continuously heated by the microwave radiation field, with much of the resulting thermal energy spent to heat methylcyclohexane.

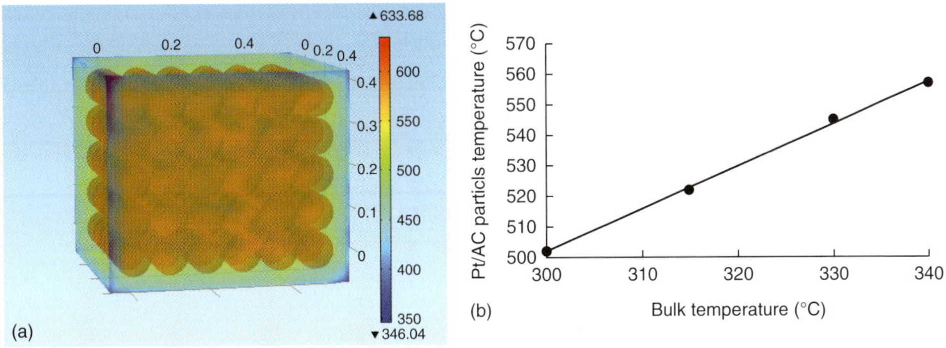

Figure 13.10 Relation between the Pt/AC particle temperatures and bulk temperatures. (a) Simulation image and (b) plot of temporal change on Pt/AC particle temperatures and bulk temperatures. From Ref. [28].

13.4
Dehydrogenation of Methane with Microwaves in a Heterogeneous Catalytic System

Methane is the major component of natural gas (97% by volume). Transformation of methane to other high-value products has been widely investigated: for example, (i) methane reforming to hydrogen and (ii) oxidative coupling of methane to form higher hydrocarbons such as C_2 products. In this regard, Marún and coworkers [29] examined the catalytic oligomerization of methane to C_2–C_6 hydrocarbons (e.g., acetylene to benzene) under MWH in a continuous flow reactor. Several catalysts were employed for the reaction activated by microwave radiation: activated carbon, MnO_2, TiO_2, Fe, and Ni. Of these, activated carbon exhibited efficient activity on the reaction under MWH by rapid heating to over 1000 °C at an applied microwave power of 600 W. Selectivity depended on the applied power and on the diluting gas used in the reaction. The major C_2 products were in the order ethylene > acetylene > ethane, while 33% selectivity of benzene product was achieved under 378-W microwave irradiation. However, when the applied microwave power was increased to 754 W, the reaction generated only CO_2. The proposed steps in the overall mechanism of methane oligomerization are reported below for the formation of the C_2 products (Eqs. (13.2)–(13.5)):

$$CH_4 \rightarrow\ :CH_2 + H_2 \tag{13.2}$$

$$2CH_4 \rightarrow\rightarrow\rightarrow C_2H_4 + 2H_2 \tag{13.3}$$

$$2CH_4 \rightarrow\rightarrow\rightarrow C_2H_2 + 3H_2 \tag{13.4}$$

$$C_2H_4 + H_2 \rightarrow C_2H_6 \tag{13.5}$$

Another viable mechanism suggested for ethylene and acetylene formation involved the dehydrogenation of secondary hydrocarbon products (Eqs. (13.6) and (13.7)):

$$C_2H_6 \rightarrow\rightarrow\rightarrow C_2H_4 + H_2 \tag{13.6}$$

$$C_2H_4 \rightarrow\rightarrow\rightarrow C_2H_2 + H_2 \tag{13.7}$$

The reaction that yielded the higher hydrocarbons most likely proceeded via formation of free radical intermediates. Thus, formation of propane, propene, and benzene was thought to originate via Eqs. (13.8)–(13.10), respectively [29]:

$$C_2H_6 +\ :CH_2 \rightarrow\rightarrow\rightarrow C_3H_8 \tag{13.8}$$

$$C_3H_8 \rightarrow C_3H_6 + H_2 \tag{13.9}$$

$$3C_2H_2 \rightarrow C_6H_6 \tag{13.10}$$

The microwave-assisted dry reforming of methane over activated carbon under MWH on a fixed-bed reactor (Eq. (13.11)) was reported by Fidalgo *et al.* [30], who used activated carbon made from coconut shells as both the catalyst and the microwave absorber:

$$CH_4 + CO_2 \leftrightarrows 2CO + 2H_2 \tag{13.11}$$

Results showed that the conversions of CO_2 and CH_4 were clearly greater under MWH. Moreover, on stream for 5 h and at a reaction temperature of 800 °C at the catalyst bed (activated carbon), the reported conversion yield under MWH was 70%, while the conversion yield under conventional heating was 40%. The microwave effect on the dry reforming of methane was described by the Fidalgo *et al.* [30] in terms of more uniform heating of the catalyst bed under MWH when compared to conventional (electrical) heating. Additionally, they observed the formation of a micro-plasma within the bed of activated carbon upon MWH that yielded a temperature greater than the average temperature of the catalyst bed:

$$C_2H_6 \xrightarrow[800\,°C]{AC} C_2H_4 + H_2 \tag{13.12}$$

The dehydrogenation of ethane to ethylene (Eq. (13.12)) by MWH was investigated by Kim and coworkers [31] using activated carbon produced from Korean raw anthracite coals as the catalyst/absorber that was subjected to MWH and conventional heating in a fixed-bed reactor system. Based on the same bed temperature of 800 °C, ethane conversion under MWH exhibited a better performance than did conventional heating: about 13% higher conversion with the microwaves (73% vs. 60%); product yield with microwave irradiation was about 64% while with conventional heating the product yield was about 52%. The temperature of the outlet gas product was much lower when microwave irradiation was used (range 45–50 °C), whereas the outlet temperature under conventional heating was 835 °C.

13.5
Problems and Improvements of Microwave-Assisted Heterogeneous Catalysis

The heat distribution along the vertical positions of the Pd/AC catalysts in the reactor has been examined under microwave heating and, for comparison, also under conventional heating using a ceramic heater. An optic fiber thermometer was used to measure the temperature from the bottom to the top at the center of the reactor in the presence and absence of the organic hydride MCH (flow rate of liquid MCH was 0.2 mL min^{-1}). This thermometer was limited to recording a maximum temperature of 220 °C. Under such conditions, the temperature recorded at the 3-cm position from the bottom was 174 ± 2 °C; under conventional heating, the temperature at the same location was 181 ± 1 °C [28].

The temperature distribution within the catalyst particles under microwave heating is displayed in Figure 13.11, which shows a rather large distribution of temperatures: 176–177 °C between 3.0 and 3.5 cm from the bottom of the

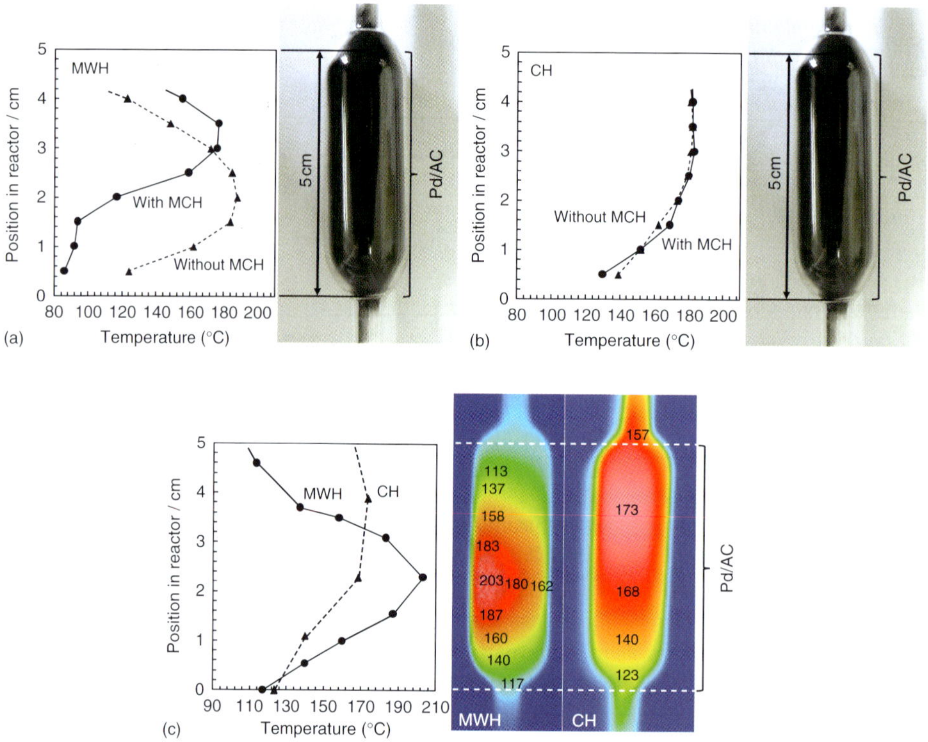

Figure 13.11 Temperature distribution for lengthwise location of Pd/AC catalyst in the reactor with and without methylcyclohexane. (a) Microwave heating, (b) conventional heating with a ceramic heater, (c) temperature distribution for a reactor containing Pd/AC catalysts without methylcyclohexane under microwave heating (MWH) and conventional heating (CH). From Ref. [28].

reactor in the presence of liquid MCH. However, at 4.0 cm from the bottom of the reactor, the temperature decreased to 155 °C, and even more below the 3.0-cm mark: 94 7°C at 1.5 cm and well below the boiling point of MCH (101 °C) and of the dehydrogenation reaction temperature (340 °C). Thus, with an almost 50% drop in the temperature, the catalytic powder located below 1.5 cm are not likely to participate in the dehydrogenation reaction [28].

The temperatures of the dry Pd/AC catalyst powder (in the absence of MCH) were 188 °C and 185 °C at 2.0 cm and 1.5 cm from the bottom, respectively. Perusal of the distribution in Figure 13.11a showed that the temperatures at the top and bottom of the reactor were significantly lower. The distribution of the microwaves' electric field in the waveguide was likely at the origin of such variations in the heat distributions. Significant selective heating of the Pd/AC particulates occurred primarily where the electric field density was greatest, exactly at the middle (height) position of the reactor. Figure 13.11c summarizes the temperature distribution at the inner walls of the Pd/AC catalyst reactor in the absence of MCH (dry Pd/AC catalysts) observed by thermography immediately (5 sec) after the reactor was taken out of the setup. A rather large distribution was observed under microwave heating, with a temperature of 203 °C at 2.5 cm from the bottom of the reactor and decreasing to 180 °C and 162 °C horizontally away from the inset of the microwave radiation. The lower temperatures at the top and at the bottom of the reactor in the MW method were attributed to the influence of the distribution of the microwaves' electric field. Consideration of the these results led the authors [28] to deduce that the temperature distribution under microwave radiation is the consequence of several factors: (i) the distribution of the microwaves' electric field, (ii) the temperature of the cooler MCH liquid fed to the reactor, and (iii) the liquefied toluene gas exiting the upper part of the reactor.

Under conventional heating, no changes were observed in the distribution of temperature in the presence or absence of MCH, although the temperature decreased gradually toward the lower portions of the reactor (13.11b). Subsequent to bringing the temperature at the middle position of the Pd/AC catalyst layer to 180 °C under conventional heating with the ceramic heater, the temperatures were also measured at various positions in the surrounding atmosphere of the reactor (13.11c). Outside the reactor, the temperature was ambient (26 °C) at the bottom of the reactor, whereas in the upper part of the setup it was 63 °C. On comparing, the temperature inside the setup was 65 °C (bottom), which increased to 82 °C in the upper portion as a result of air convection occurring from the bottom-up [28].

Studies reported so far have demonstrated that selective heating of a solid catalyst in a heterogeneous reaction is an ideal occurrence when using microwaves. The combination of the catalyst and/or catalyst support with high absorption of microwaves and a nonpolar solvent satisfy this condition. A fixed-bed reactor is also an effective component of a reaction system. The temperature distribution peculiar to microwave irradiation can enhance adsorption and desorption (contact efficiency) of the substrate on the catalyst surface and can also lead to energy savings during the heating stage(s). However, the heat developed on the catalyst can escape to the atmosphere through the reactor walls. Therefore, in order to avoid such heat losses, insulation of the reaction container is most desirable.

Usually, one makes use of heat insulating materials (e.g., fiberglass) wrapped around the external walls of the reactor to prevent escape of heat to the surroundings. Unfortunately, such materials can also attenuate the microwave radiation reaching the reacting substrates in the reaction vessel. Accordingly, a novel reactor design was conceived that avoided heat losses and attenuation of the microwaves (Figure 13.12a,b) [26]. The proposed reactor system consisted of a double-walled glass vessel in which the double wall was vacuum-filled that typifies a Dewar-like structure used to preserve cold fluids (e.g., liquid helium). In this case, microwave irradiation of the sample was not obstructed by the vacuum-filled double walls, while any heat emanating from the heated sample was intercepted by the reactor's double walls.

This double-walled glass vessel also indicates the effect already in organic synthesis with heterogeneous catalysis reaction. For example, the synthesis of

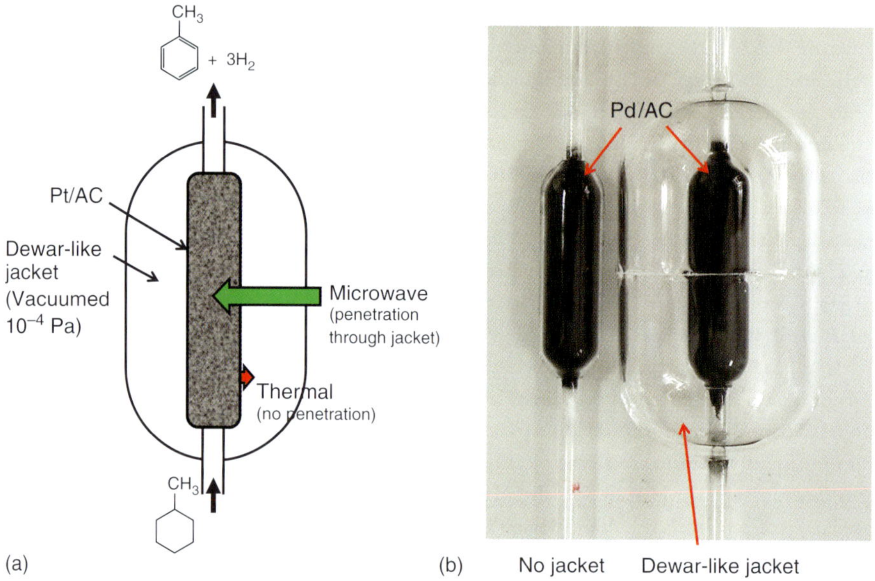

Figure 13.12 (a) Sketched image and (b) photograph of the conventional microwave reactor and microwave Dewar-like vacuum reactor. From Horikoshi *et al.*, unpublished data [26].

biphenyl by the Pd/AC Suzuki–Miyaura cross-coupling reaction was progressed with a Dewar-like microwave vessel [32]. The Pd/AC catalyst, phenylboronic acid, bromobenzene, K_2CO_3 as the base, and toluene solvent were mixed and subsequently added under an argon atmosphere to the double-walled microwave vessel (Figure 13.13a,b). The syntheses were performed by placing the respective reactors in a commercial multimode microwave chemical apparatus (Tokyo Rikakikai Co. Ltd, MWO-1000S system) equipped with a fiber-optic thermometer (Anritsu Meter Co., Ltd, FL-2000).

The temperature distributions in the Dewar-like (doubled-walled) microwave vessel doubled-walled and in the conventional single-walled vessels were examined by recording the increase in the temperature of water (20 ml) at various times following continuous irradiation of the dielectric medium with 120-W microwaves in the commercial multimode microwave chemical apparatus. The temperature gap of center position and inner wall side position under nonstirring conditions was <1 °C for doubled-walled microwave vessel and 7 °C for conventional single-walled vessel. The potential of the double-walled vacuum-filled vessel in the synthesis of biphenyl was enhanced significantly (six- to sevenfold) with this double-walled vessel relative to the conventional single-walled reactor vessel currently used in commercial microwave chemical devices. Dewar-like microwave vessel can make a microwave input power fall remarkably because keeping temperature of a sample is excellent. Exactly, its possible to achieve energy saving in chemistry process.

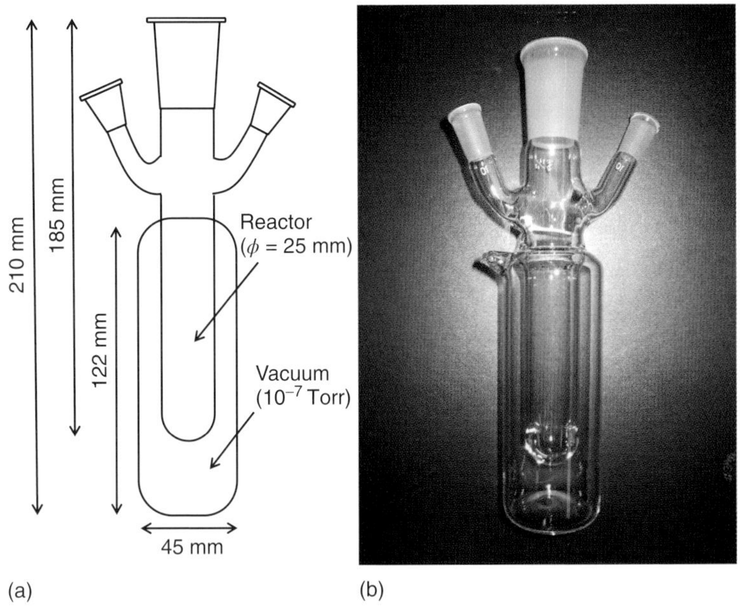

Figure 13.13 (a) Sketched image and dimensions of the double-walled reactor and (b) actual photograph of the microwave Dewar-like vacuum vessel. Reproduced in part from Ref. [32]. Copyright 2010 by the American Chemical Society.

Acknowledgments

We are particularly grateful to Dr Suttisawat (Ph.D. researcher at the Tokyo University of Science and Sophia University, 2010–2012) and to Miss Kamata (Master course student at Sophia University) for their assistance in the studies on the dehydrogenation of tetralin and methylcyclohexane.

References

1. Winter, C.-J. (2005) Into the hydrogen energy economy – milestones. *Int. J. Hydrogen Energy*, **30**, 681–685.
2. see Panasonic Corporation *http://panasonic.co.jp/ap/FC/en_index.html* (accessed 7 May 2015).
3. Züttel, A. (2004) Hydrogen storage methods. *Naturwissenschaften*, **91**, 157–172.
4. Gretz, J., Baselt, J.P., Ullmann, O., and Wendt, H. (1990) The 100 MW Euro-Quebec hydro-hydrogen pilot project. *Int. J. Hydrogen Energy*, **15**, 419–424.
5. Wang, Y., Chen, K.S., Mishler, J., Cho, S.C., and Adroher, X.C. (2011) A review of polymer electrolyte membrane fuel cells: technology, applications, and needs on fundamental research. *Appl. Energy*, **88**, 981–1007.
6. Sakintuna, B., Lamari-Darkrim, F., and Hirscher, M. (2007) Metal hydride materials for solid hydrogen storage: a review. *Int. J. Hydrogen Energy*, **32**, 1121–1140.
7. Makowski, P., Thomas, A., Kuhn, P., and Goettmann, F. (2009) Organic materials for hydrogen storage applications: from physisorption on organic solids to chemisorption in organic molecules. *Energy Environ. Sci.*, **2**, 480–490.

8. Zhu, Q.-L. and Xu, Q. (2015) Liquid organic and inorganic chemical hydrides for high-capacity hydrogen storage. *Energy Environ. Sci.*, **8**, 478–512.
9. United States Department of Energy (2014) *http://www.hydrogen.energy.gov/annual_progress14.html* (accessed August 2015).
10. Christian, M.L. and Aguey-Zinsou, K.-F. (2012) Core–shell strategy leading to high reversible hydrogen storage capacity for $NaBH_4$. *ACS Nano*, **6**, 7739–7751.
11. NEEL Institute (Highlights 2007) Magnesium for Reversible Hydrogen Storage, The NEEL Institute, Grenoble. *http://neel.cnrs.fr/spip.php?article1281&lang=fr* (accessed 11 February 2015).
12. Ichikawa, M. (2007) in *Organic Storage Materials and Nano Techology for Hydrogen Society*, Chapter 3 (ed M. Ichikawa), CMC Publisher, pp. 127–169.
13. Suttisawat, Y., Horikoshi, S., Sakai, H., and Abe, M. (2010) Hydrogen production from tetralin over microwave-accelerated Pt-supported activated carbon. *Int. J. Hydrogen Energy*, **35**, 6179–6183.
14. Stuerga, D. (2006) in *Microwaves in Organic Synthesis*, Chapter 1 (ed A. Loupy), Wiley-VCH Verlag GmbH, Weinheim, pp. 1–61.
15. Suttisawat, Y., Horikoshi, S., Sakai, H., Rangsunvigit, P., and Abe, M. (2012) Enhanced conversion of tetralin dehydrogenation under microwave heating: effects of temperature variation. *Fuel Process. Technol.*, **95**, 27–32.
16. Saito, Y., Aramaki, K., Hodoshima, S., Saito, M., Shono, A., Kuwano, J., and Otake, K. (2008) Efficient hydrogen generation from organic chemical hydrides by using catalytic reactor on the basis of superheated liquid-film concept. *Chem. Eng. Sci.*, **63**, 4935–4941.
17. Tsukahara, Y., Higashi, A., Yamauchi, T., Nakamura, T., Yasuda, M., Baba, A., and Wada, Y. (2010) In situ observation of non-equilibrium local heating as an origin of special effect of microwave on chemistry. *J. Phys. Chem. C*, **114**, 8965–8970.
18. Zhang, X., Hayward, D.O., and Mingos, D.M.P. (2003) Effect of microwave dielectric heating on heterogeneous catalysis. *Catal. Lett.*, **88**, 33–38.
19. Perreux, L. and Loupy, A. (2006) in *Microwave in Organic Synthesis* (ed A. Loupy), Wiley-VCH Verlag GmbH, Weinheim, pp. 134–218.
20. Djenni, Z.C., Hamada, B., and Chemat, F. (2007) Atmospheric pressure microwave assisted heterogeneous catalytic reactions. *Molecules*, **12**, 1399–1409.
21. Suttisawat, Y., Sakai, H., Abe, M., Rangsunvigit, P., and Horikoshi, S. (2012) Microwave effect in the dehydrogenation of tetralin and decalin with a fixed-bed reactor. *Int. J. Hydrogen Energy*, **37**, 3242–3250.
22. Horikoshi, S., Matsubara, A., Takayama, S., Sato, M., Sakai, F., Kajitani, M., Abe, M., and Serpone, N. (2009) Characterization of microwave effects on metal-oxide materials: zinc oxide and titanium dioxide. *Appl. Catal. B: Environ.*, **91**, 362–367.
23. Semenov, V.E. and Rybakov, K. II, (2004) What type transport phenomena can be induced by microwave field in solids and how these phenomena contribute to materials processing. Proceedings of the International Symposium on Novel Materials Processing by Advanced Electromagnetic Energy Sources, Osaka, Japan, March 19–22, 2004, pp. 111–117.
24. Antonio, C. and Deam, R.T. (2007) Can microwave effects be explained by enhanced diffusion? *Phys. Chem. Chem. Phys.*, **9**, 2976–2982.
25. Dincov, D.D., Parrott, K.A., and Pericleous, K.A. (2004) Heat and mass transfer in two-phase porous materials under intensive microwave heating. *J. Food Eng.*, **65**, 403–412.
26. Horikoshi, S. and Serpone, N., *et al.*, (2015) to be submitted for publication.
27. Horikoshi, S., Kamata, M., Sumi, T., and Serpone, N. (2015) Selective heating of Pd/AC catalyst in heterogeneous systems for the microwave-assisted continuous hydrogen evolution from organic hydrides: Temperature distribution in the fixed-bed reactor. *ACS Catal.*, submitted.

28. Hodoshima, S., Shono, A., Kuwano, J., and Saito, Y. (2008) Chemical recuperation of low-quality waste heats by catalytic dehydrogenation of organic chemical hydrides and its energy analysis. *Energy Fuels*, **22**, 2559–2569.
29. Marún, C., Conde, L.D., and Suib, S.L. (1999) Catalytic oligomerization of methane via microwave heating. *J. Phys. Chem. A*, **103**, 4332–4340.
30. Fidalgo, B., Domínguez, A., Pis, J.J., and Menéndez, J.A. (2008) Microwave-assisted dry reforming of methane. *Int. J. Hydrogen Energy*, **33**, 4337–4344.
31. Kim, D.K., Cha, C.Y., Lee, W.T., and Kim, J.H. (2001) Microwave dehydrogenation of ethane to ethylene. *J. Ind. Eng. Chem.*, **7**, 363–374.
32. Horikoshi, S., Osawa, A., Suttisawat, Y., Abe, M., and Serpone, N. (2010) A novel Dewar-like reactor for maintaining constant heat and enhancing product yields during microwave-assisted organic syntheses. *Org. Process Res. Dev.*, **14**, 1453–1456.

Part VI
Applications – Oil Refining

Microwaves in Catalysis: Methodology and Applications, First Edition.
Edited by Satoshi Horikoshi and Nick Serpone.

14
Microwave-Stimulated Oil and Gas Processing

Leonid M. Kustov

14.1
Introduction

Existing technologies in oil and gas processing mostly reached the saturation level in their development and further progress may be possible, if at all, by using non-traditional approaches, such as nonequilibrium conditions, gradient technologies, combination of endothermic and exothermic processes, *ex situ* or *in situ* application of electromagnetic activation modes, such as microwave activation and low-temperature plasma, among others.

Microwave technologies have been introduced as benign and robust approaches in many areas, including in industrial processes [1]. Microwave irradiation (in the heating or plasma option) is widely explored in processes of heavy oil and gas processing. It is obvious that in this case the dipolar polarization mechanism does not work and a conduction mechanism or Maxwell–Wagner effects (Sections 17.1 and 17.8) may be responsible for the microwave heating.

14.2
Early Publications

One of the first patents related to the effects of microwave (MW) activation in catalysis was issued in 1985, which focused on the catalytic transformation of methane into ethylene on nickel and iron catalysts activated by microwave field pulses [2]. Although initially the focus was on supported metallic catalysts, efforts are currently extended to mixed oxides (e.g., perovskites and transition metal oxides) because of their capability to absorb the MW energy.

Wan [3] was the first to use microwave heating in heterogeneous catalysis with ferro- and para-magnetic metallic particles dispersed in a liquid reaction medium. Wan's team discovered that the use of the pulsed microwave technique allowed for the accurate control of catalyst temperature and process selectivity. Apart from metallic powders, Wan *et al.* studied the commercially available materials containing nickel and copper wires and gauzes, as well as dispersed

Microwaves in Catalysis: Methodology and Applications, First Edition.
Edited by Satoshi Horikoshi and Nick Serpone.

metallic particles supported on nonconducting carriers. The early studies were focused on the reactions of hydrogenation and hydrocracking of hydrocarbons [4, 5], on methane decomposition [6], on the oxidation of hydrocarbons and the synthesis of acetylene [7–10], on the conversion of olefins [11], and on oil sands (Province of Alberta, Canada) and bitumens [12].

Roussy *et al.* [13–15] used highly dispersed platinum, that is, species that absorb microwave radiation very well, on an Al_2O_3 support that is transparent to microwaves in the oxidation of ethylene, in the isomerization of hexane and 2-methylpentane, and in the hydrogenolysis of methylcyclopentane. The authors also studied the isomerization of 2-methyl-2-pentene on acidic oxide catalysts, as well as the oxidative condensation of methane on transition metal oxides. They found that the use of microwave fields *in situ* could lead to results quite different from those obtained using thermal activation. These authors reached an important conclusion in that the temperature dependencies of the conversion are not informative because of strong differences in the methods of heating and difficulties in determining the true catalyst temperature under microwave field conditions. They suggested a comparison of the conversion versus selectivity dependencies rather than the conversion versus temperature curves.

Discussion of the contributions of purely thermal and nonthermal effects in the case of microwave-assisted processes has been a subject of debate for more than 30 years. Klimov *et al.* studied the epoxidation of ethylene while separating thermal from nonthermal effects [16]. The catalyst was heated not only by a microwave field but also by hot air. To decrease the convective heat loss, the reactor and part of the air supply line located in the microwave furnace were placed into an evacuated jacket. The specific action of the microwave field was only observed for the $Ag/\alpha\text{-}Al_2O_3$ catalyst subjected to a reducing treatment before measurements. In an oxidative atmosphere, no differences were observed between the experiments performed under microwave and thermal conditions. The effect was discussed taking into account another observation: formation of defects in the bulk of silver particles and on the surface is suppressed in a reducing atmosphere.

According to Roussy, structural changes or "electron behavior" effects are one of several possible reasons behind the increase of selectivity in the isomerization of 2-methylpentane on Pt/Al_2O_3 catalysts, including 0.3% Pt/Al_2O_3 (EuroPt-3) and on 0.3% Pt–0.3% Re/Al_2O_3 (EuroPt-4) [17]. In the case of the EuroPt-3 catalyst containing 0.95% Cl, the selectivity to branched isomers was enhanced twofold under MW conditions. In the case of EuroPt-4, which is a bimetallic catalyst, the product distributions differed for the MW-assisted process and conventional heating.

Another nonthermal effect was revealed for the $BaBiO_{3-x}$ catalyst in the oxidative condensation of methane [18]. An increase in selectivity with respect to C_2 products under microwave activation was accounted for by a decrease in the rate of carbanion oxidation because of a decrease in the concentration of oxygen adsorbed by the catalyst. The different behaviors of catalysts activated under the microwave and thermal conditions were confirmed by the measurements of the real and imaginary complex permittivity components under microwave conditions (single-mode, high-power microwave field heating catalysts) and in a classic resonator (multi-mode conductivity measurements [18] with a reduced microwave power supply after the thermal treatment of the sample). When the catalyst reached a temperature higher than 500 °C, the imaginary permittivity component increased in oxygen and air. Conversely, the dielectric loss decreased in an inert atmosphere (helium). The authors explained these effects in terms of suppression of redox processes under microwave conditions. It was noted that the metallic behavior corresponded to a maximum degree of $BaBiO_{3-x}$ system reduction ($x = 0.49$), and it was concluded that the electromagnetic field decreased the rate of oxygen adsorption.

14.3 Use of Microwave Activation in Catalytic Processes of Gas and Oil Conversions

Some examples of the application of microwave radiation in processes of gas processing, including dry methane reforming and oxidative coupling of methane, are given in the review by Zhang and Hayward [19]. A few applications of microwave activation in oil sands and oil processing are given in the review by Mutyala and coworkers [20]. The examples include, *inter alia*, bitumen extraction, upgrading heavy oils, removing heteroatoms, and the underground heating of oil sands to reduce bitumen viscosity and allow it to be pumped to the surface.

Among the gas-phase catalytic reactions that most frequently were subjected to stimulation by microwave fields, the following processes are worth noting: (i) CO_2 reforming of methane, (ii) oxidative coupling of methane, and (iii) partial oxidation of higher hydrocarbons. In the first two reactions, which are of great importance for practical applications, the key efficiency factor is the achievement of high selectivity with respect to the desired products at equal reagent conversions in comparing thermal and MW conditions.

14.3.1 Hydrogen Production

One of the reactions considered with respect to hydrogen production from natural gas is the water/gas shift reaction (Eq. (14.1)) [21]. When the steam/CO ratio is low (about 1), the curve of CO concentration has a different shape in the case of the thermal and microwave treatments with the performance of the catalyst being more efficient under microwave irradiation. With an increase in reaction temperature or steam/CO ratio, a less significant increase in CO conversion is observed. This was attributed to the more efficient energy absorption by steam and the catalyst under microwave irradiation. The kinetics of MW-assisted water gas shift reaction on a Cu–Zn-based catalyst has also been examined [22]. The efficiency of the low-temperature water/gas shift reaction increases with increasing temperature and steam/CO molar ratio. The heat generation from microwave irradiation was always larger than that from the chemical reaction:

$$CO + H_2O \rightarrow CO_2 + H_2 \tag{14.1}$$

14.3.2
CO_2 Conversion

The reaction of CO_2 reduction leading to formate was explored under MW irradiation [23]. Neither formaldehyde nor methanol was found in the products. $NaBH_4$ was the reducing agent. The reaction duration was 5 min.

Carbon dioxide reforming of methane (Eq. (14.2)) on carbon-based and Ni/Al_2O_3 catalysts was investigated from the point of view of energy consumption, since this is the most exothermic reaction among the catalytic conversions of methane [24–27]. The mixture of carbon and Ni/Al_2O_3 was a more efficient system than the individual catalysts at a flow rate of $1\,m^3\,h^{-1}$ of introduced CH_4. The catalytic activity of the mixture (carbon + Ni/Al_2O_3) was also better than the Ni/carbon catalyst. The energy consumed was $44.4\,kW\,h\,m^{-3}$ of H_2 produced for the carbon as a catalyst, and $4.6\,kW\,h\,m^{-3}$ of H_2 produced for the mixture with Ni/Al_2O_3. The latter value is comparable to the energy consumption for conventional steam methane reforming ($1.2\,kW\,h\,m^{-3}$ of H_2 produced). Thus, the microwave-assisted CO_2 reforming of CH_4 on a mixture of carbon and metal-based catalysts is a benign alternative to steam reforming. The role of porosity and pore size in dry reforming was estimated. The presence of micropores was a necessary prerequisite for the efficient performance of the carbon catalysts. The availability of oxygen surface groups in carbon materials was a negative factor, especially under microwave heating:

$$CO_2 + CH_4 \rightarrow 2CO + 2H_2 \tag{14.2}$$

Research into the microwave plasma used to activate CO_2 with the formation of hydrocarbons was reviewed in Ref. [28].

14.3.3
Synthesis Gas (Syngas) Production

Syngas ($CO + H_2$) is a valuable feedstock for methanol and Fischer–Tropsch processes that can be obtained by dry reforming of methane with CO_2 using 2.45-GHz microwave plasma conditions [29]. The presence of a Ni catalyst resulted in a lower ratio of syngas compared to the plasma reaction without a catalyst. The increase in the $CH_4{:}CO_2$ ratio led to an increase in selectivity toward ethane and hydrogen; thus, oxidative coupling of methane may occur under plasma conditions.

The syngas can also be produced by partial methane oxidation with air (or oxygen) – see Eq. (14.3) [30, 31]. Comparison with thermal process demonstrated

Table 14.1 Effect of microwave activation on the partial oxidation of methane into syngas.

Catalyst	Method of activation	Temperature (K)	CH_4 conversion (%)	Selectivity (%)	
				CO	CO_2
10% Co/ZrO_2	MW	723	48	79	21
	MW	1073	100	99	1
	Thermal	873	63	66	34
	Thermal	1073	94	93	7
10% Ni/La_2O_3	MW	673	58	74	26
	MW	973	100	100	<1
	Thermal	873	40	77	23
	Thermal	1073	91	92	8

obvious advantages of the microwave method (Table 14.1), including the higher conversion of methane at a lower temperature and better selectivity to CO versus CO_2. The authors explained the difference of 50–250 K between the temperatures of reaching similar conversions under microwave and thermal conditions by the formation of hot spots on oxygen defects:

$$2CH_4 + O_2 \rightarrow 2CO + 4H_2 \tag{14.3}$$

14.3.4 Methane Decomposition

Methane decomposition into char and hydrogen occurring on a carbon product from the pyrolysis of sewage sludge was studied in a multi-mode microwave reactor [32]. At the same MW power, the stable temperatures of the catalysts in hydrogen, nitrogen, and methane followed the sequence: $T_{\text{nitrogen}} > T_{\text{hydrogen}} > T_{\text{methane}}$.

A similar approach was described by Domínguez *et al.* [33]. The activated carbon was used as an MW-absorbing material to serve for CH_4 decomposition into hydrogen and carbon deposits onto the activated carbon. Methane conversion was greater under microwave conditions than with conventional heating, especially at low reaction temperatures. An increase in the space velocity showed a negative effect on CH_4 conversion, especially for microwave heating. In some experiments, the authors observed the formation of carbon nanofibers on the surface of activated carbon.

Acetylene is one of the intermediates formed in the conversion of methane to carbon deposits and can be obtained as a predominant product in the MW-assisted conversion of CH_4 [34]. Nickel, copper, and tungsten-based catalysts showed fairly high activity. The decrease in the temperature of the catalyst surface was shown to increase the extent of conversion and acetylene selectivity, as well as the stability.

14.3.5 Methane Steam Reforming

Methane steam reforming is an industrial process (Eq. (14.4)). It has been investigated under MW plasma at atmospheric pressure conditions [35]. Nanocarbon powder, CO_2, C_2H_2, C_2H_4, and HCN were the side products. Intermediate species, such as OH, NH, CH, were identified by optical emission spectroscopy. The selectivity to H_2 was as high as 92.7% for a H_2O/CH_4 molar ratio over 0.5. Methane conversion reached 91.6% at a H_2O/CH_4 ratio of 1. The specific energy consumption of H_2 formation was optimal at $[CH_4]_{in} = 5\%$, 1.0 kW, and $12\,l\,min^{-1}$. Ni/Al_2O_3 catalysts loaded in the discharge zone turned out to be inefficient to enhance methane conversion and hydrogen selectivity because of the rapid catalyst deactivation by carbon deposits:

$$CH_4 + H_2O \rightarrow CO + 3H_2 \tag{14.4}$$

14.3.6 Oxidative Coupling of Methane

The oxidative condensation of methane, the main desired products of which are ethylene and ethane, was described for the first time in the early 1980s (Eq. (14.5)). Since then, this reaction has been the object of serious scientific interest. Currently, the practical importance of this process again manifested because of the necessity of the utilization of natural and strand oil gases:

$$2CH_4 + O_2 \rightarrow C_2H_4(\text{or } C_2H_6) + 2H_2O \tag{14.5}$$

Numerous studies of the process illustrated by reaction (14.5) showed that there is a kinetic limit of the yield of the desired products (20–28%). This limit is determined by the ratio between methane and C_2 hydrocarbon reactivities. At comparatively low conversions (≤10–15%), an 80–85% selectivity could be reached. However, there is a trade-off between the selectivity and the conversion of methane. Free radical reactions substantially influence the process, because the primary reaction product (ethane) is formed via the recombination of methyl radicals generated as a result of methane activation with the participation of the catalyst surface [36].

According to several authors, the use of microwave activation can cause a decrease in the reaction temperature and a suppression of complete oxidation. The first studies in this direction were performed by Bond *et al.* [37] with the use of sodium aluminate as the catalyst. The selectivities with respect to C_2 hydrocarbons obtained under thermal and microwave heating conditions did not differ significantly, but the reaction temperatures differed by more than 400 K. The authors reported on the changes of the composition of reaction products, the decrease in the selectivity to CO_2, and the increase in the selectivity to CO. Under microwave conditions, the ethylene/ethane ratio increased, whereas ethane was the main product under thermal conditions.

Roussy also examined several mixed oxides and compared their activity in a microwave versus thermal conditions [38]. The catalysts were La_2O_3, $La_2Zr_2O_7$, $SmLiO_2$, $(SmLiO_2)_{0.8}(CaOMgO)_{0.2}$, Li/MgO, and $BaBiO_{3-x}$. No differences between thermal and microwave regimes were found for $La_2Zr_2O_7$, whereas La_2O_3 showed enhanced selectivity to ethylene under thermal heating. $SmLiO_2$ and $(SmLiO_2)_{0.8}(CaOMgO)_{0.2}$ showed opposite dependencies of the selectivity versus conversion: at low conversions, the selectivity to C_2 hydrocarbons reached 100% under microwave activation and was close to zero under thermal conditions. At high methane conversions (30–40%), the selectivities were comparable (about 50–60%). The authors explained the observed effects by different ratios between the rates of homogeneous and heterogeneous stages leading to the formation of the desired products and side products under different heating conditions. The rate constants for the reactions of complete oxidation of ethane and ethylene into CO_2 were substantially higher under the thermal conditions, in agreement with the observed dramatic difference in the selectivity to the C_2 products. The authors assumed that the reasons were the process of complete oxidation of C_2 hydrocarbons predominated in the gas phase and the substantially lower temperature of the gas phase under microwave conditions. It was also likely that active centers responsible for the formation of $^{\bullet}CH_3$ radicals were selectively activated by the microwave field in the case of the samples characterized by high dielectric loss values for $SmLiO_2$ and $(SmLiO_2)_{0.8}(CaOMgO)_{0.2}$.

The influence of temperature of the gas phase was investigated in detail for the Li/MgO and $BaBiO_{3-x}$ catalysts [39]. Different mechanisms of the formation of methyl radicals capable of dimerization to yield ethane were assumed for these catalysts. In the case of $BaBiO_{3-x}$, the presence of surface oxygen ions O^{2-} was assumed. These active sites can cause the rupture of the C–H bond in the methane molecule with the formation of the surface anion CH_3^-. The second mechanism was established by Ito *et al.* [40] for the Li/MgO catalyst. It was based on the existence of O^- ions on the surface of the catalyst. These ions could abstract a hydrogen atom from methane to produce methyl radicals, which further could be dimerized in the gas phase to yield ethane. It was assumed that the formation of undesirable products (carbon oxides) occurred either by a heterogeneous or homogeneous mechanism when the reaction was governed by the first and second mechanisms, respectively. The influence of the gas-phase oxidation of methyl radicals was demonstrated for the Li/MgO catalyst by diluting it with a material (silica) inert with respect to microwave radiation. Microwave activation of the diluted sample resulted in a selectivity with respect to C_2 products comparable to that obtained via thermal heating. This result was explained by the positive effect of the microwave activation increasing the selectivity to C_2 hydrocarbons as a result of a decrease in the rate of the gas-phase oxidation of methyl radicals, because the gas phase over the catalyst remained relatively cold. For the $BaBiO_{3-x}$ catalyst, dilution with an MW-inert material did not influence significantly the selectivity to C_2 products because of a lower contribution of gas-phase reactions on this catalyst.

Ni and coworkers [41] studied the effect of the oxygen to methane ratio in the feed on the yields and selectivities of the methane oxidative coupling process on $La_2O_3/CeO_2/Al_2O_3$ as a catalyst. It was found that the selectivities for ethane and ethylene increased when oxygen was absent compared to the oxygen/methane mixtures. Modeling of the process parameters allowed the authors to deduce that the temperature at the center of the catalyst bed was 85 °C higher than that at the periphery when the catalyst was MW-heated in a gas mixture with an oxygen concentration of 12.5%; the temperature difference was 168 °C in the absence of oxygen.

Catalysts with perovskite structures were studied by Chen *et al.* [42–44]. Depending on their composition and structure, perovskites can exhibit very different electromagnetic properties; they can be insulators, metallic conductors, superconductors, materials with superparamagnetic resistance, and dielectrics [45]. Studies were performed for oxides serving as oxygen-conducting materials, which included $SrCe_{0.95}Yb_{0.05}O_3$, $BaCe_{0.93}La_{0.07}O_3$, and $Li_2SO_4/BaCe_{0.93}La_{0.07}O_3$. The abstraction of hydrogen from methane presumably played a key role in the oxidative condensation of methane. It is noteworthy that the perovskite samples can be easily heated by a microwave field to reaction temperatures. As in the case of other MW-assisted reactions, a ~200 K decrease in the reaction temperature was revealed (Table 14.2). Differences in product distribution for microwave and thermal processes are insignificant.

The reaction of methane oxidative coupling on mixed oxide catalysts $(Bi_2O_3)_{1-x}(WO_3)_x$ ($x = 0.2–0.4$) was explored in Ref. [44]. Higher conversions were achieved with microwave activation at the same temperatures in comparison with the thermal process (Table 14.3). The selectivity to C_2 products

Table 14.2 Comparison of MW and thermal activation of methane oxidative coupling on perovskites.

Catalyst	Temperature (°C)	CH_4 conversion (%)	Selectivity (%)					Method of heating
			C_2H_2	C_2H_4	C_2H_6	CO	CO_2	
$BaCe_{0.93}La_{0.07}O_3$	590	25	2	30	29	6	33	MW
$BaCe_{0.93}La_{0.07}O_3$	825	25	0	38	26	1	35	Thermal

Table 14.3 Methane oxidative coupling on mixed oxide catalysts $(Bi_2O_3)_{1-x}(WO_3)_x$.

Heating mode	Temperature (°C)	Methane conversion (%)	Selectivity (%)			C_2H_4/C_2H_6
			C_2^+	CO	CO_2	
Thermal	840	20	62	20	18	2.8
MW	525	20	72	11	17	1.3
MW	580	30	65	12	23	1.8

increased under microwave conditions because of suppression of the contribution of complete ethane oxidation to CO_x. The ethane and ethylene ratios were substantially different for the two methods of activation. The results were interpreted on the basis of the relationship between the mobility of lattice oxygen and catalytic properties [46, 47], and the influence of microwave activation on the transport of ions [48, 49].

It should be noted that the results of the studies of the microwave activation in the oxidative coupling of methane are quite contradictory. In particular, some of the results are in conflict with the existence of the kinetic limit independent of the particular mechanism and localization of its separate stages. Nevertheless, the decrease in the reaction temperature by ~200 K under microwave conditions with some increase in the selectivity to C_2 hydrocarbons seems to be well established. It is hardly probable that the use of microwave radiation changes the mechanism of methane activation involving the formation of primary methyl radicals.

The role of MW plasma was also examined in the application to the methane oxidative coupling using metals supported on ZSM-5 zeolite, although this catalyst was never considered as an active or selective catalyst for this reaction [50]. The metals included Fe, Ni, Co, and Cu. Methane conversion to C_2 products was found to be enhanced under MW-plasma conditions compared to the thermal mode. The catalysts at the input power of 120 W, pressure of 5–10 Torr, flow rate of 125 ml min^{-1}, and a methane/oxygen ratio of 4 : 1 were ranked in terms of the conversion as follows: Co–ZSM-5 > Fe–ZSM-5 ~ Cu–ZSM-5 > Ni–ZSM-5. The selectivity to ethylene was not very high (29.8%) compared to that of ethane (about 10%); acetylene was the major product (49.9%). The highest conversion was 54.9% for Co–ZSM-5. It was established that the higher the frequency, the higher the efficiency of the process.

The same MW-plasma approach was also used in the conversion of methane to a mixture of C_2 hydrocarbons, H_2, and carbon black on a Pd–NiO/γ-Al_2O_3 catalyst [51]. The yield of C_2 hydrocarbons increased by 63.7% in the microwave plasma. The Pd–NiO/γ-Al_2O_3 catalyst turned out to be more active than Pt–Sn/γ-Al_2O_3, owing to deposition of carbon on the surface of the first catalyst. In the case of radio-frequency plasma, the C_2 yield was 72%.

14.3.7 Partial Oxidation and Other Hydrocarbon Conversion Processes

The catalytic partial oxidation of organic compounds of various classes is a basic reaction of many fundamental organic syntheses and petrochemical processes. In the past decades, such processes with the participation of hydrocarbons (basically saturated hydrocarbons) have attracted special attention because of the opportunities of production of valuable chemical products.

The first experiments to use microwave fields in such processes were described by Wan [5] in the steam conversion of hydrocarbons (Table 14.4). Microwave activation was assumed to initiate the dissociation of adsorbed water into $OH^\bullet$ and $H^\bullet$

Table 14.4 MW activation of partial oxidation of C_1–C_6 hydrocarbons.

Hydrocarbon	Catalyst	Process conditions	Products
Methane	Ni	H_2O (steam)	Methanol, acetone, dimethyl ether
Methane	Ni	Pulse MW activation, H_2O (steam)	Acetone, C_2, C_3, methanol, ethanol
Propane	Ni	H_2O (steam)	Methanol, propanols, butanols
Propene	CuO	H_2O (steam)	Propanol, ethanol, acetone, propylene oxide
n-Hexane	CuO	H_2O (steam)	Methanol, propanol, hexanone
Cyclohexane	V_2O_5	H_2O_2	Cyclohexanol, cyclohexanone

radicals, albeit the yields of oxygen-containing compounds were very low (~5%) under microwave methane activation.

Conde *et al.* [52] showed that the microwave frequency affects substantially the catalytic process. While studying the oligomerization of methane on nickel catalysts, the authors showed that the variation of the microwave frequency changed the composition of reaction products. In the entire range of frequencies studied, an increase in the MW radiation power increased the conversion of methane and decreased the selectivity to C_2 hydrocarbons. At a 4.6 GHz frequency, an increase in input power resulted in a decrease in the selectivity to ethane and ethylene with a simultaneous increase in the selectivity to acetylene and benzene. Noteworthy, at 2.45 GHz, acetylene was not formed at all. An increase in the frequency at a constant input power caused an increase in the selectivity to benzene. The use of an inert diluent (helium) also changed the reaction pattern and the selectivity to different products under microwave conditions. These data led the authors to conclude that the reason behind the changes in the reaction pattern was the variation of the real temperature of the process rather than some specific microwave activation of the reactants or the catalyst. The ability of the catalyst to transform the microwave energy into heat demonstrated the dependence with a maximum versus the MW frequency. Thus, the operating frequency is a critical parameter to influence the amount of heat released and the real temperature of the catalyst, and ultimately the product distribution.

Microwave heating has also been used to perform other reactions, namely, (i) the isomerization of 2-methylpentane on Pt/Al_2O_3 [53], (ii) the epoxidation of ethylene on Ag/Al_2O_3 [54], and (iii) the hydrocracking of cyclic hydrocarbons [3].

Mora *et al.* studied the conversion of C_5–C_7 paraffins under MW-plasma conditions (argon microwave discharge, 2.45 GHz, reduced pressure) [55]. Optical emission spectroscopy was used to estimate the plasma temperature. The microwave power and the feed flow rate affected both the conversion and the selectivity, with the main products being hydrogen and ethylene. The selectivity to hydrogen was quite high at a low input power (100–150 W). However, at a higher power (>300 W) or increased contact time, ethylene was the major product. Carbon deposits also formed.

Syngas was produced by MW-plasma conversion (reforming) of diverse hydrocarbons that include methane, isooctane, and gasoline [56]. The rates of both hydrogen and carbon monoxide formation increased with increased power and decreased with diminishing total flow rate. The maximum efficiency was only 3.12% with isooctane at a power consumption of 28.8 W, O/C ratio of 1, and 0.1 g min^{-1} of fuel supply. The better performance was observed with the use of liquid fuels rather than methane.

Reforming of *n*-hexane into $CO + H_2$ was studied under atmospheric steam plasma using a microwave discharge [57]. Emission spectra showed that O, H, and OH were present in the plasma. The conversion of the hydrocarbon and steam depended on the plasma power, as well as the ratio and total flow rate of the feed components.

Styrene is one of the large-scale chemicals whose further conversion into valuable products has been widely studied. Epoxidation of styrene by O_2 proceeds quite effectively under MW conditions (40 W) on sulfated Co–Y-doped ZrO_2 as a catalyst [58]. The same conversion can be achieved at shorter times under microwave conditions in comparison with thermal heating. The selectivity to the target product diminishes at a higher input power (800 W) and styrene glycol is formed. The best solvent for this reaction was dimethylformamide, one of the best MW-absorbing organic solvents. The maximum selectivity for styrene oxide was 91% at 120 °C.

Aliphatic alkenes were also studied as substrates of the MW-stimulated reactions, including partial oxidation [59]. For instance, cyclohexene can be selectively converted either into the epoxide with a 65% conversion and 75% selectivity or into the enol with 70% conversion and 80% selectivity, or into the enone with 99% conversion and 89% selectivity. The reaction time ranges from 1 to 20 min. A Co-salen/SBA-15 catalyst was used without any solvents.

Effective use of microwave heating in the oxidation of alkenes and alcohols under the action of hydrogen peroxide was described in Ref. [60]. The catalyst was a MCM-41 type mesoporous titanium silicate modified by organic compounds. It was recycled several times without activity loss. Another example of alkene oxidation under MW activation is reported by Gourari *et al.* with ethylene as the substrate and 1% Pt/Al_2O_3 as the catalyst [15].

The effect of microwave radiation in the oxidation of benzene to phenol was examined with the use of nitrous oxide as a mild oxidizer, a very promising one-stage method for the production of phenols [61]. Microwave activation of the zeolite catalyst of this reaction (Fe–ZSM-5) was used to decrease the contribution of coking. Although some increase in selectivity with respect to phenol and a decrease in the rate of catalyst deactivation were observed, no substantial improvement of process characteristics was obtained.

Vanadium-containing catalysts have been shown to be suitable in MW-assisted oxidation processes [62], because of the good MW absorption capacity of V_2O_5.

This was demonstrated in the reaction of *o*-xylene oxidation with air into phthalic anhydride with high yield (67%) at 563 K on V_2O_5/SiO_2. The dispersion of V_2O_5 on SiO_2 was more homogeneous when the catalyst was prepared by the microwave method. The temperature of the MW process was about 100 °C lower compared to the thermal conditions.

Zeolites modified with transition metals were used as catalysts under microwave irradiation [63]. Two main dielectric mechanisms of microwave heating were determined in zeolites: rotational polarization phenomenon and interfacial polarization. The selective catalytic oxidation of styrene with H_2O_2 resulted in the formation of benzaldehyde.

Positive effects of the *in situ* microwave activation were reported by Wan and coworkers in the hydrocracking of naphthenes [64].

Regioselective alkylation of naphthalene with alcohols occurs on H-mordenite ($SiO_2/Al_2O_3 = 240$) under microwave irradiation [65]. For the microwave-assisted reaction, high reaction rates and high selectivity for 2,6-dialkylnaphthalenes were found. In the reaction of 2-isopropylnaphthalene with isopropyl alcohol, the conversion and the selectivity were 43.5% and 66.4%, respectively. In di-*tert*-butylation of naphthalene with *tert*-butyl alcohol, the conversion and the selectivity reached 86.5% and 70.4%, respectively. Both conversions and selectivities were enhanced by microwave heating.

14.3.8 Oxidative Dehydrogenation

Two oxidative dehydrogenation reactions have been studied from the viewpoint of the *in situ* application of microwave radiation for the conversion of ethane to ethylene and ethylbenzene to styrene. In this regard, Nigrovski *et al.* explored the styrene production on multiwalled carbon nanotubes modified by iron oxide [66]. The optimal catalyst composition was 3% Fe_2O_3/multi-wall carbon, nanotubes (MWCNT) which operated at 380–450 °C. The oxide Fe_2O_3 was reduced into Fe nanocrystals encapsulated by polyhedral graphite shells only under microwave conditions.

Microwave-assisted oxidative dehydrogenation of ethane into ethylene was studied using a kinetic approach consisting of the comparison of kinetic dependencies (shape of kinetic equations, "selectivity or yield vs. conversion" curves) and apparent parameters (activation energies) obtained in thermal and MW modes [67]. Three types of catalysts were studied: $VSbO_x$, $VMoNbO_x$, and $VMoO_x$. Note that the ability to absorb MW energy changes in the order: $VSbO_x > VMoNbO_x > VMoO_x$; the first catalyst could be heated up to about 800 °C within 2 min in a mono-mode MW unit with a 100% efficiency (the entire MW energy is absorbed by the catalyst). In the case of VMo and VMoNb oxides, a distinct difference between ethane yields was observed at given conversion of limiting reactant (oxygen). The Arrhenius plots obtained for these catalysts under thermal and MW regimes are shown in Figure 14.1. It is clear that the largest effect of the MW radiation is found for the $VMoO_x$ catalyst (curves 5 and 6).

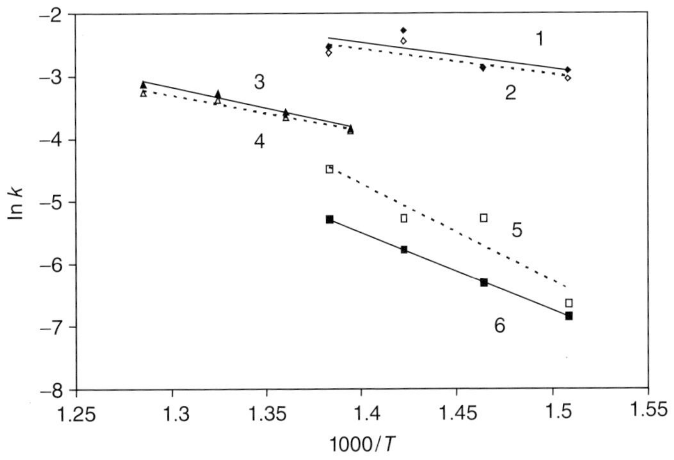

Figure 14.1 Arrhenius plots (k is the overall ethane conversion apparent rate constant): 1 – $VSbO_x$ microwave, 2 – $VSbO_x$ thermal, 3 – $VMoNbO_x$ microwave, 4 – $VMoNbO_x$ thermal, 5 – $VMoO_x$ microwave, 6 – $VMoO_x$ thermal.

14.3.9 Oil Processing

Desulfurization of crude oil and dibenzothiophene was explored under MW activation [68]. Iron powder proved to be an efficient catalyst, though no comparison was made with the thermal regime.

The hydrocatalytic process was developed in a reverse microemulsion using microwave irradiation [69]. The mixed oxides $CoMo/Al_2O_3$ and $NiMo/Al_2O_3$ were used as catalysts. The demetallization and desulfurization processes were studied using such model compounds as nickel(II) phthalocyanine tetrasulfonate tetrasodium salt, nickel tetraphenylporphyrin, and thiophene. The phthalocyanine and porphyrin were demetallized to about 85% and 60%, respectively. The extent of desulfurization of thiophene was close to 92–93%. These conversions exceed most reported data. The catalytic hydrodesulfurization was also used with a real sample of Venezuelan heavy oil; a rather high desulfurization activity (59%) was achieved at a short reaction time. The extent of hydrodemetallization of the heavy oil containing nickel and vanadium was over 35% and 47% for Ni and V, respectively.

14.4 Prospects for the Use of Microwave Radiation in Oil and Gas Processing

The majority of results available in the literature on microwave effects in catalytic processes of oil and gas conversion provides convincing evidence that MW activation can enhance the reaction rate and change product selectivity, while decreasing the reaction temperature. However, the barriers to overcome on the way to industrial application of MW irradiation in such catalytic processes are related to the problem of scaling up. Moreover, there are several examples on the positive effects of microwave plasma; however, unlike corona discharge and some other types of plasma, microwave plasma requires the use of reduced pressures (typically 10–50 Torr) that significantly suppresses the space-time yield of the desired products.

In mastering the technology for MW-assisted oil and gas processing, one should take into account that oil components and most gases are not absorbing the microwave irradiation, so the only option is to use the catalysts that are capable of strongly interacting with the microwave radiation. The Maxwell–Wagner interphase polarization mechanism can efficiently operate in mixed oxide catalysts: for instance, catalysts of oxidative dehydrogenation of ethane or ammoxidation of propane or conversion of propane into acrylics that consist of several well-characterized phases of vanadium and molybdenum oxides with additives of tellurium, antimony, or niobium.

It is important to emphasize that microwave activation is widely used in the preparation of heterogeneous catalysts. In fact, microwave activation of catalysts either at the stage of decomposition of precursors or further reduction allows skillful scientists to prepare unique materials with very uniform distribution of supported metal or metal oxide nanoparticles in diverse (MW absorbing or not) matrices. The MW-assisted preparation of catalysts allows one to accelerate the procedure. In the case of catalysts consisting of several phases, the replacement of traditional heating with microwave heating can contribute to the preferable formation of specific phases.

Special attention should be paid to bifunctional catalysis, in which the different stages of the process take place on different components of the catalytic system. A typical example is the transformation of hydrocarbons on metallic catalysts deposited on acid carriers. A metallic component (typically a Pt family metal) activates the initial molecule by breaking bonds and/or by charge transfer (dehydrogenation, hydrogenolysis), whereas carbon skeleton transformations (alkylation and isomerization) occur on acid centers of the carrier. Simultaneous hydrogenation/dehydrogenation, cyclization, aromatization (and other processes) can occur on both complex catalyst components in various combinations. Microwave radiation generates temperature gradients in a heterogeneous system whose components interact differently with the electromagnetic fields that can then cause substantial changes in the rate of the process and, which is especially important, in the distribution of reaction products [70].

Although oil processing under MW radiation is unlikely to occur in the near future because of upscaling problems, the use of MW technologies for some gas processing remains an attractive perspective, because the volume of the reactors in these processes is much smaller compared to the oil processing and petrochemistry. Of course, this can be considered only for processes and catalysts that have demonstrated significant advantages over conventional thermal processes through optimization of MW conditions that include the operating frequency, input power, gas mixture flow rate, contact time, and the configuration of the microwave electromagnetic field. With regard to the latter, a multi-mode reactor, or a mono-mode resonator reactor can be used. An appropriate comparison should be made to determine the best reactor design.

Acknowledgment

The financial support of the Ministry of Education and Science of Russian Federation (project RFMEFI61314X0012) is acknowledged.

References

1. Ishii, T.K. (ed) (1995) *Industrial Applications of Microwaves, Handbook of Microwave Technology*, Academic Press, San Diego, pp. 277–307.
2. Wan, J.K.S. and Kriz, J.F. (1985) Hydrodesulphurization of hydrocracked pitch. US Patent 4,545,879.
3. Wan, J.K.S. (1993) Microwaves and chemistry: the catalysis of an exciting marriage. *Res. Chem. Intermed.*, **19**, 147–153.
4. Wan, J.K.S., Wolf, K., and Heyding, R.D. (1984) in *Catalysis on the Energy Scene* (eds S. Kaliaguine and A. Mahay), Elsevier, Amsterdam, pp. 439–455.
5. Wolf, K., Choi, H.K., and Wan, J.K.S. (1986) Microwave assisted catalytic conversion of cyclohexene. *Aust. J. Res.*, **3**, 53–59.
6. Wan, J.K.S., Tse, M., and Husby, H. (1990) High-power pulsed microwave catalytic processes: decomposition of methane. *J. Microwave Power Electromagn. Energy*, **25**, 32–38.
7. Bamwenda, G., Depew, M.C., and Wan, J.K.S. (1991) Microwave induced catalytic reactions of carbon dioxide and water: mimicry of photosynthesis. *Res. Chem. Intermed.*, **16**, 241–255.
8. Bamwenda, G., Moore, E., and Wan, J.K.S. (1992) Production of acetylene by a microwave catalytic reaction of water and carbon. *Res. Chem. Intermed.*, **17**, 243–262.
9. Ioffe, M.S., Pollington, S.D., and Wan, J.K.S. (1995) High-power pulsed radiofrequency and microwave catalytic processes: selective production of acetylene from the reaction of methane over carbon. *J. Catal.*, **151**, 349–355.
10. Bamwenda, G., Depew, M.C., and Wan, J.K.S. (1993) Acetylene from benzene: a novel and selective microwave assisted cracking synthesis. *Res. Chem. Intermed.*, **19**, 553–564.
11. Cameron, K.L., Depew, M.C., and Wan, J.K.S. (1991) Pulsed microwave catalytic decomposition of olefins. *Res. Chem. Intermed.*, **16**, 57–70.
12. Depew, M.C., Lem, S., and Wan, J.K.S. (1991) Microwave induced catalytic decomposition of some alberta oil sands and bitumens. *Res. Chem. Intermed.*, **16**, 213–223.
13. Roussy, G., Thiebaut, J.M., and Mdejaram, M.S. (1993) Durable changes of the catalytic properties of alumina-supported platinum induced by microwave irradiation. *Catal. Lett.*, **21**, 133–138.
14. Seyfried, L., Garin, F., Maire, G., Thiebaut, J.M., and Roussy, G. (1994) Microwave electromagnetic-field effects on reforming catalysts: 1. Higher selectivity in 2-methylpentane isomerization on alumina-supported Pt catalysts. *J. Catal.*, **148**, 281–287.
15. Gourari, S., Roussy, G., Thiebaut, J.M., and Zoulalian, A. (1992) Etude d'une réaction catalytique en présence d'un rayonnement microonde. *Chem. Eng. J.*, **49**, 79–88.
16. Klimov, A.Y., Bal'zhinimaev, B.S., Makarshin, L.L., Zaikovskii, V.I., and Parmon, V.N. (1998) Effect of microwave

irradiation on the catalytic properties of silver in ethylene epoxidation. *Kinet. Catal.*, **39**, 511–515.
17. Roussy, G., Hilaire, S., and Thiebaut, J.M. (1997) Permanent change of catalytic properties induced by microwave activation on 0.3% Pt/Al_2O_3 (EuroPt-3) and on 0.3% Pt-0.3% Re/Al_2O_3 (EuroPt-4). *Appl. Catal. A: Gen.*, **156**, 167–180.
18. Roussy, G., Marchal, E., Thiebaut, J.M., Ename-Obiang, F., and Marchal, E. (2000) Microwave broadband permittivity measurement with a multimode helical resonator for studying catalysts. *Meas. Sci. Technol.*, **12**, 542–547.
19. Zhang, X. and Hayward, D.O. (2006) Applications of microwave dielectric heating in environment-related heterogeneous gas-phase catalytic systems. *Inorg. Chim. Acta*, **359**, 3421–3433.
20. Mutyala, S., Fairbridge, C., Paré, J.R.J., Bélanger, J.M.R., Ng, S., and Hawkins, R. (2010) Microwave applications to oil sands and petroleum: A review. *Fuel Process. Technol.*, **91**, 127–135.
21. Chen, W.-H., Jheng, J.-G., and Yu, A.B. (2008) Hydrogen generation from a catalytic water gas shift reaction under microwave irradiation. *Int. J. Hydrogen Energy*, **33**, 4789–4797.
22. Chen, W.-H., Cheng, T.-C., Hung, C.-I., and Lin, B.-J. (2012) Chemical reactions and kinetics of a low-temperature water gas shift reaction heated by microwaves. *Int. J. Hydrogen Energy*, **37**, 276–289.
23. Pulidindi, I.N., Kimchi, B.B., and Gedanken, A. (2014) Selective chemical reduction of carbon dioxide to formate using microwave irradiation. *J. CO_2 Util.*, 7, 19–22.
24. Fidalgo, B. and Menéndez, J.A. (2012) Study of energy consumption in a laboratory pilot plant for the microwave-assisted CO_2 reforming of CH_4. *Fuel Process. Technol.*, **95**, 55–61.
25. Fidalgo, B., Arenillas, A., and Menéndez, J.A. (2011) Mixtures of carbon and Ni/Al_2O_3 as catalysts for the microwave-assisted CO_2 reforming of CH_4. *Fuel Process. Technol.*, **92** (8), 1531–1536.
26. Fidalgo, B., Arenillas, A., and Menéndez, J.A. (2010) Influence of porosity and surface groups on the catalytic activity of carbon materials for the microwave-assisted CO_2 reforming of CH_4. *Fuel*, **89**, 4002–4007.
27. Fidalgo, B., Domínguez, A., Pis, J.J., and Menéndez, J.A. (2008) Microwave-assisted dry reforming of methane. *Int. J. Hydrogen Energy*, **33** (16), 4337–4344.
28. Liu, C.-J., Xu, G.-H., and Wang, T. (1999) Non-thermal plasma approaches in CO_2 utilization. *Fuel Process. Technol.*, **58**, 119–134.
29. Cho, W., Ju, W.-S., Lee, S.-H., Baek, Y.-S., and Kim, Y.C. (2004) Investigation of synthesis gas production from natural gas and CO_2 by microwave plasma technology. *Stud. Surf. Sci. Catal.*, **153**, 205–208.
30. Chen, W.-H., Chiu, T.-W., and Hung, C.-I. (2010) Enhancement effect of heat recovery on hydrogen production from catalytic partial oxidation of methane. *Int. J. Hydrogen Energy*, **35**, 7427–7440.
31. Bi, X.-J., Hong, P.J., Xie, X.G., and Dai, S.-S. (1999) Microwave effect on partial oxidation of methane to syngas. *React. Kinet. Catal. Lett.*, **66**, 381–386.
32. Deng, W., Su, Y., Liu, S., and Shen, H. (2014) Microwave-assisted methane decomposition over pyrolysis residue of sewage sludge for hydrogen production. *Int. J. Hydrogen Energy*, **39**, 9169–9179.
33. Domínguez, A., Fidalgo, B., Fernández, Y., Pis, J.J., and Menéndez, J.A. (2007) Microwave-assisted catalytic decomposition of methane over activated carbon for CO_2-free hydrogen production. *Int. J. Hydrogen Energy*, **32**, 4792–4799.
34. Radoiu, M.T., Chen, Y., and Depew, M.C. (2003) Catalytic conversion of methane to acetylene induced by microwave irradiation. *Appl. Catal. B: Environ.*, **43** (2), 187–193.
35. Wang, Y.-F., Tsai, C.-H., Chang, W.-Y., and Kuo, Y.-M. (2010) Methane steam reforming for producing hydrogen in an atmospheric-pressure microwave plasma reactor. *Int. J. Hydrogen Energy*, **35** (1), 135–140.
36. Arutyunov, V.S. and Krylov, O.V. (1998) *Oxidative Conversion of Methane*, Nauka, Moscow.
37. Bond, G., Moyes, R.B., and Whan, D.A. (1993) Recent applications of microwave

heating in catalysis. *Catal. Today*, **17**, 427–437.
38. Roussy, G., Thiebaut, J.M., Souiri, M., Marchal, E., Kiennemann, A., and Maire, G. (1994) Controlled oxidation of methane doped catalysts irradiated by microwaves. *Catal. Today*, **21**, 349–355.
39. Roussy, G., Marchal, E., Thiebaut, J.M., Kiennemann, A., and Maire, G. (1997) C_2^+ selectivity enhancement in oxidative coupling of methane over microwave-irradiated catalysts. *Fuel Process. Technol.*, **50**, 261–274.
40. Ito, T., Wang, J., Lin, C.H., and Lunsford, J.H. (1985) Oxidative dimerization of methane over a lithium-promoted magnesium oxide catalyst. *J. Am. Chem. Soc.*, **107**, 5062–5068.
41. Ni, B.H., Lee, C., Sun, R.C., and Zhang, X.L. (2009) Microwave assisted heterogeneous catalysis: effects of varying oxygen concentrations on the oxidative coupling of methane. *React. Kinet. Catal. Lett.*, **98** (2), 287–302.
42. Chen, C.-L., Hong, P.-J., Dai, S.S., and Kan, J. (1997) Microwave effects on the oxidative coupling of methane over proton conductive catalysts. *J. Chem. Soc., Faraday Trans.*, **91**, 1179–1180.
43. Fu, M.S., Ni, L., and Chen, X.M. (2013) Abnormal variation of microwave dielectric properties in A/B site co-substituted $(Ca_{1-0.3x}La_{0.2x})[(Mg_{1/3}Ta_{2/3})_{1-x}Ti_x]O_3$ complex perovskite ceramics. *J. Eur. Ceram. Soc.*, **33** (4), 813–823.
44. Chen, C.-L., Hong, P.-J., Dai, S.S., Zhang, C.-C., and Yang, X.-Y. (1997) Microwave effects on the oxidative coupling of methane over Bi_2O_3-WO_3 oxygen ion conductive oxides. *React. Kinet. Catal. Lett.*, **61**, 175–180.
45. Bhalla, A.S., Guo, R., and Roy, R. (2000) The perovskite structure – A review of its role in ceramic science and technology. *Mater. Res. Innovations*, **4**, 3–17.
46. Doornkamp, C. and Ponec, V. (2000) The universal character of the Mars and Van Krevelen mechanism. *J. Mol. Catal., A: Chem.*, **162** (1-2), 19–32.
47. Gellings, P.J. and Bouwmeester, H.J.M. (2000) Solid state aspects of oxidation catalysis. *Catal. Today*, **58**, 1–53.
48. Freemann, S.A., Booske, J.H., Cooper, R.T., Meng, B., Kieffer, J., and Rerdon, B.J. (1993) Studies of microwave field effects on ionic transport in ionic crystalline solids. *Ceram. Trans.*, **36**, 213–220.
49. Kenkre, V.M., Kus, M., and Katz, J.D. (1992) Explanation of the barrier-depression effect in ceramic. *Phys. Rev. B*, **46**, 13825–13831.
50. Cho, W., Baek, Y., Moon, S.-K., and Kim, Y.C. (2002) Oxidative coupling of methane with microwave and RF plasma catalytic reaction over transitional metals loaded on ZSM-5. *Catal. Today*, **74**, 207–223.
51. Cho, W., Kim, Y.C., and Kim, S.-S. (2010) Conversion of natural gas to C_2 product, hydrogen and carbon black using a catalytic plasma reaction. *J. Ind. Eng. Chem.*, **16**, 20–26.
52. Conde, L.D., Marun, C., Suib, S.L., and Fathi, Z. (2001) Frequency effects in the catalytic oligomerization of methane via microwave heating. *J. Catal.*, **204**, 324–332.
53. Ringler, S., Garin, F., Maire, G. *et al.* (1997) Proceedings of Europacat-3, Krakow, Poland, August 31–September 6, 1997, European Congress on Catalysis, p. 124.
54. Klimov, A.Yu, Bal'zhinimaev, B.S., Makarshin, L.L. *et al.* (1997) Proceedings of Europacat-3, Krakow, Poland, August 31–September 6, 1997, European Congress on Catalysis, p. 189.
55. Mora, M., del Carmen García, M., Jiménez-Sanchidrián, C., and Romero-Salguero, F.J. (2010) Transformation of light paraffins in a microwave-induced plasma-based reactor at reduced pressure. *Int. J. Hydrogen Energy*, **35**, 4111–4122.
56. Kim, T.-S., Song, S., Chun, K.-M., and Lee, S.H. (2010) An experimental study of syn-gas production via microwave plasma reforming of methane, *iso*-octane and gasoline. *Energy*, **35**, 2734–2743.
57. Sekiguchi, H. and Mori, Y. (2003) Steam plasma reforming using microwave discharge. *Thin Solid Films*, **435**, 44–48.
58. Tyagi, B., Shaik, B., and Bajaj, H.C. (2009) Microwave-assisted epoxidation

of styrene with molecular O_2 over sulfated $Co-Y-ZrO_2$ solid catalyst. *Catal. Commun.*, **11**, 114–117.
59. Luque, R., Badamali, S.K., Clark, J.H., Fleming, M., and Macquarrie, D.J. (2008) Controlling selectivity in catalysis: Selective greener oxidation of cyclohexene under microwave conditions. *Appl. Catal. A: Gen.*, **341**, 154–159.
60. Balu, A.M., Hidalgo, J.M., Campelo, J.M., Luna, D., Luquea, R., Marinas, J.M., and Romero, A.A. (2008) Microwave oxidation of alkenes and alcohols using highly active and stable mesoporous organotitanium silicates. *J. Mol. Catal. A: Chem.*, **293**, 17–24.
61. Gopalakrishnan, S., Munch, J., Herrmann, R., and Schwieger, W. (2006) Effects of microwave radiation on one-step oxidation of benzene to phenol with nitrous oxide over Fe-ZSM-5 catalyst. *Chem. Eng. J.*, **120**, 99–105.
62. Liu, Y., Lu, Y., Liu, S., and Yin, Y. (1999) The effects of microwaves on the catalyst preparation and the oxidation of o-xylene over a V_2O_5/SiO_2 system. *Catal. Today*, **51** (1), 147–151.
63. Jury, F.A., Polaert, I., Estel, L., and Pierella, L.B. (2013) Synthesis and characterization of MEL and FAU zeolites doped with transition metals for their application to the fine chemistry under microwave irradiation. *Appl. Catal. A: Gen.*, **453**, 92–101.
64. Wan, J.K.S., Wolf, K., and Heyding, R.D. (1984) Some chemical aspects of the microwave assisted catalytic hydro-cracking processes. *Stud. Surf. Sci. Catal.*, **19**, 561–568.
65. Yamashita, H., Mitsukura, Y., Kobashi, H., Hiroki, K., Sugiyama, J., Onishi, K., and Sakamoto, T. (2010) Microwave-assisted regioselective alkylation of naphthalene compounds using alcohols and zeolite catalysts. *Appl. Catal., A: Gen.*, **381**, 145–149.
66. Nigrovski, B., Zavyalova, U., Scholz, P., Pollok, K., Müller, M., and Ondruschka, B. (2008) Microwave-assisted catalytic oxidative dehydrogenation of ethylbenzene on iron oxide loaded carbon nanotubes. *Carbon*, **46** (13), 1678–1686.
67. Sinev, I., Kardash, T., Kramareva, N., Sinev, M., Tkachenko, O., Kucherov, A., and Kustov, L.M. (2009) Interaction of vanadium containing catalysts with microwaves and their activation in oxidative dehydrogenation of ethane. *Catal. Today*, **141**, 300–305.
68. Leadbeater, N.E. and Rashid Khan, M. (2008) Microwave-promoted desulfurization of heavy and sulfur-containing crude oil. *Energy Fuels*, **22**, 1836–1839.
69. Madriz, L., Carrero, H., Ramon Dominguez, J., and Vargas, R. (2013) Lenys Fernández, Catalytic hydrotreatment in reverse microemulsions under microwave irradiation. *Fuel*, **112**, 338–346.
70. Kustov, L.M. and Sinev, I.M. (2010) Microwave activation of catalysts and catalytic processes. *J. Phys. Chem. (Russ.)*, **84**, 1676–1694.

Part VII
Applications – Biomass and Wastes

Microwaves in Catalysis: Methodology and Applications, First Edition.
Edited by Satoshi Horikoshi and Nick Serpone.

15
Algal Biomass Conversion under Microwave Irradiation

Shuntaro Tsubaki, Tadaharu Ueda, and Ayumu Onda

15.1
Introduction

Effective utilization of biomass is important for chemical industry and energy production to reduce their strong dependency on fossil resources and promote effective use of renewable biomass. Exploitation of local biomass is also beneficial for regional economy in the rural areas. Algae have been expected as the third-generation biomass next to the starch-based biomass (the first generation) and the lignocellulose (the second generation). Alage generally grow fast with high production at unit area. Cultivations of algae are less competitive with food production. They can also contribute to bioremediation of waste water and eutrophic ocean by capturing nitrogen, phosphorus, and heavy metals. Moreover, algae can be more easily converted to useful chemicals than lignocellulose since they do not contain lignin.

Macroalgae accumulate hydrocolloids (also called as *phycocolloids*), such as alginate, carrageenan, agar, and fucoidan, which are usable as feedstock for food ingredients, health-promoting foods, polymer materials and so on. For instance, macroalgae can be converted to biofuels by genetically modified microorganisms (*Escherichia coli* and yeast) that is capable of utilizing brown algae alginates [1, 2]. Microalgae, on the other hand, particularly accumulate oils (triacylglycerols and hydrocarbons), which are applicable as biodiesel, gasoline, naphtha, and jet fuels [3]. Proteins and pigments are also available from microalgae [4].

Hydrothermal treatment is one of the effective conversion processes of algal biomass, since they do not require energy-consuming drying process of wet algae. Hydrothermal treatment is a collective term of reactions in water medium at high temperature under above saturated vapor pressure. Increase in temperature decreases ion product of water under hydrothermal condition and initiates autohydrolysis of substrate without addition of catalysts. At the same time, decrease in the dielectric constant increases solubility of less polar compounds [5]. These unique properties of hydrothermal reaction have been widely utilized in synthesis of inorganic materials and treatment of organic wastes as a green

Microwaves in Catalysis: Methodology and Applications, First Edition.
Edited by Satoshi Horikoshi and Nick Serpone.

processes. Hydrothermal treatment has also been used as pretreatment of lignocellulose before enzymatic hydrolysis [6]. In the case of microalgae treatments, hydrothermal treatment is applied to liquefaction of microalgae to produce biocrude oils [6]. Hydrothermal liquefaction of algae at around 300–350 °C converts oils, carbohydrates, and proteins into biocrude oils, and then they are transformed to naphtha, gasoline, and jet fuels via hydrogenation.

The microwave irradiation technology has been applied to accelerate hydrothermal treatment of biomass. Microwave irradiation promotes molecular motion of dipolar molecules and electrolytes contained in a solution, generating heat by dielectric and conduction losses. Microwave irradiation can directly heat irradiated materials implying that it should degrade biomass from inside. At the same time, microwave irradiation can directly activate catalysts composed of microwave-absorbing materials (e.g., carbon-based catalysts) and promote a variety of organic chemical reactions [7]. Previously, we have demonstrated that microwave irradiation and hydrothermal reaction synergistically works together toward extracting polysaccharides and polyphenols from food waste biomass and agricultural residues (such as tea residue, soybean residue, corn fibers, and cassava residues) [8]. Microwave irradiation is also effective for promoting delignification of lignocellulose to promote enzymatic susceptibility of woods [9]. In the case of algal biorefinery, microwave irradiation have been utilized for extraction of pigments [10], polysaccharides [11], oils [12], transesterification of oils [13], as well as solvothermal treatments of macroalgae and microalgae [14].

In the present chapter, we will summarize three topics in our recent works regarding microwave-assisted treatments especially focusing on conversion of macroalgae: (i) the effects of microwave irradiation on hydrothermal treatment of biomass by using model biomass compounds; (ii) synergistic effects of ionic conductive catalyst and microwave irradiation on the hydrolysis of polysaccharides and fast-growing macroalgae (*Ulva* spp.); (iii) the dielectric property of algal polysaccharides in water to study how microwave energy affect algal substrates.

15.2 Microwave Effect on Hydrothermal Conversion – Analysis Using Biomass Model Compounds

15.2.1 Degradation Kinetics of Neutral Sugars under Microwave Heating

Hydrothermal reaction is one of the environmentally friendly technologies however, the selectivity of the product is sometimes limited. In the case of biorefinery process, it is important to avoid degradation of sugars into by-products such as furfurals and organic acids to improve sugar yields. Therefore, the microwave effects on stability of neutral sugars were investigated by comparing

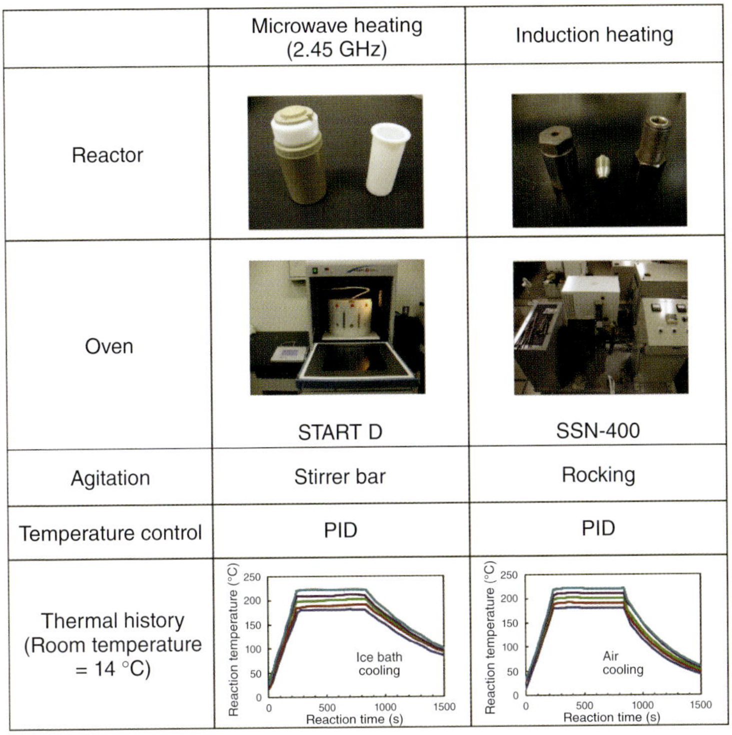

Figure 15.1 The descriptions of microwave and induction ovens and their thermal histories.

with the external heating [15]. Induction oven was used as an external heating system to generate identical thermal histories between microwave and external heating. The description of microwave and induction ovens was summarized in Figure 15.1.

Microwave heating was performed by START D microwave oven (Multimode, Frequency; 2.45 GHz, maximum output; 1 kW, Milestone Inc., Shelton, USA) and HPR-100 TFM^{TM} reactor (100-ml closed reactor made of Teflon). Reaction temperature was controlled by PID (Propotional-integral-derivative) control with direct temperature measurement using the thermocouple thermometer. Distributions of microwave and reactant were kept homogeneous by using diffuser, rotation table, and stirrer bar. After the reaction, the reactor was cooled in ice bath. Induction heating was conducted by using SSN-400 (Shikoku Rika., Co., Kochi, Japan) equipped with a hastelloy-C lined iron autoclave (volume; 50 ml). The temperature of the reactor was measured by the thermocouple thermometer and controlled by PID to trace the same thermal history as the microwave experiments. The reactant was mixed by rocking the whole induction coil unit and the autoclave units. After the reaction, the autoclave was immediately air cooled by fan.

Table 15.1 Kinetic parameters of decomposition of neutral monosaccharides under hydrothermal water (200 °C) using microwave and induction heating.

Neutral monosaccharide	Microwave heating		Induction heating	
	k ($\times 10^{-4}$ s^{-1})	R^2	k ($\times 10^{-4}$ s^{-1})	R^2
Glucose	2.31	0.953	3.02	0.933
Galactose	2.13	0.974	2.39	0.987
Mannose	2.86	0.979	4.28	0.983
Arabinose	2.61	0.996	3.00	0.988
Xylose	3.87	0.998	4.38	0.998

Reproduced with permission from Ref. [15]. Copyright 2013 Elsevier.

Table 15.1 shows the degradation kinetics of five kinds of neutral sugars by microwave and induction heating at 200 °C [15]. Microwave heating exhibited $2.13 \times 10^{-4} - 3.87 \times 10^{-4}\ s^{-1}$ in degradation rate depending on the neutral sugars. Hexoses were relatively stable while pentoses, especially xylose, were very fragile to hydrothermal treatment. Although degradation behavior of each neutral sugars to hydrothermal treatments were the same, induction heating exhibited higher degradation rate by 1.1–1.5-fold for microwave heating. Similar to monosaccharides, degradations of disaccharides were also suppressed by microwave heating [16]. Figure 15.2 shows saccharification rates of maltose and cellobiose as well as selectivities to glucose by microwave and induction heating. The results were expressed as a function of severity parameter ($\log R_0$) calculated from the following equation:

$$\log R_0 = \log t \exp\left(\frac{T - 100}{14.75}\right) \tag{15.1}$$

where t is the reaction time (min) and T is the temperature (°C), respectively [17]. Microwave showed higher stability of glucose even at severe conditions at $R_0 \geq 4$. In addition, glucose selectivity was also higher for microwave than induction heating.

The degradation behavior of neutral sugars under acidic condition was summarized in Table 15.2. Although sugar degradations were promoted by addition of sulfuric acid, the order of degradation rate of neutral sugars were almost the same as the hydrothermal degradation. Induction heating showed higher degradation rate than microwave by 1.54–2.23-fold.

These results suggested that microwave heating can promote stability of sugars under hydrothermal treatment. This is probably because microwave directly heats water molecules by dielectric loss, and the water catalyzes sugar hydrolysis with lower possibility of side reactions (Scheme 15.1a). In the case of external heating, the heats were nonselectively input via reactor wall and facilitated both hydrolysis and degradation of sugars with lower glucose selectivity (Scheme 15.1b), a phenomenon well known as the *wall effect*.

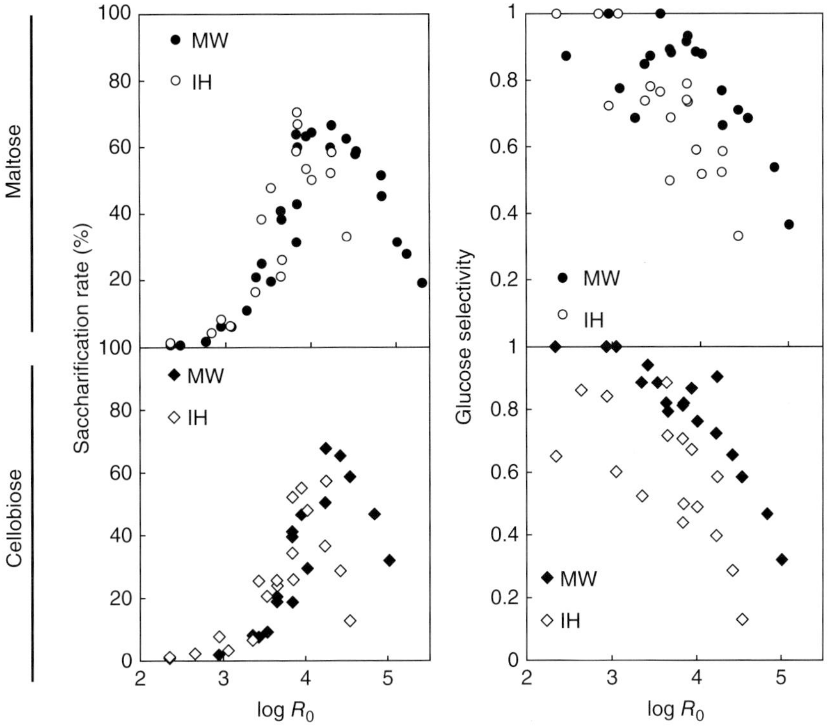

Figure 15.2 Saccharification rate and glucose selectivity of the hydrothermal hydrolysis of maltose (top) and cellobiose (bottom) by using microwave (MW) and induction heating (IH) as a function of severity parameter (log R_0). Reproduced with permission from Ref. [16]. Copyright 2012 Elsevier.

Table 15.2 Kinetic parameters of decomposition of neutral monosaccharides under dilute sulfuric acid (180 °C, 0.8% w/w) using microwave and induction heating.

Neutral monosaccharides	Microwave heating		Induction heating	
	k ($\times 10^{-3}$ s^{-1})	R^2	k ($\times 10^{-3}$ s^{-1})	R^2
Glucose	0.61	0.999	1.07	0.979
Galactose	0.75	0.998	1.23	0.971
Mannose	1.09	0.981	2.05	0.831
Arabinose	1.18	0.998	2.63	0.949
Xylose	2.00	1.000	3.08	0.995

Reproduced with permission from Ref. [15]. Copyright 2013 Elsevier.

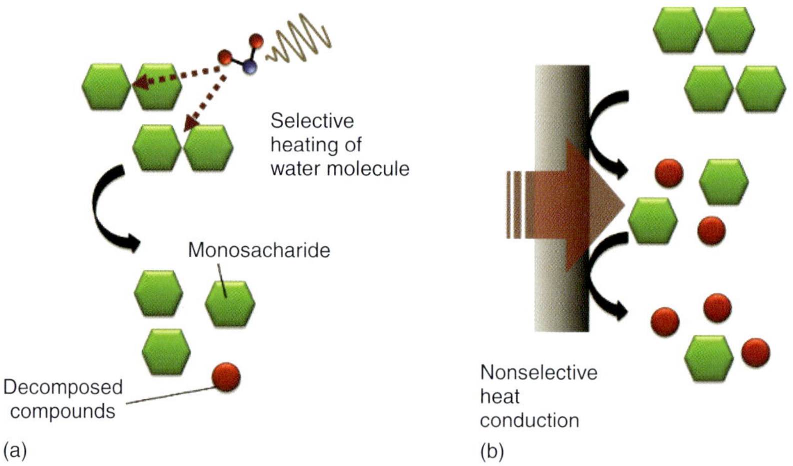

Scheme 15.1 Proposed degradation mechanism of disaccharides by hydrothermal hydrolysis using (a) microwave and (b) induction heating.

15.2.2 Effects of Ionic Conduction on Hydrolysis of Disaccharides under Hydrothermal Condition

Dielectric loss of water decreases with increase in the temperature due to increase in molecular mobility of water molecules. Okada *et al.*, have reported that relative permittivity and dielectric loss of water at hydrothermal condition significantly decreased and the dielectric dissipation peak shifted to higher frequency [18]. Therefore, water is less affected by microwave as they are heated. On the other hand, electrolyte solutions such as sea water become more easily heated by microwave at elevated temperature due to an increase in ionic conduction. For instance, foods containing a lot of salts are more heated when the get hotter [19].

The effects of electrolytes on hydrothermal hydrolysis of cellobiose were shown in Figure 15.3a. By addition of small amount of NaCl, saccharification rate effectively increased with reduction in energy consumption by 30–40%. The comparison of Arrhenius Plot demonstrated that the addition of electrolytes was effective for improving frequency factor (Figure 15.3b). The valence of the electrolyte also affected hydrolytic behavior. Divalent cations showed higher saccharification rate at lower reaction temperature than monovalent cations (Figure 15.3c). The effect of ionic conduction of salt is also effective for hydrothermal treatment of salty food waste biomass. Previously, we have shown that pickled stones of Japanese apricot (*Prunus mume*) are more easily heated and hydrolyzed by microwave than untreated ones [20]. The synergistic effects of salt and citric acid in the pickling juice promoted higher microwave absorption of the reaction medium.

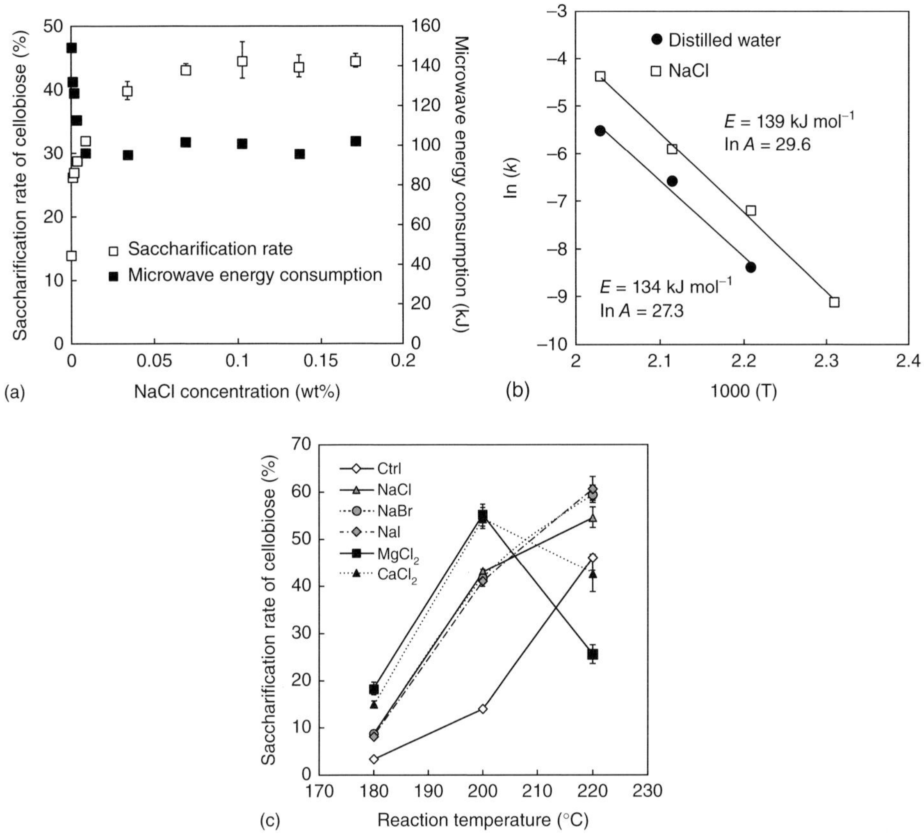

Figure 15.3 The effects of electrolytes on hydrothermal hydrolysis of cellobiose: (a) effect of concentration of NaCl (cellobiose 200 mg/100 ml, reaction temperature 200 °C, reaction time 5 min with 4 min of heating up time). (b) Arrhenius plot of hydrothermal hydrolysis of cellobiose with and without addition of NaCl (0.17 M). (c) Effects of monovalent and divalent cations (0.17 M) on hydrothermal hydrolysis of cellobiose. Reproduced with permission from Ref. [16]. Copyright 2012 Elsevier.

15.3 Hydrolysis of Biomass Using Ionic Conduction of Catalysts

15.3.1 Hydrolysis of Starch and Crystalline Cellulose Using Microwave Irradiation and Polyoxometalate Cluster

Homogeneous acids effectively work as both acid catalyst and microwave sensitizer through ionic conduction. We have been applied polyoxometalate (POM) cluster (heteropoly acids) as a super strong acid and less corrosive catalyst for hydrolysis of biomass [21]. POM is a high molecular weight anion composed of phosphate and silicate as hetero ions and tungstate and molybdate as addenda ions. The strong acidity of POM has been used in wide variety of organic chemical reaction as well as biomass hydrolysis [22]. In the present study, the effects of phosphotungstic acid ($H_3PW_{12}O_{40}$; PW) and silicotungstic acid ($H_4SiW_{12}O_{40}$; SiW) were investigated for hydrolysis of corn starch and crystalline cellulose (Avicel cellulose, cotton linter, and α-cellulose) as model biomass substrates [23]. Relatively high saccharification rates (≥80.4% for corn starch and ≥39.0% for Avicel cellulose) were obtained at short duration of reaction (5–10 min) with low concentration of POMs (2–14 mM). POM was stable enough under microwave irradiation (160-200 C) and they could be recycled at least four times by diethyl ether extraction. Microwave produced relatively higher amount of reducing sugars with lower selectivity to glucose than induction oven. Microwave irradiation exhibited higher selectivity to oligosaccharides (Figure 15.4a), indicating that microwave irradiation could promote hydrolysis of sugar chain from inside. In addition, aqueous solutions of POMs are easily heated by microwave irradiation due to their high ionic conduction. The energy consumption required for hydrolysis is, therefore, reduced from the normal hydrothermal hydrolysis or strong acid resins which are less affected by microwave (Figure 15.4b).

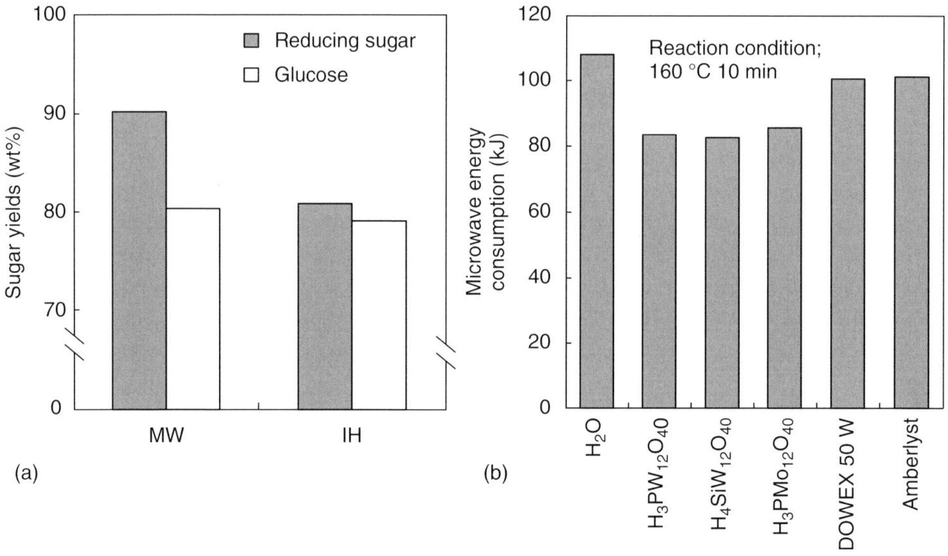

Figure 15.4 Effects of polyoxometalate cluster and microwave irradiation on hydrolysis of corn starch and Avicel cellulose: (a) comparison of microwave (MW) and induction heating (IH) on yields of reducing sugar and glucose and (b) comparison of POMs and strong cation exchange resins on microwave energy consumption. Reproduced with permission from Ref. [23]. Copyright 2013 Elsevier.

15.3.2
Hydrolysis of Fast-Growing Green Macroalgae Using Microwave Irradiation and Polyoxometalate Cluster

The effects of POM were further investigated for hydrolysis of *Ulva* spp. as a real biomass. *Ulva* spp. is a kind of green macroalgae and often brings about green tide at eutrophic ocean. The stranded seaweeds deteriorate and release a strong odor, so they are routinely collected and burned. The green-tide is becoming a big issue in Japan, China, and Europe. On the other hand, the fast growing rate of *Ulva* spp. is attractive for production of algal biomass. Recently, Hiraoka and his coworkers found that *Ulva meridionalis* [24] even doubles a day (Figure 15.5, [25]), and they are attractive biomass resource for capturing carbon dioxide.

The effects of POMs and microwave irradiation were, then, investigated on the hydrolysis of polysaccharides in *Ulva meridionalis* [26]. *Ulva* spp. predominantly contains sulfated glucuronoxylorhamnan (ulvan, Figure 15.6a) as a major polysaccharide and they are potentially used as a food ingredients as well as biomedical applications [27]. *Ulva* spp. also accumulates starch and cellulose as storage and structural polysaccharides. The sugar yields (neutral sugar and uronic acids) attained maximum value of 35% at 160 °C and 10 min of hydrolysis by using 2 mM of PW (Figure 15.7). The monosaccharide analysis showed that major hydrolysates were consisting of glucose and rhamnose, which are originated from ulvan and starch. Glucose is one of the most important feedstock to produce building block compounds such as ethanol and lactic acids. Rhamnose is a kind of rare sugar, which is a raw material as sweetener and flavoring ingredients [28]. PW and SiW exhibited higher activity than sulfuric and hydrochloric acids due to their stronger acidity at the same concentration (50 mM, Figure 15.8a). The size exclusion chromatograms showed that POMs could degrade acidic polysaccharides into monosaccharides; however, catalysis of sulfuric and hydrochloric acids

Figure 15.5 *Ulva meridionalis*. The photograph was kindly supplied from Dr Masanori Hiraoka at Kochi University.

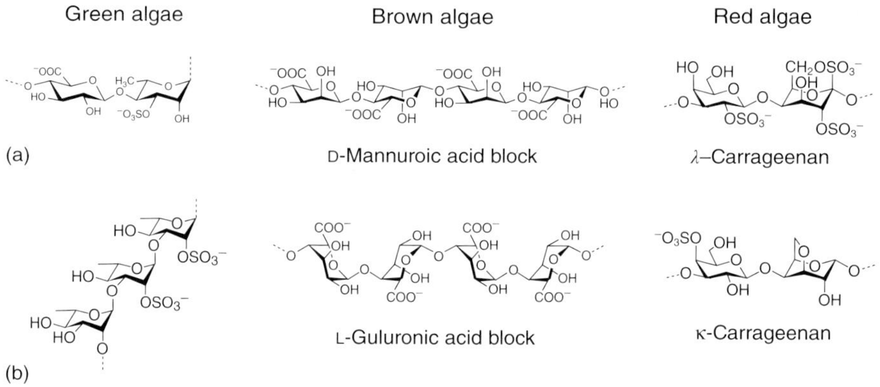

Figure 15.6 Typical chemical structures of algal hydrocolloids of green, brown, and red algae: (a) ulvan, (b) rhamnan sulfate, (c) alginates, and (d) carrageenans.

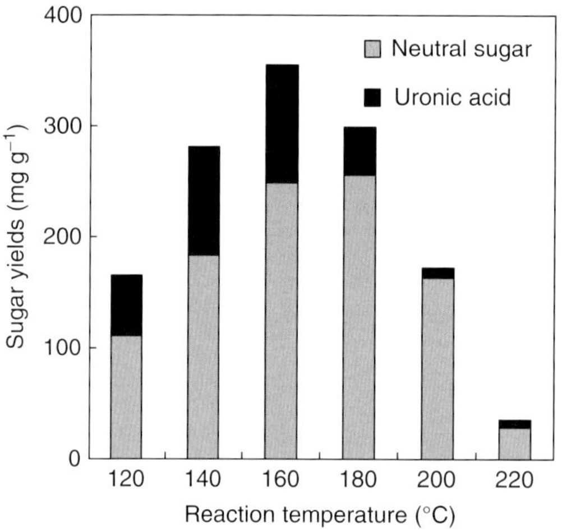

Figure 15.7 Effects of reaction temperature on polyoxometalate cluster (2 mM) and microwave irradiation on hydrolysis of *Ulva meridionalis*. Reproduced with permission from Ref. [26]. Copyright 2014. The Royal Society of Chemistry.

were not enough to obtain monosaccharides (Figure 15.8b). Microwave heating showed higher sugar yields than induction heating by 9% (neutral sugar) and 33% (uronic acid) at the same thermal history (Figure 15.8c). Since uronic acids are fragile to heat, microwave could improve their stability under hydrothermal condition by avoiding the wall effect. Microwave should also affected catalysis of POMs by strong microwave absorption due to ionic conduction. These synergistic effects of microwave irradiation and POM catalysis worked together toward obtaining useful sugars from fast-growing *Ulva* spp.

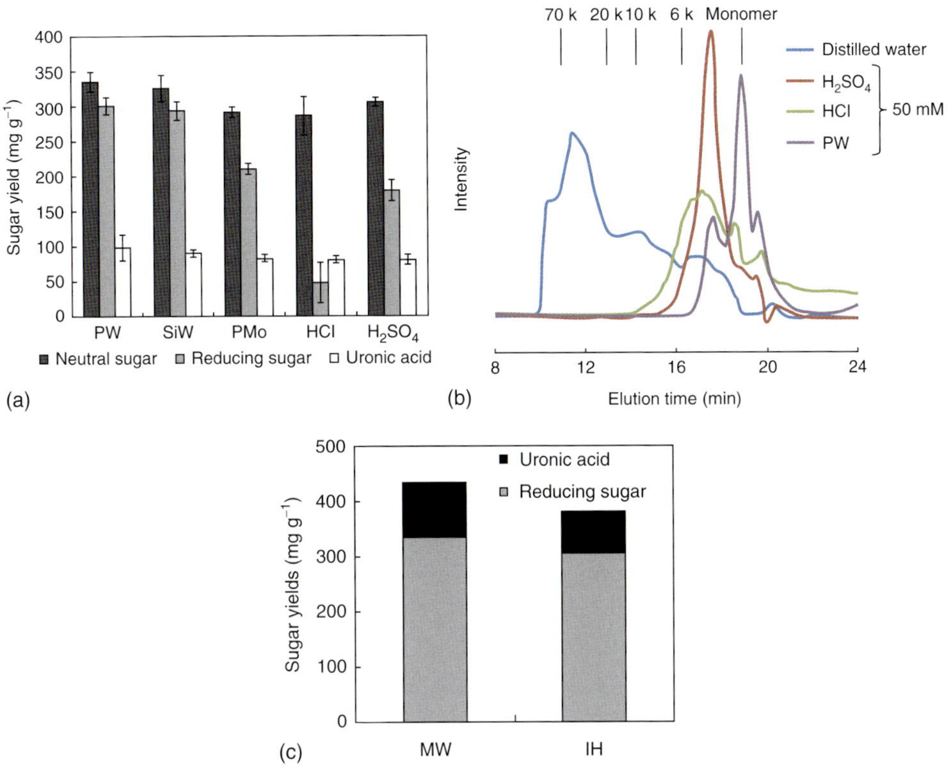

Figure 15.8 Effects of polyoxometalate cluster and microwave irradiation on hydrolysis of *Ulva meridionalis*: (a) comparison of POMs (PW; phsophotungstic acid, SiW; silicotungstic acid, PMo; phosphomolybdic acid), HCl, and H_2SO_4 on yields of neutral sugar, reducing sugar, and uronic acids, (b) size exclusion chromatograms of the hydrolysates, and (c) comparison of microwave and induction heating. Reproduced with permission from Ref. [26]. Copyright 2014. The Royal Society of Chemistry.

15.4
Dielectric Property of Algal Hydrocolloids in Water

15.4.1
Comparison of Dielectric Property of Aqueous Solution of Hydrocolloids Obtained from Algae and Land Plants

The heating behavior of reactants under microwave irradiation is strongly dependent on dielectric property of the irradiated materials. The aqueous electrolyte solution is, for instance, heated by dielectric and conduction losses due to rotation of dipolar molecules and ionic conduction of electrolytes, respectively. Particularly, hydrocolloids in macroalgae are a kind of polyelectrolyte substantially substituted by acidic functional groups such as carboxylic and sulfate groups. For example, alginates, extracted from brown algae, are a block copolymers composed of D-mannuronic acid blocks (M) and L-guluronic acid block (G) as well as MG blocks. Carrageenans and agar, extracted from red algae, are composed of sulfated and anhydrous galactose residues. These acidic functional groups were expected to affect dielectric property of the aqueous solution of the hydrocolloids; therefore, the relationship of chemical structure of the algal hydrocolloids (e.g., ulvan, rhamnan sulfate, sodium alginate, and carrageenan, Figure 15.6a,b,c,d) and their dielectric property were analyzed between 100 MHz and 20 GHz using the dielectric probe kit with a network analyzer (Keysight Technologies 5242A Network Analyzer) [29, 30].

Figure 15.9 shows typical dielectric spectra of sodium alginate and κ-carrageenan. Corn starch and *Citrus* pectin originated from land plants were also presented as controls. All the spectra showed peaks due to dielectric relaxation of water at around 20 GHz at room temperature (γ relaxation process). The spectra of sodium alginate and κ-carrageenan also showed significant increase in the lower frequency due to ionic conduction. Ulvan and rhamnan sulfates showed similar dielectric property as carrageenan [30]. Corn starch did not show any ionic conduction because they do not have acidic functional groups. *Citrus* pectin, on the other hand, showed slight ionic conduction because it

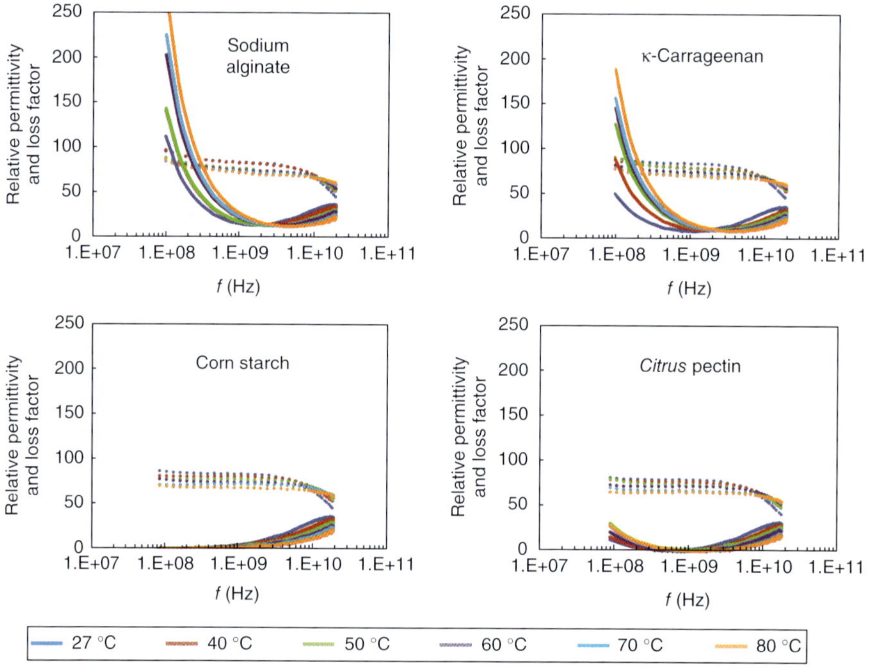

Figure 15.9 Temperature dependency of the dielectric spectra of native (a) sodium alginates, (b) κ-carrageenans, (c) corn starch, and (d) *Citrus* pectin in water (2.0 wt%). The dashed lines represent relative permittivity, whereas the solid lines represent the loss factors. Reproduced with permission from Ref. [29]. Copyright 2015 Elsevier.

retains carboxylic groups on galacturonic acid residues. These results indicated that aqueous solutions of algal hydrocolloids are significantly dependent on ionic conduction of acidic functional groups. Although increase in temperature decreased dielectric loss of water, ionic conduction pronounced with increase in temperature. The phenomenon coincided with the hydrothermal reaction with addition of electrolytes (Section 15.2.2) in that salt solution become more easily heated by microwaves at elevated temperature.

15.4.2
The Effects of the Degree of Substitution of Acidic Functional Groups on Dielectric Property of Aqueous Solution of Algal Hydrocolloids

The effects of degree of substitution of acidic functional groups on dielectric property were investigated by dielectric measurement of partially reduced sodium alginates and desulfated carrageenans, ulvan, and rhamnan sulfate (Scheme 15.2). Sodium alginates were reduced using 1-ethyl-3-(3-dimethylaminopropyl) carbodiimide (EDC) and sodium borohydride ($NaBH_4$) (Scheme 15.2a) treatments [31]. Sulfated hydrocolloids were first protonated by the addition of strong ion-exchange resin to produce H^+-form hydrocolloids, and then desulfated by solvolysis (Scheme 15.2b) [32]. The typical dielectric spectra of the reduced sodium alginate and desulfated carrageenan were shown in Figure 15.10. Reduction of sodium alginates (reduction rate = 43%) significantly decreased ionic conduction at lower

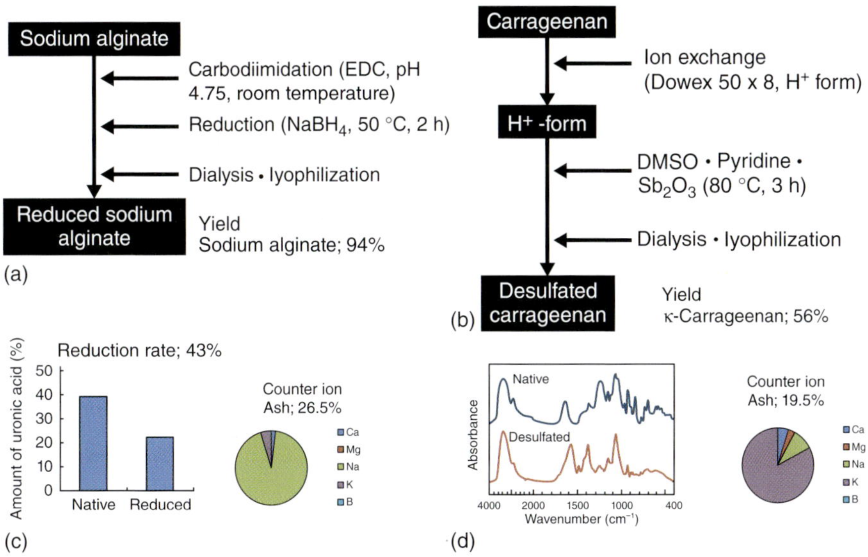

Scheme 15.2 Preparation procedures of partially reduced (a) sodium alginate [31] and (b) desulfated carrageenan [32a]. Their effects on the degree of substitution and compositions of counter cations of (c,d) sodium alginates and (e-f) κ-carrageenan. The desulfation procedures was conducted according to Nakagawa *et al.* [32b].

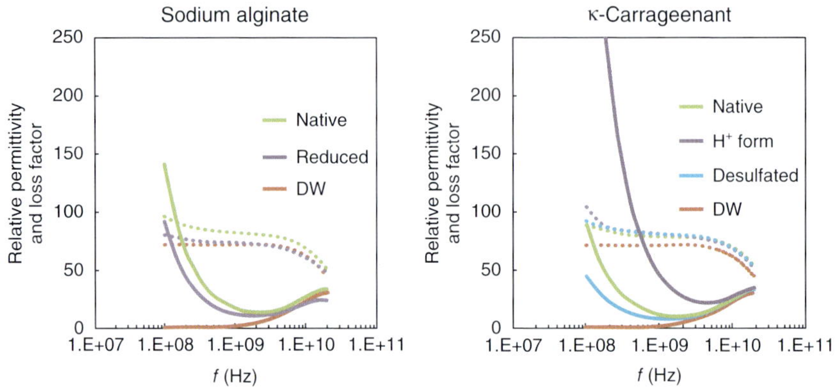

Figure 15.10 Dependency of the dielectric spectra on acidic functional groups of (a) sodium alginates, (b) κ-carrageenan in water (2.0 wt%). The dashed lines represent relative permittivity, whereas the solid lines represent loss factors. DW; distilled water. Reproduced with permission from Ref. [29]. Copyright 2015 Elsevier.

frequency; however, this did not affect dielectric loss of water. Desulfation of carrageenan also showed decrease in ionic conduction without change in dielectric loss at higher frequencies. Desulfation of ulvan and rhamnan sulfate also showed the similar behavior as carrageenans [29]. In addition, the exchange in countercation of sulfate group has led to significant increase in ionic conduction, because proton in aqueous solution shows higher conductivity than other ions. The phenomenon is well known as the *Grotthus mechanism* or *proton relay* [33]. The high conductivity of proton was also effective for promoting dielectric property. These results confirmed that the acidic functional groups and their counter cations generate ionic conduction, which strongly affects the dielectric property of the solution.

Relative permittivity of aqueous solution of hydrocolloids was higher than that of water (Figure 15.10a,b), indicating that hydrocolloids behave as a water-structure-maker (kosmotropes) [19]. Basically hydrocolloids are hydrophilic compounds, however, they also retain hydrophobic surface in their ring structure. On the other hand, inorganic ions, such as K^+ and Cl^-, behave as a water-structure-breaker (chaotropes) [19]. Therefore, contamination of salts in hydrocolloid solution decreases relative permittivity [34].

15.4.3 The Correlation of Loss Tangent at 2.45 GHz and Other Physical Properties of Sodium Alginates and Carrageenans in Water

The correlation between loss tangent at 2.45 GHz and conductivity, pH, non-freezing bound water, relaxation time of free water, and bulk viscosity of the hydrocolloids were shown in Figure 15.11 [29]. Loss tangent of the algal

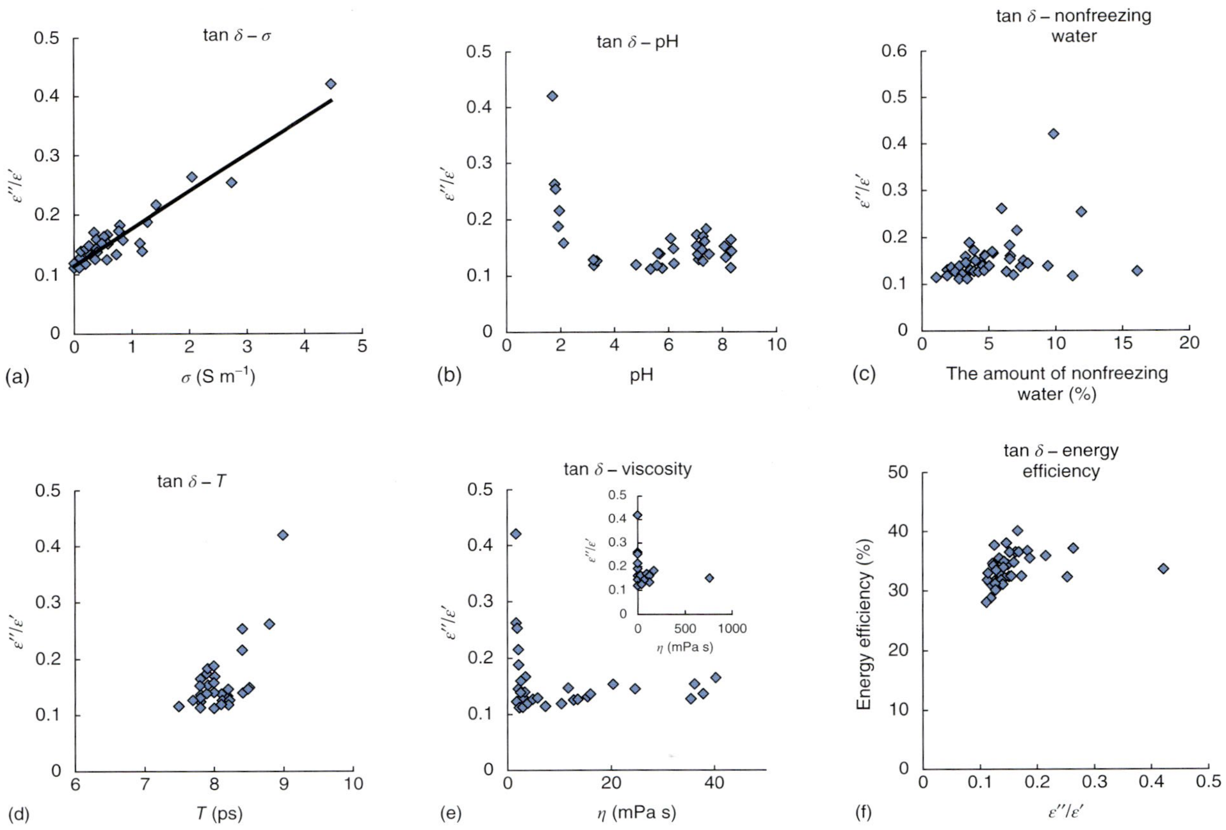

Figure 15.11 Correlations between tan δ ($\varepsilon''/\varepsilon'$) at 2.45 GHz with other physical properties of hydrocolloids in water including (a) conductivity (σ), (b) pH, (c) nonfreezing water, (d, e) viscosity, and (f) energy efficiency. Reproduced with permission from Ref. [29]. Copyright 2015 Elsevier.

hydrocolloids strongly correlated with conductivity (Figure 15.11a) and drastically increased at proton-rich condition below pH 2 (Figure 15.11b). This indicates that proton is significantly important for loss tangent. Loss tangent of hydrocolloids rich in acidic functional groups increased with an increase in formation of bound water (Figure 15.11c). Contrary, *Citrus* pectin also produced bound water up to 16% due to branch structure of the sugar chain; however, the loss tangent did not increase. The results suggested that both proton concentration and formation of bound water may be important to initiate the proton conduction.

The relaxation time of free water was, then, calculated from the Cole–Cole equation as shown below:

$$\varepsilon^* = \varepsilon_\infty + \frac{\varepsilon_s - \varepsilon_\infty}{1 + (j\omega\tau)^{1-\alpha}} \tag{15.2}$$

where ε^*, ε_s, ε_∞, j, ω, τ, and α are the relative permittivity and loss factor, the static frequency, the infinite frequency, an imaginary unit, the angular frequency, the relaxation time, and the distribution parameter, respectively [35]. The relaxation times of water were mostly around 8 ps, however, the longer relaxation time brought larger loss tangent (Figure 15.11d). This could be due to restriction of water molecule by hydration of proton at low pH conditions. The high loss tangent was obtained at very low bulk viscosity (Figure 15.11e). The phenomenon may be due to higher ionic conduction with lower restriction of ions by water molecule at low viscosity. Finally, microwave heating efficiency was calculated from the ratio of microwave energy used (Qmw) and the theoretical energy (Qth) calculated from the following Eq. (15.3):

$$Q_{TH} = mc_p\Delta T \tag{15.3}$$

where m and c_p are the mass of the irradiated medium (kg) and the specific heat capacity, respectively [36]. The relationship between heating efficiency and loss tangent was summarized in Figure 15.11f. Although high loss tangent could be obtained from the high ionic conductive solution of algal hydrocolloids, energy efficiency did not significantly change since high loss tangent brings low penetration depth. In this case, lower frequency can be one of the solutions to obtain good microwave penetration into aqueous solution with high loss tangent.

The dielectric property of the aqueous solution of algal polysaccharide was found to be determined by the dynamics of water molecules and electrolytes under electromagnetic field. Especially, ionic conduction generated by interactions among acidic functional groups, counter cations, and water molecules were significantly important for the dielectric property. The regulation of dielectric property of reaction system and selection of microwave frequency will be important for efficient utilization of microwave energy for algal conversion process to produce useful chemicals.

15.5 Summary and Conclusions

Microwave heating was effective for hydrothermal treatment of biomass by promoting stability of sugars under high-temperature water with avoidance of the wall effect. The ionic conduction by the addition of electrolytes was very effective for hydrothermal treatment of biomass under microwave irradiation with efficient microwave energy use. Since algal biomass predominantly contains water and electrolytes, they are highly compatible with microwave-assisted conversion processes. Addition of electrolyte catalyst was also effective for improving both hydrolytic activity and dielectric property of the reaction system due to ionic conduction. POM showed strong catalysis on hydrolysis of biomass model polysaccharides as well as green macroalgae.

Microwave heating also has a lot of advantages as chemical engineering processes. Microwave is capable of providing compact reactor size, short duration of conversion process, and low severity of the reaction. Therefore, it could be applied as small-scale and distributed biorefinery system in the rural areas where most part of the biomass is produced. Microwave process is, therefore, a promising technology to achieve efficient recycle of algal biomass feedstock.

Acknowledgments

Part of this study was performed through Program to Disseminate Tenure Tracking System of the Ministry of Education, Culture, Sports, Science and Technology (Kochi University). Dielectric measurement was conducted through Analysis and Development System for Advanced Materials (ADAM), RISH, Kyoto University. Authors are also deeply grateful to Prof. Jun-ichi Azuma at Osaka University; Dr Tomohiko Mitani and Dr Hiroshi Nishimura at RISH Kyoto University; Dr Keiichiro Kashimura at Chubu University; Prof. Kazumichi Yanagisawa, Dr Masanori Hiraoka, Dr Kei Okamura, Dr Minoru Ikehara, Dr Shingo Hadano, and Ms. Kiriyo Oono at Kochi University.

References

1. Wargacki, A.J., Leonard, E., Win, M.N., Regitsky, D.D., Santhos, C.N.S., Kin, P.B., Cooper, S.R., Raisner, R.M., Herman, A., Sivitz, A.B., Lakshmanaswamy, A., Kashiyama, Y., Baker, D., and Yoshikuni, Y. (2012) *Science*, **335**, 308–313.
2. Enquist-Newman, M., Faust, A.M.E., Bravo, D.D., Santos, C.N.S., Raisner, R.M., Hanel, A., Sarvabhowma, P., Le, C., Regitsky, D.D., Cooper, S.R., Peereboom, L., Clark, A., Martinez, Y., Goldsmith, J., Cho, M.Y., Donohoue, P.D., Luo, L., Lamberson, B., Tamrakar, P., Kim, E.J., Villari, J.L., Gill, A., Tripathi, S.A., Karamchedu, P., Paredes, C.J., Rajgarhia, V., Kotlar, H.K., Bailey, R.B., Miller, D.J., Ohler, N.L., Swimmer, C., and Yoshikuni, Y. (2014) *Nature*, **505**, 239–243.
3. Chisti, Y. (2007) *Biotechnol. Adv.*, **25**, 294–306.
4. Spolaore, P., Joannis-Cassan, C., Duran, E., and Lambert, A. (2006) *J. Biosci. Bioeng.*, **101**, 87–96.

5. Akiya, N. and Savage, P.E. (2002) *Chem. Rev.*, **102**, 2725–2750.
6. (a) Sun, Y. and Cheng, J. (2002) *Bioresour. Technol.*, **83**, 1–11; (b) Savage, P.E. (2012) *Science*, **338**, 1039–1040.
7. Horikoshi, S. and Serpone, N. (2014) *Catal. Sci. Technol.*, **4**, 1197–1210.
8. (a) Tsubaki, S., Iida, H., Sakamoto, M., and Azuma, J. (2008) *J. Agric. Food. Chem.*, **56**, 11293–11299; (b) Tsubaki, S., Nakauchi, M., Ozaki, Y., and Azuma, J. (2009) *Food Sci. Technol. Res.*, **15**, 307–314; (c) Yoshida, T., Tsubaki, S., Teramoto, Y., and Azuma, J. (2010) *Bioresour. Technol.*, **101**, 7820–7826; (d) Tsubaki, S., Sakamoto, M., and Azuma, J. (2010) *Food Chem.*, **123**, 1255–1258; (e) Tsubaki, S. and Azuma, J. (2011) in *Advances in Induction and Microwave Heating* (ed S. Grundas), I-Tech Education and Publishing, Vienna, pp. 697–722; (f) Hermiati, E., Azuma, J., Tsubaki, S., Djumali, M., Candra, S.T., Ono, S., and Bambang, P. (2012) *Carbohydr. Polym.*, **87**, 939–942;(g) Tsubaki, S. and Azuma, J. (2013) *Bioresour. Technol.*, **131**, 485–491.
9. Azuma, J., Tanaka, F., and Koshijima, T. (1984) *Mokuzai Gakkaishi*, **30**, 501–509.
10. Pasquet, V., Chérouvrier, J.R., Farhat, F., Thiéry, V., Piot, J.M., Bérard, J.B., Kaas, R., Serive, B., Patrice, T., Cadoret, J.P., and Picot, L. (2011) *Process Biochem.*, **46**, 59–67.
11. (a) Rodriguez-Jasso, R.M., Mussatto, S.I., Pastrana, L., Aguilar, C.N., and Teixeira, J.A. (2011) *Carbohydr. Polym.*, **86**, 1137–1144; (b) Quitain, A.T., Kai, T., Sasaki, M., and Goto, M. (2013) *Ind. Eng. Chem. Res.*, **52**, 7940–7946.
12. (a) Balasubramanian, S., Allen, J.D., Kanitkar, A., and Boldor, D. (2011) *Bioresour. Technol.*, **102**, 3396–3403; (b) Lee, J.Y., Yoo, C., Jun, S.-Y., Ahn, C.-Y., and Oh, H.-M. (2010) *Bioresour. Technol.*, **101**, S75–S77; (c) Prabhakaran, P. and Ravindran, A.D. (2011) *Lett. Appl. Microbiol.*, **53**, 150–154; (d) Biller, P., Friedman, C., and Ross, A.B. (2013) *Bioresour. Technol.*, **136**, 188–195; (e) McMillan, J.R., Watson, I.A., Ali, M., and Jaafar, W. (2013) *Appl. Energy*, **103**, 128–134.
13. (a) Patil, P.D., Gude, V.G., Mannarswamy, A., Cooke, P., Munson-McGee, S., Nirmalakhandan, N., Lammers, P., and Deng, S. (2011) *Bioresour. Technol.*, **102**, 1399–1405; (b) Patil, P.D., Guide, V.G., Mannarswamy, A., Cooke, P., Nirmalakhandan, N., Lammers, P., and Deng, S. (2012) *Fuel*, **97**, 822–831; (c) Patil, P.D., Reddy, H., Muppaneni, T., Mannarswamy, A., Schaub, T., Holguin, F.M., Lammers, P., Nirmalakhandan, N., Cooke, P., and Deng, S. (2012) *Green Chem.*, **14**, 809–818; (d) Patil, P., Reddy, H., Muppaneni, T., Ponnusamy, S., Sun, Y., Dailey, P., Cooke, P., Patil, U., and Deng, S. (2013) *Bioresour. Technol.*, **137**, 278–285; (e) Koberg, M., Cohen, M., Ben-Amotz, A., and Gedanken, A. (2011) *Bioresour. Technol.*, **102**, 4265–4269; (f) Cheng, J., Yu, T., Li, T., Zhou, J., and Cen, K. (2013) *Bioresour. Technol.*, **131**, 531–535.
14. (a) Guo, J., Zhuang, Y., Chen, L., Liu, J., Li, D., and Ye, N. (2012) *Bioresour. Technol.*, **120**, 19–25; (b) Zhuang, Y., Guo, J., Chen, L., Li, D., Liu, J., and Ye, N. (2012) *Bioresour. Technol.*, **116**, 133–139.
15. Tsubaki, S., Oono, K., Onda, A., Yanagisawa, K., and Azuma, J. (2013) *Carbohydr. Res.*, **375**, 1–4.
16. (a) Tsubaki, S., Oono, K., Onda, A., Yanagisawa, K., and Azuma, J. (2012) *Bioresour. Technol.*, **123**, 703–706; (b) Tsubaki, S., Onda, A., Yanagisawa, K., and Azuma, J. (2012) *Proc. Chem.*, **4**, 288–293.
17. Overend, R.P. and Chornet, E. (1987) *Philos. Trans. R. Soc. London, Ser. A*, **321**, 523–536.
18. Okada, K., Yao, M., Hiejima, Y., Kohno, H., and Kajihara, Y. (1999) *J. Chem. Phys.*, **110**, 3026–3036.
19. Chaplin, M. (2006) Water Structure and Sciences, *http://www1.lsbu.ac.uk/water/* (accessed 07 May 2015).
20. Tsubaki, S., Ozaki, Y., and Azuma, J. (2010) *J. Food Sci.*, **75**, C152–C159.
21. Kozhevnikov, I.V. (1998) *Chem. Rev.*, **98**, 171–198.
22. (a) Shimidzu, K.I., Furukawa, H., Kobayashi, N., Itaya, Y., and Satsuma, A. (2009) *Green Chem.*, **11**, 1627–1632; (b) Tian, J., Wang, J., Zhao, S., Jiang, C.,

Zhang, X., and Wang, X. (2010) *Cellulose*, **17**, 587–594; (c) Tian, J., Fan, C., Cheng, M., and Wang, X. (2011) *Chem. Eng. Technol.*, **34**, 482–486; (d) Cheng, M., Shi, T., Guan, H., Wang, S., Wang, X., and Jiang, Z. (2011) *Appl. Catal., B*, **107**, 104–109.

23. Tsubaki, S., Oono, K., Ueda, T., Onda, A., Yanagisawa, K., Mitani, T., and Azuma, J. (2013) *Bioresour. Technol.*, **144**, 67–73.
24. Horimoto, R., Masakiyo, Y., Ichihara, K., and Shimada, S. (2011) *Bull. Natl. Sci. Mus. (Jpn.)*, **37**, 155–167.
25. Hiraoka, M. (2012) *J. Jpn. Inst. Energy*, **91**, 1154–1160.
26. Tsubaki, S., Oono, K., Hiraoka, M., Ueda, T., Onda, A., Yanagisawa, K., and Azuma, J. (2014) *Green Chem.*, **16**, 2227–2233.
27. Lahaye, M. and Robic, A. (2007) *Biomacromolecules*, **8**, 1765–1774.
28. Alves, A., Sousa, R.A., and Reis, R.L. (2013) *J. Appl. Phycol.*, **25**, 407–424.
29. Tsubaki, S., Hiraoka, M., Hadano, S., Okamura, K., Ueda, T., Nishimura, H., Kashimura, K., and Mitani, T. (2015) *Carbohydr. Polym.*, **115**, 78–87.
30. Tsubaki, S., Hiraoka, M., Hadano, S., Nishimura, H., Kashimura, K., and Mitani, T. (2014) *Carbohydr. Polym.*, **107**, 192–197.
31. Fujihara, M. and Nagumo, T. (1986) *Nihon Kagaku Kaishi*, **4**, 576–579.
32. (a) De Araújo, C.A., Noseda, M.D., Cipriani, T.R., Gonçalves, A.G., Duarte, M.E.R., and Ducatti, D.R.B. (2013) *Carbohydr. Polym.*, **91**, 483–491; (b) Nakagawa, K., Inoue, Y., and Kamata, T. (1977) *Carbohydr. Res.*, **58**, 47–55.
33. Bonin, J., Costentin, C., Louault, C., Robert, M., and Savéant, J.M. (2011) *J. Am. Chem. Soc.*, **133**, 6668–6674.
34. Kishikawa, Y., Seki, Y., Shingai, K., Kita, R., Shinyashiki, N., and Yagihara, S. (2013) *J. Phys. Chem. B*, **117**, 9034–9041.
35. Gabriel, C., Gabriel, S., Grant, E.H., Halstead, B.S.J., and Mingos, D.M.P. (1998) *Chem. Soc. Rev.*, **27**, 213–224.
36. Moseley, J.D. and Kappe, C.O. (2011) *Green Chem.*, **13**, 794–806.

16 Microwave-Assisted Lignocellulosic Biomass Conversion

Tomohiko Mitani and Takashi Watanabe

16.1 Introduction

Scale-up of microwave-assisted chemical processing is a long-standing problem for commercialization of microwave chemistry. The wavelength of 2.45 GHz microwave is 122 mm in vacuum and it is shortened in a dielectric medium in inverse proportion to the square root of permittivity. In addition, microwave energy is absorbed in a lossy medium and eventually converted to heat. These natures of microwave, which are usually expressed as "penetration depth," make it difficult to scale-up a processing system. Unlike conventional chemical processing such as steam heating system, a simple expansion of reactor size leads to uneven heating because microwave energy does not reach to the center of the reactor.

A continuous flow microwave reactor is expected as an alternative way to increase productivity. An attractive feature of microwave-assisted chemical processing is shortening of chemical reaction time. Since productivity is dependent on system size and required processing time, "speed-up" of microwave-assisted chemical processing will make up for limitation of "scale-up." Hence, a continuous flow system has a greater possibility to improve productivity compared with a batch system.

A review paper [1] summarized various types of continuous flow microwave reactors, and comparison of seven commercially available microwave reactors was reported [2]. As an example of scale-up approach, a commercial continuous flow microwave reactor with a capacity of 200 ml achieved the production rate between 0.5 and 3.0 mol h^{-1} (1–6 l h^{-1}) [3]. However, the production rate is still insufficient for a plant-scale facility.

In this section, continuous flow microwave reactors that were developed for lignocellulosic biomass conversion are described. After conversion of lignocellulosic biomass by pretreatment and subsequent enzymatic saccharification and fermentation process is overviewed in Section 16.2, a multi-mode continuous flow microwave reactor developed in 1980s is explained in Section 16.3. Direct-irradiation continuous flow microwave reactors developed in 2000s is described

Microwaves in Catalysis: Methodology and Applications, First Edition.
Edited by Satoshi Horikoshi and Nick Serpone.

in Section 16.4. In Section 16.5, a pilot-plant-scale continuous flow microwave reactor is introduced as a scale-up of the direct-irradiation reactor.

16.2 Lignocellulosic Biomass Conversion

Lignocellulosic biomass, which is derived from lignified plant cell wall components, is the most abundant renewable organic resources on the Earth. Widely available in large quantity and noncompeting with food supply, lignocellulosic biomass is expected to serve as a platform for green chemicals and fuels, and effectively mitigate greenhouse gas emission effects. Major components of plant cell walls comprise cellulose (30–50%), hemicelluloses (10–30%), and a heterogeneous aromatic polymer, lignin (10–35%). The separation and conversion of plant cell wall components have emerged as important technologies for the sustainable production of renewable fuels and chemicals. Cell wall polysaccharides are coated with lignin; thus, pretreatment disintegrating the cell walls is necessary for conversion of the polysaccharides into bioethanol and chemicals by enzymatic saccharification and fermentation. In conversion of storage polysaccharides such as starch (e.g., from corn) and sucrose (e.g., from sugarcane), the pretreatment process to remove the lignin coverage is not required, which gives higher feasibility in terms of cost balance. However, negative effects on food supply and emission of higher CO_2 in overall process from cultivation to bioethanol has been pointed out [4].

A typical bioethanol production process from lignocellulosic biomass by separate hydrolysis and fermentation (SHF) is shown in Figure 16.1. The first step is pretreatment of biomass to increase susceptibility of cellulolytic and hemicellulolytic enzymes to their substrate by disrupting the lignin coating. The pretreated biomass is hydrolyzed with enzymes, and the monosaccharides formed are converted to bioethanol by fermentation. The enzymatic hydrolysis and fermentation

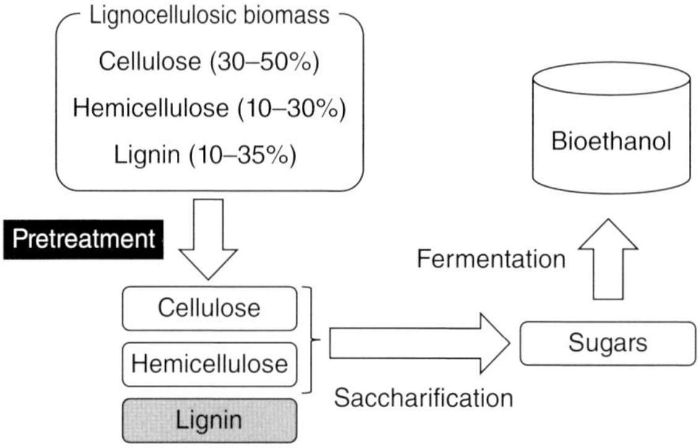

Figure 16.1 Bioethanol production from lignocellulosic biomass by separate hydrolysis and fermentation (SHF) process.

can be carried out simultaneously, which is called simultaneous saccharification and fermentation (SSF), to suppress inhibition of glycosyl hydrolases by the sugars. The most simple conversion process is a one-step consolidated bioprocess (CBP), in which production of glycosyl hydrolases, enzymatic saccharification and fermentation can be performed in a single reactor where a single micro-organism or microbial consortium converts the pretreated biomass into bioethanol without added enzymes. In all the conversion processes, efficient pretreatment system is required to increase the hydrolyzability of polysaccharides with production of a less amount of inhibitors for hydrolysis and fermentation. A number of pretreatments including mechanical, thermal, and chemical methods have been studied [5]. Recently, microwave pretreatment has drawn attention from the aspects of rapid and efficient process, and pretreatments with acids, alkali, and oxidative catalysts in aqueous and organic solvents have been applied to hardwood, softwood, and herbaceous biomass [6].

16.3 Multi-mode Continuous Flow Microwave Reactor

A multi-mode continuous flow microwave reactor was developed in 1980s [7]. A conceptual image of the microwave reactor is shown in Figure 16.2. Lignocellulosic biomass slurry soaked in water or organic solvents flows in a dielectric tube, which passes through a metal vessel. Microwave is radiated from the irradiation port to the metal vessel. As you know, electromagnetic wave is reflected on the metal plate; whereas it can penetrate through dielectric materials with dielectric loss. Lignocellulosic biomass slurry is therefore irradiated with microwave continuously while passing through the microwave irradiation region. The metal

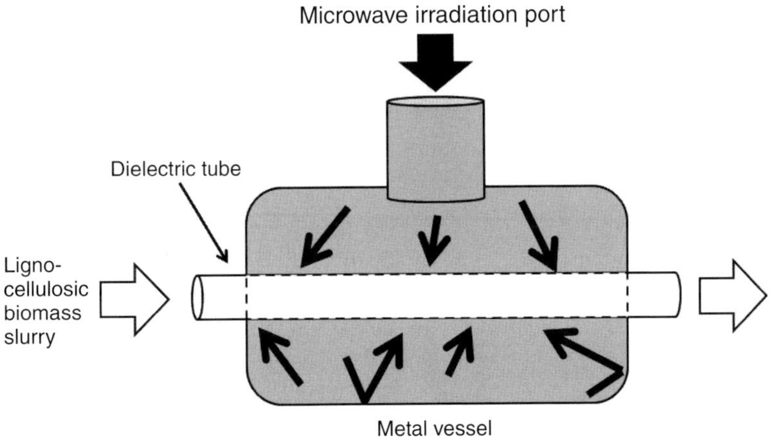

Figure 16.2 Conceptual image of multi-mode continuous flow microwave reactor.

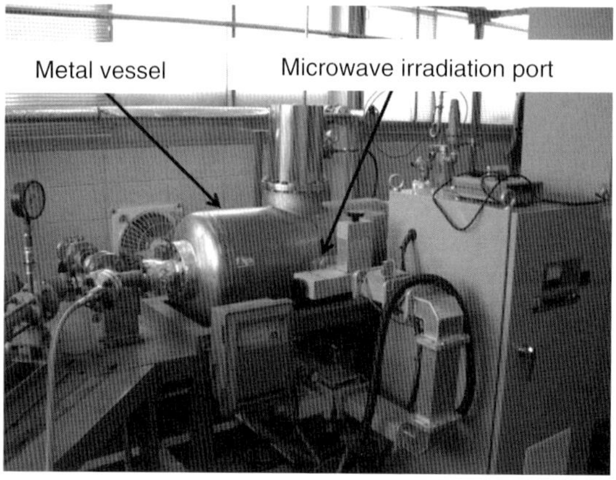

Figure 16.3 Photograph of the multi-mode continuous flow microwave reactor with a 4.9 kW microwave generator.

vessel also plays a role of fail-safe against pressure. Even though the dielectric pipe is burst into fragments due to high pressure, the vessel prevents the high-temperature slurry and shattered dielectric pipe from being splattered around.

A photograph of a developed multi-mode continuous flow microwave reactor is shown in Figure 16.3. A 2.45 GHz and 4.9 kW microwave generator is attached to the metal vessel. Ceramic is used as material of the dielectric pipe from viewpoints of strengths in temperature and pressure. Experimental results show that reducing sugar was obtained in a 55.7% yield from *Akamatsu* (Japanese red pine) and in a 61.0% yield from *Buna* (Japanese beech), respectively, through the microwave pretreatment process using the reactor shown in Figure 16.3 [7]. These woody biomass slurries were made from 1 kg of wood ground by a Wiley mill and 10 l of water. The flow rate was $10-15\,\mathrm{l\,h^{-1}}$. More than 90% of potential glucose was obtained at the pretreatment temperatures over 200 °C for softwood, hardwood and herbaceous plant resources such as rice straw, rice hulls, and sugar cane bagasse [8]. These slurries were made from 1 kg of lignocellulosic biomass in 15 l of 0.5% acetic acid solution. The flow rate was $8-20\,\mathrm{l\,h^{-1}}$.

A multi-mode continuous flow microwave reactor looks like a large microwave oven, and it is simple in design. As microwave does not leak out of the vessel, most of radiated microwave energy is eventually consumed by lignocellulosic biomass slurry and dielectric pipe, except for backward reflection to a microwave generator. However, scale-up of the reactor is still difficult because temperature of lignocellulosic biomass slurry is hardly controlled. Microwave absorptivity depends on the permittivity, which varies as a function of temperature. If the absorptivity of slurry increases with a rise in temperature, the reactor operation will result in thermal runaway. In the reverse case, the slurry temperature may not reach to the desired temperature for a successful microwave pretreatment or other microwave-assisted chemical reactions.

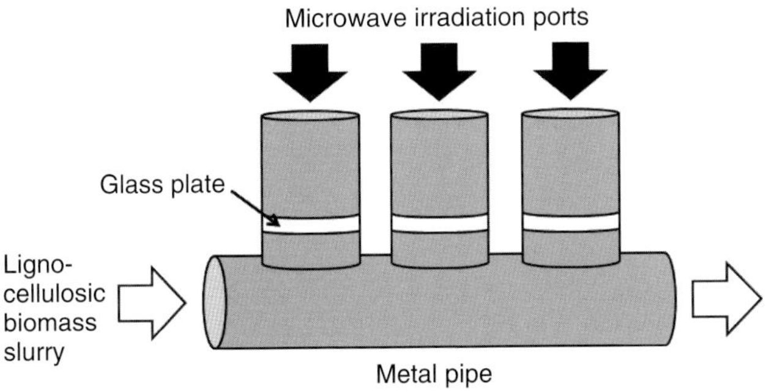

Figure 16.4 Conceptual image of direct-irradiation continuous flow microwave reactor.

16.4 Direct-Irradiation Continuous Flow Microwave Reactor

16.4.1 Concept of Reactor

A direct-irradiation continuous flow microwave reactor was developed in 2000s [9]. A conceptual image of the microwave reactor is shown in Figure 16.4. A big difference from the multi-mode reactor described in Section 16.3 is a method of microwave irradiation. As shown in Figure 16.4, woody biomass slurry flows through a metal pipe and it is directly irradiated with microwave at T-junction metal pipe sections. A glass plate is placed to prevent the slurry from leakage. Microwave can penetrate through the glass plate and heat up the slurry.

A great feature of the direct-irradiation continuous flow microwave reactor is flexibility to chemical reaction time and temperature. The T-junction sections as well as microwave generators can be added to or removed from the whole system depending on reaction time and flow rate. Also microwave irradiation power can be changed at each T-junction section. For instance, high-power microwave can be used at the first T-junction section for rapid heating; whereas moderate-power microwave can be independently used at the latter sections for keeping the slurry temperature.

16.4.2 Designing of Microwave Irradiation Section

The microwave irradiation section (T-junction section) of a direct-irradiation continuous flow microwave reactor was designed with 3D electromagnetic simulation. Computer-aided designing software is a powerful tool for designing a microwave reactor in the light of not only cost and time but also visualization of electromagnetic waves. However, it is necessary to know dielectric properties of materials taken into simulation. Note that permittivity of materials depends

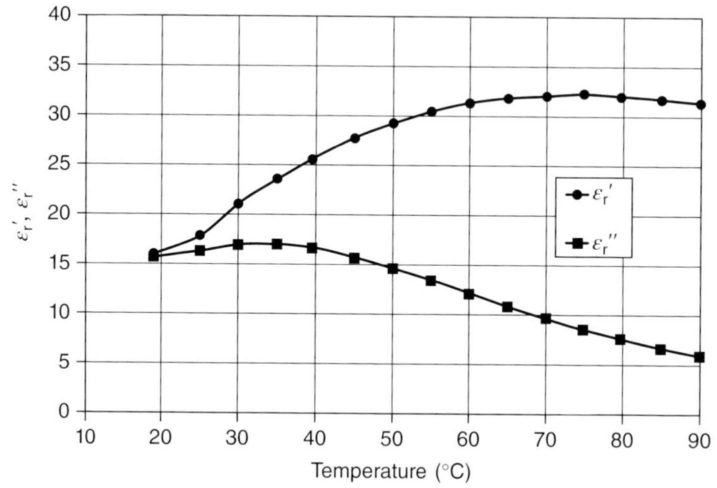

Figure 16.5 Measured permittivity characteristics of ethylene glycol at 2.45 GHz (ε_r': real part of relative permittivity, ε_r'': imaginary part of relative permittivity).

on frequency and temperature. Permittivity data of some materials are obtained from the handbook [10]; otherwise one should measure permittivity for accurate and meaningful simulation.

As an example of organic solvents, Figure 16.5 shows measurement results of permittivity under temperature from 19 (room temperature) to 90 °C at the frequency of 2.45 GHz [6]. The real part ε_r' and imaginary part ε_r'' of relative permittivity was measured with coaxial probe method [11]. As shown in Figure 16.5, the measured permittivity is drastically affected by changes in temperature.

Main evaluation items of electromagnetic simulations are reflection coefficient at the microwave irradiation port, and power density distribution or power absorption distribution in the analytical space. For high-power microwave irradiation, it is also important to confirm if there is an excessive concentration of electromagnetic field in the analytical space. Figure 16.6 shows simulation results of power density distribution in the cross section of the direct-irradiation continuous flow microwave reactor, when 1 kW of 2.45 GHz microwave was input from the microwave irradiation port. The dotted rectangle shows the position of ethylene glycol placed in the simulation model. The measured permittivity data of ethylene glycol at 25 and 80 °C were introduced in Figure 16.6a and b, respectively. As seen in Figure 16.6a, microwave is absorbed around the surface of ethylene glycol at 25 °C. On the other hand, microwave propagates deeply inside of ethylene glycol at 80 °C as shown in Figure 16.6b. This propagation difference is attributed to the permittivity change with temperature. It is therefore important to understand permittivity properties of microwave-heated materials before conducting simulations and experiments.

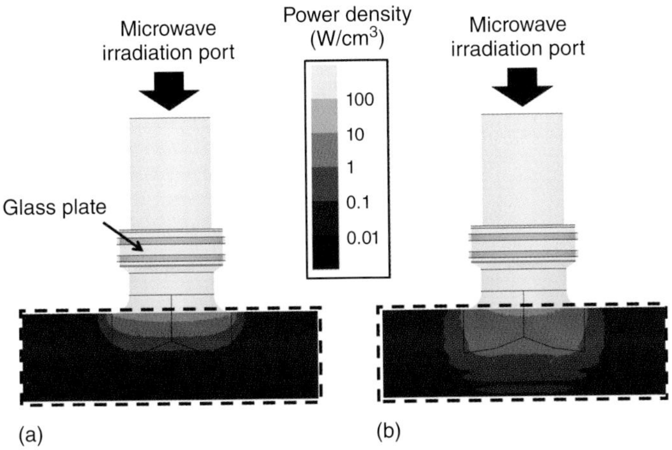

Figure 16.6 Power density distribution in the cross section of the direct-irradiation continuous flow microwave reactor at the ethylene glycol temperatures of (a) 25 °C and (b) 80 °C.

16.4.3 Prototypes of Reactors

Figure 16.7 shows the first prototype of direct-irradiation continuous flow microwave reactor [9]. The reactor was designed for microwave pretreatment of woody biomass slurry with organic solvents. The metal pipe size is the Japanese Industrial Standard (JIS) 80A, and its inner diameter is 78.1 mm. Three 2.45 GHz and 1.2 kW microwave generators are mounted in line on the metal pipe via waveguide. Each microwave generator has an isolator for protecting the generator from reflected microwave, a three-stub tuner for impedance matching, and a

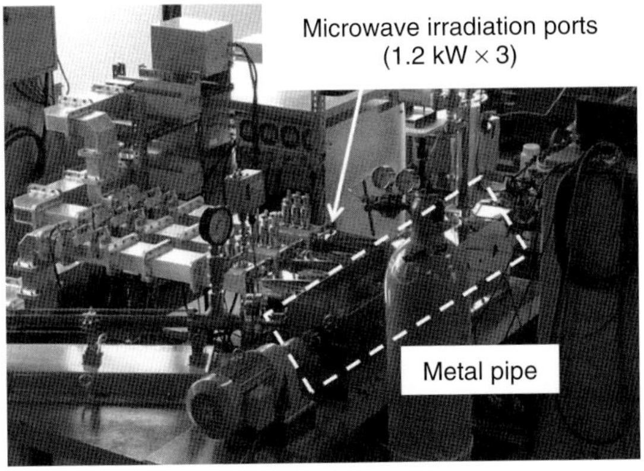

Figure 16.7 A photograph of the first prototype of direct-irradiation continuous flow microwave reactor with three 1.2 kW microwave generators.

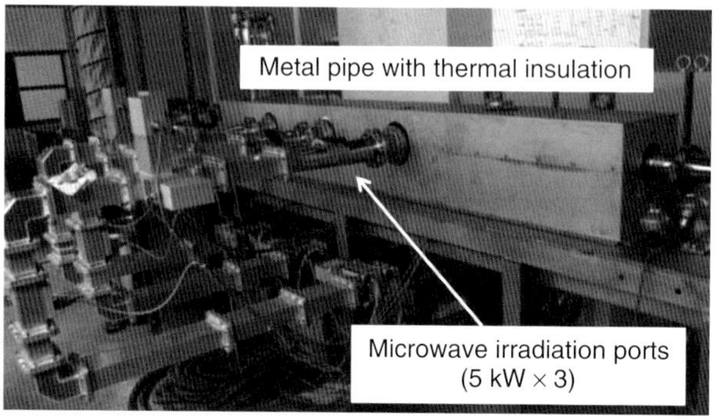

Figure 16.8 Photograph of the second prototypes of direct-irradiation continuous flow microwave reactor with three 5 kW microwave generators.

directional coupler for input and reflected power monitoring. Woody biomass slurry flows through the metal pipe by a pump while being agitated for uniform heating.

Figure 16.8 shows the second prototype of direct-irradiation continuous flow microwave reactor [6]. The reactor was designed for microwave pretreatment of water-based woody biomass slurry, and its size and microwave power were enhanced in comparison with the first prototype. The metal pipe size is JIS 100A, and its inner diameter is 106.3 mm. Three 2.45 GHz and 5 kW microwave generators are mounted in line on the metal pipe, which is covered with thermal insulator for prevention from thermal dissipation.

Experiments of microwave pretreatment and microwave-assisted reaction in aqueous and organic solvents are described in Ref. [6]. Pretreatment in solvents with high dielectric loss factor such as ethylene glycol and glycerol is advantageous for conversion of microwave energy to heat. Microwave-assisted pretreatment in glycerol-containing organic and inorganic acids was effective for enzymatic saccharification of lignocellulosic biomass including recalcitrant softwood. We showed that the microwave-assisted glycerolysis with acids degraded lignin and exposed crystalline and noncrystalline cellulose [12]. In reactions with catalysts having microwave sensitizer effects, the microwave pretreatments shows higher sugar yield than pretreatment by conventional external heating. For instance, significant pretreatment effects were found in microwave reaction with ammonium molybdate and H_2O_2 [13]. Maximum sugar yield of 59.5% was obtained by microwave irradiation of beech wood in aqueous solutions containing ammonium molybdate and H_2O_2 at 140 °C for 30 min, while external heating in an autoclave gave a sugar yield of 41.8%. Selection of solvents and catalyst is important to increase the efficiency of the pretreatment system.

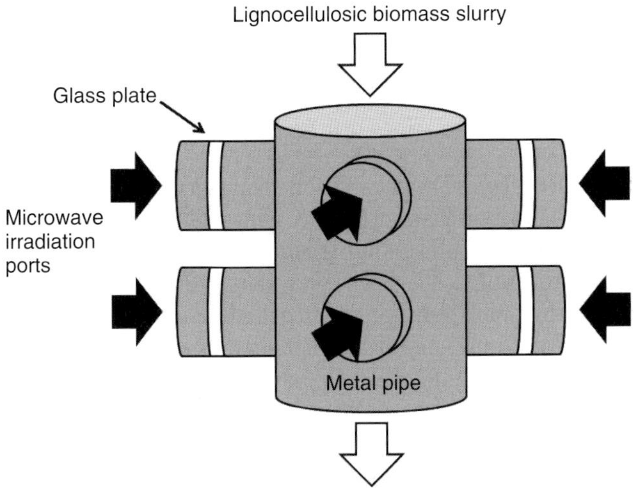

Figure 16.9 Conceptual image of pilot-plant-scale continuous flow microwave reactor.

16.5 Pilot-Plant-Scale Continuous Flow Microwave Reactor

16.5.1 Concept of Reactor

A pilot-plant-scale continuous flow microwave reactor was developed as a scale-up of the direct-irradiation continuous flow microwave reactor [14]. A conceptual image of the reactor is shown in Figure 16.9. As mentioned in Section 16.1, a simple expansion of the metal pipe poses uneven heating arose from penetration depth. The number of microwave irradiation ports is therefore increased in the circumferential direction of the metal pipe, in order to enhance the total microwave power and to avoid temperature imbalance of lignocellulosic biomass slurry in the metal pipe. Then the metal pipe diameter of the pilot-plant-scale reactor can become much larger than that of the direct-irradiation reactor. Although lots of microwave generators need to be mounted on the reactor, microwave power density can be dispersed in the metal pipe. In terms of reaction time and temperature, the pilot-plant-scale reactor inherits the flexibility originated from the direct-irradiation reactor.

16.5.2 Designing of Microwave Irradiation Section

The microwave irradiation section of a pilot-plant-scale continuous flow microwave reactor was designed with 3D electromagnetic simulation, as well as the direct-irradiation reactor. Figure 16.10 shows measured permittivity

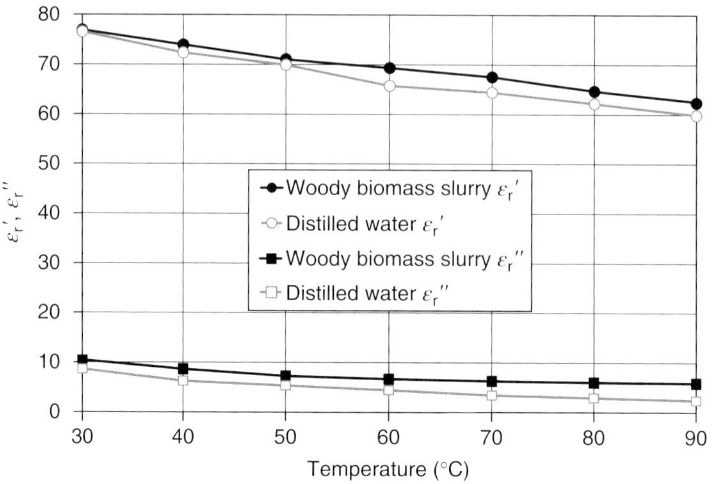

Figure 16.10 Measured permittivity characteristics of woody biomass slurry and distilled water at 2.45 GHz (ε_r': real part of relative permittivity, ε_r'': imaginary part of relative permittivity).

characteristics of the woody biomass slurry, which consisted of 85 w/w% water, 0.03 w/w% organic acid, and 15 w/w% particles of a fast growing wood, *Eucalyptus globulus,* where w/w% indicates the percent by weight. For comparison, measured permittivity characteristics of distilled water are inserted in Figure 16.10. The relative permittivity of the woody biomass slurry decreases as temperature rises. This trend predicts that the penetration depth becomes shorter with a rise in temperature.

A simulation model of the microwave irradiation section is shown in Figure 16.11. Microwave is radiated at 2.45 GHz from four directions in every 90° toward the metal pipe. A glass plate was set between the metal pipe and each microwave irradiation port. The metal pipe is filled with woody biomass slurry whose permittivity data are taken into simulation. Also a metal cylinder, which takes the place of agitator, is coaxially put in the metal pipe. Absorbing boundary condition is set on the top and bottom surfaces of woody biomass slurry.

Figure 16.12 shows simulation results of power absorption distribution in the cross section of the direct-irradiation continuous flow microwave reactor, when 1.5 kW of 2.45 GHz microwave was input from each microwave irradiation port. The measured permittivity data of woody biomass slurry at 40 and 80 °C were introduced in Figure 16.12a and b, respectively. As seen in Figure 16.12, microwave is absorbed inside of the woody biomass slurry more deeply at 40 °C than at 80 °C due to the difference of penetration depth. Reflection coefficient at the microwave irradiation port was calculated to be less than −30 dB [14], which indicates that less than 0.1% of microwave power is only reflected backward to the microwave irradiation port. Isolation between microwave irradiation ports was also calculated to be less than −110 dB [14], which is too small to interfere with each other.

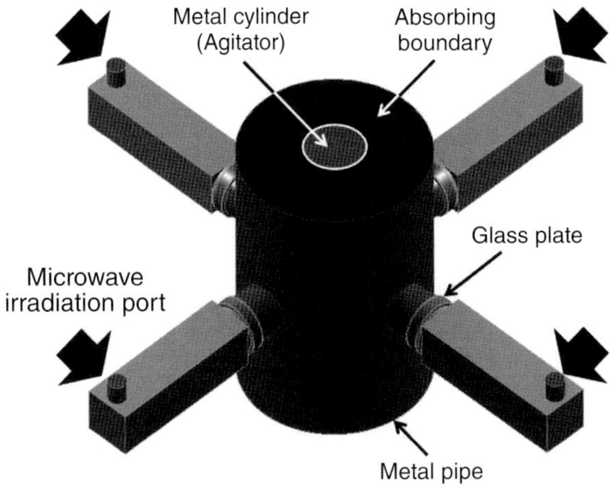

Figure 16.11 Simulation model of the pilot-plant-scale continuous flow microwave reactor. Absorbing boundary was set at the top and bottom planes of the metal pipe.

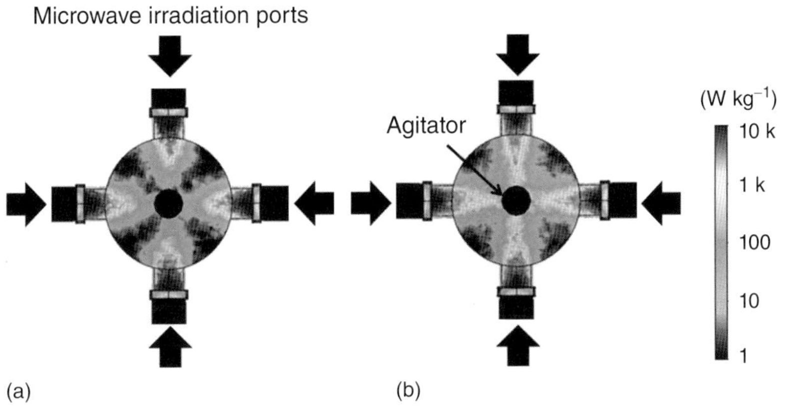

Figure 16.12 Power absorption distribution in the cross section of the designed pilot-plant-scale continuous flow microwave reactor at the slurry temperatures of (a) 40 °C and (b) 80 °C.

16.5.3
Demonstration Experiments of Microwave Pretreatment

The developed pilot-plant-scale continuous flow microwave reactor is shown in Figure 16.13. The metal pipe size is JIS 250A, and its inner diameter is 254.2 mm. Eight 2.45 GHz and 1.5 kW microwave generators are mounted on the metal pipe. Four generators are on the same level and arranged in every 90°. Then another four generators are aligned in a two-tiered manner. The volume of microwave irradiation section is about 50 l. Woody biomass slurry is poured from the top of the

Figure 16.13 Photograph of the pilot-plant-scale continuous flow microwave reactor.

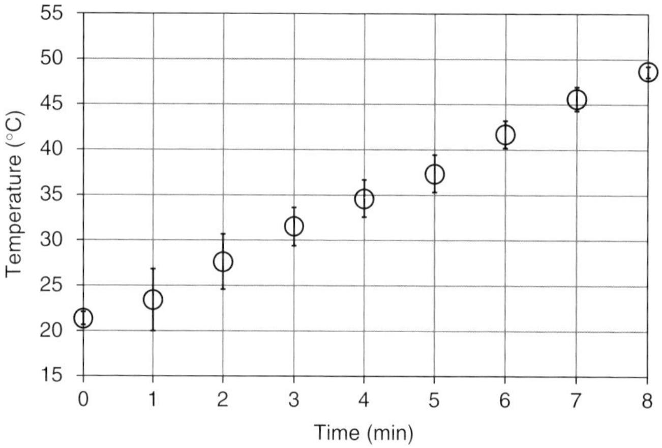

Figure 16.14 Microwave heating characteristics by the pilot-plant-scale continuous flow microwave reactor.

reactor, irradiated with microwave while being agitated, and taken out from the bottom after microwave pretreatment. Pressure resistance of the reactor is up to 2.0 MPa.

Microwave heating test was conducted for investigating the reactor performance. Microwave heating characteristics are shown in Figure 16.14, when the metal pipe was filled with woody biomass slurry. The total microwave power was 12 kW (1.5 kW × 8). The temperature data shown in Figure 16.14 is the average of four measurement points inside of the reactor. From the experimental results, the heating rate of the reactor is about 0.056 K s^{-1}. Then the microwave absorption efficiency of the reactor is estimated to be 79% from thermal calculation [14]. Efficiency degradation is mainly attributed to heat loss by thermal conduction and radiation. As a feature of microwave heating is direct heating to materials, it is inevitable that a wide temperature gap is created between the heated materials and surrounding environment. Thermal insulation is therefore mandatory to improve the microwave absorption efficiency. Heat recovery system is another option for effective energy utilization.

Fast growing wood *E. globulus* was pretreated with aqueous maleic acid in the pilot-plant-scale microwave reactor at 190 °C for 30 min. The pretreated biomass was separated by centrifuge into soluble and insoluble fractions. The soluble fraction was detoxified with lignin-derived adsorbent. The soluble and insoluble fractions were mixed with cellulolytic and hemicellulolytic enzymes and hydrolyzed for 48 h. To the prehydrolyzed mixture, a seed culture of ethanologenic bacterium *Zyomomonas mobilis* and nutrients for the bacterium were added to start the ethanol fermentation. The bacterium used was genetically transformed to utilize xylose [15]. After 48 h, ethanol was produced in a yield exceeding 6% concentration. Thus, the woody biomass was pretreated in the pilot-plant-scale microwave reactor, and bioethanol was successfully produced [16].

16.6
Summary and Conclusions

Three types of continuous flow microwave reactors were explained. For further scale-up, microwave frequency is an important element. Although most of microwave heating apparatuses are designed at 2.45 GHz, electromagnetic wave irradiation at lower frequency, for example, at 915 MHz, has a great possibility to increase in capacity of microwave reactor as the penetration depth becomes deeper. Another important aspect is best use of rapid heating. Lots of studies claim that microwave-assisted chemical reaction shows a surprising reduction in reaction time. Nevertheless, no "speed-up" technologies on microwave-assisted chemical reaction have been seen yet, except for a continuous flow system. Considering the limitation of penetration depth, "speed-up" (not "scale-up") will be a direction for progress toward commercialization of microwave chemistry.

References

1. Glasnov, T.N. and Kappe, C.O. (2007) *Macromol. Rapid Commun.*, **28**, 395–410.
2. Moseley, J.D., Lenden, P., Lockwood, M., Ruda, K., Sherlock, J., Thomson, A.D., and Gilday, J.P. (2008) *Org. Process Res. Dev.*, **12**, 30–40.
3. Bergamelli, F., Iannelli, M., Marafie, J.A., and Moseley, J.D. (2010) *Org. Process Res. Dev.*, **14**, 926–930.
4. Watanabe, T. (2013) in *Lignocellulose Conversion: Enzymatic and Microbial Tools for Bioethanol Production* (ed V. Faraco), Springer, New York, pp. 1–20.

5. Hendriks, A.T.W.M. and Zeeman, G. (2009) *Bioresour. Technol.*, **100**, 10–18.
6. Watanabe, T. and Mitani, T. (2013) in *The Role of Green Chemistry in Biomass Processing and Conversion*, Chapter 9 (eds H. Xie and N. Gathergood), John Wiley & Sons, Inc., pp. 281–291.
7. Magara, K., Ueki, S., Azuma, J., and Koshijima, T. (1988) *Mokuzai Gakkaishi*, **34**, 462–468.
8. Magara, K., Azuma, J., and Koshijima, T. (1989) *Wood Res.*, **76**, 1–9.
9. Mitani, T., Oyadomari, M., Suzuki, H., Yano, K., Shinohara, N., Tsumiya, T., Sego, H., and Watanabe, T. (2011) *J. Jpn. Inst. Energy*, **90**, 881–885.
10. Komarov, V.V. (2012) *Handbook of Dielectric and Thermal Properties of Materials at Microwave Frequencies*, Artech House.
11. Keysight Technologies (2014) Basics of Measuring the Dielectric Properties of Materials, Application Note, pp. 17–20.
12. Liu, J., Takada, R., Karita, S., Watanabe, T., Honda, Y., and Watanabe, T. (2010) *Bioresour. Technol.*, **101**, 9355–9360.
13. Verma, P., Watanabe, T., Honda, Y., and Watanabe, T. (2011) *Bioresour. Technol.*, **102**, 3941–3945.
14. Hasegawa, N., Mitani, T., Shinohara, N., Daidai, M., Katsura, Y., Sego, H., and Watanabe, T. (2014) *IEICE Trans. Electron.*, **E97-C**, 986–993.
15. Yanase, H., Miyawaki, H., Sakurai, M., Kawakami, A., Matsumoto, M., Haga, K., Kojima, M., and Okamoto, K. (2012) *Appl. Microbiol. Biotechnol.*, **94**, 1667–1678.
16. Watanabe, T., Mitani, T., and Yanase, H. (2013) Microwave processing for the production of bioethanol and aromatic chemicals from woody biomass. 8th PRWAC and IAWS 2013, Nanjing, China, October 17–21, 2013, Abstract, p. 113.

17
Biomass and Waste Valorization under Microwave Activation

Leonid M. Kustov

17.1
Introduction

Diverse sources of radiation (γ-irradiation, electron and neutron beams, radio frequency, microwave, or ultrasound, irradiation by UV, and visible light) are used for the initiation and stimulation of chemical processes, including catalytic reactions. Of particular interest are the effects caused by the microwaves, that is, electromagnetic waves with the frequency range from 200 MHz to 300 GHz. The 915 MHz frequency (33.3 cm) is most frequently used in industrial devices, whereas the frequency 2.45 GHz (12.2 cm) is used in microwave laboratory setups.

The important advantage of microwave activation is the transfer of energy through radiation rather than heat transfer or convection. This ensures fast penetration of energy into the volume of materials transparent to microwave fields, that is, instantaneous heating.

The most important effects related to the appearance of temperature gradients and nonequilibrium conditions are observed when a reaction medium or material (for instance, a catalyst) consists of several phases with different microwave radiation absorption coefficients.

The microwave treatment allows one to decrease the reaction duration, increase yields and selectivity, and decrease energy and reagent consumption. The use of microwave energy pulses provides additional reduction in the energy consumption.

Microwave technologies can be used for both catalyst preparation, including preliminary (*ex situ*) activation, and direct (*in situ*) activation of catalytic processes. In this chapter, the use of microwave activation directly in catalytic processes of biomass and conversion of renewables is discussed.

One of the possible mechanisms of the microwave action, which is widely realized in the case of biomass conversion due to the presence of water and other polar oxygenated compounds in the feed, is dielectric polarization. The second mechanism involves currents of free charges excited in solids, contributing to heating because of Ohmic losses; this mechanism is characteristic of catalysts themselves that possess a substantial conductivity (metals and semiconductors).

Microwaves in Catalysis: Methodology and Applications, First Edition.
Edited by Satoshi Horikoshi and Nick Serpone.

The third mechanism that should also be taken into account is caused by the loss of vortex currents excited by magnetic fields.

Certainly, microwave fields cannot initiate chemical reactions by exciting and, especially, breaking chemical bonds, as in photochemical processes under the action of electromagnetic radiation. The energy of microwave photons is ~1 J mol^{-1}, which is far below the activation energies of chemical processes. So in the case of biomass conversion all three mechanisms can be realized if the system contains polar molecules and solid catalysts containing supported metals or transition metal oxides. Attention should be given to materials and oxides used as components of carriers and catalysts. Carbon and transition metal oxides reach high temperatures, whereas titanium and zirconium oxides are heated to moderate temperatures, and aluminum, magnesium, and silicon oxides cannot be heated above 100–150 °C. A disperse system consisting of conducting particles is a nonuniform material whose dielectric properties exhibit pronounced frequency and phase composition dependencies. The dielectric loss in such materials is related to the aggregation of charges at interphase boundaries and is known as the *Maxwell–Wagner effect*.

Numerous data show that, under microwave activation conditions, higher reaction rates are obtained than in thermally activated processes, and the yields of final products increase. These advantages are often ascribed to the so-called microwave effects, without a detailed analysis of the reaction conditions. It is important to separate thermal and nonthermal (electromagnetic) microwave effects. As such, let us now examine some of the most impressive examples of the use of microwave activation in the conversion of renewable resources and biomass.

17.2 Vegetable Oil and Glycerol Conversion

The influence of microwave radiation on the formation of acrylonitrile in the ammoxidation of glycerol (a side product of the production of biodiesel fuel from vegetable oils) under the action of hydrogen peroxide and ammonia on V–Sb–Nb oxide catalysts (Eq. (17.1)) has been investigated in Refs. [1, 2] (Eq. (17.1)). The selectivity to acrylonitrile under microwave heating reached 83.8% at a 46.8% conversion. For comparison, the reaction under thermal activation gives a selectivity of only ~60% at a conversion not exceeding 15%:

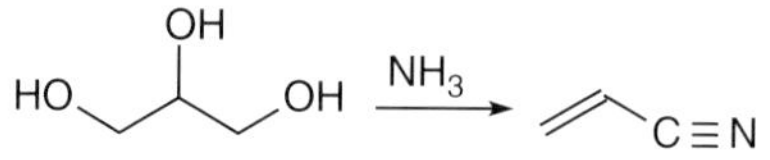

(17.1)

Drago *et al.* [3] reported on the significant enhancement of the etherification reaction of triglycerides under microwave activation. The derivatives of this reaction served as final potential glycerol-free biofuel. The process is catalyzed by silica functionalized with 10 wt% of sulfonic groups. The highest activity was observed when the mixture of vegetable oil (soybean oil) and *tert*-butyl-methyl ether in a 1:10 ratio and 1% of the catalyst was MW-treated at 20 W. The developed process provides advantages since the step of removing free glycerol is eliminated.

Microwave-assisted pyrolysis of rapeseed oil was proposed for biofuel production [4]. The pyrolysis of rapeseed oil and waste oil was performed using a batch reactor either thermally at 500–600 °C or in a MW-induced process at a 300–500 W microwave power. The effect of HSZM-5 zeolite was also studied, though the dielectric loss tangents of both oil and catalyst are very low: range, 0.001–0.12. Rather high conversions (70–97%) of rapeseed oil and waste oil into gas and liquid products were observed under thermal heating. Light bio-oil predominates in the products at 550 °C. The catalyst helped to increase the yield of light bio-oil. Organic products representing hydrocarbons in the diesel range were also found (alkanes and alkenes C_9–C_{26}) as well as some fatty acids. The microwave pyrolysis produced aromatics, C_7–C_{29} alkanes, cycloalkanes, alkenes, cycloalkenes, dienes, and acids. The formation of aromatic compounds was a distinctive feature of the microwave pyrolysis.

Epoxidation of vegetable oil under MW irradiation is another example of conversion of renewables [5]. The reaction was carried out in an isothermal batch reactor by peroxyacetic and peroxypropionic acids formed *in situ* from carboxylic acid and hydrogen peroxide. Microwave irradiation was shown to accelerate the epoxidation reaction rate in the case when the continuous phase was aqueous.

The epoxidized vegetable oil can be further transformed into carbonated oil under MW activation with tetrabutyl ammonium bromide as a catalyst [6]. This reaction can be considered as a means of utilization of CO_2, which is incorporated into vegetable oils. The reaction time with MW activation was 25–33% of that required for the thermal process. The introduction of water accelerates the reaction, while the selectivity to cyclocarbonate was enhanced under MW conditions. Conversions as high as 90% were found with the reaction times being 70 and 40 h for the thermal and microwave heating, respectively.

17.3 Conversion of Carbohydrates

5-Hydroxymethylfurfural produced from biomass is currently considered as one of the so-called platform chemicals. The MW-stimulated fast conversion of carbohydrates into this product catalyzed by self-assembled mesoporous TiO_2 nanoparticles was reported by Dutta *et al.* [7]. Monosaccharides (D-fructose, D-glucose) and disaccharides (sucrose, maltose, cellobiose) were transformed into 5-hydroxymethylfurfural in aqueous or organic media. The authors studied the influence of the solvent nature, microwave absorbing coefficient, catalyst mass, reaction time, and substrate nature on the yields of the target product. The high surface area, the presence of Lewis acid sites, and uniform morphology of titania nanoparticles were assumed to be beneficial for the high activity in the dehydration of carbohydrates. It is noteworthy that the commercial sample of TiO_2 is poorly active in this reaction. The higher value of $\tan\delta$ for the solvent dimethylsulfoxide (DMSO) provided higher yield of the product in organic media than in water. The catalyst could be recycled four times without the loss of activity.

Hydrolysis of xylan, epimerization of D-xylose and D-lyxose, and furfural production were found to proceed more efficiently under MW irradiation [8]. Hydrolysis of xylan with sodium molybdate as a catalyst and subsequent epimerization resulted in a mixture of D-xylose and D-lyxose (1.6 : 1) that can be reduced to D-xylitol and D-lyxitol (yields: 88% and 85%) or can be dehydrated to furfural (53%). Under MW irradiation, the reaction time decreased 400 times compared to the thermal mode.

The MW-stimulated hydrolysis of polysaccharides was also performed onto polyoxometalate acid catalysts [9]. Polyoxometalates are capable of absorbing microwaves, and this property was used in the hydrolysis of corn starch and crystalline cellulose. Phosphotungstic and silicotungstic acids demonstrated high activity, whereas phosphomolybdic acid was much inferior. The active catalyst can be recycled four times without loss of activity by ether extraction. The catalyst reduced the energy consumption in the hydrolysis process by 17–23%.

Fructose was shown to be transformed into 5-hydroxymethylfurfural in ionic liquids under MW activation in the absence of catalysts [10]. Yields of up to 97% could be obtained after 3 min in the reaction, which proceeds via the cyclic fructofuranosyl intermediate.

17.4
Cellulose Conversion

Value-added chemicals can be produced selectively from cellulose under MW irradiation [11]. Microwave-assisted depolymerization yields liquid products from cellulose reaching 87% using acid catalysts. The selectivities to glucose or levulinic acid as the main products may be as high as 50% and 69%, respectively. Spherical carbonized residues are formed as a side product. MW-stimulated depolymerization of cellulose yielded 55% glucose in just 20 min in the presence of F^-/Cl^- as reported by Shaveta *et al.* [12].

Pyrolysis of cellulose at temperatures as low as 200–280 °C under MW conditions was also studied by Al Shra'ah *et al.* [13]. The effects of temperature, cellulose structure (crystalline or amorphous), and microwave absorbing agents were determined. The maximum yield of bio-oil (45%) was found in the case of amorphous cellulose at 260 °C. The presence of water enhanced the yield to 52%. On the contrary, the introduction of activated carbon as a MW-absorbing material resulted in an increase in the yield of gas products. Levoglucosan was produced with a high yield by pyrolysis of amorphous cellulose at 260 °C (Eq. (17.2)). In comparison, the thermal pyrolysis proceeded at a significantly higher temperature (400 °C):

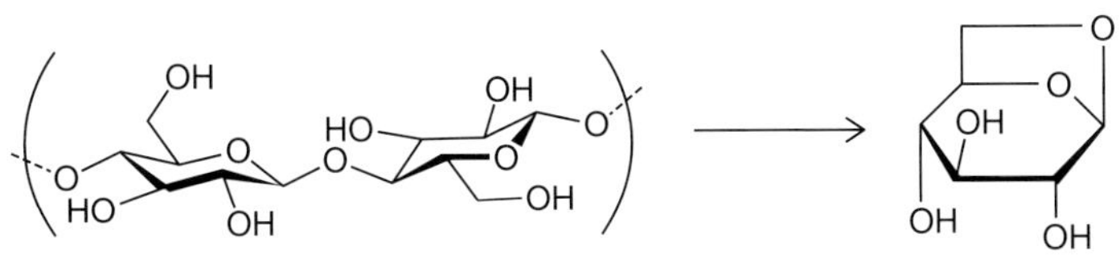

(17.2)

Microcrystalline cellulose can be converted into glucose under MW irradiation with NaOH completely (Eq. (17.3)) [14]:

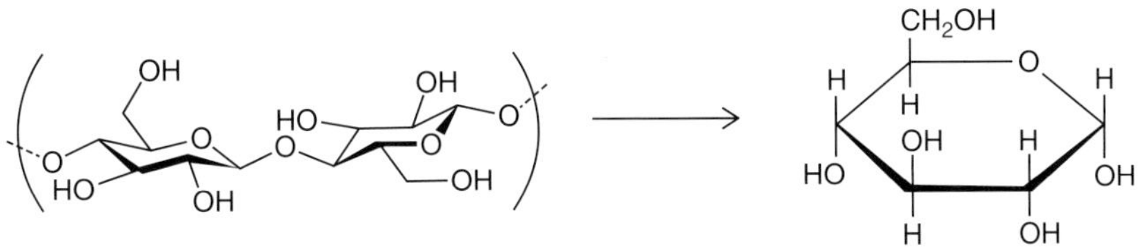

(17.3)

Cellulose can be transformed selectively into levulinic acid if the MW reaction is carried out in SO_3H-functionalized ionic liquids (Eq. (17.4)) [15]. The highest yield was 55%. The anion of the ionic liquid plays a crucial role and the activity decreases in the order: $HSO_4^- > CH_3SO_3^- > H_2PO_4^-$, in agreement with the acidity strength of the acid. Alkylated cellulose can be synthesized in an aqueous medium using an easy protocol [16]. The reaction takes less than 30 min with the conversion reaching 87%:

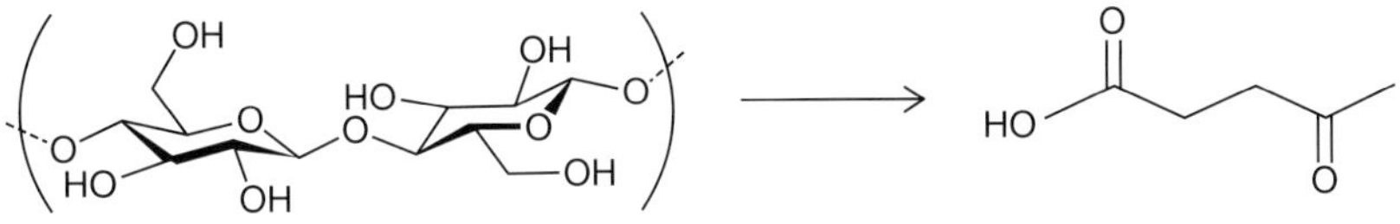

(17.4)

Microwave-assisted conversion of glucose and cellulose directly into 5-hydroxymethyl-furfural proceeds in ionic liquids with $CrCl_3$ or $ZrCl_4$ as catalysts (Eq. (17.5)) [17, 18]. The yields are as high as 60% and 90% in the case of cellulose and glucose, respectively:

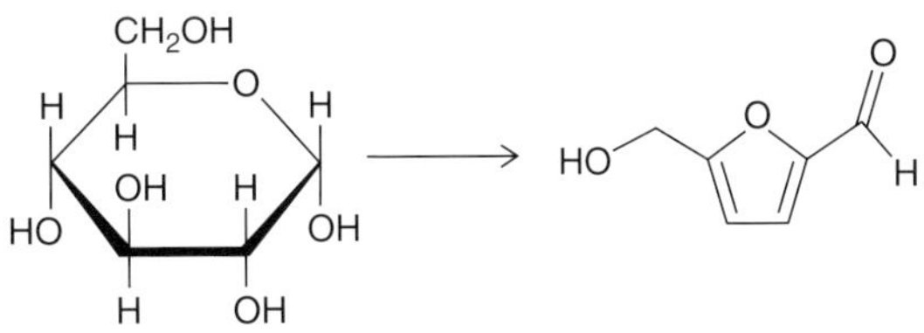

(17.5)

Ionic liquids may be used not only as a polar and MW-absorbing media but also as a catalyst (reduced amounts) in the conversion of microcrystalline cellulose into 5-hydroxymethylfurfural [19]. The 1,1,3,3-tetramethylguanidine (TMG)-based ionic liquids, including tetrafluoroborate ([TMG][BF_4]) and lactate ([TMG]L), were used with [TMG]BF_4 being the most active (yield of product: 28.6%). Cellulose esterification by long-chain chlorides with *N,N*-dimethylaminopyridine as a catalyst under MW conditions was studied in Ref. [20]. One minute of MW irradiation was sufficient to achieve a significant degree of esterification.

Quaternized ethylenediamine-functionalized cellulose prepared under MW irradiation was used for the solvent-free synthesis of cyclic carbonates from CO_2 and epoxides [21]. A selectivity of 99% was found at a CO_2 pressure of 1.2 MPa. BET surface areas up to 71 $m^2\,g^{-1}$ were determined for the quaternized cellulose.

Several publications have been devoted to MW pretreatment of cellulose preceding the fermentation [22, 23] in order to facilitate the following enzymatic conversion owing to the decreased degree of polymerization. The rate of enzymatic hydrolysis of cotton cellulose was enhanced 50-fold after such a treatment in ionic liquids [23].

An MW reactor was designed for the hydrolysis of cellulosic biowaste [24]. The kinetic analysis showed that the advantages of microwave heating were related to the achievement of a higher rate constant at moderate temperatures and prevention of "hot spot" formation within the reactor.

Heterogeneous catalysts such as the H-forms of zeolites also turned out quite efficient in the MW-stimulated hydrolysis of cellulose in ionic liquids [25]. The zeolites demonstrated a better performance than the sulfated ion-exchange resin (Eq. (17.6)).

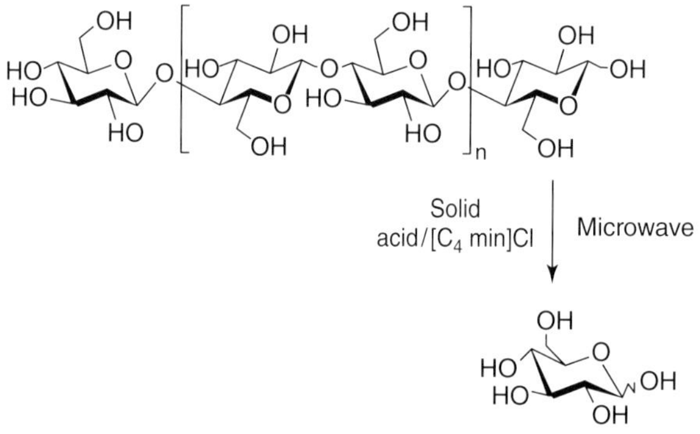

(17.6)

Mixed lignocellulosic material was also successfully converted in cholinium ionic liquid for further efficient enzymatic transformation [26]. About 60–90% of cellulose was hydrolyzed to glucose after the MW pretreatment in the ionic liquid at 110 °C for 20 min.

17.5 Lignin Processing

Since lignin is a branched and cross-linked polymer formed by phenolic and aliphatic fragments, it is more difficult to depolymerize. Nevertheless, attempts to use MW radiation turned out to be successful for lignin pyrolysis into phenolic moieties and fuel components. Bio-oil produced by microwave pyrolysis of kraft lignin [27] is affected by the contents of microwave-absorbing materials (20–40 wt%) and MW input power (1.5–2.7 kW). The temperatures measured under MW conditions for 800 s were 900–1240 K. The yields of the aqueous phase, oil phase, gas products, and char were 17–21%, 15–20%, 21–27%, and 32–40%, respectively.

Further attempts to stimulate the lignin conversion using activated carbon as a MW-absorbing solid were made in Refs. [28, 29]. Phenols, guaiacols, hydrocarbons, and esters comprising 71–87% of bio-oil were produced. The char produced from lignin has the higher calorific value (20.4–24.5 MJ kg^{-1}) in comparison with raw lignin (19 MJ kg^{-1}). A high concentration of long-chain fatty acid esters (42.2% in bio-oil) was obtained when Zn powder was used as the catalyst.

17.6 Waste and Renewable Raw Material Processing

Several examples of the use of microwave heating in biocatalysis, in the first place, in the hydrolysis of cellulosic wastes for plant raw material processing were described in Ref. [30]. Among diverse reactions of side product and waste processing, the pyrolysis of glycerol into synthesis gas [31] and the oxidation of compounds modeling waste lignin on the Co(salen)/SBA-15 mesoporous catalyst [32] are worth noting. In the latter case, complete conversion of apocynol (model compound) under the action of hydrogen peroxide was obtained in 40 min, whereas the usual heating regime gave conversions no higher than 60% and only in ~24 h.

Reviews of the methods for the utilization of wastes with the use of microwave technologies are given in Refs. [33, 34]. Processes considered included reprocessing of old tires and plastic wastes and restoration of (removal of harmful contaminants from) soils and underground water. It is, however, noted that there are upscaling problems when the microwave technologies are transferred from laboratory studies to an industrial scale. The use of microwave methods for the production of biogas with a low content of CO_2 and CH_4 by processing dead-water precipitates was described [35]. Compared with thermal pyrolysis, the microwave method allows one to obtain a higher yield (~94%) of syngas ($CO + H_2$), with the CO_2 and CH_4 concentrations about 50% and 70%, respectively, much better than in the thermal process.

A review [36] considered the use of microwave technology in processing tar sands and oil products, in particular, for the extraction of bitumens, conversion of heavy residues, the removal of sulfur and nitrogen compounds, and heating tar sands for decreasing the viscosity of bitumens. One more example of the useful conversion of wastes is the biotechnological production of hydrogen under microwave processing conditions [37].

Recently, microwave technologies have been applied for the liquefaction of coal [38]. Microwave-assisted pyrolysis of coal and biomass in the presence of microwave-absorbing materials provides significant advantages over thermal processing [39].

Cotton fabric can be effectively converted into cellulose esters under MW heating in an *N,N*-dimethylacetamide/lithium chloride mixture [40]. The 4-(*N,N*-dimethylamino)pyridine served the catalyst.

Gasification of palm shell char using CO_2 was found to be enhanced by MW activation [41]. The Boudouard reaction ($C + CO_2 \leftrightarrows 2CO$) was the main process. Addition of an Fe catalyst into the char allowed the authors to obtain promising results: a CO_2 conversion of 99% could be reached within 60 min of the reaction before the deactivation.

There are several publications devoted to the production of biofuels (bio-oil) from biomass under MW conditions, including the review in Ref. [42].

Syngas ($CO + H_2$) is a valuable product that can be produced from biomass (sewage sludges, coffee hulls, and glycerol) and further used for the production of methanol or hydrocarbons [43]. The highest yield of syngas and the lowest H_2/CO ratio were found in the case of glycerol. The lowest syngas yield with the highest H_2/CO ratio was observed for sewage sludges. Microwave heating was characterized by enhanced syngas yields compared to thermal pyrolysis. The microwave process was shown to be less time and energy consuming in comparison to the thermal process.

Xie *et al.* reported on the accelerated syngas production from biomass under MW conditions with Fe, Co, and Ni on Al_2O_3 as catalysts, although alumina poorly absorbed the MW radiation [44]. Another task was tar removal. Ni/Al_2O_3 was quite stable and demonstrated superior activity in syngas production and tar removal. The ($CO + H_2$) yield was as high as 80%. A fairly high ratio of catalyst to biomass is required to reach the high activity: 1 : 5–1 : 3. The introduction of water vapor enhanced the syngas yield owing to the occurrence of steam reforming. A fluidized bed gasifier reactor was proposed.

Diversified wastes of packing materials deserve special attention. Fuel can be produced by MW pyrolysis of a beverage packing waste [45]. The raw material represented a composite system made of layers of paper, polyethylene, and aluminum. Although the applied MW power was too high (3 kW), the use of different MW-absorbing solids (e.g., chopped tire, carbon, and iron powder) turned out to be efficient. Paper was converted into bio-oil composed of alcohols, aldehydes, acids, and anhydrosugars. The polymer layer was transformed into a viscous liquid, with linear alkanes, alkenes, cyclic, and aromatic hydrocarbons being present in the mixture. Aluminum can be recovered.

An agricultural waste (e.g., corn stover) has been used as a feed in the catalytic MW-assisted pyrolysis process [46]. A microwave power of 500 W was applied and the treatment time was 30 min. The use of CO_2 instead of N_2 as a carrier gas showed a negative effect because of the MW-energy consumption for heating CO_2 molecules. The presence of metal oxides as the catalysts enhanced the efficiency but reduced the calorific value of the solid product. The catalyst also reduced the formation of polyaromatic hydrocarbons, which are known to be eco-toxicants. The yield of liquid products was higher, and the yield of gas products was lower in the case where the catalysts were used owing to the occurrence of the Fischer–Tropsch reaction.

If the MW treatment is performed at a reduced pressure, microwave plasma can be generated. There are a few interesting studies related to the conversion of biomass in the MW plasma. Hydrogen can be produced at high yield from rice straw by MW-plasma pyrolysis [47]. Though a rather high applied power was used (800–1000 W), the concentrations of hydrogen in the gas products reached 48–56% such that about 40 mg of hydrogen per gram of rice straw was obtained at the highest conversion of 67.5%.

Oil (triolein, soybean oil, microalgae biomass of *Nannochloropsis*, and castor and jatropha seeds), cooked oil, or microalgae can be converted into biodiesel using MW treatment [48] with SrO as the catalyst.

Microwave pyrolysis of wheat straw was investigated at 500 °C in Refs. [49, 50]. The yield of solid products diminished when the microwave power increased, unlike the yield of gas products that included CO and H_2. The minimal yield of CO_2 was found with CaO as an additive. Additives of CuO and Fe_3O_4 catalyze the formation of liquid products. Pyrolysis occurs most intensively if carbon is added to the mixture, with the CO content being very high, and the maximum content of H_2 exceeding 35%.

Sugarcane bagasse is an abundant by-product of sugar production and can serve as a benign raw material for MW pyrolysis [51]. Metal oxides catalyzed the conversion and decreased the maximum temperature. NiO or CaO increased the formation of H_2, whereas CuO or MgO acted in the opposite manner. Introduction of CaO or MgO resulted in the increase in the gas yield, while NiO or CuO stimulated the formation of liquid products. Surprisingly, catalyst addition increased the activation energy, but increased the preexponential factor of the reaction.

It should be noted that most of the studies so far available in the literature are related to the use of solids that are usually not considered as good catalysts for the reactions occurring in the system (reforming, cracking, etc.). There are only a few reports on the effect of the nature of the catalyst, though this issue may be equally important in the optimization of the microwave regime. One of such paper [52] describes the effect of such catalysts as CaO, $CaCO_3$, NiO, Ni_2O_3, γ-Al_2O_3, and TiO_2 on the efficiency of sewage sludge pyrolysis. Wet sewage sludge was applied, so that the dipolar polarization mechanism predominated in the MW process. The products included char, water and water-soluble compounds, oil, and noncondensable gas. The catalysts changed dramatically the product distribution. In most cases, the presence of a catalyst resulted in an increase in the temperature rise of sewage sludge, and the catalysts tested can be arranged in the following order for the temperature rise: $Ni_2O_3 \approx \gamma\text{-}Al_2O_3 > TiO_2 > NiO > CaCO_3$. Ni-containing catalysts demonstrated the highest activities toward the decomposition of organic compounds in the sludge and enhanced the yields of bio-oil and gas products. The use of CaO was beneficial for the formation of H_2-enriched syngas, whereas

CO-enriched syngas was formed on Ni-catalysts. The metal oxides γ-Al_2O_3 and TiO_2 catalyzed the decomposition of organic compounds in the sludge; however, the composition of gas products was unaffected.

The techniques of MW-initiated pyrolysis have been reviewed by Motasemi *et al.* [53]. The upscaling of the pyrolysis technique is one of the major barriers for the commercialization of the MW pyrolysis. The microwave-assisted technology is an effective method to reduce the pyrolysis reaction time and increase the yields of value-added products from diverse feedstocks. A very important advantage of the MW technology is that it requires no pretreatment or shredding of the raw materials.

The use of supercritical media is considered these days as a benign method of performing chemical processes, because it allows increasing dramatically the productivity (space-time yield) of the processes. The combination of the supercritical and microwave technologies would probably result in a synergy and additional benefits, though the use of the microwave technology is limited in terms of the applied pressure. Such a combined MW and SC approach was discussed for the transesterification of algal biomass in supercritical ethanol [54]. The lipid-rich, dry algae (*Nannochloropsis salina*) was converted successfully into fatty acid ethyl esters. A short reaction time, reduction of energy consumption, and high conversion are the benefits of the integrated approach; SiC was used as an inert additive.

A similar process was performed in supercritical methanol, where the MW approach was compared to the supercritical approach [55]. The microwave approach improves the efficiency of extraction of algae, reduced extractive-transesterification time, and increased the yield. The supercritical method gave purified extracts and reduced energy consumption.

Yet another study disclosed the bio-syngas generation from municipal solid waste treatment [56]. The yield of syngas in the MW process was rather high. The syngas fraction with a high H_2 content (50–55 vol%) was obtained, and a positive effect of the moisture on syngas production was found that explained the gasification of char.

Hydrogen can also be produced from banyan leaves in the MW plasma under atmospheric pressure [57]. The applied microwave power was proportional to the rate of production of hydrogen, hydrogen productivity per gram of banyan leaves, and amount of hydrogen formed with respect to the H-atom content of converted banyan leaves. Increasing the microwave power led to an increase in H_2 and to a decrease in CO_2 concentrations in the products as a result of the water–gas shift reaction: $CO_2 + H_2 \rightarrow CO + H_2O$.

Automotive engine oil is another waste that needs to be utilized. The efficient MW-assisted method of conversion of such a raw material into H_2 was proposed in Ref. [58]. Pyrolysis was performed in a continuous stirred bed reactor. The gas products (41 wt% yield) contained substantial amounts of light aliphatic hydrocarbons (up to 86 vol%) that could find application as a fuel. H_2 (up to 19 vol%) and CO were also formed (the $H_2 + CO$ content about 35 vol%). The presence of carbon improved the process efficiency.

As an example of another approach to the utilization of wastes, eggshell wastes can be used as a catalyst for the MW-stimulated biodiesel process [59]. The catalytic activity was tested in biodiesel synthesis by the transesterification of palm oil with methanol under MW conditions. The catalyst present was mainly CaO (99.2 wt%) with a high concentration of strong basic sites. Biodiesel production was enhanced under microwave versus thermal heating. The yield of fatty acid esters achieved 96.7% with the reaction time of 4 min at an input MW power of 900 W; alcohol-to-oil ratio was 18 : 1; and catalyst loading was 15%.

17.7 Carbon Gasification

Carbon gasification or liquefaction is another challenge in view of the shortage of fossil fuels. One approach was proposed in Ref. [60]. The process is based on reaction 17.7; MW irradiation enhanced the efficiency of the conversion. Tests with different carbon materials demonstrated that charcoal conversion is characterized by the least consumption of energy. Comparison of the single-mode and multi-mode regimes showed that the latter is more efficient from the point of view of energy consumption and recovery. The energy efficiency could reach 45%:

$$C + CO_2 \rightarrow 2CO \qquad (17.7)$$

Coal and charcoal gasification into hydrogen-enriched syngas with varying the O_2/fuel ratio from 0 to 0.544 was demonstrated under MW-plasma activation [61]. Plasma-producing gases at a 5 kW input power were steam and air. With increasing O_2/fuel ratio, the H_2 and CH_4 concentrations in the syngas diminished, whereas those of CO and CO_2 increased. With steam plasma and the O_2/fuel ratio equal to zero, the syngas with a high content of hydrogen (>60%) was produced with an H_2/CO ratio over 3. The conversion of the carbon material increased with increasing O_2/fuel ratio, and the cold gas efficiency reached the maximum at the O_2/fuel ratio of 0.272. Charcoal demonstrated a lower carbon conversion and cold gas efficiency than the coal samples.

One more example does not belong to any of the categories of biomass valorization, but the tire wastes represent a growing problem that need to be solved [62]. Microwave pyrolysis of this kind of waste allows the conversion into three types of products: a solid, a liquid, and a gas. The operating parameters were as follows: the microwave frequency was 2.45 GHz and the energy output was up to 6 kW. Further fractionation makes it possible to obtain liquid products with a low density (~0.9 g cm^{-3}) and viscosity (1.25–3.9 cP). Rather high yields of aromatic and olefinic compounds can be obtained. It is noteworthy that tires containing substantial proportion of aromatics were converted faster than the samples composed of natural rubber.

Although the extraction of lipids does not fall in any of the topics discussed earlier, it is relevant to also consider the data on MW-assisted lipid extraction. MW-enhanced lipid extraction from microalgae with the goal of biodiesel synthesis was reported in Ref. [63]. The research was conducted on extraction of algae oil from microalgae using microwave fragmentation technology. The benefit of this technology is the reduction of the production costs of microalgae-based biodiesel. Microwaves assisted the lipid extraction of microalgae by appropriate solvents. Extracted microalgae is characterized by the largest extracted quantity of microalgae lipid (30 wt%), and the transesterification conversion was 76.2% at 68 °C with Li_4SiO_4 (3 wt%) as an additive; the oil/methanol ratio was 1 : 18 for 4 h.

17.8
Prospects for the Use of Microwave Irradiation in the Conversion of Biomass and Renewables

To summarize, there are good reasons in certain instances for believing that catalytic conversion processes of biomass and renewables occur under the action of microwave fields differently than under traditional thermal activation with convective or conductive heating. The necessary conditions for the efficient action of a microwave field on a process is the strong interaction of a solid (a catalyst or its separate components, carbon species present in the biomass) or dipolar liquid components with this field.

Among various mechanisms of the action of microwave field on solids, several classes of catalysts and processes that offer promise for use in practice can be identified. For instance, the Maxwell–Wagner interphase polarization mechanism can likely operate in mixed oxide catalysts (such as catalysts of oxidation based on metal oxides). For such catalysts, "nontrivial" and "nonthermal" effects can most probably be expected; these effects are of obvious interest for the fundamental science of catalysis.

Among heterogeneous catalysts, there are also many systems representing massive conductors (metals and activated carbons) and systems containing a deposited active metallic phase. Strictly speaking, the Maxwell–Wagner mechanism may also be valid for metallic catalysts deposited on a carrier inert with respect to microwave fields. Heat exchange between a deposited conducting phase and a carrier transparent to microwave radiation is of great importance for heterostructures of the type "metal + carrier" transparent to the electromagnetic field. If the characteristic time of thermal relaxation is larger than the reaction duration, such a system is of great interest for bifunctional catalysis with the separation of functions between a metal and a carrier active in the given reaction – for instance, a carrier possessing acidic or basic properties or acting as a donor of active lattice oxygen. The main special feature of heating conductors is the propagation of the electromagnetic field. This can be of obvious advantage for processes that require fast and selective catalyst heating. In these instances, the microwave field can be used for the valorization of biomass and renewables.

Currently, the procedure for selecting an appropriate catalyst for a heterogeneous catalytic process of valorization of biomass and renewables ideally suitable for activation in microwave fields is not quite clear. The selection is complicated because the catalyst should first have substantial catalytic activity, and secondly possess certain electrophysical properties (dielectric/magnetic properties and conductivity). To obtain a maximum positive effect of microwave radiation on a catalytic system, comparatively inert components strongly interacting with electromagnetic fields may be introduced into catalysts.

The advantages of microwave technologies can more effectively be used if the catalytically and electromagnetically active catalyst components are placed on a substrate transparent to microwave fields. This is done to form superheated regions (hot spots) only in the volume or on the surface of the active catalyst phase, which decreases the temperature of the catalyst. This offers the possibility of suppressing side reactions and decreasing energy costs for heating inert materials that do not directly participate in the catalytic process.

The development of additional criteria (except for temperature-conversion dependencies) for revealing specific effects that appear under the action of microwave fields on reacting systems is an important problem. This is related, in particular, to the changes in the kinetic characteristics and dependences (reaction orders with respect to components and activation energies) that are less sensitive to errors in temperature measurements than a comparison of activities or transformation rates. Changes in process selectivity with respect to certain products at equal conversions and different methods of reacting system activation are most interesting and informative.

It is also necessary to track changes in the state of a solid catalyst under the action of microwave fields. In certain instances, the appearance of metastable structures that are not typically formed in the absence of such action can be expected. If they have catalytic properties and/or reactivity different from those usually observed, the appearance of the corresponding nontrivial effects in chemical reactions is possible.

Last but not least, bifunctional catalysis in which the occurrence of various stages of the process takes place on various catalytic components should be specifically considered. A typical example is transformations of renewables on metal nanoparticles supported on acid or basic carriers. A metallic component (as a rule, a Pt family metal) activates the initial molecule by breaking bonds and/or charge transfer (dehydrogenation, hydrogenolysis), whereas carbon skeleton transformations (depolymerization, alkylation, isomerization) occur on acid/basic centers. Microwave radiation generating temperature gradients in a heterogeneous system whose components interact differently with electromagnetic fields can then cause substantial changes in the reaction rates of the entire process and different stages and, especially important, may change the distribution of reaction products.

The advantages of the use of microwave activation of heterogeneous catalytic reactions of the valorization of biomass and renewables compared with thermal activation become obvious when fast heating or cooling and pulsed energy supply are required. Apart from the search for and synthesis of suitable catalysts, the development of new microwave equipment for applications on pilot and industrial scales is necessary, as well as process upscaling.

Effects observed in catalysis under the action of microwaves are still not well understood and have been poorly studied, although it is clear that irradiation of microwave-absorbing materials (catalysts, carriers, and reaction medium) can cause rapid volume heating, effectively remove moisture from solids, and modify surface properties. This distinguishes the action of microwaves from traditional thermal treatments.

Solid catalytic materials can be divided into three groups according to their character of interaction with microwave radiation. The first group includes metals whose smooth surface fully reflects microwave radiation. Metals then are not heated because there is almost no microwave radiation loss into their volume. However, if the surface is rough, microwave radiation can cause an arc discharge on it. The second group includes dielectrics transmitting microwave radiation almost unchanged through their volume. These are aluminum and silicon oxides, fused quartz, various glasses, porcelain and clays, and polymers. The third group includes dielectrics that absorb microwave radiation, which is particularly accompanied by sample heating.

In practice, mixtures containing substances that weakly and strongly absorb microwave radiation are often used for microwave heating. By varying the composition of such mixtures, it is possible to control the maximum temperature of mixture heating and the composition of reaction products.

Acknowledgment

The financial support of the Ministry of Education and Science of Russian Federation (project RFMEFI61614X0012) is acknowledged.

References

1. Calvino-Casilda, V., Guerrero-Perez, M.O., and Banares, M.A. (2010) Microwave-activated direct synthesis of acrylonitrile from glycerol under mild conditions: effect of niobium as dopant of the V–Sb oxide catalytic system. *Appl. Catal., B*, **95** (3–4), 192–196.
2. Calvino-Casilda, V., Guerrero-Perez, M.O., and Banares, M.A. (2009) Efficient microwave-promoted acrylonitrile sustainable synthesis from glycerol. *Green Chem.*, **11**, 939–941.
3. Drago, C., Liotta, L.F., La Parola, V., Testa, M.L., and Nicolosi, G. (2013) One-pot microwave assisted catalytic transformation of vegetable oil into glycerol-free biodiesel. *Fuel*, **113**, 707–711.
4. Omar, R. and Robinson, J.P. (2014) Conventional and microwave-assisted pyrolysis of rapeseed oil for bio-fuel production. *J. Anal. Appl. Pyrolysis*, **105**, 131–142.
5. Leveneur, S., Ledoux, A., Estel, L., Taouk, B., and Salmi, T. (2014) Epoxidation of vegetable oils under microwave irradiation. *Chem. Eng. Res. Des.*, **92** (8), 1495–1502.
6. Mazo, P.C. and Rios, L.A. (2012) Improved synthesis of carbonated vegetable oils using microwaves. *Chem. Eng. J.*, **210**, 333–338.

7. Dutta, S., De, S., Patra, A.K., Sasidharan, M., Bhaumik, A., and Saha, B. (2011) Microwave assisted rapid conversion of carbohydrates into 5-hydroxymethylfurfural catalyzed by mesoporous TiO_2 nanoparticles. *Appl. Catal. A: Gen.*, **409–410**, 133–139.
8. Hricovíniová, Z. (2013) Xylans are a valuable alternative resource: production of d-xylose, d-lyxose and furfural under microwave irradiation. *Carbohydr. Polym.*, **98** (2), 1416–1421.
9. Tsubaki, S., Oono, K., Ueda, T., Onda, A., Yanagisawa, K., Mitani, T., and Azuma, J. (2013) Microwave-assisted hydrolysis of polysaccharides over polyoxometalate clusters. *Bioresour. Technol.*, **144**, 67–73.
10. Li, C., Zhao, Z.K., Cai, H., Wang, A., and Zhang, T. (2011) Microwave-promoted conversion of concentrated fructose into 5-hydroxymethylfurfural in ionic liquids in the absence of catalysts. *Biomass Bioenergy*, **35** (5), 2013–2017.
11. Hassanzadeh, S., Aminlashgari, N., and Hakkarainen, M. (2014) Chemo-selective high yield microwave assisted reaction turns cellulose to green chemicals. *Carbohydr. Polym.*, **112**, 448–457.
12. Shaveta, S., Bansal, N., and Singh, P. (2014) F^-/Cl^- mediated microwave assisted breakdown of cellulose to glucose. *Tetrahedron Lett.*, **55** (15), 2467–2470.
13. Al Shra'ah, A. and Helleur, R. (2014) Microwave pyrolysis of cellulose at low temperature. *J. Anal. Appl. Pyrolysis*, **105**, 91–99.
14. Peng, H., Chen, H., Qu, Y., Li, H., and Xu, J. (2014) Bioconversion of different sizes of microcrystalline cellulose pretreated by microwave irradiation with/without NaOH. *Appl. Energy*, **117**, 142–148.
15. Ren, H., Zhou, Y., and Liu, L. (2013) Selective conversion of cellulose to levulinic acid via microwave-assisted synthesis in ionic liquids. *Bioresour. Technol.*, **129**, 616–619.
16. Biswas, A., Kim, S., Selling, G.W., and Cheng, H.N. (2013) Microwave-assisted synthesis of alkyl cellulose in aqueous medium. *Carbohydr. Polym.*, **94**, 120–123.
17. Li, C., Zhang, Z., and Zhao, Z.K. (2009) Direct conversion of glucose and cellulose to 5-hydroxymethylfurfural in ionic liquid under microwave irradiation. *Tetrahedron Lett.*, **50** (38), 5403–5405.
18. Liu, B., Zhang, Z., and Zhao, Z.K. (2013) Microwave-assisted catalytic conversion of cellulose into 5-hydroxymethylfurfural in ionic liquids. *Chem. Eng. J.*, **215–216**, 517–521.
19. Qu, Y., Wei, Q., Li, H., Oleskowicz-Popiel, P., Huang, C., and Xu, J. (2014) Microwave-assisted conversion of microcrystalline cellulose to 5-hydroxymethylfurfural catalyzed by ionic liquids. *Bioresour. Technol.*, **162**, 358–364.
20. Satgé, C., Verneuil, B., Branland, P., Granet, R., Krausz, P., Rozier, J., and Petit, C. (2002) Rapid homogeneous esterification of cellulose induced by microwave irradiation. *Carbohydr. Polym.*, **49** (3), 373–376.
21. Roshan, K.R., Jose, T., Kathalikkattil, A.C., Kim, D.W., Kim, B., and Park, D.W. (2013) Microwave synthesized quaternized celluloses for cyclic carbonate synthesis from carbon dioxide and epoxides. *Appl. Catal. A: Gen.*, **467**, 17–25.
22. Budarin, V.L., Clark, J.H., Lanigan, B.A., Shuttleworth, P., and Macquarrie, D.J. (2010) Microwave assisted decomposition of cellulose: a new thermochemical route for biomass exploitation. *Bioresour. Technol.*, **101** (10), 3776–3779.
23. Ha, S.H., Mai, N.L., An, G., and Koo, Y.-M. (2011) Microwave-assisted pretreatment of cellulose in ionic liquid for accelerated enzymatic hydrolysis. *Bioresour. Technol.*, **102** (2), 1214–1219.
24. Orozco, A., Ahmad, M., Rooney, D., and Walker, G. (2007) Dilute acid hydrolysis of cellulose and cellulosic bio-waste using a microwave reactor system. *Process Saf. Environ. Prot.*, **85** (5), 446–449.
25. Zhang, Z. and Zhao, Z.K. (2009) Solid acid and microwave-assisted hydrolysis of cellulose in ionic liquid. *Carbohydr. Res.*, **344** (15), 2069–2072.
26. Ninomiya, K., Yamauchi, T., Ogino, C., Shimizu, N., and Takahashi, K. (2014) Microwave pretreatment of lignocellulosic material in cholinium ionic liquid

for efficient enzymatic saccharification. *Biochem. Eng. J.*, **90**, 90–95.

27. Farag, S., Fu, D., Jessop, P.G., and Chaouki, J. (2014) Detailed compositional analysis and structural investigation of a bio-oil from microwave pyrolysis of kraft lignin. *J. Anal. Appl. Pyrolysis*, **109**, 249–257.
28. Bu, Q., Lei, H., Wang, L., Wei, Y., Zhu, L., Zhang, X., Liu, Y., Yadavalli, G., and Tang, J. (2014) Bio-based phenols and fuel production from catalytic microwave pyrolysis of lignin by activated carbons. *Bioresour. Technol.*, **162**, 142–147.
29. Bu, Q., Lei, H., Ren, S., Wang, L., Zhang, Q., Tang, J., and Ruan, R. (2012) Production of phenols and biofuels by catalytic microwave pyrolysis of lignocellulosic biomass. *Bioresour. Technol.*, **108**, 274–279.
30. Zhu, S., Wu, Y., Yu, Z., Zhang, X., Li, H., and Gao, M. (2006) The effect of microwave irradiation on enzymatic hydrolysis of rice straw. *Bioresour. Technol.*, **97**, 1964–1968.
31. Fernández, Y., Arenillas, A., Díez, M.A., Pis, J.J., and Menéndez, J.A. (2009) Pyrolysis of glycerol over activated carbons for syngas production. *J. Anal. Appl. Pyrolysis*, **84**, 145–150.
32. Badamali, S.K., Luque, R., Clark, J.H., and Breeden, S.W. (2009) Microwave assisted oxidation of a lignin model phenolic monomer using Co(salen)/SBA-15. *Catal. Commun.*, **10**, 1010–1013.
33. Appleton, T.J., Colder, R.I., Kingman, S.W., Lowndes, I.S., and Read, A.G. (2005) Microwave technology for energy-efficient processing of waste. *Appl. Energy*, **81**, 85–113.
34. Jones, D.A., Lelyveld, T.P., Mavrofidis, S.D., Kingman, S.W., and Miles, N.J. (2002) Microwave heating applications in environmental engineering—A review. *Resour. Conserv. Recycl.*, **34**, 75.
35. Dominguez, A., Fernández, Y., Fidalgo, B., Pis, J.J., and Menéndez, J.A. (2008) Bio-syngas production with low concentrations of CO_2 and CH_4 from microwave-induced pyrolysis of wet and dried sewage sludge. *Chemosphere*, **70**, 397–403.
36. Mutyala, S., Fairbridge, C., Pare, J.R.J., Bélanger, J.M.R., Ng, S., and Hawkins, R. (2010) Microwave applications to oil sands and petroleum: a review. *Fuel Process. Technol.*, **91**, 127–135.
37. Guo, L., Li, X.M., Bo, X., Yang, Q., Zeng, G.-M., Liao, D.-X., and Liu, J.-J. (2008) Impacts of sterilization, microwave and ultrasonication pretreatment on hydrogen producing using waste sludge. *Bioresour. Technol.*, **99**, 3651–3658.
38. Wang, T.X., Zong, Z.M., Zhang, J.W., Wei, Y.-B., Zhao, W., Li, B.-M., and Wei, X.-Y. (2008) Microwave-assisted hydroconversions of demineralized coal liquefaction residues from Shenfu and Shengli coals. *Fuel*, **87**, 498–507.
39. Mushtaq, F., Mat, R., and Ani, F.N. (2014) A review on microwave assisted pyrolysis of coal and biomass for fuel production. *Renewable Sustainable Energy Rev.*, **39**, 555–574.
40. Ratanakamnuan, U., Atong, D., and Aht-Ong, D. (2012) Cellulose esters from waste cotton fabric via conventional and microwave heating. *Carbohydr. Polym.*, **87** (1), 84–94.
41. Lahijani, P., Zainal, Z.A., Mohamed, A.R., and Mohammadi, M. (2014) Microwave-enhanced CO_2 gasification of oil palm shell char. *Bioresour. Technol.*, **158**, 193–200.
42. Yin, C. (2012) Microwave-assisted pyrolysis of biomass for liquid biofuels production. *Bioresour. Technol.*, **120**, 273–284.
43. Fernández, Y. and Menéndez, J.A. (2011) Influence of feed characteristics on the microwave-assisted pyrolysis used to produce syngas from biomass wastes. *J. Anal. Appl. Pyrolysis*, **91** (2), 316–322.
44. Xie, Q., Borges, F.C., Cheng, Y., Wan, Y., Li, Y., Lin, X., Liu, Y., Hussain, F., Chen, P., and Ruan, R. (2014) Fast microwave-assisted catalytic gasification of biomass for syngas production and tar removal. *Bioresour. Technol.*, **156**, 291–296.
45. Undri, A., Rosi, L., Frediani, M., and Frediani, P. (2014) Fuel from microwave assisted pyrolysis of waste multilayer packaging beverage. *Fuel*, **133**, 7–16.

46. Huang, Y.-F., Kuan, W.-H., Chang, C.-C., and Tzou, Y.-M. (2013) Catalytic and atmospheric effects on microwave pyrolysis of corn stover. *Bioresour. Technol.*, **131**, 274–280.
47. Lin, Y.-C., Wu, T.-Y., Liu, W.-Y., and Hsiao, Y.-H. (2014) Production of hydrogen from rice straw using microwave-induced pyrolysis. *Fuel*, **119**, 21–26.
48. Koberg, M. and Gedanken, A. (2013) in *New and Future Developments in Catalysis* (ed S.L. Suib), Elsevier, Amsterdam, pp. 209–227.
49. Zhao, X., Wang, M., Liu, H., Zhao, C., Ma, C., and Song, Z. (2013) Effect of temperature and additives on the yields of products and microwave pyrolysis behaviors of wheat straw. *J. Anal. Appl. Pyrolysis*, **100**, 49–55.
50. Zhao, X., Wang, W., Liu, H., Ma, C., and Song, Z. (2014) Microwave pyrolysis of wheat straw: product distribution and generation mechanism. *Bioresour. Technol.*, **158**, 278–285.
51. Kuan, W.-H., Huang, Y.-F., Chang, C.-C., and Lo, S.-L. (2013) Catalytic pyrolysis of sugarcane bagasse by using microwave heating. *Bioresour. Technol.*, **146**, 324–329.
52. Yu, Y., Yu, J., Sun, B., and Yan, Z. (2014) Influence of catalyst types on the microwave-induced pyrolysis of sewage sludge. *J. Anal. Appl. Pyrolysis*, **106**, 86–91.
53. Motasemi, F. and Afzal, M.T. (2013) A review on the microwave-assisted pyrolysis technique. *Renewable Sustainable Energy Rev.*, **28**, 317–330.
54. Patil, P.D., Reddy, H., Muppaneni, T., Ponnusamy, S., Cooke, P., Schuab, T., and Deng, S. (2013) Microwave-mediated non-catalytic transesterification of algal biomass under supercritical ethanol conditions. *J. Supercrit. Fluids*, **79**, 67–72.
55. Patil, P.D., Gude, V.G., Mannarswamy, A., Cooke, P., Nirmalakhandan, N., Lammers, P., and Deng, S. (2012) Comparison of direct transesterification of algal biomass under supercritical methanol and microwave irradiation conditions. *Fuel*, **97**, 822–831.
56. Beneroso, D., Bermúdez, J.M., Arenillas, A., and Menéndez, J.A. (2014) Influence of the microwave absorbent and moisture content on the microwave pyrolysis of an organic municipal solid waste. *J. Anal. Appl. Pyrolysis*, **105**, 234–240.
57. Lin, Y.-C., Wu, T.-Y., Jhang, S.-R., Yang, P.-M., and Hsiao, Y.-H. (2014) Hydrogen production from banyan leaves using an atmospheric-pressure microwave plasma reactor. *Bioresour. Technol.*, **161**, 304–309.
58. Lam, S.S., Russell, A.D., and Lee, C.L. (2012) Production of hydrogen and light hydrocarbons as a potential gaseous fuel from microwave-heated pyrolysis of waste automotive engine oil. *Int. J. Hydrogen Energy*, **37** (6), 5011–5021.
59. Khemthong, P., Luadthong, C., Nualpaeng, W., Changsuwan, P., Tongprem, P., Viriya-empikul, N., and Faungnawakij, K. (2012) Industrial eggshell wastes as the heterogeneous catalysts for microwave-assisted biodiesel production. *Catal. Today*, **190** (1), 112–116.
60. Bermúdez, J.M., Ruisánchez, E., Arenillas, A., Moreno, A.H., and Menéndez, J.A. (2014) New concept for energy storage: microwave-induced carbon gasification with CO_2. *Energy Convers. Manage.*, **78**, 559–564.
61. Yoon, S.J. and Lee, J.-G. (2012) Hydrogen-rich syngas production through coal and charcoal gasification using microwave steam and air plasma torch. *Int. J. Hydrogen Energy*, **37** (22), 17093–17100.
62. Undri, A., Rosi, L., Frediani, M., and Frediani, P. (2014) Upgraded fuel from microwave assisted pyrolysis of waste tire. *Fuel*, **115**, 600–608.
63. Dai, Y.-M., Chen, K.-T., and Chen, C.-C. (2014) Study of the microwave lipid extraction from microalgae for biodiesel production. *Chem. Eng. J.*, **250**, 267–273.

Part VIII
Applications – Environmental Catalysis

Microwaves in Catalysis: Methodology and Applications, First Edition.
Edited by Satoshi Horikoshi and Nick Serpone.

18
Oxidative and Reductive Catalysts for Environmental Purification Using Microwaves

Takenori Hirano

18.1
Introduction

Control of emission of air pollutants such as CO, VOCs (volatile organic compounds), NO_x, and SO_2 is high on the agenda, and heterogeneous catalysis plays a key role. One of major group of emission controlled by environmental legislation comprises VOCs and odorants released from all industrial and refinery effluents, that is, generated by a wide range of industrial operations, such as chemical processing, waste water treatment, coating operations, and paints. These compounds are also of concern in controlling indoor air quality. VOCs removal from wide range of process effluents becomes necessary. The main classifications of VOCs include aliphatic and aromatic hydrocarbons; organic oxygen and nitrogen compounds; and organic halide compounds. Compared to thermal combustion, catalytic oxidation procedure has the advantage of operating at high conversion at much lower temperatures and high space velocities. This means that temperature control is easier and that the smaller reactor leads to a reduction in the capital cost [1, 2]. The most active catalysts are usually derived from supported platinum group metals, however, in the presence of poisons or oxidizing chlorinated hydrocarbons, catalysts composed of oxides such as copper chromite have properly been used.

Another major group of emission comprises NO_x and SO_2 released from power plants and automobiles. High-temperature combustion of coal and oil provides heat and generates electricity that results in huge volumes of NO_x, SO_2, and CO being vented to the atmosphere. These compounds form acid rain resulting in devastating forests or affecting public health. Selective catalytic reduction (SCR) of NO_x with ammonia using specially developed catalysts is now the most important process for removing NO_x from the effluent gas of power plants and many other stationary sources [3, 4]. In automobile exhaust emissions, high concentration levels of CO, hydrocarbons, and NO_x are included. Since NO_x is difficult to remove in conjunction with the other pollutants, particularly in the presence of oxygen, controlling the air/fuel ratio in the engine, efficient three-way catalysts incorporating several platinum group metals have been introduced and operated

Microwaves in Catalysis: Methodology and Applications, First Edition.
Edited by Satoshi Horikoshi and Nick Serpone.

under conditions that are effective for the oxidation of CO and hydrocarbons and the reduction of NO_x [5]. SO_2 is also removed from the flue gas in a small contact process unit after the usual SCR procedure.

Much attention has been devoted to the application of microwave heating method to the catalytic processes for environmental purification [6]. Microwave-induced catalysis differs in several ways from conventional catalytic methods, that is to say, firstly the catalyst may serve a dual purpose both as a promoter of the reactions occurred on the surface and as an efficient converter of the incident microwave into thermal energy required for the reaction to proceed. In addition, the fact that the incident microwave energy can be switched on and off extremely rapidly combined with the almost instantaneous heating that occurs in lossy oxides of catalyst upon absorption of microwave, allowing a high degree of control of reaction conditions including temperature. During the on cycle, the incident microwave pulses supply the required energy for the catalyst not only to reach the activation temperature but also sometimes to form an active phase in the irradiated catalyst. While during the off cycle, between each pulse packet, the catalyst surface would conduct the heat away from the hot spots and thus decrease the temperature. These properties of microwave catalysis are well suited for the environmental catalytic processes, which require quick heating time and treatment of high effluent gas flow rate.

The technology of microwave-induced catalysis has apparent benefits in the catalytic system of environmental purification where oxidative and reductive reactions such as elimination of VOCs, NO_x, and SO_2 occur on the solid surface of oxides. Using appropriate devices, the heating can be concentrated on the catalytic mass, resulting in downsizing of the catalytic reactors and shortening of the heating time with very little energy absorption by the reactants on supports [7]. It is worthy to apply the microwave heating method to the catalytic processes for the environmental purification.

18.2 Microwave Heating of Catalyst Oxides Used for Environmental Purification

The microwave dielectric heating effect uses the ability of some solid materials of absorbents or catalysts to transform electromagnetic energy into heat via their dielectric loss and thereby drive catalytic reactions [8]. Both conduction and dielectric polarization are sources of microwave heating. Dielectric loss at microwave frequencies may be composed of migration loss and conduction loss from lower frequencies; moreover, vibration and deformation loss from higher frequencies for the oxides are used as catalyst.

Increasing temperature of catalyst oxides would result in shift of migration loss to higher frequencies and increase in conduction loss due to increasing ionic conductivity. These would contribute to increase in dielectric loss at 2.45 GHz for the oxides resulting in increasing temperature again. This successive process of increase in temperature and dielectric loss would result in thermal runaway [9]. The ratio of the dielectric loss and the dielectric constant of catalysts

Table 18.1 Effect of microwave heating on the temperature reached of pure oxides at 2.45 GHz and 25 °C.

Oxide	*T* (°C)
CaO	58
BaO	57
CeO_2	46
TiO_2	40
ZrO_2	47
V_2O_5	221
Nb_2O_5	51
Ta_2O_5	55
Cr_2O_3	54
MoO_3	41
WO_3	188
MnO_2	194
Fe_2O_3	165
Co_2O_3	272
NiO	260
CuO	341
ZnO	131
γ-Al_2O_3	85
SiO_2	106
SnO_2	143
Bi_2O_3	59

Microwave pulse irradiation conditions: 2.45 GHz, 25 °C, 5 s ON–10 s OFF, 8 cycles.

define the dielectric loss tangent $= \varepsilon''/\varepsilon' = \tan\delta$, which defines the ability of the catalyst oxides to convert electromagnetic energy into heat energy at a given frequency and temperature. Several workers have reported the temperature reached by pure oxides, which could be able to be one of the components of practical catalysts for environmental purification, when placed in conventional microwave oven with pulse microwave irradiation at 2.45 GHz and room temperature [10, 11].

In Table 18.1, typical temperature data measured for a number of pure oxides with microwave heating are exhibited [11]. Oxides such as CuO, Co_2O_3, NiO, V_2O_5, and MnO_2 are susceptible to incident microwave, rapidly heated to reach about 200 °C or above after eight cycles of 5 s ON–10 s OFF microwave irradiation. Similar results on NiO, Co_2O_3, V_2O_5, and CuO have also been reported [10]. The reason why these oxides are easily heated under microwave irradiation has not been interpreted; however, it is speculated that the increase in conduction with temperature of these oxides would be associated with the thermal activation of the electrons that pass from the oxygen 2p valence band to the 3s3p conduction band [12]. In addition, electrical conduction would be enhanced by the oxide defects formed at higher temperatures, which sharply decreases the energy gap between the vacancy and conduction bands.

Comparing the heating properties of oxides measured, most of transition metal oxides can absorb microwave to be heated; however, rare earth oxides, alkaline earth group oxides, SiO_2, and γ-Al_2O_3 appear to absorb weakly exhibiting little rising in temperature. From a practical viewpoint as VOCs oxidation or NO_x reduction catalyst, the most active catalysts have been derived from supported platinum group metals, where γ-Al_2O_3 is usually used as a support, that is, in the case of automobile emission control catalysts, supported platinum group metals of Pt, Rh, and Pd are usually considered to be the most efficient and active catalysts available. However, considering the effectiveness of microwave irradiation, catalysts composed of transition metal oxides would be more suitable for the reactor using microwave than the supported metal catalysts, because materials normally used for supports such as γ-Al_2O_3, SiO_2, and CeO_2 have not so high susceptibility to incident microwave as understood by Table 18.1.

Several kinds of commercial catalyst consisting of mixed oxides including transition metal elements, which are frequently used for the purpose of improving or protecting our environment by their high activities for oxidation or reduction, were submitted to measuring their heating properties under microwave irradiation [13] (T. Hirano and T. Matoba, 1994, unpublished results). The submitted catalysts are G-3A comprised of 90% Fe_2O_3 and 10% Cr_2O_3, G-132 of 40% CuO and 60% ZnO, G-53D of 70% NiO and 30% SiO_2, N-140 of 40% CuO and 60% MnO_2, N-150 of 40% Fe_2O_3 and 60% MnO_2, and NF-9 of 100% NiO from Nissan Girdler. Since every catalysts are shaped tablets, small amounts of graphite are added at concentrations between 0.5% and 1% to achieve high stability of the tablets; nevertheless, these amounts are confirmed to be too small to influence on the microwave heating. These catalysts of mixed oxides used for environmental purification were repeatedly pulse heated by microwave irradiation of 2 s ON – 10 s OFF, and the resulting temperatures reached after several times of pulse cycles are given in Table 18.2, where measured dielectric constants of the catalysts at 2.45 GHz and 25 °C are also shown (T. Hirano and T. Matoba, 1994, unpublished results). Among the catalysts tested, NF-9, N-140, and N-150 show rapid temperature rising, on the other hand, G-132 and G-53D show a slow temperature rise.

Table 18.2 Dielectric constant ε', loss factor ε'', and effect of microwave heating on the temperature reached of complex oxides of commercial catalyst.

Catalyst	Surface area ($m^2\,g^{-1}$)	Dielectric constant ε'	Dielectric loss factor ε''	tan δ	T (°C)	t (min)
G-3A (Fe_2O_3–Cr_2O_3)	78	2.9	0.38	0.13	61	5
G-132 (CuO–ZnO)	35	2.9	0.35	0.12	49	5
G-53D (NiO–SiO_2)	193	3.1	0.45	0.15	48	8
N-140 (CuO–MnO_2)	133	3.3	0.63	0.19	96	6
N-150 (Fe_2O_3–MnO_2)	175	3.8	0.49	0.13	79	5
NF-9 (NiO)	152	3.8	0.70	0.18	101	5

Microwave pulse irradiation conditions: 2.45 GHz, 25 °C, 2 s ON – 10 s OFF.

Comparing the temperature rising rates of the mixed oxides with those of pure oxides, little correlation seems to be found, that is to say, commercial catalysts of N-140 and NF-9 containing CuO and NiO, respectively, show relatively higher heating rates, on the other hand, in spite of containing CuO and NiO, the heating rates of G-132 and G-53D are slow under microwave irradiation. It is of interest that N-150 consisting of Fe_2O_3 and MnO_2, despite both oxides have independently shown slower rising rates, shows relatively higher rising rate. These results seem to indicate that the coexistence of different oxides in a catalyst changes the microwave susceptibility from those possessed by each oxides separately, that is, the dielectric loss in the mixed oxides would be significantly altered by the mixing of different oxides consequently. It is reported that the existence of two types of sites in perovskite-type oxides has been shown to have different dipole strengths, resulting in hot spot formation for perovskites during dielectric heating [14].

18.3 Microwave-Assisted Catalytic Oxidation of VOCs, Odorants, and Soot

Removal of VOCs including odorants, major contributors to atmospheric pollution, at source is ideal. Catalytically aided oxidation is a preferred method over thermal combustion owing to lower processing temperatures and also a higher selectivity to CO_2 [1a,b]. Lowering the temperature at which these processes occur with great efficiency can consequently reduce energy cost. On this point of view, it is of interest to utilize the microwave-assisted catalyst for these processes because of some advantages, such as volumetric and high heating rates, indirect contact, easy control for heating, and the minimization of equipment size.

For the purposes of oxidative removal of environmental pollutants, microwave-heated catalytic processes have been extensively attempted. Oxidations of propane as a model of aliphatic hydrocarbons which are more difficult to be oxidized than aromatic hydrocarbons such as benzene, toluene, and xylene, and CO over doped perovskite catalysts such as $La_{0.5}Co_{0.2}MnO_3$ were investigated under microwave irradiation [15]. Comparing microwave heating to conventional heating, there is no indication that the two heating methods lead to different reaction pathway, because only small difference in selectivity is observed. However, the reaction mechanism of propane oxidation and that of CO oxidation are considered to be different, that is, CO is oxidized by reaction with adsorbed oxygen at low temperature, on the other hand, propane is considered to be oxidized by reaction with lattice oxygen at high temperature according to a Mars–van Krevelen mechanism. The use of microwave energy is assumed to enable fast and efficient heating of the perovskite catalysts of complex oxides, where only the catalyst is selectively heated to form hot spots enabling lower bulk temperature. Thus, direct heating with microwave irradiation is rapid and effective, and for the periodic operation, it is much more suitable than conventional heating.

As a test VOC molecule, acetylene in dry air was used, and oxidation over perovskites of $LaCoO_3$ and $BaTiO_3$ was studied by using pulsed microwave [16]. The oxidation efficiency is found to increase linearly with the pulse repetition rate, leading to very promising results for the oxidation of VOCs even at very low energy cost. *n*-Hexane oxidation over hopcalite of a mixture of copper and manganese oxides assisted by microwave was studied and the effectiveness of hopcalite and microwave for the oxidation is confirmed [17].

In the case of microwave-assisted methane oxidation including oxidative coupling, numerous studies have been conducted [18]. The results of the oxidation of methane over Ni-based and Co-based catalysts exhibit that the temperature of the catalyst bed required for the oxidation with microwave irradiation is much lower than that with conventional heating. The heating effect is more emphasized when the irradiated catalyst has a strong dipole character. The effect of microwave heating on methane oxidation over $La_{0.8}Ce_{0.2}MnO_2$ in the presence of SO_2 was also studied [14]. Microwave irradiation decreases the temperature of methane conversion by about 200 °C than that reached by conventional heating. It is proposed that local hot spots on the catalyst surface would explain the enhanced activity toward methane oxidation.

Six commercial catalysts of G-3A, G-132, G-53D, N-140, N-150, and NF-9, whose heating properties under microwave irradiation are displayed in previous section, were subjected to *n*-pentane oxidation test as a model reaction of VOCs removal under microwave irradiation and conventional heating [13, 19] (T. Hirano and T. Matoba, 1994, unpublished results). On all catalysts, the activity with microwave heating is always superior to that with conventional heating as shown in Figure 18.1, where *n*-pentane conversions of N-150 and G-3A measured in heating and cooling process are shown. It is noteworthy that under microwave irradiation, the performance of N-150 in cooling process is extremely higher than that in heating process. It is guessed that new active sites for *n*-pentane oxidation would be formed on N-150 at high temperatures reached by microwave

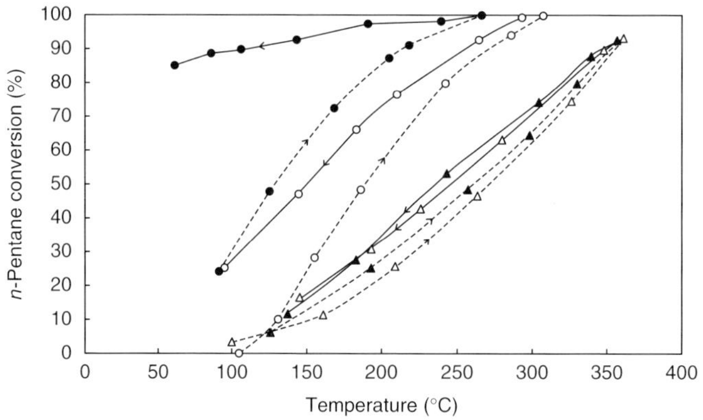

Figure 18.1 *n*-Pentane oxidation over N-150 and G-3A under microwave irradiation (MW) and conventional heating (CH) measured with increasing (dotted line) and decreasing (full line) temperature; •: N-150 MW, o: N-150 CH, ▲: G-3A MW, △: G-3A CH; 800 ppm *n*-pentane in air, GHSV: 1000 h^{-1}.

heating, even though similar tendency is observed with the conventional heating processes, the extent is remarkably smaller compared with microwave heating. On the other hand, *n*-pentane conversion over G-3A is inferior to N-150, and moreover the effect of microwave irradiation on G-3A is much less than that on N-150. These results should be explained not only by hot spots but also by the new active sites formed by microwave irradiated. Comparison of the activation energies of *n*-pentane oxidation (except for G-3A) reveals that all the catalysts exhibit lower values under microwave irradiation than those from conventional heating.

As another model reaction of VOCs removal, oxidation of acetaldehyde was also carried out over the same series of commercial catalysts of mixed oxides under microwave irradiation and conventional heating [19] (T. Hirano and T. Matoba, 1994, unpublished results). Different from *n*-pentane, little difference in activity is observed between microwave heating and conventional heating as shown in Figure 18.2. Even though only G-3A shows higher conversion in cooling process than heating process regardless of microwave or conventional heating method, other catalysts including N-150 show opposite behavior that the conversion in cooling process is always somewhat inferior to that in heating process different from the case of *n*-pentane oxidation.

It is known that there is an order of destructiveness of organic compounds which runs as follows: alcohols > aldehydes > ketones > alkenes > alkanes [20]. From the analogy to the proposed reaction mechanisms of CO and propane oxidation [15], it is speculated that acetaldehyde-containing oxygen would react with dissociatively adsorbed oxygen on the catalyst surface; on the other hand, *n*-pentane would react with lattice oxygen according to so-called Mars–van Krevelen mechanism. The higher activity of Mn-based catalysts, such as N-140 and N-150, for *n*-pentane oxidation under microwave irradiation could be explained by the ease of reoxidation of the reduced surface giving up lattice oxygen. On the contrary, reoxidation by the conventional heating would be much

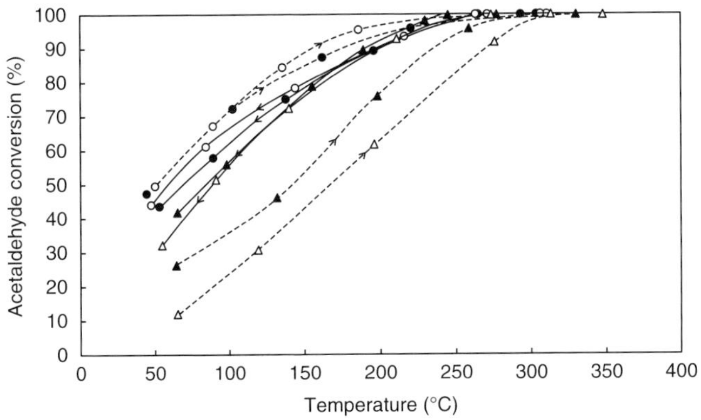

Figure 18.2 Acetaldehyde oxidation over N-150 and G-3A under microwave irradiation (MW) and conventional heating (CH) measured with increasing (dotted line) and decreasing (full line) temperature; •: N-150 MW, o: N-150 CH, ▲: G-3A MW, △: G-3A CH; 1400 ppm CH_3CHO in air, GHSV: 9000 h^{-1}.

slow comparing with the microwave heating, resulting in the lower performance for *n*-pentane oxidation caused by the lower oxidized states of surface where small amounts of lattice oxygen would exist. On the other hand, in the case of acetaldehyde oxidation, amounts of dissociatively adsorbed oxygen on the surface under both heating methods would be almost similar resulting in the small differences in the conversion.

A possibility of forming new active phase by microwave irradiation at high temperatures would not be excluded, on the basis of the appearance of higher conversion in cooling process after the exposure to high temperatures in *n*-pentane oxidation. By XRD measurements, it is confirmed that the oxidation states of oxides in the tested commercial catalysts are more reduced by the conventional heating than by the microwave heating. It is guessed that the microwave irradiation would keep the surface of catalyst in more oxidized state than the conventional heating in the oxidation process of *n*-pentane. On the other hand, in the case of acetaldehyde oxidation, only hematite is observed in N-150 as an oxidized state of iron oxide, whereas maghemite in addition to hematite is observed in G-3A after oxidation process with both microwave and conventional heating. Several studies have pointed to the oxygen species including overabundance of oxygen in the surface layers, which would be capable of extracting hydrogen from a methylene C–H bond of *n*-pentane [21].

Microwave-assisted oxidation or decomposition of other VOCs such as organic halide compounds and organic nitrogen compounds were studied using appropriate catalysts. A series of model organic halide compounds of polychlorinated hydrocarbons are decomposed by the technique of microwave catalysis, using metal powders of Ni, Fe, and Co as catalysts [22]. Primary free radical mechanisms of the decomposition by breaking selectively C–X bond are established. Oxidative decomposition of trimethylamine under microwave irradiation over mixed catalyst of zeolite and MnO_2 was investigated, and the effect of microwave irradiation is confirmed comparing with conventional heating [23].

In addition to VOCs components, diesel particulate matter consists mainly of highly agglomerated solid carbonaceous material, the so-called soot. A major problem is the fact that exhaust temperatures are generally low as compared to soot light-off temperatures. Hence, apart from additional filter, exhaust gas preheating is usually required. Alternatively, microwave heating has been applied, whose advantage is a bulk heating technique that is fast and selective. For microwave-assisted soot filters, regeneration efficiency close to 80% can be reached by using $La_{0.8}Ce_{0.2}MnO_3$ perovskite [24].

18.4 Microwave-Assisted Reduction of NO_x and SO_2

NO_x and SO_2 are emitted from fossil fuel combustion in mobile and stationary sources and are the major air pollutants. NO_x contribute a lot of photochemical smog, acid rain, ozone depletion, and greenhouse effect. SO_2 is also generally accepted to be the most important precursor to acid rain [7]. Microwave-induced catalysis has been applied to topics of environmental concern including degradation of NO_x and SO_2 [25]. Among many technologies to reduce the emission of NO_x and SO_2, the catalytic technology of direct and reductive decomposition of these oxides is more attractive because of the compact apparatus without reducing agent and high efficiency. It is reported that the reduction of NO_x could be up to 98% conversion when microwave energy is applied continuously [26]. On the other hand, as a case of using reducing agent, the use of both catalyst and ammonium bicarbonate combined with microwave energy could increase the catalytic reductions of NO_x and SO_2 significantly [27].

As an application of microwave dielectric heating in environment-related heterogeneous gas-phase catalytic systems, SO_2 reduction by CH_4 has been reported whereby the amount of microwave power absorbed by the dielectric depends not only on the electromagnetic field applied but also on the dielectric property of the heated catalyst [28]. Microwave-assisted catalytic reduction of SO_2 with CH_4 over Al_2O_3-supported MoS_2 catalyst is similarly reported [29]. The existence of hot spots in the catalysts heated by incident microwave has been verified by the detection of α-Al_2O_3 at a recorded temperature some 200 °C lower than the temperature at which the phase transition of γ-Al_2O_3 to α-Al_2O_3 is usually observed. In the case of microwave-assisted purification of automobile emission, NO_x reduction temperature is lower shifted by the microwave irradiation and the reason could be attributed to a kind of internal heating which possesses many characteristics such as uniformity, quickness, and high-energy efficiency [30].

Wan *et al.* report a decomposition of NO_x over C-21, a commercial catalyst comprising of Fe_2O_3 and Cr_2O_3, is markedly accelerated under microwave irradiation [31]. Similarly, the NO_x decomposition activity of commercial catalysts of G-3A, NF-9, G-53D, G-132, N-140, and N-150, whose temperatures were maintained around 900 °C by controlling microwave irradiation conditions in a stream of NO–H_2O–He, was also studied [13, 32]. Among the catalysts tested, only G-3A, N-150, and G-132 exhibit an apparent activity of NO_x decomposition, and it is noteworthy that G-3A shows 100% NO_x conversion, and 50% conversion is kept even at 450 °C as shown in Figure 18.3. On the other hand, with the conventional heating, G-3A shows low conversion of less than 10% even at 800 °C. Addition of CH_4 to the NO_x stream increases the performance of G-3A remarkably to shift the temperature of giving 90% conversion to 170 °C lower comparing with the temperature obtained without CH_4 [33]. On the contrary, addition of O_2 to the NO_x stream decreases the performance of G-3A to give 40% conversion at 900 °C under microwave irradiation, even though no conversion is observed with the conventional heating. Approximately all conditions, the obtained activation energies of microwave heating, are always lower than those of conventional heating. Using XRD, it is found that the crystal state of G-3A exposed to higher temperatures under microwave irradiation in NO–H_2O–He stream is mainly magnetite, whereas that heated by conventional method is composed of a mixture of hematite and magnetite. Also AFM 3D-height images indicate that the surface of microwave

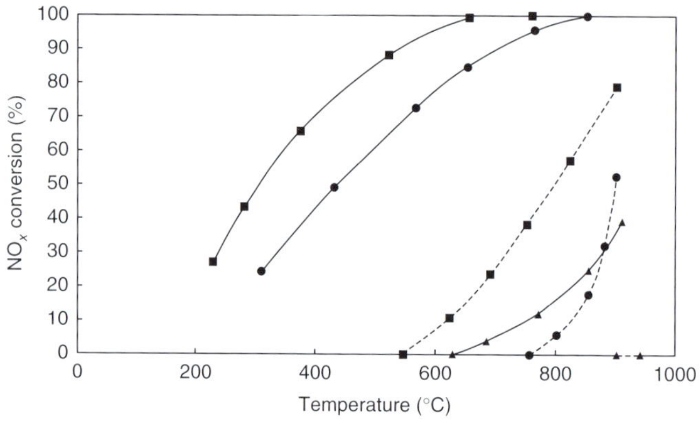

Figure 18.3 NO_x reductive reaction over G-3A under microwave irradiation (full line) and conventional heating (dotted line) measured with decreasing temperature; •: 630 ppm NO–2.6% H_2O in He, GHSV: 4800 h^{-1}, ▲: 630 ppm NO–5% O_2–2.6% H_2O in He, GHSV: 4800 h^{-1}, ■: 1600 ppm NO–6000 ppm CH_4–3% H_2O in He, GHSV: 4800 h^{-1}.

irradiated seems to be coarsely granular, on the contrary, that of conventionally heated seems to be almost smooth as shown in Figure 18.4a,b [11, 13].

During conventional heating, where heat is transformed to the exterior of the catalyst particles based on thermal conduction, the exterior of the particles would be highest in temperature. Therefore, sintering would predominantly take place at the exterior of the particles through a shell-progressive mechanism. In contrast, as microwave heating is a kind of internal heating, the interior of the particles might be highest in temperature during dielectric heating based on its bulk heating property, by creating an effectively inverse temperature gradient throughout the particles. For this reason, superheating and sintering would occur predominantly in the interior of the catalyst particles as a result of the high temperature at the interior of the particles; thus, observed changes in the surface state of catalyst during dielectric heating proceed through a growing core mechanism, different

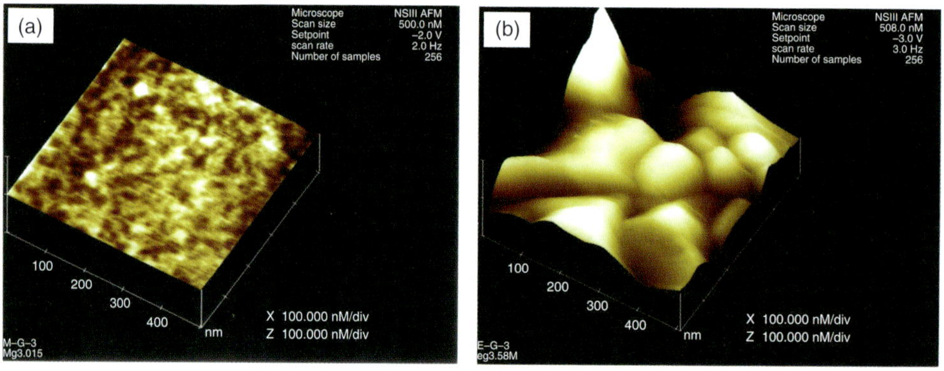

Figure 18.4 AMF 3D-height images of G-3A used for NO_x reductive reactions under microwave irradiation (a) and conventional heating (b).

from the growing shell mechanism of the conventional heating. As a result of such microwave heating mechanism, the reduced surface would be formed, which is appropriate to reduce or decompose NO_x, and the decomposition of SO_2 under microwave irradiation might also proceed according to the similar mechanism.

18.5 Conclusions

Real benefits of microwave heating of catalysts for the removal of environmental pollutants would be summarized as follows. That is to say, very high rates of temperature rising of catalyst oxides of pure and mixed oxides can be achieved even in the shaped pellets, extrusions, and monoliths. Considerable bulk superheating can be achieved to form hot spots, new active phase, and specific surface state different from those formed by conventional heating, resulting in higher activities of catalysts for environmental purification. In heterogeneous catalyst systems, there may be significant microwave heating effects due to both the heterogeneity of the dielectric and the presence of interface regions in the complex oxides of commercial catalysts.

In microwave-irradiated oxidative and reductive catalysts, it is quite possible for some regions to be hotter than others on a macro- or micro-scale, being said as “hot spots,” to form new active sites that could not be easily formed by the conventional heating. That is to say, the formation of oxidative or reductive active sites would be caused by the reoxidation of the surface giving up lattice oxygen under oxidative conditions or reduction of the surface under reductive conditions induced by the microwave irradiation. Many literatures have stated that the applications of microwave heating to catalytic systems would possibly shorten process time and reduce the required process energy consumption, apart from being effective for the removal of pollutants such as VOCs, NO_x, and SO_2. Apart from highly specialized applications using dielectric catalysts, little commercialization has been seen in the field of environmental purification engineering. Following reasons would be considered. Firstly, little fundamental data exist regarding the dielectric properties of oxidative or reductive catalysts used actually. The second reason is a lack of knowledge concerning the design of microwave heating apparatus. In spite of such problems, microwave-assisted catalytic processes for the environmental purification, where a number of benefits including process time saving and increased process conversion are apparent, may be considered to be likely candidates for further development.

References

1. (a) Spivey, J.J. (1987) *Ind. Eng. Chem. Res.*, **26**, 2165–2180; (b) Chu, W. and Windawi, H. (1996) *Chem. Eng. Prog.*, **92**, 37–43.
2. Armor, J.N. (1992) *Appl. Catal.*, **1**, 221–256.
3. Bosch, H. and Janssen, F. (1987) *Catal. Today*, **2**, 369–532.

4. Janssen, F. (1993) *Catal. Today*, **16**, 155–287.
5. (a) Acres, G.J.K. (1971) Improvements in and Relating to Catalysis 1390182; (b) Acres, G.J.K. and Cooper, B.J. Catalyst, 3951860.
6. Wan, J.K.S. (1993) *Res. Chem. Intermed.*, **19**, 147–158.
7. Jones, D.A., Leyveld, T.P., Mavrofidis, S.D., Kingman, S.W., and Miles, N.J. (2002) *Resour. Conserv. Recycl.*, **34**, 75–90.
8. Liu, C.C., Na, B.K., Walters, A.B., and Vannie, M.A. (1994) *Catal. Lett.*, **26**, 9–24.
9. (a) Kozuka, H. (1995) 115th Gathering for Discussion on New Ceramics, pp. 1–12; (b) Kozuka, H., Tsai, C.L., and Mackenzie, J.D. (1991) International Conference on Science and Technology of New Glasses, October 16–17, 1991, pp. 311–316.
10. Mignos, D.M.P. and Baghurst, D.R. (1911) *Chem. Soc. Rev.*, **20**, 1–47.
11. Hirano, T. (2002) *Hyoumen*, **40**, 1–14.
12. Tamura, H. (1994) *Am. Ceram. Soc. Bull.*, **73**, 92–95.
13. Hirano, T. and Matsuda, Y. (1997) Proceedings of the 2nd Kyushu-Taipei International Congress on Chemical Engineering, pp. 335–338.
14. Steenwinkel, Y.Z., Castricum, H.L., Beckers, J., Eiser, E., and Blieke, A. (2004) *J. Catal.*, **221**, 523–531.
15. Beckers, J., Zande, L.M., and Rothemberg, G. (2006) *ChemPhysChem*, **7**, 747–755.
16. Roussean, A., Guaitella, O., Ropcke, J., Gatilova, L.V., and Tolmachev, Y.A. (2004) *Appl. Phys. Lett.*, **85**, 2199–2201.
17. Hartmann, I. and Einicke, W.D. (2007) *Chem. Ing. Tech.*, **79**, 1205–1211.
18. Bi, X.J., Hong, P.J., Xie, X.G., and Dain, S.S. (1999) *React. Kinet. Catal. Lett.*, **66**, 381–386.
19. Hirano, T., Matoba, T., and Tomiyama, Y. (1997) Treatment of Organic Pollutant, JP, 09-075670, A.
20. Hermia, J. and Vigneron, S. (1993) *Catal. Today*, **17**, 349–358.
21. Centi, G., Trifiro, F., Ebner, J.R., and Franchetti, V.M. (1988) *Chem. Rev.*, **88**, 55–80.
22. Dinesen, T.R.J., Tse, M.Y., Depew, M.C., and Wan, J.K.S. (1991) *Res. Chem. Intermed.*, **15**, 113–127.
23. Kamitani, Y., Yanase, S., Koide, S., and Hirano, T. (1998) Deodorizing Method, JP, 10-137330, A.
24. Steenwinkel, Y.Z., Zande, L.M., Castricum, H.L., Bliek, A., Brink, R.W., and Elzinga, G.D. (2005) *Chem. Eng. Sci.*, **60**, 797–804.
25. Wan, J.K.S., Tse, M.Y., Husby, H., and Depew, M.C. (1990) *J. Microwave Power Electromagn. Energy*, **25**, 32–38.
26. Zang, D.X., Yu, A.M., and Jin, Q.H. (1997) *Chem. J. Chin. Univ.*, **18**, 1271–1280.
27. Wei, Z., Lin, Z., Niu, H., He, H., and Ji, Y. (2009) *Hazard. Mater*, **162**, 837–841.
28. Zhang, X. and Hayword, D.O. (2006) *Inorg. Chim. Acta*, **359**, 3421–3433.
29. Zhang, X., Hayword, D.O., Lee, C., and Mingos, D.M.P. (2001) *Appl. Catal., B*, **33**, 137–148.
30. (a) Tang, J., Zhang, T., Ma, L., Li, N., Liang, D., and Lin, L. (2002) *J. Catal.*, **221**, 560–564; (b) Turner, M.D., Laurence, R.L., Ynguesson, K.S., and Cornner, W.C. (2001) *Catal. Lett.*, **71**, 133–138; (c) Zhenming, Y., Jinsong, Z., Xiaoning, C., Qiang, L., Zhijum, X., and Zhimin, Z. (2001) *Appl. Catal., B*, **34**, 129–135.
31. Tse, M.Y., Depeu, M.C., and Wan, J.K.S. (1990) *Res. Chem. Intermed.*, **13**, 221–236.
32. (a) Hirano, T., Matoba, T., and Tomiyama, Y. (1996) Method for Removing Gaseous Contaminant in Gas, JP, 08-299762, A. (b) Hirano, T., Matoba, T., and Tomiyama, Y. (1996) Treatment of Gas Containing Nitrogen Oxides, JP, 08-323150, A
33. Hirano, T., Matoba, T., and Tomiyama, Y. (1996) Treatment Method for Gas Containing Nitrogen Oxides, JP, 08-309153, A.

19
Microwave-/Photo-Driven Photocatalytic Treatment of Wastewaters

Satoshi Horikoshi and Nick Serpone

19.1
Situation of Wastewater Treatment by Photocatalytic Classical Methods

The photoassisted oxidative (and reductive) decomposition of pollutants by means of TiO_2 photocatalyst nanoparticulates is an effective and attractive oxidation (reduction) method in the general area of advanced oxidation technologies. Several review articles have appeared that summarize environmental protection using TiO_2 materials as the photocatalysts [1–4]. Applications of photoassisted treatments to air pollution have been developed by TiO_2 fixation on several suitable substrate supports, such as in filters found in air conditioners [5]. However, this photoassisted degradation methodology is not suitable for large-scale wastewater treatment because the rates of degradation of organic compounds dissolved in wastewaters tend to be rather slow. In this regard, relatively little has been done in this area in the last decade as large-scale treatments of organic pollutants in aquatic environments have not been without some problems, not least of which is the low photodegradation efficiency, a result of several factors, most notably: (i) the poor adsorption of wastewater organic pollutants on the TiO_2 surface; (ii) the penetration of UV light tends to be shallow in turbid wastewaters; (iii) the need for dissolved oxygen in the photoassisted degradations; and (iv) the need to immobilize TiO_2 nanoparticles in their use in aquatic ecosystems. Also relevant, the processing time has been the principal problem in actual wastewater treatments that have used TiO_2 nanomaterials. Many of the above problems could be resolved if the activity of the photocatalysts were improved. To achieve such an objective, Horikoshi and coworkers proposed sometime ago [6] the coupling of both microwave radiation and ultraviolet radiation (UV/MW) to enhance the activity of photocatalysts. With the latter coupled (i.e., integrated) methodology, it was possible to enhance the photoassisted degradation processes in TiO_2 dispersions by the added assistance of microwave radiation in the remediation of wastewaters contaminated with such pollutants as dyes, polymers, surfactants, herbicides, and endocrine disruptors, among others.

Microwaves in Catalysis: Methodology and Applications, First Edition.
Edited by Satoshi Horikoshi and Nick Serpone.

19.2
Experimental Setup of an Integrated Microwave/Photoreactor System

Continuous microwave irradiation of a wastewater sample can typically be achieved in a single-mode applicator using a 2.45-GHz microwave generator, a power monitor (to assess incident and reflected microwave power), a three-stub tuner, and an isolator (an air cooling device) such as the one fabricated by the Hitachi Kyowa Engineering Co. Ltd (Figure 19.1a,b) [7]. A typical microwave-photoreactor (MPR) setup might contain a model wastewater sample (30 ml) containing TiO_2 particles (Evonik P25; 60 mg loading) introduced into a closed high-pressure 150-ml Pyrex glass cylindrical reactor. Subsequent irradiation of the reactor with UV light from a super high-pressure Hg lamp can be achieved through the means of a light guide. The solution temperature is usually measured with an optical fiber thermometer.

Three different methodologies can be used to examine the photodecomposition of aqueous samples of pollutants in aqueous TiO_2 dispersions. The first is the photo-/microwave-assisted method using UV light and microwave irradiation in the presence of TiO_2 (UV/MW). The second method entails UV irradiation alone (UV), whereas the third involves a thermal-assisted photodegradation of the TiO_2 dispersions using UV light and externally applied conventional heat (UV/CH). In the present context, the external heat can be supplied by coating one part of the cylindrical photoreactor with a thin metallic film on one side at the bottom of the reactor, whereas the uncoated side permits the UV radiation to reach the dispersion. No differences in temperature profiles were observed when using both UV/MW and UV/CH methods.

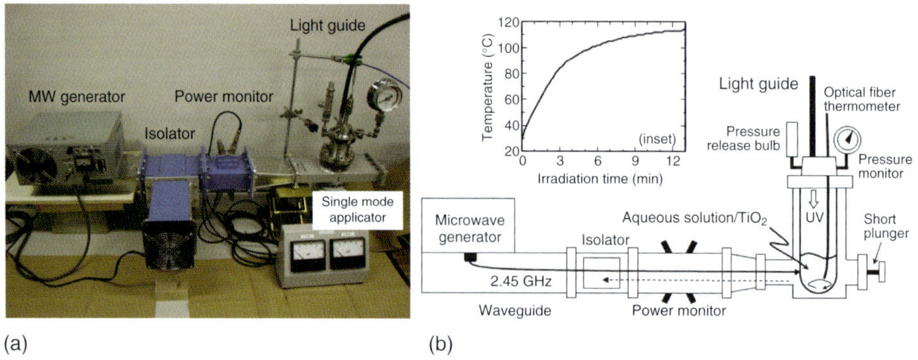

Figure 19.1 (a) Photograph of an integrated microwave/photoreactor system having a single-mode applicator and (b) schematic of the system and a typical plot (inset) of the change of temperature with irradiation time for an aqueous TiO_2 dispersion under microwave irradiation. Reproduced from Ref. [7]. Copyright 2014 by Elsevier B.V.

19.3 Microwave-/Photo-Driven Photocatalytic Wastewater Treatment

19.3.1 Degradation of Rhodamine B Dye

The photooxidative remediation of wastewaters with TiO_2 as the photocatalyst has been reported in many international journals (see, e.g., Ref. [8]). In the degradation of a dye substrate, the photoassisted reaction can be accomplished by means of electron transfer from the excited dye that is exposed to visible light. In this regard, it is noteworthy that the Japanese Industrial Standards (JIS) includes the methylene blue (MB) dye as a standard wastewater substance to ascertain the activity of photocatalysts. Some dyes are poorly photodecomposed and thus are not useful in ascertaining the photoactivity of a metal-oxide photocatalyst. For example, the rate of photodegradation of the cationic rhodamine B (RhB) dye is slow in acidic aqueous media because the surface of the TiO_2 particles is positively charged ($Ti\text{-}OH_2^+$; pI = 6.3). However, RhB has proven an interesting model compound to examine the microwave effect. In earlier studies, the major focus was on the degradation of organic pollutants, as exemplified by the degradation of the RhB dye catalyzed by TiO_2 semiconductor particles under both UV and microwave irradiation [6]. Changes in color intensity of the RhB dye solutions occurring under various conditions are illustrated in Figure 19.2 [9]. The photodegradation of RhB is clearly evident on using the TiO_2-assisted UV/MW method. These observations demonstrate that a method that can treat large quantities of pollutants in wastewaters by a hybrid combination of microwaves and TiO_2-photoassisted technologies is conceivable. The photodegradation of RhB by this metal oxide is unaffected by

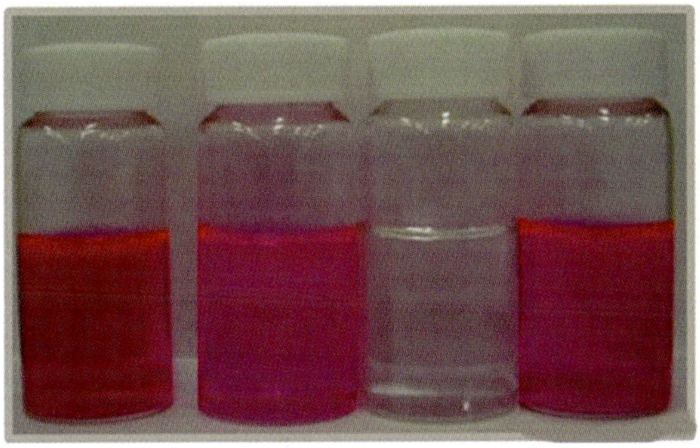

Figure 19.2 Visual comparison of color fading in the degradation of RhB solutions (0.05 mM) subsequent to being subjected to various degradation methods for 150 min. From left to right: initial RhB solution; RhB subjected to photoassisted degradation (UV); RhB subjected to integrated microwave-/photo-assisted degradation (UV/MW); and RhB subjected to thermal- and photo-assisted degradation (UV/CH). Reproduced from Ref. [9]. Copyright 2009 by Elsevier B.V.

conventional heating (CH) – compare, for example, the results from the UV and the UV/CH methods in the presence of TiO_2 (Figure 19.2).

Microwave effects were examined by the temporal decay of total organic carbon (TOC) in the degradation of aqueous RhB solutions under four different methodologies. Results displayed in Figure 19.3a show that for a TiO_2 loading of 60 mg (volume 30 ml) there is no distinction between the efficacy of the UV and UV/CH methods [6, 7]. In the absence of the metal-oxide TiO_2, no changes in TOC occurred when the RhB dye solution was irradiated only by the microwaves, even after 3 h. At a TiO_2 loading of 30 mg (volume 30 ml), the UV/MW method proved very efficient in decreasing the TOC by nearly a factor of 6 from 18.6 mg l^{-1} (ppm) to 3 mg l^{-1} after 3 h.

The degradation of RhB was also examined at two different UV-light irradiances (0.3 and 2.0 mW cm^{-2}) to assess the microwave effects through the loss of TOC as depicted in Figure 19.3b. Clearly, the degradation of RhB by the UV/MW method was faster even at the lower irradiance of 0.3 mW cm^{-2} than occurred by the UV method alone at the higher irradiance of 2.0 mW cm^{-2}. Evidently, the situation at the lower UV-light irradiance accentuated the effect of the microwave radiation.

The disappearance of TOC in aqueous RhB solutions under various atmospheric conditions (air, oxygen gas, and helium gas) in the presence of TiO_2 is illustrated in Figure 19.3c. Recombination of photogenerated valence band holes with conduction band electrons after UV irradiation of TiO_2 particles is known to compete with formation of reactive oxygen species (e.g., $^{\bullet}OH$ and $^{\bullet}OOH$ radicals) that typically lead to degradative processes. The degradation of an organic pollutant via a TiO_2-photoassisted reaction typically follows the order oxygen gas > air > inert gas (e.g., He). In the present instance, integrating microwave irradiation to UV irradiation for an oxygen-saturated solution led to a twofold enhancement in the decrease in TOC [6, 7]. Moreover, the decrease in TOC in the air-equilibrated RhB solution by the UV/MW method was enhanced relative to the degradation of oxygen-saturated RhB solutions by the UV method alone. Within experimental error, the decomposition rate of the He-purged RhB solution by the UV/MW method was nearly the same as the rate for an oxygen-purged dye solution by the UV method alone, again indicating the influence of microwave factors in the degradation process.

The impact of microwave power on the degradation of RhB solutions is illustrated in Figure 19.3d [6, 7]. The temperature rise from ambient (25 °C) at power levels of 150, 225, and 300 W was 62, 75, and 138 °C, respectively, within the first 20 min of irradiation. The increase in microwave power enhanced the degradation of RhB solutions, as witnessed by the decrease in TOC for an irradiation period of about 2 h: the decrease in TOC by the UV/MW was 61% at a microwave power of 150 W, 82% for 225 W, and 92% at 300 W. In the absence of MW irradiation, the extent of TOC decrease by the UV method alone was only 14%. Clearly, under the conditions used, microwave power output enhanced the decomposition dynamics.

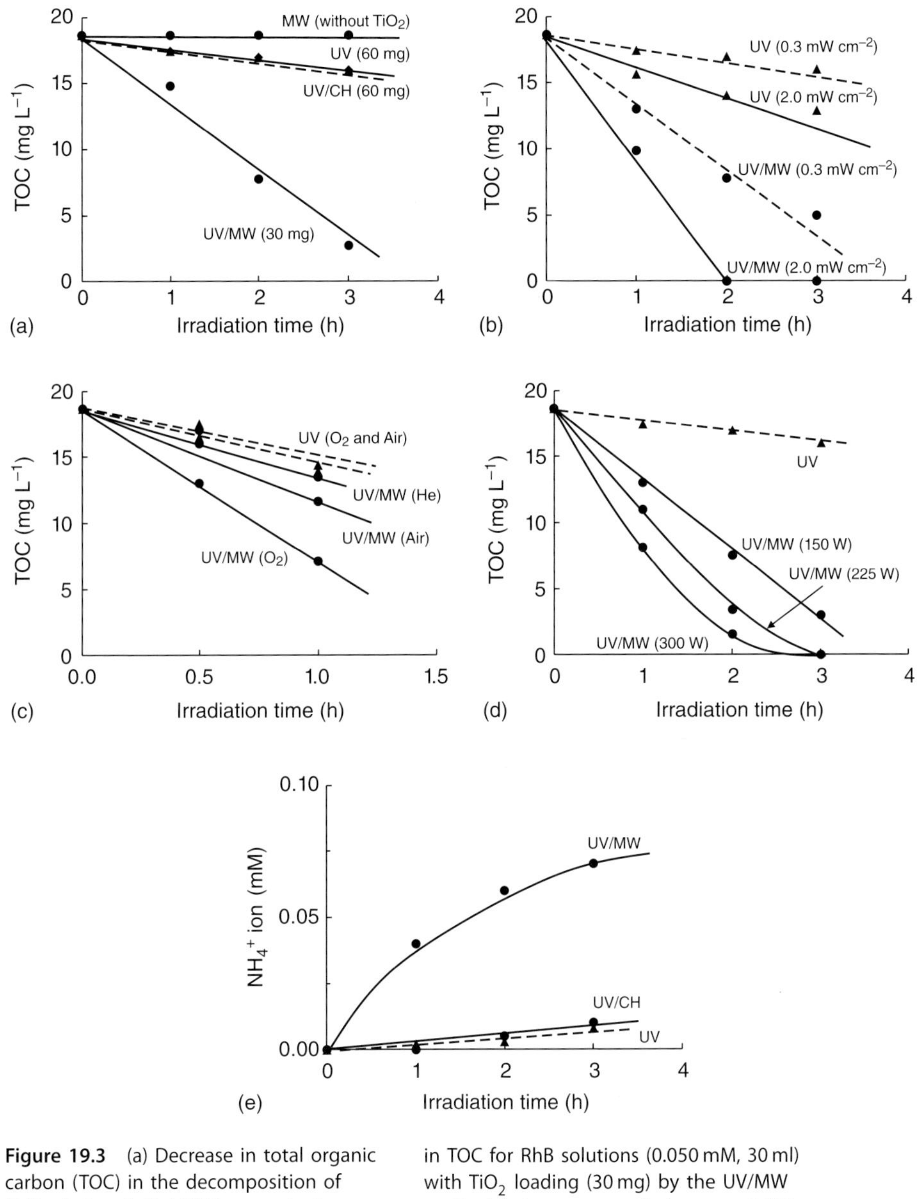

Figure 19.3 (a) Decrease in total organic carbon (TOC) in the decomposition of RhB solution (initial TOC concentration, 18.6 mg l^{-1}; 30 ml) by MW (without TiO$_2$), UV (60 mg), UV/CH (60 mg), and UV/MW (30 mg); (b) temporal evolution of the decrease in TOC during the degradation of RhB solution (0.050 mM, 30 ml) at a radiance of 0.3 and 2.0 mW cm^{-2}; (c) decrease in TOC values for the influence of different added gases on the degradation of RhB (0.050 mM); (d) decrease in TOC for RhB solutions (0.050 mM, 30 ml) with TiO$_2$ loading (30 mg) by the UV/MW method (microwave applied power at 150, 225, and 300 W); and (e) temporal evolution of the formation of NH$_4^+$ ions in the decomposition of RhB (0.050 mM) using the UV, UV/CH, and UV/MW methods; the radiance was 0.3 mW cm^{-2}. Reproduced from Ref. [7]; Copyright 2014 by Elsevier B.V. and in part from Ref. [6]; Copyright by the American Chemical Society.

Formation of NH_4^+ (and NO_3^-) ions in the decomposition of RhB was also investigated by the UV, UV/CH, and UV/MW methods in the presence of TiO_2 nanoparticulates (only data of NH_4^+ are shown in Figure 19.3e) [6, 7]. Again, the UV/MW method led to more significant changes than either the UV or UV/CH methods alone in converting the two nitrogen atoms in the RhB structure. In this regard, the mineralization yields of the two nitrogen atoms, given as the sum of the yields of NH_4^+ and NO_3^- ions, respectively, after 3 h of irradiation were 77% (70% + 7%) for the UV/MW procedure, 12.8% (10% + 2.8%) for the UV method, and 8% (8% + 0%) for the UV/CH method. The increased formation yield of NH_4^+ ions by the UV/MW method was significant by comparison with the UV and UV/CH methods. These features inferred that the mechanistic details in the degradation of RhB in aqueous media in the presence of TiO_2 differed under UV/MW irradiation relative to the UV-induced degradation.

The initially formed intermediates from the degraded RhB were identified by electrospray ionization (ESI) ionization mass spectra and subsequently confirmed by HPLC/absorption spectroscopy (HPLC: high pressure liquid chromatography) [10]. An initial adsorption model of RhB molecule on the TiO_2 surface was proposed by computer simulations that led to estimates of frontier electron densities of all atoms in the RhB structure, which afforded inferences as to the position of $^{\bullet}OH$ (or $HOO^{\bullet}$) radical attack on the RhB structure. Results from these simulations led to the proposed degradation mechanisms summarized in Scheme 19.1 [10]. For the UV method, RhB approaches the positively charged TiO_2 surface through the two oxygen atoms in the carboxylate function bearing the greater negative charge, with further assistance provided by the repulsion between the positively charged nitrogen atoms and the positive TiO_2 surface. Accordingly, de-ethylation of RhB by the UV method was rather inefficient as evidenced by the formation of intermediate (**I**) only (liquid chromatography/mass spectral detection analysis).

Diethylamine was formed through the cleavage of the C–N bond in the $C–N(C_2H_5)_2$ fragment of RhB. For the UV/MW method, the increase in the hydrophobic nature of TiO_2 through microwave irradiation facilitated adsorption of RhB through the aromatic rings aided by the three oxygen atoms in RhB; that is, RhB lay flat on the particle surface. The principal intermediates formed in the degradation of RhB by the UV/MW method were the N-de-ethylated species I–IV. Ultimately, the amino groups of intermediate IV were converted predominantly into NH_4^+ ions. The hydrophobic methyl component of the ethyl group could not adsorb onto the hydrophilic TiO_2 surface, so that transformation of the nitrogen atoms was not a priority event in the photoassisted process.

Generally, a quaternary cationic ammonium function on the RhB structure cannot adsorb on the positively charged TiO_2 surface, which explains why priority transformation of the nitrogens in RhB did not occur under acidic pH conditions below pH 7. The above notwithstanding, it is somewhat enigmatic that under UV/MW irradiation, the extent of NH_4^+ ion formation (and NO_3^- ions) from the nitrogens of the RhB dye was far more efficient than under UV or UV/CH irradiation (Figure 19.3e).

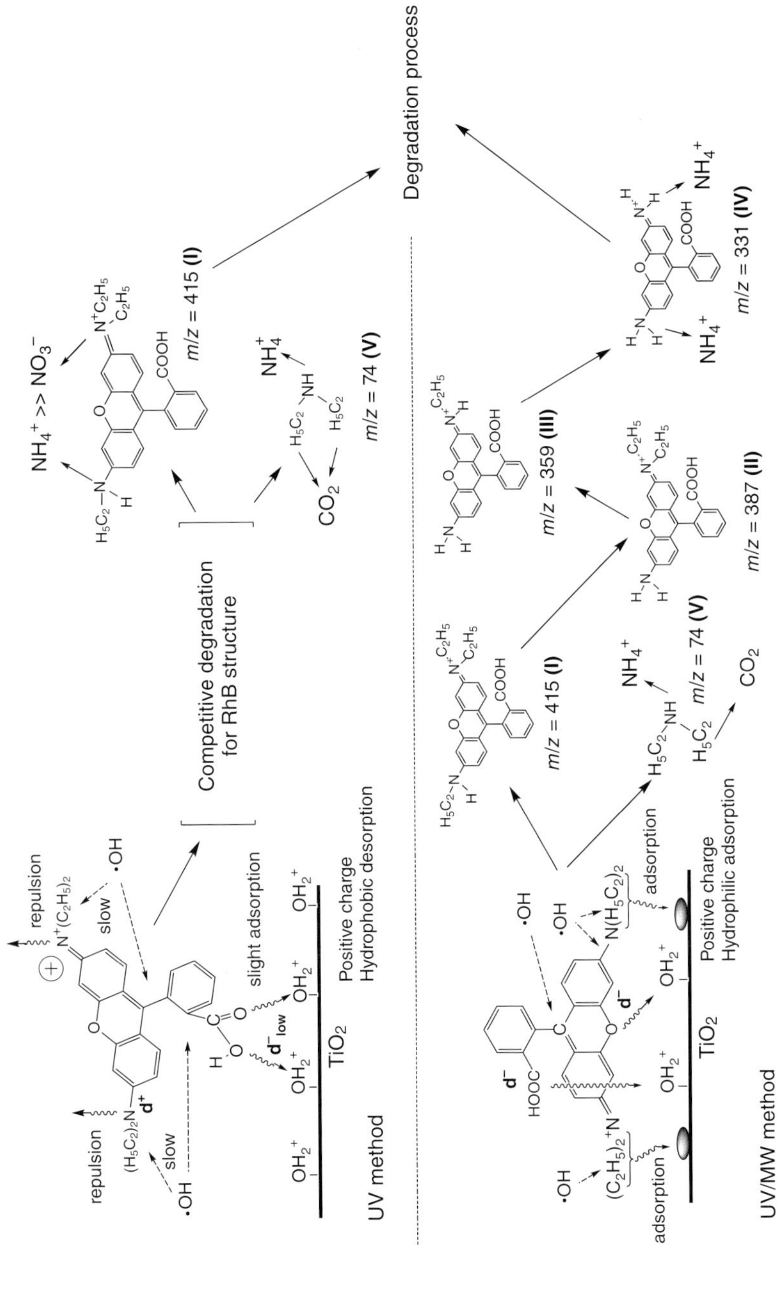

Scheme 19.1 Proposed initial mechanistic steps in the degradation of RhB dye by the UV and UV/MW methods. Reproduced from Ref. [10]. Copyright 2003 by the American Chemical Society.

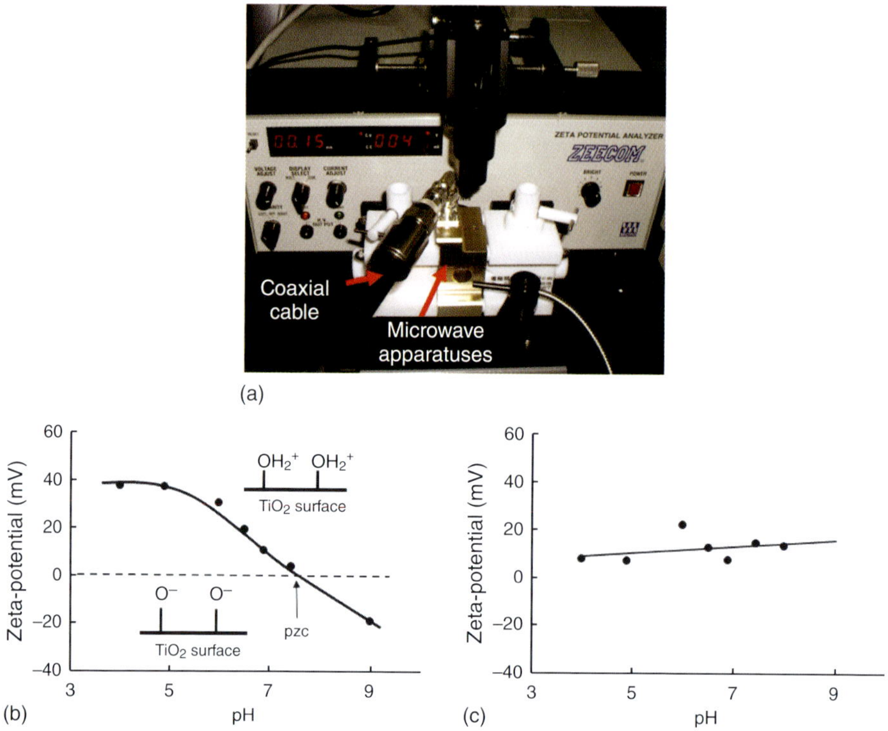

Figure 19.4 The *in situ* observation of the pH dependence of the zeta-potential on the TiO_2 particle surface (P25): (a) photograph of the experimental setup, (b) under UV irradiation alone, and (c) under simultaneous UV/MW irradiation. Reproduced from Ref. [7]. Copyright 2014 by Elsevier B.V.

19.3.2
Change of TiO_2 Surface Condition under a Microwave Field

The surface electric charge of TiO_2 particles is an important factor that impinges on the adsorption of substrates on the TiO_2 surface. Accordingly, the electric charge on the TiO_2 surface was ascertained by a zeta-potential analysis using the coupled microwave/UV irradiation system shown in Figure 19.4a [7]. Under UV-light irradiation alone, the zeta-potential was positive in acidic media and decreased with an increase in pH of the aqueous TiO_2 dispersions, Figure 19.4b. The point of zero charge (pzc) of the TiO_2 particles under UV irradiation was attained at pH ∼ 7.2. On the other hand, the typical zeta-potential curve was not evident under simultaneous UV/MW irradiation conditions, Figure 19.4c [7]. In the latter case, the zeta-potential remained somewhat positive in the 0–20 mV range for dispersions throughout the pH range 4–9. The scatter in the data of Figure 19.4c was likely caused by some unexpected convection currents originating from the microwave heating of only a portion of the dispersion in the sample cell. In all cases, the pzc measurements were repeated no less than five times at each pH examined. Evidently, the surface charge on the TiO_2 particles caused by UV irradiation was significantly perturbed when the particles were also exposed to MW irradiation, although significant scatter was seen in the data. A change of the TiO_2 surface charge by the microwave radiation changes the adsorbed state of RhB in a manner that facilitates the cationic nitrogen group to approach the TiO_2 surface.

19.3.3 Specific Nonthermal Microwave Effect(s) in TiO_2 Photoassisted Reactions

In the degradation of various pollutants in wastewaters, the effect of microwaves manifested itself in the enhancement of the reaction dynamics that otherwise could not be obtained by conventional heating. In many instances, such enhancement has been attributed to a specific effect(s) (nonthermal) imparted by the microwaves and not to microwave dielectric heating (thermal factor). As evidence for this assertion, the microwave-assisted photodegradation of bisphenol-A (BPA) was examined in the presence of TiO_2 at near-ambient temperature to unravel some details on the importance of the microwave nonthermal factor. BPA is one of many listed endocrine disruptors that tend to accumulate in the natural world with serious consequential damage to the reproductive cycle in a variety of animal species [11]. In an earlier study [12], the photodegradation process of the contaminant BPA was found to be affected only by the thermal effect of the microwave radiation that enhanced its decomposition.

To further delineate between the microwaves' thermal and nonthermal factors, the degradation of BPA was re-visited later [13] using an integrated microwave/photoreactor (MPR) system equipped with a cooling system (Figure 19.5a). The Pyrex reactor consisted of a double-layered structure with internal and external diameters of 75 and 100 mm, respectively. The silicone oil refrigerant was circulated through the inner part of this double-layered structure using a circulation cooling apparatus. Note that the silicone oil absorbed only a negligible quantity of the microwave radiation owing to its low dielectric loss ($\varepsilon_r'' = 0.009$ below 25 °C). The temperature of the silicone oil was maintained at −20 °C by the recirculation apparatus operating at the maximal flow rate possible. The UV-light source was a super high-pressure 150-W mercury lamp. The temperature of the aqueous TiO_2 dispersion was measured through a sealed optical fiber thermometer. Air-equilibrated aqueous bisphenol-A solutions (0.050 mM, 50 ml) containing TiO_2 particles (loading, 150 mg; Evonik P25) were introduced into the MPR reactor under dark conditions. The course of the BPA degradation was followed using three experimental protocols: (i) the integrated microwave-/photo-assisted method in the presence of TiO_2 particulates (UV/MW); (ii) the UV/MW method under controlled ambient temperature with the cooling system (UV/MW/Cool); and (iii) the photoassisted method alone (UV) mediated by TiO_2. The temperatures of the solution by the UV and UV/MW/Cool methods

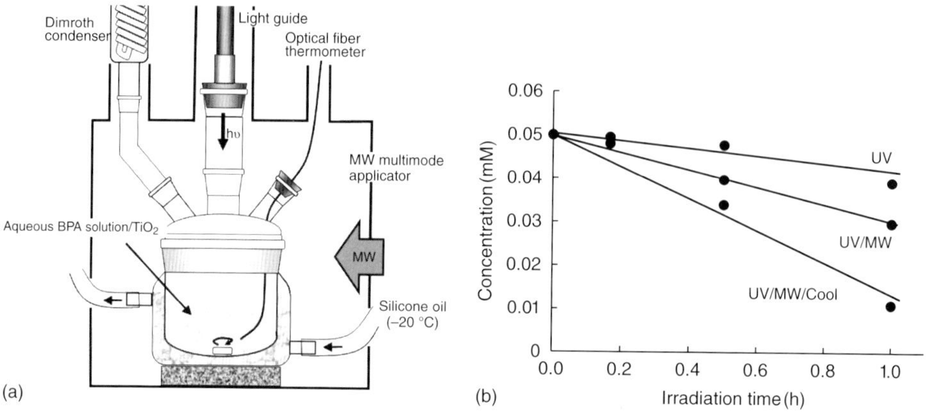

Figure 19.5 (a) Schematic illustration of the microwave-assisted photoreactor (MPR) coupled to a cooling system in the microwave multi-mode applicator. (b) Temporal decrease in the concentration of bisphenol-A (BPA: 0.050 mM) during its decomposition in aqueous media by photoassisted oxidation (UV), by the microwave-/photo-assisted oxidation (UV/MW) method, and by the integrated microwave-/photo-assisted degradation under cooling conditions (UV/MW/Cool). Reproduced from Ref. [13]. Copyright 2007 by Elsevier B.V.

were kept at 21 °C, whereas the temperature in the UV/MW method reached 85 °C after 1 h of irradiation.

The time course of the degradation of bisphenol-A by the UV, UV/MW, and UV/MW/Cool methods is reported in Figure 19.5b. The relevant degradation dynamics subsequent to UV/MW irradiation was about twofold faster than for the UV method alone (3.3×10^{-4} mM min^{-1} vs. 1.7×10^{-4} mM min^{-1}, respectively). The earlier study [12] found that the kinetics of photomineralization of BPA through loss of TOC were nearly identical for the UV/MW method and for the photoassisted method under conventional heating (UV/CH) at otherwise identical temperatures. However, when we examined the results summarized in Figure 19.5b [13], it was evident that the rate of the microwave-/photo-assisted degradation of BPA under controlled ambient conditions (UV/MW/Cool), with temperature maintained constant at about 21 °C, was twofold faster than for the degradation of BPA by the UV/MW protocol (6.5×10^{-4} mM min^{-1} vs. 3.3×10^{-4} mM min^{-1}, respectively) taking place at the higher temperature. The extent of degradation of BPA followed the order: UV/MW/Cool (78%) > UV/MW (42%) > UV (21%) after microwave and/or UV irradiation for 1 h. Clearly, under the conditions used in the later study [13], the microwave-/photo-assisted degradation of BPA was most efficient when carried out at near-ambient temperature, under which the microwave-assisted photodegradation of BPA was not only due to a microwave thermal effect but also due to a significant nonthermal effect that we inferred might implicate hot spots on the TiO_2 particle surface that led to enhanced photodegradation under constant ambient temperature. The existence of hot spots on a Pd/activated carbon catalyst has been documented by Horikoshi and coworkers [14, 15]. Other origins of the microwave nonthermal effect may include additional formation of charge carriers on the TiO_2 photocatalyst and formation of additional trap sites that may otherwise prolong the lifetimes of the charge carriers in TiO_2 through diminished recombination, and thus an ultimate increase in process kinetics.

19.3.4
Microwave Frequency Effects on the Photoactivity of TiO_2

Microwave frequency (2.45 GHz generally used vs. the 5.8 GHz frequency) effects on the photoactivity of P25 TiO_2 have been examined in the photodecomposition of methylene blue (MB) [16]. Results of the temperature change of the solution and the photoassisted degradation of MB by the UV and UV/MW methods are displayed in Figure 19.6, which shows that the temperatures of the aqueous MB solutions subjected to the 2.45 and 5.8-GHz microwaves were nearly identical on adjusting the input power of the incident microwaves (Figure 19.6a). As such, the temperature of the aqueous MB solutions can have no influence on any difference between decomposition rates when the solutions are subjected to different microwave frequencies.

The degradation of MB was considerably attenuated at the 5.8-GHz frequency relative to the 2.45-GHz microwaves (Figure 19.6b). In fact, observations for the UV/MW (5.8 GHz) method are not significantly different from the degradation of MB when using the UV method alone subsequent to irradiation for 3 h. By contrast, the extent of degradation reached 94% for the UV/MW method when using the 2.45-GHz microwaves. The frequency effect was also examined in the decomposition of BPA and 4-chlorophenol (4-CP). In both cases, the results showed similar tendencies as for the MB. Differences in the behavior of both MW frequencies could not be attributed to the chemical constitution of the substrates.

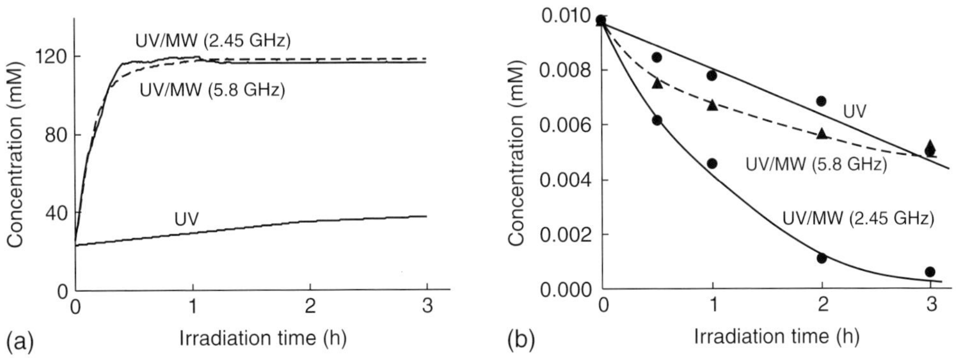

Figure 19.6 (a) Temporal changes of the temperature and (b) decrease in methylene blue (MB) concentrations by the UV and by the integrated UV/MW protocols with 2.45 and 5.8-GHz microwaves in the presence of TiO_2. Adapted from Ref. [16]. Copyright 2009 by Elsevier B.V.

The authors inferred, that loss of photoactivity of the P25 TiO_2 was likely due to some specific effect(s) imparted on this TiO_2 specimen by the 5.8 GHz microwaves [16]. These results show that the enhanced photoactivity of P25 required 2.45 GHz frequency and support the earlier assertion that the microwaves impart not only a thermal effect when used to degrade pollutants in the presence of such metal-oxide specimens, but also a significant effect of the microwaves' nonthermal factor.

19.3.5
Increase in Radical Species on TiO_2 under Microwave Irradiation

It is well known that when a TiO_2 particle is illuminated at wavelengths corresponding to photon energies equal to or larger than the bandgap energy, valence band electrons are excited to the conduction band and holes are produced in the valence band. Most of the electron/hole pairs produced recombine after some tens of picoseconds with the energy being released as emission of photons, phonons, or both [17]. Coordination defects at the surface of the particle and defects within the particle lattice trap the remaining charges. The holes trapped at the surface have a highly reactive oxidation potential, and the electrons have a highly reactive reduction potential. Thus, photogenerated holes and electrons can induce reactions at the surface that, for historical reasons, are often referred to as *photocatalytic reactions* [18]. Anatase titanium dioxide has a bandgap of 3.20 eV, which corresponds to an absorption onset at a wavelength of 387 nm. This means that electron/hole pairs are created when TiO_2 is irradiated with UV light at wavelengths shorter than 387 nm (Eq. (19.1)). To the extent that not all the photogenerated electrons and holes recombine, some of the holes can migrate to the surface and react with surface-bound OH groups and/or water molecules surrounding the particles that ultimately lead to the formation of hydroxyl radicals ($^{\bullet}OH$; Eq. (19.2)). Dissolved oxygen molecules react with conduction band electrons (e^-) to yield superoxide radical anions ($O_2^{-\bullet}$; Eq. (19.3)), which on protonation generate the hydroperoxy radicals $^{\bullet}OOH$ (Eq. (19.4)). Accordingly, the photooxidation of organic substrates with the UV/TiO_2-driven photoassisted process (Eq. (19.5)) depends on the concentration of $^{\bullet}OH$ (and/or $^{\bullet}OOH$) radicals produced by the photooxidation of surface hydroxyl groups and/or chemisorbed H_2O:

$$TiO_2 + h\nu \rightarrow TiO_2(e^- + h^+) \rightarrow e^- + h^+ \tag{19.1}$$

$$h^+ + {}^{-}OH_{surf.}(\text{and/or } H_2O) \rightarrow {}^{\bullet}OH(+H^+) \tag{19.2}$$

$$e_{cb}^- + O_2 \rightarrow O_2^{-\bullet} \tag{19.3}$$

$$O_2^{-\bullet} + H^+ \rightarrow {}^{\bullet}OOH \tag{19.4}$$

$${}^{\bullet}OH\ (\text{or } {}^{\bullet}OOH) + \text{organic pollutant} \rightarrow \text{Oxidative products} \tag{19.5}$$

The possible enhancement of the photoactivity of TiO_2 photocatalysts subsequent to being exposed to microwave radiation from the viewpoint of the amount of $^{\bullet}OH$ radicals generated was also investigated [19]. Formation of $^{\bullet}OH$ radicals during TiO_2-assisted photooxidations that were driven simultaneously

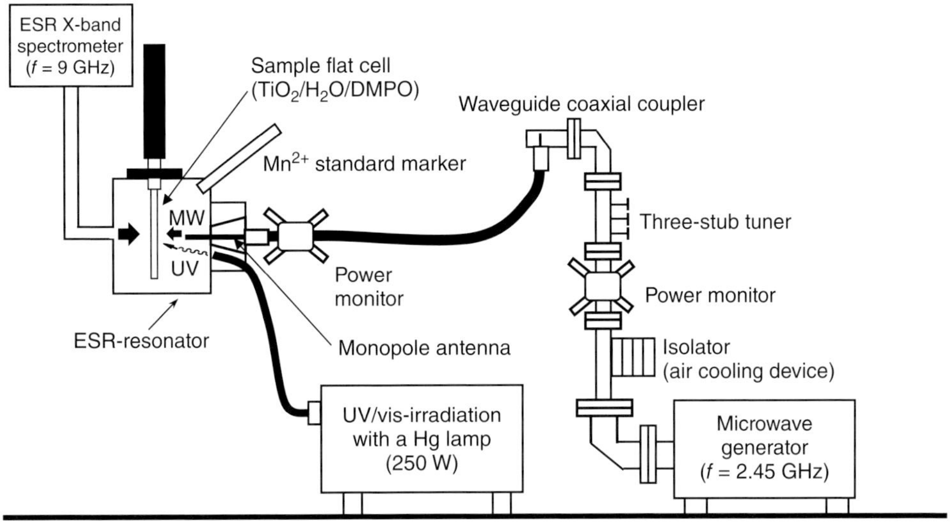

Figure 19.7 Setup used to generate •OH radicals in water alone under MW irradiation, in an aqueous TiO_2 dispersion by MW irradiation alone, and by the UV and UV/MW methods. Adapted from Ref. [19]. Copyright 2003 by Elsevier B.V.

by UV light and microwave radiation was probed by electron spin resonance spectroscopy employing a novel setup in which the ESR (electron spin resonance) sample (contained the DMPO spin-trap agent and various TiO_2 particles in aqueous media; DMPO = 5,5-Dimethyl-1-pyrroline-N-oxide) was irradiated by both UV light and microwave radiation [19]. In this case, microwave radiation was produced using a magnetron microwave generator (frequency, 2.45 GHz), a three-stub tuner, a power monitor, and an isolator (Figure 19.7). The UV irradiation source was an Ushio 250-W mercury lamp; the emitted UV light irradiated the sample at an angle to the horizontal plane using a fiber-optic light guide.

The number of •OH radicals generated under various experimental conditions is summarized in Table 19.1. For P25 titania, the number of •OH radicals produced by the UV/MW method was nearly 30% greater than the quantity generated by the UV method alone [19]. A fivefold increase in incident microwave power from 3 to 16 W led to a significant increase (about 40%) in the number of •OH radicals. Such an increase was sufficient to increase the efficiency of the photooxidation of the organic pollutant in water.

For a UV100 TiO_2 sample, the increase in the number of •OH radicals produced was only 10% greater on increasing the MW power five times. On the other hand, the number of •OH radicals generated for the pristine anatase and rutile TiO_2 samples decreased under microwave irradiation. The P25 specimen was clearly influenced by the microwaves and generated •OH radicals efficiently under the influence of microwaves. Therefore, the rate of decomposition of pollutants can be enhanced when P25 is used to decompose the wastewater sample by the

Table 19.1 Number of DMPO-•OH spin adducts produced in the various heterogeneous systems under microwave irradiation, UV irradiation, and MW/UV irradiation relative to those formed in the rutile TiO_2 specimen for the TiO_2/H_2O/MW heterogeneous system.

Methodology	P25	UV100	Anatase	Rutile
UV	182	45	110	110
UV/MW (3 W)	259	51	92	76
UV/MW (16 W)	369	—	—	—

UV/MW method. On the other hand, to the extent that the quantity of •OH radicals produced by the other TiO_2 systems (commercial anatase and rutile) does not increase even when irradiated with microwaves, the rate of decomposition is not expected to be greatly affected.

19.3.6 Microwave Nonthermal Effect(s) as a Key Factor in TiO_2 Photoassisted Reactions

A key factor in process dynamics – that is, microwave nonthermal effects – was also examined when "impurities" were present in TiO_2 (e.g., N-doped TiO_2 [20]), as well as oxygen vacancies and lattice distortions in TiO_2 crystallites [21]. Within this context, the characteristics of 2.45-GHz microwaves were examined on two second-generation titania samples, namely two N-doped TiO_2 nanomaterials prepared by annealing P25 TiO_2 and Ishihara's ST01 TiO_2 at 400 and 500 °C, respectively, in air in the presence of urea, and for comparison on the undoped pristine samples [20]. Nitrogen doping led to an increase in the size of the particles and to a decrease in the surface area. Temperature–time profiles showed that the heating efficiency of the N-doped specimens under microwave irradiation was greater, particularly significant for the N-doped P25 sample, but rather small for the N-doped ST01 sample relative to the corresponding undoped specimens. The effect of microwaves on surface optical phonons of the undoped P25 and N-doped P25 TiO_2 systems, with and without UV/vis irradiation, was negligible (Raman data). By contrast, microwave irradiation of pristine ST01 and N-doped ST01 TiO_2 samples showed significant changes in the 144 cm^{-1} optical phonons, from which we inferred a microwave thermal effect on the ST01 and N-doped ST01 specimens, whereas for the P25 samples the microwaves also imparted a specific effect (i.e., a nonthermal effect) as the microwaves influenced the N-dopant sites in contrast to the ST01 systems where the dopant sites seemed unaffected. The microwave-/photo-assisted degradation of 4-CP under various conditions of UV/vis irradiation and conventional heating, as opposed to microwave heating, confirmed the role of the specific microwave effect on the P25 systems. As a further verification of the latter, the possible effects of the 2.45-GHz microwaves on the Raman-active lattice phonons of undoped anatase and rutile specimens (Wako Pure Chemical Ind. Ltd.), and for comparison P25 TiO_2 nanoparticles, by the Raman technique when subjected to MW irradiation alone or in combination with UV illumination (UV/MW) [22]. Results showed significant changes in band intensities for anatase (E_g mode at 143 cm^{-1}) and for rutile (446 cm^{-1}), whereas only negligible changes in intensity were again observed at 144 cm^{-1} for the P25 nanoparticles exposed to MW and UV/MW irradiation. These observations confirmed that microwaves had a negligible impact on the lattice phonons of P25 titania, even though the specimen was thermally heated by the microwaves. It was deduced [22] that the microwave energy absorbed by the P25 TiO_2 was consumed to bring about changes at the interfacial boundaries between the coupled rutile and anatase polymorphic structures, which likely caused the formation of oxygen vacancies. The latter are well known to trap conduction band electrons to yield F-type color centers [23–25], thereby allowing valence band holes to produce additional $^\bullet$OH radicals as evidenced by the ESR results [22], unlike the pristine anatase and rutile samples where the number of $^\bullet$OH radicals decreased in UV/MW irradiation relative to UV irradiation alone. The resulting increased charge separation within the P25 particles and the greater number of $^\bullet$OH radicals are consistent with the increased rate enhancements and greater photoactivity, which were not otherwise seen for the anatase and rutile samples. A specific microwave effect was deduced by the available data [22].

Subjecting pristine P25 titania to various heat treatments at different temperatures (645–800 °C) resulted not only in changes in the anatase-to-rutile ratios but also to a decrease in lattice distortions (X-ray diffraction data); the decrease was greater in the rutile phase than in the anatase phase. In addition, particle sizes transmission electron microscopy (TEM) increased linearly with increase in temperature [21]. Changes in lattice distortions in the heat-treated P25 samples caused the kinetics of photodegradation of 4-CP to increase the greater was the extent of lattice distortions in the anatase and rutile phases of P25 TiO_2, at least under UV/MW irradiation, while under UV/CH and UV irradiation the kinetics increased the greater was the distortion at first and then decreased with increased distortions in both phases. The heat treatments probably caused changes in the number of defect sites (e.g., oxygen vacancies) inherently present in pristine P25 titania. Moreover, microwave irradiation for 4 min caused the Raman band intensities of pure rutile (446 cm^{-1}) and pure anatase (143 cm^{-1}), as well as those of the anatase/rutile phases in the heat-treated P25 samples, to decrease significantly; on termination of the microwave irradiation the bands tended to regain their original intensities [21]. Such changes were again ascribed to some specific (nonthermal) effect of the 2.45-GHz microwaves on the nanostructure of the samples.

Heat treatment of P25 titania and ST01 TiO_2 in the presence of H_2 caused the former to change color from white to light blue, while the latter changed from white to pale yellow with the colors being stable even after exposure of the samples to air oxygen. The change in the extent of lattice distortion of the heat/H_2-treated ST01 specimen was twofold greater than for the P25 sample under otherwise identical conditions. The UV–vis absorption spectra of both sets of specimens revealed a broad unresolved band envelope above 400 nm that was attributed [21] to the formation of F-type color centers originating from oxygen vacancies [25]. The kinetics of photodegradation of 4-CP by these heat/H_2-treated specimens were enhanced significantly under UV/MW irradiation (about threefold) relative to irradiation by the UV/CH and UV methods. In addition, the treated P25 sample was nearly 25% more efficient than the untreated sample, while the corresponding ST01 specimen was 85% more efficient. Such increased efficiency under UV/MW irradiation for the heat/H_2-treated ST-01 specimen relative to the pristine ST01 sample was not due to a microwave thermal effect but due to some nonthermal effect of the microwaves impacting on the nanostructure of the metal-oxide samples [21].

A nanosecond transient diffuse reflectance spectroscopic study of commercially available pristine Ishihara ST01 titania, which on treatment with hydrogen yielded an oxygen-vacancy rich Vo-ST01 system was undertaken to examine the effect that microwaves have on the photogenerated charge carriers when exposed to microwave radiation; for comparison, a N-doped N-ST01 sample was also examined (prepared using urea as the nitrogen donor) [26]. Time-resolved absorption measurements were carried out using the second harmonic (532 nm) output pulse from a neodymium-doped yttrium aluminium garnet (Nd:$Y_3Al_5O_{12}$ or so-called Nd:YAG) laser (Ekspla, SL311) employed as the pump light source. The repetition rate of the pump laser was 10 Hz and the pulse duration was about 150 ps. About 2-μs pulses from a Xe flash lamp (Hamamatsu, L4642) were used as the probe (~5 mm in diameter) light source focused on the sample specimen. A schematic of the setup is illustrated in Figure 19.8a. A picture of the microwave irradiation setup with the single-mode cavity is shown in Figure 19.8b; it included a short plunger, an iris, a three-stub tuner, a power monitor, and an isolator. The microwave equipment was arranged such that the waveguide was perpendicularly set so as not to interfere with the laser light. The highly precise continuous microwave radiation

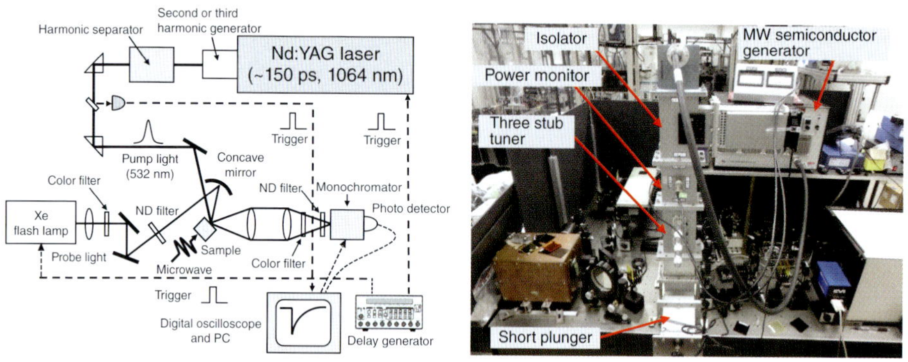

(a) (b)

Figure 19.8 (a) Image and (b) photograph of the nanosecond transient diffuse reflectance spectroscopic system. Reproduced from Ref. [26]; Copyright 2015 by the Royal Society of Chemistry.

was generated from a 2.45-GHz microwave semiconductor generator (Fuji Electronic Industrial Co. Ltd; GNU-201AA; maximal power, 200 W).

Transient absorption decay kinetics at 550 nm for all three samples (pristine ST01, oxygen-vacancy rich Vo-ST01, and N-doped ST01) in the dry and wet states were determined *in situ* using the 150-ps Nd-YAG pulsed laser system (10 Hz) and the Xe flash lamp (2-μs pulses) probe while samples were being microwave-irradiated (2.45 GHz). The transient(s) absorbing at the probe wavelength displayed double exponential decay kinetics: a *fast* decay that occurred within about 5–12 ns ascribed to recombination of photogenerated shallow-trapped or free conduction band electrons with valence band holes, and a *slower* decay that occurred within hundreds of nanoseconds to several microseconds attributable to recombination of electrons trapped in deep traps (e.g., oxygen vacancies yielding F color centers, or as Ti^{3+}) with free holes [26]. Significant differences were observed for the dry N-doped ST01 sample and for the wet pristine ST01/H_2O and the Vo-ST01/H_2O systems when subjected to microwave irradiation. In addition, the role played by microwave nonthermal factors was also examined to assess the effect of oxygen vacancies on the photodegradation of 2,4-dichlorophenoxyacetic acid (2,4-D) in aqueous TiO_2 dispersions under an integrated microwave-/UV illumination method (UV/MW) and UV irradiation with conventional heating (UV/CH) at 100 °C.

The presence of ions and/or oxygen vacancies within the crystal lattice enhances microwave heating of ceramics [27]. In fact, oxygen vacancies are considered responsible for Joule heating by microwave radiation in order to bring about changes of the microscopic electrical resistance in ceramics. In addition, although microwave energy conversion to thermal energy is affected by the presence of oxygen vacancies, it is nonetheless possible under some conditions to control the recombination of charge carriers (Figure 19.9) [26] and increase catalyst activity [28].

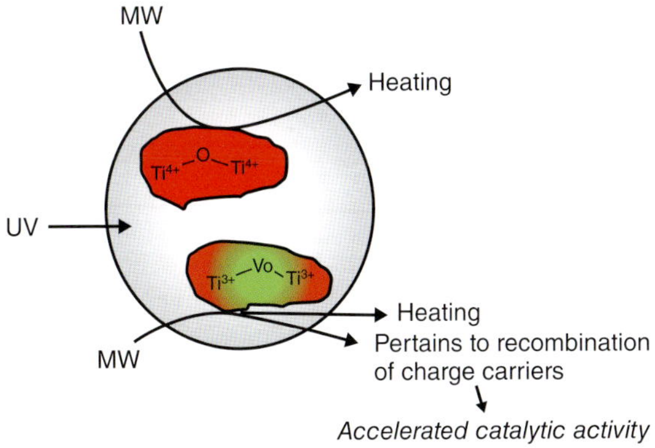

Figure 19.9 Cartoon depicting the influence of microwave radiation on a Vo-ST01 particle. Reproduced from Ref. [26]; Copyright 2015 by the Royal Society of Chemistry.

19.4
Microwave Discharge Electrodeless Lamps (MDELs)

19.4.1
The Need for More Efficient UV Light Sources

It was shown in the preceding section that microwave radiation is effective in enhancing the photocatalytic activity of metal-oxide specimens. Accordingly, a treatment method that can treat larger quantities of pollutants in wastewaters is conceivable by a hybrid combination of the microwave technology and the photocatalytic technology. However, the light source can be a problem in the scale-up of microwave-assisted TiO_2-photoassisted processes. The bandgap energy of anatase TiO_2 is 3.20 eV, so that UV light below 387 nm is necessary to activate this metal oxide. Generally, an electrode Hg lamp is used as the UV-light source in Advanced Oxidation Processes. However, it is difficult to set up a typical mercury lamp in a microwave field from the viewpoint of the electrical discharge. In addition, it is difficult to open the window for the light beam to the microwave applicator because of possible leakages of the microwave radiation. Moreover, a small hole can only irradiate a small fraction of the dispersions. Accordingly, novel microwave-driven electrodeless lamps (MDELs) were developed by us (also photoreactors) and others as UV-light sources to overcome many of these problems (see, e.g., review article in Ref. [29]).

An MDEL device has certain attractive features based on electric and magnetic fields coupling discharge that are significantly different from those of an electrode conventional Hg lamp, namely (i) the relatively long lifetime of electrodeless lamps, (ii) the lack of complications in lamp shape because the lamps are electrodeless, (iii) the absence of variations in light intensity, (iv) the shorter ignition time to lighting the MDEL device relative to a typical Hg electrode lamp, (v) the UV radiation can be supplied external to the reactor to avoid absorption of the microwave radiation by the reactor contents, (vi) the facility with which the lamps can be replaced, (vii) the filling compounds (gases) that might be deleterious to metals and plastics can be used, (viii) the electrical energy can be transmitted from outside the reactor by a wireless process, and (ix) not least both UV and microwave radiations can be used simultaneously to induce oxidative/degradative processes to destroy organics and microorganisms using only microwave energy.

A novel TiO_2-coated MDEL system was recently reported by Církva and coworkers [30], which they used to photodegrade monochloroacetic acid. They prepared thin titanium dioxide films by the sol–gel process and characterized them using several techniques. Apparently, process efficiency depended on the intensity of the light (i.e., light irradiance) and on the initial pH of the solution. Monochloroacetic acid was mineralized to Cl^- ions, CO_2, and H_2O over the irradiated titanium dioxide film in the microwave field, and was particularly enhanced in alkaline media in the presence of H_2O_2, and significantly so on increasing the light intensity.

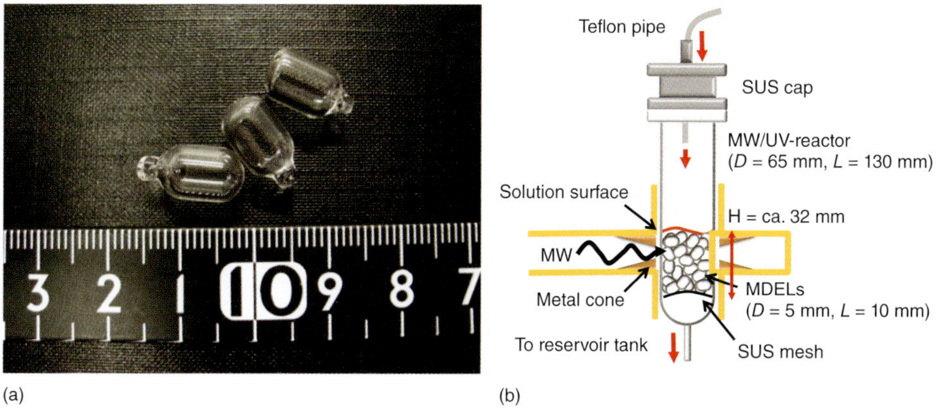

Figure 19.10 (a) Photograph of small MDELs and (b) the experimental setup of a small MDEL device in a single-mode microwave apparatus. Adapted from Ref. [31]. Copyright 2011 by Elsevier B.V.

Along the same lines, small size microwave discharge granulated electrodeless lamps (MDELs) were fabricated using vacuum-UV transparent synthetic quartz as the envelope and a mixture of Hg and Ar as the gas-fills [31]. Dimensions of the devices were 10 mm (length) by 5 mm (external diameter); see Figure 19.10a. Subsequent to evacuating the MDEL quartz envelope to 133×10^{-7} Pa, the system was purged with argon gas (133×10^{-3} Pa) after which a small quantity of liquid mercury was added. Continuous microwave radiation was produced using a Hitachi Kyowa Engineering System microwave generator (frequency, 2.45 GHz; maximal power, 800 W), an isolator, a power monitor, and a short-circuit plunger. To test the performance of such MDELs, the photodegradation of the highly fluorinated perfluorooctanoic acid (PFOA) was examined in 300-ml air-equilibrated aqueous solutions that were circulated with a peristaltic pump through the multi-pass UV/MW reactor (Figure 19.10b) containing the MDELs (20 pieces) at a flow rate of 600 ml min^{-1}. Note that no TiO_2 was used in this experiment. Figure 19.11 illustrates the results of the defluorination of perfluorooctanoic acid in aqueous media [31]. The extent of defluorination of PFOA was about 80% after 200 min of microwave irradiation, which reached 100% upon further irradiation for another 200 min.

19.4.2
Purification of Water Using TiO_2-Coated MDEL Systems in Natural Disasters

The great earthquake that occurred in eastern Japan on March 11, 2011 was soon followed by a huge tsunami; hundreds of houses either collapsed directly by the ground movement or else were destroyed during the subsequent tsunami. The demand for drinking water by sterilization of natural water (accumulated rain water) increased soon after the earthquake disaster. In connection with this, the needs of water for other uses (e.g., water to rinse off mud, water for toilets) also

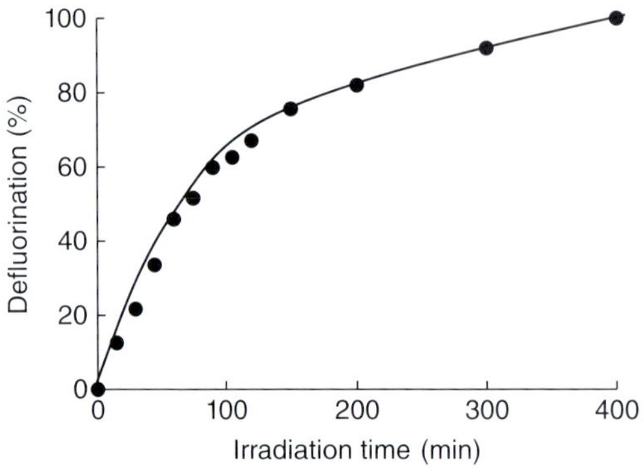

Figure 19.11 Time profiles of the extent of the photoassisted defluorination of perfluorooctanoic acid (PFOA) in aqueous solutions in the photoreactor containing the 20 MDELs systems. Adapted from Ref. [31]. Copyright 2011 by Elsevier B.V.

Figure 19.12 Water sterilization equipment used to sterilize natural water (accumulated rain water) samples using the solar cells located on the right-hand side of the photograph and the TiO_2-coated MDELs (150 pieces) [32] Copyright 2014 by S. Horikoshi and N. Serpone.

grew exponentially. Accordingly, the possible sterilization of resurgent water in the earthquake-stricken area was examined using MDELs as well as natural water collected in a pond (used as model natural water) using the setup illustrated in Figure 19.12 [32].

To the extent that an electric supply is typically unavailable in such areas after an earthquake and ensuing tsunami, the experimental setup also included a system of solar cells that were connected to the equipment so that water purification could be achieved. The sterilization equipment consisted of a reaction vessel that contained 150 pieces of bead-shaped MDELs (10 mm long by 5 mm in external diameter) positioned in the single-mode microwave applicator [32]. Sterilization was carried out by continuous introduction of the natural water (accumulated rain water) sample from the reactor located in the upper part of the setup. The equipment was so designed as to process natural water continuously on site. However, if used for long periods of time, the surface of the various MDELs tended to become dirty. To overcome this issue, a thin layer of TiO_2 nanoparticles was coated on the surface of the MDELs so that the surface of the MDELs would remain clean by the well-known self-cleaning properties of titanium dioxide [1, 5].

Preliminary experiments [32] showed that a single pass was sufficient to sterilize more than 95% of the natural water (flow rate: 0.4 ml min^{-1}) through the reactor containing the TiO_2-coated MDELs. Moreover, 100% killing of *Escherichia coli* (*E. coli*) was observed by a single pass through the reactor. In fact, sterilization of natural water was complete after only two passes through the reactor aided by an appropriate recirculating pump. More importantly, this equipment can also be used to process waters continuously that may have been contaminated with agricultural chemicals such as insecticides.

19.5
Summary Remarks

Some 15 years have passed since the discovery that led to the improvement in photocatalyst activity on exposure to microwave radiation. The photocatalyst is a material that absorbs the energy of electromagnetic waves (UV light) and changes it into chemical energy. In this regard, microwave radiation also consists of electromagnetic waves. The notion of irradiating TiO_2 with microwaves may appear strange at first because the photon energy (1×10^{-5} eV) of the microwaves of frequency 2.45 GHz is several orders of magnitude lower than the bandgap energy required (3.0–3.2 eV) to activate the TiO_2 semiconductor. Moreover, microwave radiation brings other effects to bear to a photocatalyst other than heat. Microwave nonthermal effects have been deduced to contribute significantly to the enhancement of TiO_2-photoassisted reactions, as it may affect both the surface and the crystalline structure of the metal oxide toward reactions taking place at the surface. The mechanism of the effect of microwave radiation on photocatalyzed reactions has evolved gradually. The enhanced treatment of wastewaters through improvement in photocatalyst activity when exposed to microwave radiation is now a clear possibility. In fact, coupling microwave radiation with UV light in TiO_2-photoassisted processes can contribute significantly to the treatment of wastewaters as a novel advanced oxidation technology (AOT).

References

1. Hashimoto, K., Irie, H., and Fujishima, A. (2005) TiO_2 photocatalysis: a historical overview and future prospects. *Jpn. J. Appl. Phys.*, **44**, 8269–8285.
2. Kabra, K., Chaudhary, R., and Sawhney, R.T. (2004) Treatment of hazardous organic and inorganic compounds through aqueous-phase photocatalysis: a review. *Ind. Eng. Chem. Res.*, **43**, 7683–7696.
3. Konstantinou, K. and Albanis, T.A. (2003) Photocatalytic transformation of

pesticides in aqueous titanium dioxide suspensions using artificial and solar light: intermediates and degradation pathways. *Appl. Catal., B*, **42**, 319–335.
4. Bhatkhande, D.S., Pangarkar, V.G., and Beenackers, A.A. (2002) Photocatalytic degradation for environmental applications – A review. *J. Chem. Technol. Biotechnol.*, **77**, 102–116.
5. Fujishima, A., Hashimoto, K., and Watanabe, T. (1999) *TiO_2 Photocatalysis: Fundamentals and Applications*, BKC Incorporated, Tokyo.
6. Horikoshi, S., Hidaka, H., and Serpone, N. (2002) Environmental remediation by an integrated microwave/UV illumination method. 1. Microwave-assisted degradation of rhodamine-B dye in aqueous TiO_2 dispersions. *Environ. Sci. Technol.*, **36**, 1357–1366.
7. Horikoshi, S. and Serpone, N. (2014) On the influence of the microwaves' thermal and non-thermal effects in titania photoassisted reactions. *Catal. Today*, **224**, 225–235.
8. Barbeni, M., Morello, M., Pramauro, E., Pelizzetti, E., Vincenti, M., Borgarello, E., Serpone, N., and Jamieson, M.A. (1987) Sunlight photodegradation of 2,4,5-trichlorophenoxy acetic acid and 2,4,5,trichlorophenol on TiO_2. Identification of intermediates and degradation pathway. *Chemosphere*, **16**, 1165–1179.
9. Horikoshi, S. and Serpone, N. (2009) Photochemistry with microwaves: catalysts and environmental applications. *J. Photochem. Photobiol., C*, **10**, 96–110.
10. Horikoshi, S., Saitou, A., Hidaka, H., and Serpone, N. (2003) Environmental remediation by an integrated microwave/UV illumination method. V. Thermal and nonthermal effects of microwave radiation on the photocatalyst and on the photodegradation of rhodamine-B under UV/vis radiation. *Environ. Sci. Technol.*, **37**, 5813–5822.
11. Atkinson, A. and Roy, D. (1995) In vitro conversion of environmental estrogenic chemical bisphenol-A to DNA binding metabolite(s). *Biochem. Biophys. Res. Commun.*, **210**, 424–433.
12. Horikoshi, S., Tokunaga, A., Hidaka, H., and Serpone, N. (2004) Environmental remediation by an integrated microwave/UV illumination method VII. Thermal/non-thermal effects in the microwave-assisted photocatalyzed mineralization of bisphenol-A. *J. Photochem. Photobiol., A: Chem.*, **162**, 33–40.
13. Horikoshi, S., Kajitani, M., and Serpone, N. (2007) The microwave-/photo-assisted degradation of bisphenol-A in aqueous TiO_2 dispersions revisited. Re-assessment of the microwave non-thermal effect. *J. Photochem. Photobiol., A: Chem.*, **188**, 1–4.
14. Horikoshi, S., Osawa, A., Abe, M., and Serpone, N. (2011) On the generation of hot spots by microwave electric and magnetic fields and their impact on a microwave assisted heterogeneous reaction in the presence of metallic Pd nanoparticles on an activated carbon support. *J. Phys. Chem. C*, **115**, 23030–23035.
15. Horikoshi, S., Osawa, A., Sakamoto, S., and Serpone, N. (2013) Control of microwave-generated hot spots. Part IV. Control of hot spots on a heterogeneous microwave-absorber catalyst surface by a hybrid internal/external heating method. *Chem. Eng. Process.*, **69**, 52–56.
16. Horikoshi, S., Sakai, F., Kajitani, M., Abe, M., and Serpone, N. (2009) Microwave frequency effects on the photoactivity of TiO_2: dielectric properties and the degradation of 4-chlorophenol, bisphenol-A, and methylene blue. *Chem. Phys. Lett.*, **470**, 304–307.
17. Serpone, N., Lawless, D., Khairutdinov, R., and Pelizzetti, E. (1995) Subnanosecond relaxation dynamics in TiO_2 colloidal sols (Particle sizes Rp = 1.0–13.4 nm). Relevance to heterogeneous photocatalysis. *J. Phys. Chem.*, **99**, 16655–16661.
18. Anpo, M. (2000) Utilization of TiO_2 photocatalysts in green chemistry. *Pure Appl. Chem.*, **72**, 1265–1270.
19. Horikoshi, S., Hidaka, H., and Serpone, N. (2003) Hydroxyl radicals in microwave photocatalysis. Enhanced formation of •OH radicals probed by ESR techniques in microwave-assisted photocatalysis in aqueous TiO_2 dispersions. *Chem. Phys. Lett.*, **376**, 475–480.
20. Horikoshi, S., Minatodani, Y., Sakai, H., Abe, M., and Serpone, N. (2011)

Characteristics of microwaves on second generation nitrogen-doped TiO_2 nanoparticles and their effect on photoassisted processes. *J. Photochem. Photobiol., A: Chem.*, **217**, 191–200.

21. Horikoshi, S., Minatodani, Y., Tsutsumi, H., Uchida, H., Abe, M., and Serpone, N. (2013) Influence of lattice distortion and oxygen vacancies on the UV-driven/microwave-assisted TiO_2 photocatalysis. *J. Photochem. Photobiol., A: Chem.*, **265**, 20–28.
22. Horikoshi, S., Abe, M., Sato, S., and Serpone, N. (2011) Effect of microwave radiation on the (Raman) lattice phonons in selected titanium dioxide solid specimens. *J. Photochem. Photobiol., A: Chem.*, **220**, 94–101.
23. Hurum, D.C., Gray, K.A., Rajh, T., and Thurnauer, M.C. (2005) Recombination pathways in the Degussa P25 formulation of TiO_2: surface versus lattice mechanisms. *J. Phys. Chem. B*, **109**, 977–980.
24. Emeline, A.V., Kuznetsov, V.N., Ryabchuk, V.K., and Serpone, N. (2008) Visible-light-active titania photocatalysts: the case of N-doped TiO_2s – properties and some fundamental issues. *Int. J. Photoenergy*, **2008**, 258394.
25. Kuznetsov, V.N. and Serpone, N. (2009) On the origin of the spectral bands in the visible absorption spectra of visible-light-active TiO_2 specimens. Analysis and assignments. *J. Phys. Chem. C*, **113**, 15110–15123.
26. Horikoshi, S., Tsutsumi, H., Matsuzaki, H., Furube, A., and Serpone, N. (2015) In-situ picosecond transient diffuse reflectance spectroscopy of opaque TiO_2 systems under microwave irradiation and influence of oxygen vacancies on the UV-driven/microwave-assisted TiO_2 photocatalysis. *J. Mater. Chem. C*, **3**, 5958–5969.
27. Kitchen, H.J., Vallance, S.R., Kennedy, J.L., Tapia-Ruiz, N., Carassiti, L., Harrison, A., Whittaker, A.G., Drysdale, T.D., Kingman, S.W., and Gregory, D.H. (2014) Modern microwave methods in solid-state inorganic materials chemistry: from fundamentals to manufacturing. *Chem. Rev.*, **114**, 1170–1206.
28. Nozaki, H. and Iida, T. (1964) The physical properties of titanium oxide, especially electronic structure of coloring. *Seisan Kenkyu*, **16**, 201–207.
29. Horikoshi, S., Abe, M., and Serpone, N. (2009) Novel designs of microwave discharge electrodeless lamps (MDEL) in photochemical applications. Use in advanced oxidation processes. *Photochem. Photobiol. Sci.*, **8**, 1087–1104.
30. Církva, V., Žabová, H., and Hájek, M. (2008) Microwave photocatalysis of mono-chloroacetic acid over nanoporous titanium(IV) oxide thin films using mercury electrodeless discharge lamp-sOriginal. *J. Photochem. Photobiol., A: Chem.*, **198**, 13–17.
31. Horikoshi, S., Tsuchida, A., Sakai, H., Abe, M., and Serpone, N. (2011) Microwave discharge electrodeless lamps (MDELs). VI. Performance evaluation of a novel microwave discharge granulated electrodeless lamp (MDGEL) – Photoassisted defluorination of perfluoroalkoxy acids in aqueous media. *J. Photochem. Photobiol., A: Chem.*, **222**, 97–104.
32. Horikoshi, S., Nishimura, T., Tsutsumi, H., and Serpone, N. (2015) Microwave discharge electrodeless lamps (MDELs). Part VIII. Continuous on-site solar energy remediation of contaminated. *Chem. Eng. Technol., inpress.*.

Index

Microwaves in Catalysis: Methodology and Applications, First Edition.
Edited by Satoshi Horikoshi and Nick Serpone.

n